# Fundamentals ar
# Characteristics of Renewable
# Energy Systems

T0073247

# Nano and Energy Series

SERIES EDITOR

Sohail Anwar, Pennsylvania State University, Altoona College, USA

**Nanotechnology: Business Applications and Commercialization**
Sherron Sparks

**Nanotechnology: Ethical and Social Implications**
Ahmed S. Khan

**A History of the Workplace: Environment and Health at Stake**
Lars Bluma, Judith Rainhorn

**Alternative Energy Technologies: An Introduction with Computer Simulations**
Gavin Buxton

**Computational Nanotechnology: Modeling and Applications with MATLAB®**
Sarhan M. Musa

**Advanced Nanoelectronics**
Razali Ismail, Mohammad Taghi Ahmadi, Sohail Anwar

For more information about this series, please visit:
https://www.crcpress.com/Nano-and-Energy/book-series/NANANDENE

# Fundamentals and Source Characteristics of Renewable Energy Systems

Radian Belu

CRC Press
Taylor & Francis Group
Boca Raton London New York

CRC Press is an imprint of the
Taylor & Francis Group, an **informa** business

CRC Press
Taylor & Francis Group
6000 Broken Sound Parkway NW, Suite 300
Boca Raton, FL 33487-2742

First issued in paperback 2022

ISBN-13: 978-0-367-26139-9 (hbk)
ISBN-13: 978-1-03-233794-4 (pbk)
DOI: 10.1201/9780429297281

**Library of Congress Cataloging-in-Publication Data**

Names: Belu, Radian, author.
Title: Fundamentals and source characteristics of renewable energy systems
    / Radian Belu.
Description: Boca Raton : Taylor & Francis, a CRC title, part of the Taylor
    & Francis imprint, a member of the Taylor & Francis Group, the academic
    division of T&F Informa, plc, 2020. | Series: Nano and energy series |
    Includes bibliographical references and index.
Identifiers: LCCN 2019021396 | ISBN 9780367261399 (hardback : acid-free
    paper) | ISBN 9780429297281 (ebook)
Subjects: LCSH: Electric power systems. | Renewable energy sources.
Classification: LCC TK1001 .B428 2020 | DDC 333.79/4--dc23
LC record available at https://lccn.loc.gov/2019021396

**Visit the Taylor & Francis Web site at**
**http://www.taylorandfrancis.com**

**and the CRC Press Web site at**
**http://www.crcpress.com**

**eResource material is available for this title at**
**https://www.crcpress.com/9780367261399**

# Contents

# *Preface*

In the last three decades our society and every country are facing energy challenges, such as energy conservation, use of increased energy efficient equipment, devices or systems, security of the energy supply, energy portfolio diversification, sustainability, or pollution reduction and control. Public electricity supply was originally developed in the form of local generation feeding local loads, the individual systems being built and operated by independent companies. This led to large centrally located generation stations supplying the loads via transmission and distribution systems. This trend may well have continued but for the need to minimize the energy generation and use environmental impacts, particularly pollutant emissions and concerns over the energy supply security of often imported fossil fuels. Consequently, governments, researchers, and energy planners are now actively developing alternative and cleaner forms of energy production, these being dominated by renewables (e.g., wind, solar, water bioenergy), distributed generation, local combined heat and power generation, and the use of waste products for energy production. Renewable energy sources with few exceptions, e.g., geothermal, tide energy originate from the Sun, are the oldest and the most modern energy forms and technologies used by humans. Conventional energy supplies are predicted to be exhausted sometimes in some future, and as energy demands keep increasing, it is expected that balancing the supply and demand becomes increasingly challenging. Renewable energy sources, present in large quantities in most of the part of the world, in contrast to the highly localized fossil fuels or mineral resources, are part of the today national energy policies with significant goals for increases into the generation energy portfolio. Consequently, it is expected that non-conventional energy sources and renewable energy resources will play a greater role in addressing the imbalance between energy supplies and demands. However, harnessing the renewable energy as important alternative energy sources is sometimes expensive and difficult to apply fully in particular sectors of society because of the location, variability intensity, and nature of the applications; therefore, specific matching of the renewable energy source to the application is a very important aspect of maximizing the utilization of renewable energy. In this regards, good resource characterization, assessment, analysis, the technology selections are critical factors in the renewable energy projects and developments. Many universities are now offering courses on renewable energy, distributed generation, and how sustainable, low pollutant emission technologies can be integrated effectively into the energy mix and power distribution systems. There is a great demand from employers for graduates who have studied such courses but, in almost all countries, a critical shortage of students with a strong education in energy and power engineering. The major aims of this book are to put the readers in contact with modern industry experiences, technologies, current and future trends of the use, exploitation, operation, and maintenance of renewable energy systems. Today there are constant increases of the production plants of renewable energy, guided by important social, economic, environmental, and technical considerations and aspects. The substitution of conventional energy production methods is a challenge in the current global and socio-economic context. New strategies of exploitation, energy uses of energy or maintenance procedures are emerging naturally as actions for solving the integration of these new aspects in the current systems of energy production.

This book's goals and objectives are to provide the students, instructors, as well as engineers, researchers, technicians, and others with technical interests in renewable energy systems with comprehensive knowledge of the renewable energy technologies. The book's first volume gives the essential knowledge of the major renewable energy sources and technologies, fundamental principles, source characteristics, assessment and analysis, and how they work, and are evaluated in order to properly select the optimum system or equipment for specific applications. Hydropower is well-established technology, while marine energy is still in the development phase, with almost no extended commercial applications. Overall Book Two covers major renewable energy sources, conversion systems, energy storage systems and technologies, grid integration, economic aspects, and environmental implications of the renewable energy systems, from basic and fundamental knowledge to resource assessment and design aspects. The first volume contains comprehensive descriptions of the major renewable energy sources, conversion systems, operation and control, while the second volume focuses on the major energy storage technologies, bioenergy, hydrogen economy, an introduction of the microgrids, distributed generation, environmental impacts of renewable energy sources, economics, and energy management. The first volume also gives comprehensive descriptions, theory, and assessment and analysis methods of solar and wind energy resources, photovoltaic systems, wind turbines, solar thermal systems, geothermal energy, and hydropower systems. Generation electricity from the Sun or wind can be achieved through photovoltaic systems, solar thermal energy systems and wind turbines, respectively. However, regardless the solar or wind conversion technology, good knowledge and an accurate assessment of the solar or wind energy resources are critical for any project or development. Chapter 1 of this volume introduces the basic energy concepts, physics background, energy concepts, a brief discussion of the energy trends and environmental impacts, fundamental and energy units and the system of units. Chapter 2 discusses the solar radiation concepts, solar equations, relationships, solar energy assessment, modeling and measurement methods, while Chapter 3 gives a good presentation of the solar thermal energy, operation, characteristics and main types of solar collectors, direct solar energy applications and solar thermal electricity generation. Chapter 4 is reserved to the photovoltaic systems, physics of the photovoltaic cells, basic cell models, theory and performances, cell efficiencies, types, materials, photovoltaic system control, operation, and applications. Chapters 5 and 6 focus on wind energy, wind velocity statistics, wind energy measurement techniques and assessment methods, wind turbine types, characteristics, turbine aerodynamics, as well as wind turbine components, control and operation, electric aspects of wind energy conversion systems, wind energy project and wind farm setting, design and development. Chapter 7 introduces the geothermal energy sources, Earth internal structure source origin, characterization and types, direct geothermal energy uses, and brief presentation of the geothermal electricity generation. The last volume chapter, Chapter 8 focuses on the hydropower systems, characterization and theory, with a larger presentation of the small hydro-electric systems, types of the turbine used in such systems, control, and generators. Two sections of Chapter 8 discuss the wave and tide energy system, theory, characteristics, and issues, followed by a brief presentation of the ocean thermal energy systems. This book originates from courses that the author taught in the areas of energy and power engineering, renewable energy systems, distributed generation, industrial energy systems, and energy management, as well as from the research projects that the author was involved in the last 20 years. Students or readers using this book must have only fundamental knowledge of mathematics, physics, and chemistry as usually expected for students enrolled in programs of engineering or sciences. Likewise, this book assumes no specific knowledge of power and energy engineering; it guides the reader through basic understanding from topic to topic and inside of each topic. Over 300 questions and problems are included in this volume. Each chapter also contains several solved examples to help the readers and instructors to understand and/or to teach the materials. A rich, comprehensive, and up-to-date literature is also included at the end of each chapter for professionals, engineers, students, and interested readers in the renewable energy topics and problems. The book is intended both as a textbook and as reference book for students, instructors, engineers, and professionals interested in renewable energy systems, technologies, operation, use, and design. The author is fully indebted to the students, colleagues, and co-professionals for their feedback and suggestions over the years, and last but not least to the editorial technical staff for support and help.

# *Acknowledgments*

To my wife Paulina Belu, my best friend and partner in life, for her patience, understanding, and support. The book is also dedicated to my children, Alexandru, Ruxandra, Mirela and my grandchildren Stefan-Ovdiu and Ana-Victoria for their support and life enjoinment.

# *Author*

**Radian Belu**, PhD, is an associate professor in the Electrical Engineering Department at Southern University and A&M College, Baton Rouge, Louisiana. He has a PhD in power engineering and a PhD in physics. Before joining Southern University, Dr. Belu held faculty and research positions at universities and research institutes in Romania, Canada, and United States. He also worked for several years in industry as project manager, senior engineer, and consultant. His research focuses on energy conversion, renewable energy, microgrids, power electronics, climate, and extreme event impacts on power systems. He taught and developed courses in power engineering, renewable energy, smart grids, control, electric machines and environmental physics. His research interests include power systems, renewable energy systems, smart microgrids, power electronics and electric machines, energy management and engineering education. Dr. Belu has published 1 book, 15 book chapters, and over 200 papers in referred journals and in conference proceedings. He has been PI or Co-PI for various research projects in the United States and abroad.

# 1

## Energy, Environment, and Renewable Energy

### 1.1 Introduction, Basic Energy Concepts, Fossil Fuels Sources

Energy conversion, such as that which is concerned mainly with converting direct or indirect forms of solar radiant energy to electrical, mechanical, or chemical energy, is the main topic of this chapter, as well as the book. Energy cannot be created or destroyed, but is only transformed from one form to another. The sole exception is nuclear, which is derived from a reduction in mass of the fuel but this, however, is mostly handled by physicists. Thus, strictly speaking, energy engineering is simply the engineering of transformation of energy between its different forms. One can distinguish between primary forms of energy that are found naturally, and secondary forms into which they can be transformed to enable easier transportation, usage, and storage. It is possible to necessitate more than one transformation to get it into a final useful form. There also energy forms that need less processing since they are closer to the form in which they will be finally used, which is the final state in which it can be used for some purpose. An example of a primary form is petroleum, which is found in nature, and a secondary form is electricity, which is not found in nature but which can be obtained from it. The useful form may be work that is needed to perform a certain mechanical task. Energy is an essential component of our daily lives and a vital source of economic development and national welfare. Humans discovered and learned how to make fire to cook meat, to deter predators, and to make tools and deadly weapons from metals, shifting the energy balance nature into irreversible one. During Industrial Revolution methods were discovered to convert heat into electricity, the most versatile and convenient form of energy. Electricity enabled astonishing scientific and technological advances, transforming our civilization and way of life. However, it comes with an unprecedented fossil fuel use and adverse environmental impacts, making us also dependent on a complex energy infrastructure for transportation, communication, heating, lighting, industrial processes and distribution of goods. Energy conversion is a multidisciplinary and disparate subject of applied sciences, requiring an understanding of physical and engineering principles. Energy issues also tend to be open and controversial issues, need to be address with an open an independent mind, being a rewarding and intellectually stimulating exercise. Throughout our history, the energy harnessing in its various forms presented great challenges and stimulated scientific and technological discoveries, while our energy usage has increased at an accelerate pace.

Modern energy technologies are the results of centuries of advances in science, technology, and gradual design improvements. To meet the global needs of economic growth there is a dramatic increase in the energy demand. Unlike the developed countries the developing countries are struggling to meet the increasing energy demands and challenges. Increased energy needs put considerable stresses on the Earth's resources and have had increasingly adverse environmental impacts. We are at acrossroads; our energy use must be critically analyzed to determine more sustainable, environmentally friendly, and efficient approaches for generation, distribution, and use of this vital component of our lives. The twenty-first-century economies face a two-fold energy challenge: meeting the needs of billions of people who still lack access to energy services while participating in a global transition to clean and low-carbon energy systems. Both demand urgent attention. The access to reliable, affordable, and socially acceptable energy services is a pre-requisite for alleviating extreme poverty and meeting societal development goals. The second one because of emissions from developing countries are growing rapidly and are contributing to environmental problems, putting the health and prosperity of people at risk. Historically, the energy use is marked by four trends: (1) rising consumption and transition from traditional energy sources to commercial energy forms (e.g., electricity or fossil fuels), (2) steady improvement in the power and efficiency of energy technologies, (3) a fuel diversification and de-carbonization trend for electricity production; and (4) improved pollution control and lower emissions. These trends have largely been positive. However, the technology improvement rate has not been sufficient to keep pace with the rapid demand growth negative consequences. The challenge is not so much to change course as it is to accelerate progress, toward increased energy efficiency and lower-carbon energy sources. This would have many concurrent benefits for developing countries in terms improving public health, making feasible a broader access to basic energy services and future economic growth. Moreover, to the extent that sustainable energy policies promote the development of indigenous renewable-energy industries, having additional benefits of creating new economic opportunities, reducing exposure to volatility of energy markets, and conserving resources for internal investments, by reducing overall energy costs. In this context, renewable energy is becoming more relevant part of the energy solutions, included in national policies, with goals to be a significant part of generated energy in the near future.

As a comprehensive overview of renewable energy systems, we are exploring the use of the Sun, wind, biomass, geothermal resources, and water to generate more sustainable energy. Taking a multidisciplinary approach, this book explains the renewable energy fundamentals. Starting with solar power, wind, small hydro power, other renewable energy sources, energy storage devices, grid integration issues, environmental and economic issues are also discussed. We have to keep in mind that there is a strong correlation between standard of living, as measured by the gross domestic product (GDP) per capita and the energy consumption per capita (see Table 1.1). It is natural that less developed countries will seek to increase GDP and thereby increase their energy consumption per capita. While in the developed countries the population size is almost constant, in the developing counterpart there are net increases in the population size. The net effect is a significant increase in global energy demand. The forecasts are that are no significant changes in the fossil fuel-based energy generation, over the next five years, either it is expected increase in the renewable energy sources. Oil will still be used mainly for transportation. Oil, coal, and natural gas are representing about 86% of all primary energy production.

Most of the fossil fuels (coal, oil, and natural gas) were formed from the ancient life stock remains over the course of millions of years. Some energy sources are either stores

**TABLE 1.1**

Correlation between Human Welfare Indicator and Energy Consumption

| Indicator of Human Welfare | Commercial Share of Total Energy | | |
|---|---|---|---|
| | 0%–20% | 21%–40% | 41%–100% |
| Life expectations (Years) | 59.8 | 69.0 | 69.5 |
| Probability not surviving to 40 | 21.7 | 9.4 | 9.1 |
| School enrollment (%) | 52.4 | 65.4 | 76.9 |
| Children underweight (%) | 40.9 | 15.1 | 11.9 |
| No access to clean Water (%) | 22.8 | 20.9 | 12.8 |

(repositories) of energy, usually chemical and nuclear that can be liberated for use. Other energy sources are in the form of energy flows through the natural environment, presented in varying degrees at a particular place and/or time. Examples of the first type are coal, oil, uranium, or natural gas, while wind and solar energy are of the second type. Electricity created in an energy conversion system requires energy input greater than the electricity itself, since some energy always is lost to the environment. Notice that electricity is not an energy source, but a product of whatever energy source was used. Fossil fuels are highly concentrated stores of energy (higher energy density), compared with most of the renewable energy sources, being easily and cheaply to collect, store, ship, and use where and when is desired or needed. The term *energy carrier*, thus a carrier of the above defined energy is a substitute that could be used to produce useful energy, either directly or by conversion processes. According to the degree of conversion, energy carriers are classified as *primary* or *secondary energy carriers* and as *final energy carriers*. The respective energy content of these energy carriers consists of *primary energy, secondary energy*, and *final energy* defined as follows:

*Primary energy carriers* are substances which have not yet undergone any technical conversion, whereby term primary refers to the energy content of the energy carriers and the *primary* energy flows. From primary energy (e.g., wind energy, solar radiation) or primary energy carriers (e.g., coal, crude oil, or biomass), secondary energy or secondary energy carriers can be either produced directly or by conversion energy steps.

*Secondary energy carriers* are energy carriers, produced from primary or other energy carriers directly or by one or several technical conversion processes (e.g., gasoline, heating oil, electrical energy), whereby the term secondary energy refers to the energy content of the carrier and the corresponding energy flow. This processing of primary energy is subject to conversion and distribution losses. Secondary energy carriers and secondary energies are available to be converted into other secondary or final energy carriers or energies by the consumers.

*Final energy carriers* and *final energy* respectively are energy streams directly consumed by the final user (e.g., light, fuel oil, wood chips or district heating at the building substation). They result from secondary and possibly from primary energy carriers, or energies, minus conversion and distribution losses, self-consumption of the conversion system, and non-energetic consumption. They are available for the conversion in useful energy.

*Useful energy* refers to the energy available to the consumer after the last conversion step to satisfy the respective requirements or energy demands (e.g., space heating, food preparation, information processing and distribution, transportation). It is produced from final energy carriers or final energies, reduced by losses of this last conversion (e.g., losses due to dissipation by a light bulb to generate light, losses of wood chip fired stove to provide heat). *The entire energy quantity available to humans, to be used is referred to as energy basis.* It is composed by the energy of the predominantly exhaustible resources and the largely renewable energy sources. In terms of energy resources, generally fossil and recent are distinguished, as: Fossil energy resources are energy stocks that have formed during ancient geologic ages by biologic and/or geologic processes, subdivided into fossil biogenous energy resources (i.e., stocks of energy carrier of biological origin) and fossil mineral (i.e., stocks of energy carrier of mineral or non-biological origin). The former includes hard coal, natural gas, and crude oil deposits, whereas the latter comprises, for instance, the energy contents of uranium deposits and resources to be used for nuclear fusion processes. Recent resources are energy resources that are currently generated, for instance, by biological, geophysical, and atmospheric processes, including, among others, the biomass energy content and the potential energy of a natural reservoir.

Energy sources, by contrast, provide energy systems over a long period of time; they are thus regarded as almost *inexhaustible* in terms of human times. But these energy flows are released by natural and technically uncontrollable processes from exhaustible fossil energy resources (like fusion processes within the Sun). The available energies or energy carriers can be further subdivided into fossil biogenous, fossil mineral and renewable energies or fossil biogenous, fossil mineral and renewable energy carriers.

- Fossil biogenous energy carriers primarily include the energy carriers, such as coal (lignite and hard coal) as well as liquid or gaseous hydrocarbons (such as crude oil and natural gas). A further differentiation can be made between fossil biogenous primary energy carriers (e.g., lignite) and fossil biogenous secondary energy carriers (e.g., gasoline, diesel fuel).

- Fossil minerals energy carriers comprise all substances that provide energy derived from atomic fission or nuclear fusion (such as uranium, thorium, or hydrogen).

- The term *renewable energy* refers to primary energies that are regarded as inexhaustible in terms of human (time) dimensions. They are continuously generated by the energy sources, such as solar energy, tidal energy, or geothermal energy. The energy produced within the Sun is responsible for a multitude of other renewable energies (such as wind and hydropower) as well as renewable energy carriers (such as solid or liquid bio-fuels).

- The energy content of the waste can only be referred to as renewable if it is of non-fossil origin (e.g., organic domestic waste, waste from the food processing industry). Properly speaking, only naturally available primary energies or primary energy carriers are renewable but not the resulting secondary or final energies or the related energies carriers. For instance, the electric energy generated from renewable energy sources by means of technical conversion processes itself is not renewable, since it is only available as long as the respective technical conversion systems are operated.

## 1.2 Renewable Energy Sources

Renewable energy is generally defined as energy coming from naturally replenished energy resources on a human timescale, such as sunlight, wind, rain, tides, waves, and geothermal heat. Renewable energy can replace conventional fossil fuels in three areas: electricity generation (including off-grid energy service), heating and cooling, and engine fuels. In all forms, renewable energies derive directly from the Sun, or from heat generated deep within the Earth. Included in this definition is electricity and heat generated from solar, wind, ocean, hydropower, biomass, geothermal resources, biofuels and hydrogen derived from renewable resources. Vast majority of renewable energy sources derive from solar radiation. Direct solar energy refers to solar thermal energy conversion and photovoltaics, while indirect solar energy includes wind and wave powers, hydropower, and biofuels. Other types of renewable energy are the tides (mainly due to moon attraction) and geothermal power. The magnitude of renewable energy sources is huge and may easily supply all our present and future energy needs. Non-solar renewable energies are those that do not depend direct on solar radiation. One type of renewable energy, not derived from natural resources, is the conversion the wastes into energy, often by producing methane that can be used to generate electricity or to fuel engines. It is very important to highlight the difference between primary energy sources and energy storage medium. Major advantages of the renewable energy include: sustainable, virtually inexhaustible and nonpolluting fuel is free, ideal for off-grid and distributed generation, while the main drawbacks are: variability and intermittence, lower energy density, generation sites often located far from populated areas, large initial investment, high maintenancecosts and sometimes adverse environmental impacts. It is out of question that in the future, large proportions of energy are from renewable energy sources. Renewable energy can be regarded as one of the fundamental premises for building a sustainable global society. A truly sustainable energy source is not only renewable, but also sustainable, meaning that energy production and use is endured by the biosphere, the biodiversity, and the social stability of the human culture.

Renewable energy resources exist over wide geographical areas, in contrast to conventional energy resources, concentrated often in a very limited number of countries. Rapid deployment of renewable energy and energy efficiency are resulting in significant energy security, climate change mitigation and economic benefits. There is also strong public support for promoting renewable energy. At the national level, many nations around the world already have renewable energy contributing significantly to their energy supply mixture. Renewable energy markets are projected to continue to grow strongly in the coming decades and beyond. For example, solar energy, the most important renewable energy source, is plentiful, having the highest availability compared to other energy sources. The amount of solar energy supplied to the Earth in one day is sufficient to supply the world total energy needs for an entire year. Solar energy is clean and almost free of emissions, it does not produce pollutants or harmful by-products. Solar energy conversion into electrical energy has many application fields. Residential, vehicular, space, aircraft and naval applications are the main solar energy applications. Sunlight has been used as an energy source by ancient civilizations to ignite fires and burn enemy warships using "burning mirrors." Until the eighteenth century, solar power was used for heating and lighting purposes. During the 1800s, Europeans started to build solar-heated greenhouses and conservatories.

The main renewable energy sources give rise to a multitude of very different energy flows and carriers due to various energy conversion processes occurring in nature. In this respect, wind energy, hydropower, ocean current energy, bio-fuels all represent conversion of the solar energy. The energy flows available on Earth resulting from renewable energy sources vary tremendously, in terms of energy density or with regard to spatio-temporal variations. Major renewable energy sources include solar radiation, wind energy, hydropower (in all forms), photo-synthetically fixed energy and geothermal energy. Appropriate techniques permit the exploitation and conversion of different renewable energy carriers into secondary or final energy, energy carriers or useful energy, respectively. Currently, there are tremendous variations in terms of utilization methods, technology, and given perspectives. However, not all options are possible for every site or location. The most promising renewable energy options include:

- Solar heat provision by active systems (i.e., solar thermal collector systems),
- Solar thermal electric provision (i.e., solar tower plans),
- Photovoltaic energy conversion systems,
- Power generation by wind energy conversion systems,
- Power generation by hydropower to provide electric energy,
- Utilization of ambient air and shallow geothermal energy for heat provision,
- Utilization of deep geothermal energy resources for heat and/or power provision,
- Wave and tidal energy for electrical energy provision, and
- Utilization of photo-synthetically fixed energy for heat, power, and transportation fuels.

Solar radiation represents the electromagnetic energy emitted by the Sun, while the terrestrial solar energy is the portion of the total solar radiation reaching the Earth's surface. The terms *insolation* and *irradiance* are used interchangeably to define solar radiation incident on the Earth per unit of area per unit of time as measured in W/m² or Wh/m²/day. The solar radiation amount, reaching the Earth's surface at any given location and time depends on many factors, including time of day, season, latitude, surface albedo, the atmosphere translucence and the weather conditions. The primary causes of winds (horizontal atmospheric motions) are the solar radiation uneven Earth and atmosphere heating and Earth's rotation. The atmosphere reflects about 43% off the incident solar radiation back into space, absorbs about 17% of it in the lower troposphere and transmits the remaining 40% to the surface of the Earth, where much of it is then reradiated into the atmosphere. The Sun short wavelength radiation (0.15–4 mm) passes readily through the atmosphere, while the longer wavelengths (5–20 mm) are absorbed by the atmospheric water vapors. Thus, Earth radiation is primarily responsible for the warmth of the atmosphere near the Earth's surface. Heat is also transferred from the Earth's surface to the atmosphere by conduction and convection. On average, the total amount of energy radiated to space from the Earth and its atmosphere must be equivalent to the total solar radiation amount absorbed, or the temperature of the Earth and its atmosphere would steadily increase or decrease. The more nearly perpendicular the Sun's rays strike the Earth, the more solar radiation is transferred through the atmosphere. During the year, tropical regions receive significantly more solar energy than the polar areas. Winds and ocean currents level out this thermal imbalance, preventing the tropical regions from getting progressively hotter and the polar regions from getting progressively colder. In addition, Earth's surface inhomogeneity—land, water, desert, forest, rocks, sands and so on—leads to differences in solar radiation absorption and reflection back to the atmosphere, creating large differences in atmospheric temperatures, densities and pressures, which in turn are shaping the local wind regimes. It may be worth noting that the power densities per-area by kinetic wind energy and solar radiation are the same order of magnitude, when considering exploitable values in a given area. At 20 m/s the wind exerts on a vertical plane 1.04 kW/m², while the solar radiant flux density on a horizontal surface on June 21st at noon, 50 N latitude is 1.05 kW/m². However, the efficiency conversion is about 40% for wind energy conversion systems and about 15% for photovoltaic generators.

## 1.3 Physics and Energy Fundamentals

The term *energy* is often used without a great deal of thought and is applied in different context and situations. Scientific energy concept is serving to reveal the common features in processes such as burning fuels, charging batteries, propelling machines and equipment, etc., so it is worthy to have physical meaning of the energy applied in the technical context, describing the utilization of the energy resources. Since often many people have the tendency to mix energy, work and power, this chapter subsection will define these quantities and point out the differences among them. *Energy* is broad concept, defined as the capacity or potential of systems to perform useful work. However, the ability to perform work from chemical, nuclear, and solar energy is given only if these forms of energy are transformed into mechanical and/or thermal energy. According to the law of conservation of energy, the total energy of a system remains constant, though energy may transform into another form. Input and output work change the energy content of a system or body. Energy exists in many forms such as mechanical energy, thermal energy, electrical energy, electromagnetic energy, chemical energy, atomic energy, and nuclear energy. *Energy, E,* is defined in physics as the capacity of a system to do work, *W*. Work is the consequence of expenditure of energy and is defined as the scalar product of force, *F,* and the distance, *d*, its point of application moves during a period of time, being expressed as:

$$W = \vec{F} \cdot \vec{d} \tag{1.1}$$

The SI (International System) unit of energy is the joule (J) or newton-meter (N·m). When the force is variable, the work is computed by an integral describing the motion of force, between two points, *a* and *b*. Work and its time derivatives, the instantaneous power is expressed as:

$$W = \int_a^b \vec{F} \cdot d\vec{x}$$

and

$$\frac{dW}{dt} = \vec{F} \cdot \frac{d\vec{x}}{dt} = \vec{F} \cdot \vec{v} \tag{1.2}$$

For rectilinear motion, work is done by the application of a force. Multiplying Newton's Second Law for linear motion by the force $F$ gives:

$$W = F \cdot v = \frac{d}{dt}\left(\frac{1}{2}mv^2\right) \tag{1.3}$$

Power is the scalar product of force and velocity. Power is the rate at which energy is produced or consumed, and is expressed as:

$$P = \frac{dW}{dt} \text{ or } W = \int P dt \tag{1.4}$$

The joule is also the SI unit of work. In terms of fundamental metric units, the joule (J) is equal to $kg \cdot m^2 \cdot s^{-2}$. Assuming that power is constant in time, $t$ then the total energy utilized is:

$$W = P \cdot t \tag{1.5}$$

**Example 1.1:** What is the work done if a force of 100 N moves an object through a distance of 25 m? What is the average speed of the object if this displacement is carried out in 10 s? What is the power input?

**Solution:** Mechanical work, average speed, and input power are calculated as:

$$W = F \cdot d = 100 \cdot 25 = 25{,}000 \text{ J or } 25 \text{ kJ}$$

$$v = \frac{d}{t} = \frac{100}{10} = 10 \text{ m/s}$$

$$P = F \cdot v = 100 \cdot 10 = 1000 \text{ W or } 1 \text{ kW}$$

The kinematics and dynamics of rotating objects and the work and power in the case of ratio are very important in energy conversion; most of the energy convertors, such as electric generators or wind turbines, are rotating devices. The frequency, $f$, angular velocity, $\omega$, and angular acceleration, and $\alpha$ is expressed as:

$$\omega = \frac{d\theta}{dt} \text{ (rad/s)}$$

$$\alpha = \frac{d\omega}{dt} = \frac{d^2\theta}{dt^2} \left(\text{rad/s}^2\right) \tag{1.6}$$

Here $\theta$ is an angular displacement (measured in a suitable reference). This can be integrated as:

$$\omega = \omega_0 + \int \alpha \cdot dt$$

where $\omega_0$ is the initial angular velocity. For $\alpha$ constant, above relationship is:

$$\omega = \omega_0 + \alpha \cdot t \tag{1.7}$$

Rotational dynamics is governed by Newton's Second Law for rotation, expressed by:

$$\tau = I \cdot \alpha = I \frac{d\omega}{dt} \text{ (N·m)} \tag{1.8}$$

where $\tau$ is an applied torque (shown in Figure 1.1), and I is the mass moment of inertia. The unit for momentum of inertia is $kg \cdot m^2$, and the unit of angular velocity is rad/s. The term mass moment of inertia is not to be confused with the area moment of inertia, which for thin plates of uniform density the mass moment of inertia is the area density, i.e., mass per unit area, times the area moment of inertia. If, as a special case, $\alpha = 0$ and/or $I = 0$, then $\tau = 0$. The angular momentum of a rotating body is $I \cdot \omega$, and it can be changed only by the application of a torque. The mass moment of inertia of a concentrated mass at a distance R from the axis of rotation is $mR^2$. For any other geometrical shape, the moment of inertia can be calculated.

In a way similar to the linear motion, for the angular rotation, the mechanical energy equation is given by:

$$W = \tau \cdot \omega = \frac{d}{dt} \left( \frac{1}{2} I \cdot \omega^2 \right) \tag{1.9}$$

Similar to Equation (1.2), and by using Equation (1.4), in the case of rotation, if a torque $\tau$ rotates an object by an angle $\Delta\theta$, then the mechanical work done by the torque is:

$$W = \tau \cdot \Delta\theta \tag{1.10}$$

The mechanical power in this case is expressed as:

$$P = \frac{W}{\Delta t} = \tau \frac{\Delta\theta}{\Delta t} = \tau \cdot \omega \tag{1.11}$$

where $\Delta t$ is the time interval.

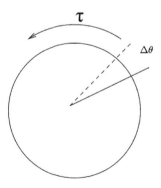

**FIGURE 1.1**
An object rated by an applied torque.

**Example 1.2:** A cylinder with mass moment of inertia of 100 kg·m² is rotating freely at 50 rad/s about its axis of symmetry. What is the constant retarding torque that must be applied to reduce the velocity to zero in 1 min.? What is the initial power of the cylinder?

**Solution:** Applying Newton Law for rotation, Equation (1.8), the constant retarding torque is:

$$\tau = I\frac{\Delta\omega}{\Delta t} = 100\frac{50-0}{60} = 83.333 \text{ N} \cdot \text{m}$$

From Equation (1.11) the initial mechanical power of this cylinder is:

$$P = \tau \cdot \omega = 83.333 \cdot 50 = 4166.7 \text{ W}$$

The total energy of a system can be subdivided and/or classified in various ways, being convenient to distinguish gravitational energy, mechanical energy, thermal energy, several types of nuclear energy (which utilize potentials from the nuclear force and the weak force), electric energy (from the electric field) and magnetic energy (from the magnetic field), among others. Many of these classifications overlap; for instance, thermal energy usually consists partly of kinetic and partly of potential energy. Some types of energy are a varying mix of both potential and kinetic energy. An example is mechanical energy which is the sum of (usually macroscopic) kinetic and potential energy in a system. Elastic energy in materials is also dependent upon electrical potential energy (among atoms and molecules), as is chemical energy, which is stored and released from a reservoir of electrical potential energy between electrons, and the molecules or atomic nuclei that attract them. Chemical energy is the energy associated with chemical bonds, the interaction energy between atomic electrons in a material. Energy can be absorbed or released during a chemical reaction as a result of changes in bonds between atoms. If during this process the energy is released (exothermic process) is of interest for energy generation. Electrical energy is associated with the electrons in conductors and is expressed in terms of voltages and currents. If a current, I, flows through a circuit with a resistance, R, with a voltage, V, across the resistance, then the power dissipated through the resistance is given by:

$$P = VI = I^2R = \frac{V^2}{R} \tag{1.12}$$

Current is measured in amperes (*A*), voltage in volts (*V*), and resistance in ohms (Ω). When the *V* and *I* are expressed in volts and amperes, *P* is expressed in watts. From Equation (1.4), electrical energy in joule is power in watts multiplied by time in seconds. More often electrical energy is expressed in kilowatt-hours (kWh) by dividing energy in joules by a conversion factor 3.6 × 10⁶ J/kWh. Energy can be transferred between systems in a variety of ways, such as the transmission of electromagnetic energy via photons, physical collisions which transfer kinetic energy, and the conductive transfer of thermal energy. Energy is strictly conserved and is also locally conserved wherever it can be defined. Classical mechanics distinguishes between kinetic energy, which is determined by an object's movement through space, and potential energy, which is a function of the position of an object within a field. Kinetic energy associated with moving objects defined for an object of mass, *m*, moving at velocity *v*, as:

$$E = \frac{1}{2}mv^2 \tag{1.13}$$

Kinetic energy is also associated with the rotational motion of an object, being given by:

$$E = \frac{1}{2} I \omega^2 \tag{1.14}$$

where again $I$ is the object momentum of inertia, and $\omega$ is the angular velocity. Potential energy is usually related to the gravitational potential terms, either the concept also applies to other situations, such energy contained by a compressed spring, the elastic potential energy. In the case of gravitational potential energy, an object of mass $m$ at a height $h$ has a potential energy given by:

$$E = mgh \tag{1.15}$$

Here $g$ is the acceleration due to the gravitation with an average value $g$ = 9.806 m/s². The mechanical energy of a body, $E$, which is a scalar, can be kinetic, due to its motion, or potential, due to its mechanical state, for example due to its height above a reference or the compression of a spring. The total mechanical energy is the sum of the kinetic and potential energies. Keep in mind that, for total energy balance, one must consider not only the work input but also the heat input. Mechanical energy is conserved only if there is no transformation from mechanical to thermal energy. Heat is transferred by a temperature difference. Thus, the conservation of mechanical energy is expressed as:

$$E_{mech} = E_{kin} + E_{pot} \tag{1.16}$$

We already discussed the kinetic energy for linear motion and rotation. To understand the potential energy concept, we consider a force being applied so that the body moves very slowly. The initial and final velocities are assumed to be zero, as are the kinetic energies. The work done by the force goes into potential energy, as we have assumed that there is no thermal energy and mechanical energy is conserved. Two examples of potential are discussed here: lifting a weight and compression of a string. For lifting of a weight, a force is needed to do work against gravity (this special force is called weight). The potential energy stored in this way is given by:

$$E_{pot} = mg(z_2 - z_1) = mgh \tag{1.17}$$

where the weight mg is assumed to be constant, and $z_1$ and $z_2$ are the initial and final heights. This potential energy only makes sense as a difference. In compression of a spring, also a force is needed to compress a spring, and the mechanical energy needed is then stored as energy of compression. In general the force and compression distance are related, such that

$$F = f(x)$$

In a linear spring, the relationship is of the form

$$F = k \cdot x \tag{1.18}$$

where $k$ is the spring constant. This is the simplest relationship for elastic potential energy, so the potential energy stored in a spring is then expressed as:

$$E_{pot} = \int_{x_1}^{x_2} kx \cdot dx = \frac{1}{2} k \left( x_2 - x_1 \right)^2$$

There are also useful relations for shafts, disks, and other rigid bodies that are rotating. The mass moment of inertia of a rotating solid cylinder of radius, $R$, and mass, $m$, is:

$$I = \frac{1}{2} mR^2$$

and its kinetic energy is given by Equation (1.14). In practice, instead of using $\omega$ in rad/s, the rotational speed is usually expressed in revolutions per minute (rpm) $n$ where

$$\omega = \frac{2\pi n}{60} \qquad (1.19)$$

Strictly speaking, mass and energy cannot be lost. However, the term is used if the mass or energy goes in a direction that is not desirable. The term is also very often used without further clarification in energy system in many different ways. (a) In one sense it is a leakage of mass. If, for example, a certain mass flow rate goes into a machine, and not all of it comes out where it should, there is a leakage. This could be due to an undesirable gap within the machine. (b) Another is a loss of energy such as heat transfer to the surroundings. (c) A third is a pressure loss, as what happens when there is a flow along a pipe. However, these are beyond our textbook and included topics, interested readers are directed to the references or elsewhere in the literature.

Chemical energy is the one associated with chemical bonds, the energy interaction between atomic electrons in a material. It can be absorbed or released during chemical reactions, due to the changes in the bonds between the atoms. Processes requiring energy input for the reaction are known as *endothermic*, while the ones releasing energy during the reaction are known as *exothermic*. The last ones are of interests in energy generation. Among the exothermic processes of great interest are the ones involving the oxidation of carbon, such as pure carbon oxidation, producing of carbon dioxide and releasing the (thermal) energy.

$$C + O_2 \rightarrow CO_2 + 32.8 \text{ MJ/kg} \qquad (1.20)$$

Pure carbon oxidation is a suitable approximation of burning coal, while other fossil fuels, such as woods, ethanol, and organic materials, involve the hydrocarbon oxidation. Some of the simple such reactions are methane (major component of natural gas) and ethanol (a common biofuel) burning. These reactions are producing steam as by-product. The energies in these reactions are referred as higher heating values (HHV) and include the latent heat of vaporization and are recovered the condensation of the produced steam. These two reactions are given below:

$$CH_4 + 2O_2 \rightarrow CO_2 + 2H_2O + 55.5 \text{ MJ/kg}$$
$$CH_6O + 3O_2 \rightarrow 2CO_2 + 3H_2O + 29.8 \text{ MJ/kg} \qquad (1.21)$$

Thermal energy of a gas results from the kinetic energy of the microscopic movement of molecules. Temperature is a characteristic of a body thermal energy, due to the internal motion of molecules. Two systems in thermal contact are in thermal equilibrium if they have the same temperature. In general, if the phase change (e.g., from solid to liquid) is not involved, the temperature of any material increases as it absorbs heat. The heat required to rise the system temperature, with *specific heat*, $c$, by an amount $\Delta T$ is expressed as:

$$\Delta Q = mc\Delta T \qquad (1.22)$$

**Example 1.3:** The specific heat of water is 4180 J/(kg·°C). Calculate the energy required to increase the temperature of 1 kg of water with 25°C.

**Solution:** By using Equation (1.11) to solve for the heat gives:

$$Q = (1 \text{ kg}) \times \left[ 4180 \text{ J}/(\text{kg}°\text{C}) \right] \times 25°\text{C} = 1.045 \times 10^5 \text{ J}$$

However, heat and temperature are different. Heat is the energy, and temperature is the potential for heat transfer from a hot to a cold place. Materials with large specific heat require a large amount of heat per unit of mass to rise their temperature by a given amount. These materials can store large amounts of thermal energy per unit of mass for a small increase in temperature, having applications in energy storage or solar thermal energy. In the heat transfer, work can also be done. Understanding of the energy concept arises from the laws of thermodynamics:

1. *Energy is conserved, cannot be created or destroyed, only transformed from one form to another.*

2. *Thermal energy, heat, cannot be transformed totally into mechanical work. Systems tend toward disorder, and in energy transformations, disorder increases.* Entropy is a measure of disorder. This means that some forms of energy are more useful than other forms. In other words: *the heat naturally flows from hot place to a cold place.*

Mathematically, the process of energy transfer is described by the *first law of thermodynamics*:

$$\Delta E = W + Q \tag{1.23}$$

where $E$ is the internal energy, $W$ represents the work done by the system, and $Q$ represents the heat flow. By convention, $Q$ and $W$ are positive if heat flows into the system and work is done by the system. When the thermal efficiency of the system is high, the heat term in Equation (1.19) is ignored:

$$\Delta E = W \tag{1.24}$$

This simplified equation is the one used to define the joule (J), for example. Errors often occur when working with energy, power, heat, or work. Units and quantities are mixed up frequently quite often. Wrong usage of these quantities and/or units can dramatically change the statements and cause misunderstandings. To have a simple understanding of this law, we can consider the behavior of system formed from one piece of hot metal and the other of cold metal that are brought into thermal contact. The system will attain thermal equilibrium by transferring heat from hot metal to cold one until the two pieces are at the same temperature. Other expression of this law is: *the entropy of the universe always increases.* Entropy is a measure of disorder. Thus, attaining thermal equilibrium increases the overall universe entropy.

Thermal energy can be used to do work only if heat flows from the hot side to the cold side of a system. An analogy is the conversion of potential energy into kinetic energy in mechanical systems. For example, if the water in a reservoir above a hydro-electric station remains in the reservoir, no electricity is generated. When the water is running down the hill and through station generators, electricity is generated. In this process, the gravitational potential energy of the stored water is converted into kinetic energy, and subsequently into electrical energy. Similarly, the thermal energy contained into hot material can be converted into other energy forms, which is the principle of heat engine operation. Besides the two laws of thermodynamics discussed above, there is a third one. It states that temperature of absolute zero cannot be attained. The third law details and its origin are not relevant to the textbook topics and are not discussed further. If the heat is moved from a hot reservoir to a cold reservoir, some of the thermal energy can be converted into mechanical work, and a device doing this is called: *heat engine*. Examples of heat engine include steam turbines, car internal combustion engines, jet engines, etc. Engine operation consists of removing heat from hot reservoir at temperature, $T_H$, while some of the heat is deposited into cold reservoir, at temperature $T_C$ and some is used to develop mechanical work. If the removed heat from the hot reservoir is $Q_H$, and the deposited heat into cold reservoir is $Q_C$, then the relationship including the developed work is:

$$Q_H = Q_C + W \qquad (1.25)$$

The terms "energy efficiency" and "energy conservation" have often been used interchangeably in policy discussions, but they do have very different meanings. Energy conservation is reduced energy consumption through lower quality of energy services, e.g., lower heating levels, through turning down thermostat levels, speed limits for cars, and capacity/consumption limits on appliances, often set by standards. Often it means doing without saving money or energy. It is strongly influenced by regulations, consumer behavior, and lifestyle changes. Energy efficiency is simply the ratio of energy services out to the energy input. It means getting the most out of every unit of energy. It is mainly a technical process caused by stock turnover where old equipment is replaced by newer more efficient ones. It is generally a by-product of other social goals: productivity, comfort, monetary savings or fuel competition. Measuring energy efficiency, particularly on a macro scale, is very difficult; there are methodological problems and it is very hard to measure between countries or sectors. *Efficiency* means different things to the two professions most engaged in achieving it. To engineers, efficiency means a physical output/input ratio. To economists, efficiency means a monetary output/input ratio, and also, confusingly, efficiency may refer to the economic optimality of a market transaction or process. Physical energy efficiency is defined as:

$$\eta = \frac{P_{OUT}}{P_{IN}} = \frac{P_{IN} - Losses}{P_{IN}} \qquad (1.26)$$

**Example 1.4:** An electric motor consumes 100 W of electricity to obtain 87 W of mechanical power. Determine its efficiency.

**Solution:** Because power is the rate of energy utilization, efficiency can also be expressed as a power ratio. The time units cancel out, and we have:

$$\eta = \frac{\text{Power Output}}{\text{Power Input}} = \frac{87 \text{ W}}{100 \text{ W}} = 0.87 \text{ or } 87\%$$

Technical systems perform the energy conversions with various efficiencies. The ratio of $Q_C$ and $Q_H$ can as the ratio the reservoir temperatures. For a heat engine, it can be written as:

$$\eta = 1 - \frac{Q_C}{Q_H} = 1 - \frac{T_C}{T_H} \qquad (1.27)$$

The efficiency form involving cold and hot reservoir temperatures is more convenient because the temperature is much more easily measured quantity than heat. The efficiency as stated by Equation (1.27) is known as *ideal Carnot efficiency*, after the name of French engineer Sadi Carnot and is the maximum efficiency attainable by a heat engine. Real heat engines typically operate at efficiencies that can be much less than the Carnot efficiency. The ideal Carnot efficiency is valid for ideal processes, taking place in either direction equally, existing in the real world only for limited cases. Time reversible ideal processes require that net entropy of the S remains constant. An alternative definition of the second law of thermodynamics is that in any real process $dS > 0$. Working on the same principles, as a heat engine is the heat pump, which uses mechanical work to transfer heat from a cold reservoir to a hot one. Similarly, conservation of energy requires that:

$$W + Q_C = Q_H \tag{1.28}$$

**Example 1.5:** What is the thermal efficiency of the most efficient heat engine that can run between a cold reservoir at 20°C and a hot reservoir at 200°C?

**Solution:** Carnot engine has the ideal or maximum efficiency, so:

$$\eta = 1 - \frac{T_C}{T_H} = 1 - \frac{273.15 + 20}{273.15 + 200} = 0.38 \text{ or } 38\%$$

Heat pumps have practical applications for the heat transfer from cold areas/reservoir to hot ones. They have applications in combined heat and power generation, in recovering the wasted heat, or moving heat from the outside to the inside in a cold day. However heat pumps may be economically attractive for heating purposes, careful considerations of cost, local climate, and other factors are necessary to assess their viability. Heat pump performance is expressed through the *coefficient of performance* (COP), the ratio of heat deposited in the hot reservoir to the work done, expressed as:

$$COP = \frac{Q_H}{W} \tag{1.29}$$

COP is a quantity greater than 1 (or, in percent, greater than 100%). Using the relationship between heat and temperature, COP can be expressed as:

$$COP = \frac{Q_H}{Q_H - Q_C} = \frac{T_H}{T_H - T_C} \tag{1.30}$$

In the case of refrigerator, the COP expression is:

$$COP = \frac{Q_C}{Q_H - Q_C} = \frac{T_C}{T_H - T_C} \tag{1.31}$$

There are six key quantities, useful in the description of a thermal system, such as thermal power plant: temperature $T$, pressure $p$, specific volume (the inverse of the density, i.e., the volume per unit of mass), specific internal energy $u$, specific enthalpy $h$ and specific entropy $s$. However, only two thermodynamic quantities are strictly needed to completely specify the thermal state of a system. We already introduced the concept of internal energy (first law of thermodynamics). Here the specific quantities are referred to the unit of mass. Specific enthalpy is defined as:

$$h = u + pv \tag{1.32}$$

Enthalpy is very useful in describing the heat transfer at constant pressure (e.g., in boilers, or condensers), where the change in the enthalpy is equal to the heat input, or in adiabatic ($Q = 0$) compression or expansion (e.g., compressors and turbines), where the network on the shaft is equal to the change in the enthalpy. The concept of entropy arises from the second and is a measure of the degree of disorder of a system. From a thermodynamic point of view there are two types of processes: reversible and irreversible processes. In the first one, the system and the surroundings can recover their initial states by changing the system slowly enough that it remains in a quasi-static thermal equilibrium throughout the process. In irreversible process the system and the surroundings are changed in such away that they are not able to return to their original states. Mathematically the entropy change is expressed as:

$$\Delta s = \frac{\Delta Q_{rev}}{T} \tag{1.33}$$

Here, $\Delta Q_{rev}$ is the heat supplied to the system reversible at absolute temperature, $T$. Notice that there is no change in entropy in a reversible adiabatic process, where $\Delta Q_{rev} \approx 0$, while in an irreversible process there is a net increase in entropy.

In some way, nuclear energy is similar with chemical energy, discussed above, being the energy associated bonds between particles inside the atoms. The most relevant nuclear energy is the one between the neutrons and protons within the nucleus rather the one to the bonds between atoms, involving the atomic electrons. Nuclear bonds represent much larger energy amounts than in chemical bonds. Energy that can be released in nuclear reactions is many order-of-magnitude higher than the one in chemical reactions. The release of the nuclear energy can be made through the fission (breaking up) of heavy nuclei like uranium and plutonium or through the fusion (bonding) of light nuclei like isotopes of hydrogen. It can be estimated through the Einstein's relationship, involving mass and the speed of light (300,000 km/s):

$$E = mc^2 \tag{1.34}$$

However, the energy released in an exothermic nuclear reaction is given in terms of changing nuclear mass, expressed as:

$$E_{exo} = \Delta mc^2 \tag{1.35}$$

*Electromagnetic radiation* is associated with the electromagnetic field under the form of waves, such as light. The electromagnetic radiation covers a very wide range of frequencies or wavelengths, X-rays, ultraviolet radiation, visible light, infrared radiation, radio waves, etc. Electromagnetic radiation for the Sun is our most important source of energy, being responsible for most other energy sources, such as fossil fuels, wind, solar energy, biomass, etc. Expressed in terms of photons or energy quanta electromagnetic energy is related to electromagnetic radiation frequency, $f$ and Planck's constant ($h = 6.626 \times 10^{-26}$ J·s), through:

$$E = hf \tag{1.36}$$

Electromagnetic radiation in the infrared through ultraviolet and X-ray regions, is produced by electron transitions between atomic energy levels, and can be either artificial or natural origin. Long wavelength radio waves may be generated by radio transmitters, radars, etc. High energy, short wavelength radiation commonly comes from excited state transitions in nuclei.

## 1.4 Heat Transfer and Essential of Fluid Mechanics

Basically, there three types of heat transfer: *conduction, convection* and *radiation*. Conduction is the thermal transfer process due to the molecule random motions. The average molecule energies are proportional to the temperature. The heat flow rate, in the steady-state along the length of bar, d of cross-sectional area, $A$ with one end at higher temperature, $T_1$ and the other at lower temperature, $T_2$ is given by the Fourier law of heat conduction:

$$Q = kA \frac{T_1 - T_2}{d} \tag{1.37}$$

where $k$ (W·m$^{-1}$·K$^{-1}$) is the thermal conduction of the bar material.

**Example 1.6:** A steel bar of 2 m and a cross-sectional area of 10 cm$^2$, has at one end a temperature of 1200°C and 200°C at the other end. Calculate the heat thermal flow along the bar in the steady-state, ignoring the heat losses from the bar surface, if the steel thermal conductivity is 50 W·m$^{-1}$·K$^{-1}$.

**Solution:** From Equation (1.32) the heat flow along the bar is calculated as:

$$Q = 50 \times 10 \times 10^{-4} \frac{1200 - 200}{2} = 25 \text{ W}$$

*Convection* represents the heat transfer through the fluid bulk motion, the actual movement or circulation of a substance. It takes place in fluids, such as water or air, where the materials are able to flow. Much of the heat transport that occurs in the atmosphere and ocean is carried by convection. However, the atmospheric circulation consists of vertical as well as horizontal components, so both vertical and horizontal heat transfer occurs. Considering a fluid of density $\rho$, temperature $T$, and moving with velocity $v$, then heat flow rate per unit of area is the product of the mass flow per unit area per second, $\rho v$ and the thermal energy per unit mass, $cT$:

$$\frac{Q}{A} = \rho v c T \tag{1.38}$$

When a cold fluid is forced to flow over a hot surface, the heat transfer rate in this forced convection from the surface to the fluid is higher than in the case of a stationary fluid. The temperature gradient at the surface is very large, and the fluid layer above the surface is rapidly heated by thermal convection, and the heat transfer rate per unit the area is often expressed as:

$$\frac{Q}{A} = Nu \frac{k(T_S - T_\infty)}{L} \tag{1.39}$$

Here, $T_S$ and $T_\infty$ are the surface and the fluid temperatures, $L$ is the characteristic length, and $Nu$ is the dimensionless Nusselt parameter. The choice of $L$ depends on the fluid-surface geometry, while the Nusselt parameter is function of two other non-dimensional parameters, the Prandtl and Reynolds numbers, depending on the fluid mechanical and thermal properties, obtained from empirical correlations.

*Radiative heat transfer* represents the energy transfer through the electromagnetic radiation, which includes the heat transfer into the vacuum. The power per unit area transferred from a surface at temperature $T$ is given by the Stefan-Boltzmann law:

$$P_{rad} = \varepsilon \sigma T^4 \qquad (1.40)$$

Here, $\varepsilon$ is the surface emissivity (a dimensionless number ranging from 0 to 1, depending on the surface nature), $\sigma \approx 5.67 \times 10^{-8}\,\text{W·m}^{-2}\text{·K}^{-4}$ is the Stefan-Boltzmann constant. Opaque surfaces absorb electromagnetic radiation from the environment. The surface absorptivity and emissivity are the same. The absorption rate per unit of area, for the environment temperature $T_0$ is:

$$P_a = \varepsilon \sigma T_0^4$$

The net emission rate per unit of area per unit of time is then given by:

$$P_{rad} = P_e - P_a = \varepsilon \sigma \left( T^4 - T_a^4 \right) \qquad (1.41)$$

A blackbody is a surface that is absorbing all incident electromagnetic radiation. The outer Sun surface determines the electromagnetic radiation flux incident to the upper Earth atmosphere. Another important relationship is the one describing mathematically the relationship between the temperature ($T$) of the radiating body and its wavelength of maximum emission ($\lambda_{max}$), the Wien's displacement law:

$$\lambda_{max} = \frac{C}{T} \qquad (1.42)$$

Here, $C$, the Wien's constant is equal to 2898 μm·K.

**Example 1.7:** By using Wien's law estimate the maximum wavelengths of the Sun and the Earth, assuming that the Sun temperature is 6000 K, the one of the Earth is 3000 K.

**Solution:** Applying Wien's law for the Sun and the Earth temperatures, we found:

$$\lambda_{max}(\text{Sun}) = \frac{2898}{6000} = 0.483\,\mu\text{m}$$

and

$$\lambda_{max}(\text{Earth}) = \frac{2898}{300} = 9.660\,\mu\text{m}$$

Radiation is often identified by the effects it produces when it interacts with an object. We are dividing radiant energy into categories based on our ability to perceive them; however, all wavelengths of radiation behave in a similar manner. An important difference among the various wavelengths is that the shorter wavelengths are more energetic. For example, the Sun emits all forms of radiation, in varying quantities with over 95% of solar radiation emitted in wavelengths between 0.1 and 2.5 μm, with much of energy concentrated in the visible and near-visible parts of the electromagnetic spectrum. The visible light spectrum band, wavelengths between 0.4 and 0.7 μm, represents over 43% of the total energy emitted, while the rest lies in the infrared (IR), about 49%, and ultraviolet (UV), about 7%. In order to have a better understanding how the Sun's radiant energy interacts with Earth's atmosphere and land-sea surface, it is helpful to have a general understanding of the basic radiation laws. However, this is beyond the scope of this book, and interested readers are direct to elsewhere in the literature. The radiation is the dominant mode energy transfer mode in the power plant furnaces. The atmosphere effects on the electromagnetic radiation transmission are very important in determining the Earth surface temperature.

### 1.4.1 Basics of Fluid Mechanics

In this section, a very brief summary of the physical properties of fluids is given, necessary to describe and understand the energy conversion in processes, such as wind energy, hydro power, or wave and tidal power. The bulk physical properties of a fluid are stated in terms of: fluid density ($\rho$), mass per unit volume; pressure ($p$), force per unit of area; and viscosity, force per unit area due to internal friction arising from the relative motion between adjacent fluid layers, acting in tangential directions. A useful concept to visualize the fluid velocity field is the notion of the *streamline*, the lines parallel to the direction of motion in all points of the fluid. Any fluid mass element flows along a stream-tube bounded by neighboring streamlines. In applications involving fluids, the forces within the fluid are often discussed more effectively in terms of fluid pressure, defined by:

$$\text{Fluid Pressure} = p \equiv \frac{\text{Force}}{\text{Area}}$$

The pressure on a surface submerged at depth, $h$ in a fluid having density $\rho$ (kg/m³), is given by:

$$p = \rho g h \tag{1.43}$$

And, the force is expressed as:

$$F = \int_A p(h) \cdot dh$$

One of the fundamental laws of fluid mechanics is the *conservation of mass* (mass continuity). If the speed of a fluid and the stream-tube cross-sectional area are, $u$ and $A$, for the fluid confined in the stream-tube, the mass flow per second is constant, expressed as:

$$\rho u A = Const. \tag{1.44}$$

In Equation (1.39), the left side represents the flow rate (in kg/s).

> **Example 1.8:** What is the mass flow rate of water at a given section of a pipe of diameter 10 cm where the fluid velocity is 30 cm/s?
>
> **Solution:** Assuming the water density to be 1000 kg/m³, the mass flow rate is:
>
> $$\dot{Q} = \rho \cdot u \cdot A = 1000 \cdot 0.3 \cdot \left( \frac{\pi}{4} 0.1^2 \right) = 2.355 \text{ kg/s}$$

In many practical applications, viscous forces are negligible than forces due to the gravity and pressure gradients over large areas of fluid flows. With this assumption the equation of energy conservation in a fluid, the Bernoulli equation (theorem), for steady flow is of the form:

$$\frac{p}{\rho} + gz + \frac{1}{2}u^2 = Const. \tag{1.45}$$

For a stationary case, when $u = 0$ everywhere in the fluid, Equation (1.45) reduces to:

$$\frac{p}{\rho} + gz = Const. \tag{1.46}$$

Equation (1.46) is the equation of hydrostatic pressure, stating that the fluid at a given depth, $z$ is all at the same pressure $p$, as expressed by the Equation (1.43). The significance of the Bernoulli Equation is that it states that the pressure in a moving fluid decreases as the fluid speed increases.

> **Example 1.9:** The atmospheric pressure on the surface of a lake is $10^5$ N/m². Calculate the pressure at depth of 10 m, assuming that the water density is $10^3$ kg/m³, and the water is stationary.
>
> **Solution:** From Equation (1.46) and the acceleration due to the gravity $g = 9.806$ m/s², $z_1 = 0$, and $z_2 = 10$ m, then:
>
> $$p_2 = p_{atm} - \rho g (z_2 - z_1) = 10^5 - (10^3)(9.806)(0-10) = 198,060 \text{ N/m}^2$$

## 1.5 Future Energy Demand and Environment

Coal and crude oil were not relevant to the energy production until the end of nineteenth century. Firewood, wind and hydropower techniques were used to provide most of the energy that was used. Windmills and watermills were the common features of the landscape for most of the medieval age. Coal become the single most important energy source, after the discovery by James Watt in 1769 of steam engine, followed of the explosive spread of its usage in industry and transportation. The steam engine, and later the internal combustion engine, replaced mechanical wind and water installations at the end of the nineteenth and beginning of the twentieth centuries. During the first part of the last century, in large part due to the discovery of huge oil reserves, the oil became a major and cheap energy source, leading also to an increased motorized transportation. During these periods firewood lost its importance as an energy source, and large hydro-electric power systems replaced the watermills. Around and after Second World War the energy demand increased dramatically. For example, the energy consumption in United States increased from 32 quads in 1950s over 100 quads in 2010s. One quad is equal to $10^{15}$ Btu. Similar trends are in all developed countries and in many developing nations.

Much of the energy used today is in the form of electricity. Figure 1.2 shows the share of energy sources used to generate electricity in the United States. The majority of electricity comes from fossil fuels, hydropower, atomic and nuclear energy, while a tiny fraction (~1%)

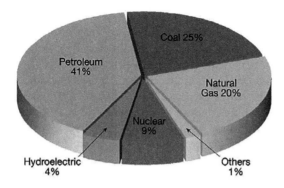

**FIGURE 1.2**
US energy mix for electricity generation.

comes from other sources. Global energy consumption in the last half of the century has increased very rapidly and is expected to continue to grow over the next 50 years, with significant differences between the last 50 years and the next. The past increase was stimulated by relatively "cheap" fossil fuels and increased rates of industrialization in developed countries yet while energy consumption in these countries continues to increase, additional factors are making the picture for the next 50 years more complex. These include the very rapid increase in energy use in China and India (countries representing about a third of the world's population), the expected depletion of oil resources sometime in the future and the effect of human activities on climate change. On the positive side, the renewable energy technologies of wind, biofuels, fuel cells, wave energy, solar thermal and photovoltaics are finally showing maturity and the ultimate promise of cost competitiveness.

Our current living standard could not be sustained without energy. The provision of energy or the related energy services (e.g., heated living spaces, information, and transportation) involves a huge variety of environmental impacts which are increasingly less tolerated by the society of the twenty-first century, making "the energy problem" a major topic in engineering education, research and policy of almost all countries. This attitude is not expected to change within the near future. On the contrary, in view of the increasing knowledge and recognition of the effects associated with energy utilization in the broadest sense of term, increased complexity has to be expected. In 2010, the worldwide consumption of fossil primary energy carriers and hydropower account to approximately 450 EJ. Roughly 28% of this overall energy consumption accounts for Europe and Eurasia, about 27% for North America, 5% for Central and South America, roughly 5% for the Middle East, 3% for Africa and 32% for Asia and the Pacific region. North America, Europe and Eurasia as well as Asia and the Pacific region consume about 90% of the currently used primary energy derived from the fossil energy carriers and hydropower. This energy consumption has increased over 2.5 times in the last four decades.

All energy sources have some environmental impacts, however the impacts vary considerably across the energy spectrum. The energy sources, such as coal or oil, in general fossil fuels, have harmful effects on the environment compared to the renewable ones, considered much more benign, even though they have some harmful impacts. Furthermore, fossil fuel and nuclear energy environmental impacts have gotten worse, due to the population growth and the increase energy use per capita, as a result of the rise in the worldwide loving standards. Living standards and the energy use per capita are strongly correlated, either way are not saying that higher per capita wealth invariably requires higher per capita energy use. Besides the consumption of fossil fuel energy and pollutant emissions, pollutants are released into environment during the operation of oil-fired or natural-gas-fired heating. These pollutants show very different environmental effects. In addition, the exploitation of fossil fuels is also associated with effects which may damage the environment, during the well drilling, oil and natural gas extraction and transportation, oil processing in refineries or transportation from refinery to the consumers. Some of environmental impacts may be ongoing and less dramatic, as for example nuclear plant or coal mine disasters, oil spills, etc., but may cost many more lives over long time periods.

Without the protection of the atmosphere, the global average temperature would be about 18°C. Some minor atmospheric constituents, such as $CO_2$, water vapor and methane, capture significant parts of the incoming solar radiation, acting similarly to a greenhouse. These atmospheric constituents have natural and artificial origins. Existing natural greenhouse effects make the life possible on Earth. Without greenhouse effects, Earth radiation would be emitted into space. Combined effects of incident solar radiation and the retention of this heating energy increases the mean global ambient temperature to about +15°C. Over millennia, a delicate balance was created in the concentration of atmospheric gases and constituents. However, several natural temperature variations have occurred, during Earth history, as evidenced by different ice ages. Additional greenhouse gases are emitted to the atmosphere as a result of energy generation and other human activities. Their effects on the climate are not fully understood and accepted by the scientific community. The reasons for climate change are controversial. As a fact, part of global temperature increase of 0.6°C during the last century is linked to natural fluctuations, while the rest is believed to be of anthropogenic origin. Detailed prediction of the anthropogenic greenhouse effect consequences is not possible. Climatic models are only giving an estimate of what would happen if the current emissions remain the same or there will be future increases.

## 1.6 Units and Dimensional Analysis

The dimension of a variable is the kind of quantity that it is, having important significance in sciences and engineering fields. Fundamental dimensions are mass, length, time, temperature, and electric charge. It is important to distinguish between the quantity, for example the mass of an object $m$, and its dimension $M$, thus we can write $[m] = M$, to be read as: "the dimension of $m$ is $M$." Two variables can have the same dimension, like for example the two sides, a and b, of a rectangle are both lengths, i.e., $[a] = L$, $[b] = L$, and so they can be added or subtracted; both a + b and a − b may have physical meanings. Variables that are of different dimensions cannot. It is meaningless for the length to be added to the area, or for a 10 km length to be added to a speed of 5 km/hr. On the other hand, two quantities of the same or different dimensions can be multiplied, and their product is that of the individual dimensions. Variables can also be non-dimensional—for examples, the efficiency and coefficient of performance. All equations must be dimensionally homogeneous, i.e., all terms in the equation must have the same dimensions. One important reason for checking the equation homogeneity of dimensions is that inhomogeneity would indicate an error. However, a homogeneous equation can still be erroneous though not because of dimensionality. There is an important exception, quite common in engineering. Sometimes it may seem that an empirically obtained equation is dimensionally inhomogeneous, but that may be because the dimensions are unknown or are suppressed. It is not enough to know the dimension of a physical quantity, in order to quantify, it also needs to know how large it is; thus every physical quantity is associated with a pure number, and corresponding units (except for dimensionless quantities). Though in principle we can use our own units, these units are only useful if they are understood by other people, though not necessarily by all. Fundamental units are those from which all other units, so-called derived, are obtained. The quantification of the units is arbitrary, as well as the choice of which units are fundamental and derived. For example, length can be a fundamental unit and area the derived, or vice versa. Dimensions and units should not be confused with each other; a variable can be described by different units that all have the same dimension. The fundamental units in the International System of Units (Système International d'Unités or SI) are mass (kilogram, kg), length (meter, m), time (second, s), temperature difference (kelvin, K), electrical current (ampere, A), luminous intensity (candela, cd) and moles (mol), see Table 1.2. Other units can be derived from these using definitions or physical laws. There are also units outside SI that are accepted for use with SI, units such as minute (min), hour (h), day (d), degree of angle (°), minute of angle ('), second of angle ("), liter (L), and metric ton (t). To make comparisons of various quantities and to quantify the magnitude of physical quantities we need a good understanding of units, their definitions and techniques of dimensional analysis. International System (SI) of units is used throughout of this book, but a comprehensive list of energy related units, including the ones used in United States, with conversions is also included. In engineering and science, dimensional analysis is the analysis of the relationships between different physical quantities by identifying their fundamental dimensions (such as length, mass, time, and electric charge) and units of measure (such as miles vs. kilometers, or pounds vs. kilograms vs. grams) and tracking these dimensions as calculations or comparisons are performed. Converting from one dimensional unit to another is often somewhat complex.

Dimensional analysis, also known as the unit-factor method, is a widely used for performing conversions using the rules of algebra. Any physically meaningful equation (and any inequality and in-equation) must have the same dimensions on the left and right sides. Checking this is a common application of performing dimensional analysis. Dimensional analysis is also routinely used as a check on the plausibility of derived equations and computations. It is generally used to categorize types of physical quantities and units based on their relationship to or dependence on other units. SI is founded on seven SI base units for seven base quantities assumed to be mutually independent, as given in Table 1.3. Other quantities, called derived quantities, are defined in terms of the seven base quantities via a system of quantity equations. The SI derived units for these derived quantities are obtained from the physics principles and equations and the seven SI base units. The fact that energy exists in many forms was one of the reasons that we have several units for

**TABLE 1.2**

SI Base Units

| Base Quantity | Name | Symbol |
| --- | --- | --- |
| Length | Meter | m |
| Mass | Kilogram | kg |
| Time | Second | s |
| Electric current | Ampere | A |
| Temperature | Kelvin | K |
| Amount of substance | Mole | Mol |
| Luminous intensity | Candela | cd |

**TABLE 1.3**

Most Common Energy Units

| Unit Name | Definition |
|---|---|
| Joule (J) | Work done by ace of 1 N acting through 1 m (also W-s) |
| Erg | Work done by 1 dyne force acting through 1 cm |
| Calorie (Cal) | Heat needed to raise the temperature of 1 g of water by 1°C |
| BTU | Heat needed to raise the temperature of 1 lb of water by 1°F |
| kWh | Energy of 1 kW of power flowing for 1 hour |
| Quad | $10^{15}$ BTU |
| Electron-Volt | Energy gained by an electron through 1 V potential difference |
| Foot-pound | Work done by 1 lb force acting through 1 ft |
| Megaton | Energy released when a million tons of TNT explodes |

this physical quantity. For example, for heat we have calories, British thermal units (BTUs), joules, ergs; foot-pound for mechanical energy; kilowatt hours (kWh) for electrical energy; and electron-Volts (eV) for nuclear and atomic energy. However, since all describe the same fundamental quantity, there are conversion relationships or factors relating them. Table 1.3 lists some of the most common energy units.

> **Example 1.10:** Show that physical dimensions of the expression of the hydrostatic pressure, $p = \rho g h$ are consistent with physical dimensions of pressure, fluid density, acceleration due to gravity, and height.
>
> **Solution:** Replacing the individual symbols in the equation of the hydrostatic pressure in terms of their fundamental physical units, we have:
>
> $$p = \rho g h$$
>
> $$[p]_{SI} = \left(kg \cdot m^{-3}\right)\left(ms^{-2}\right)(m) = \left(kg \cdot m \cdot s^{-2}\right)\left(m^{-2}\right) = N \cdot m^{-2}$$
>
> Hence the hydrostatic pressure, $p = \rho g h$ is consistent with physical dimensions of pressure, force per unit of area.

It is clear then that physical quantities that have different dimensions cannot be converted from one to the other. For instance, weight that is a force (normally gravitational) with which the one body attracts another and cannot be converted to mass though the two are frequently confused. Unfortunately, in today engineering and applied sciences areas, many power and energy units sound similar and proper care must be taken when we are dealing and using them. The terms energy and power are quite often used informally as they are synonymous (e.g., electrical energy or power, wind energy or wind power). It is important to have a clear distinction and understanding of these terms. The unit of power is Watt (W). The British (imperial) unit, still in use in United States, is the British thermal unit (Btu, 1 Btu = 1055 J). However, the Btu is commonly used in thermal processes to designate thermal energy. For a system to produce or consume power there must be forceful motion of some of its components. In practice is often convenient to express the energy in terms of power used for a specific period of time. For example, if the power of an electric motor is 1 KW, and is running for an hour, the energy consumed is one kilowatt-hour (kWh). All the systems used for the performance of several tasks desired by the human society generate work and consume energy resources. Energy is also often measured simply in terms of fuel quantities used, such as tons of coal or oil. According with the energy definition a kilogram of coal, a liter or a gallon of oil are potential sources of energy. The motion of a body is a type of work. Heat represents another form of energy. Wind or running water are able to move the blades of a rotor. Similarly, sunlight can be converted into heat, being another form of energy. Energy transfer usually refers to movements of energy between systems which are closed to transfers of matter. The portion of the energy which is transferred by conservative forces over a distance is measured as the work the source system does on the receiving system. The portion of the energy that does not work during the transfer is called heat. In other words, heat or thermal energy is a form of energy that is transferred from materials or systems at higher temperatures to ones at lower temperatures. It is usually produced by combustion of fuels, which are regarded as energy sources.

## 1.7 Summary

The chapter discusses some of the background topics of the energy, and the information introduces the purpose and scope of this book. It discusses fundamental issues of energy, fossil fuels, environmental impacts of electricity generation and energy use, and renewable energy alternatives. We are facing an increased energy demand all over the world, but especially in developing countries. The use of alternative energy and alternative use of the conventional energy sources must be considered as options of sustainable development. The challenges for the implementation of new energy technologies and sources was reviewed. Viable energy sources and technologies must be sustainable and economically competitive with conventional sources and technologies. Remaining chapters examine the major renewable energy sources, their characteristics, performances and applications. The chapter is providing a brief review of basic principles of energy conversion, conservation of energy, basic electric circuit laws, energy concepts, and units. In the chapter is also included with examples a discussion of the ways in which primary energy sources can be transformed into forms suitable for applications and direct use. These conversion processes rely on the appropriate technologies but are also governed by the fundamental laws of physics. Finally, the chapter briefly discussed the future of energy generation, conversion and use. The book is intended to be used as a required textbook for an undergraduate renewable energy course. However, it can be used to provide foundation for everyone interested in alternative energy technologies.

## Questions and Problems

1. Is the quality of life better with electricity or without? Explain.
2. Describe how a specific energy-related process can affect the environment.
3. Convert: a) 1 kJ in Btu, and b) 1000 kg crude oil equivalent per year into kW.
4. What is the difference between energy and power? Define the energy and power.
5. What is the efficiency and what is the Carnot efficiency?
6. What is quad?

7. Should 100 kWh be converted into W or J? Do the conversion.

8. Express the total USA energy usage in 2015?

9. How many kWh would generate in 30 days a 1000 MW power plant?

10. Show that the physical dimension of kinetic energy is equivalent with the one of work.

11. Using dimensional analysis verify that the hydrostatic pressure in a fluid is consistent with the dimension of pressure.

12. When the gravity is considered a source of renewable energy?

13. The power input to a machine is 150 W, and the useful output is 80 W. What is the efficiency of the machine?

14. What is the heat rate generated?

15. Besides hydro power, what are the most important renewable energy sources for your state or country?

16. In what part of the electromagnetic spectrum does solar radiation have the highest intensity?

17. A rock of 10 kg falls from 100-m high bridge. If the change in its potential energy is converted into electricity with an efficiency of 85%, how long this energy illuminate a standard 30-W light bulb?

18. Compare the energy scales associated with 1 kg of coal by computing: a) the potential energy by dropping it 100 m, b) burning it; and c) converting its mass into energy.

19. What is the power rating of your car? Convert it into kW. What is the fuel efficiency (mpg or km/L).

20. A body of 0.5 kg is moving at 20 m/s. What work is need to stop that body over 10 m distance?

21. What is the kinetic energy of a wheel of mass moment of inertia 100 kg·m$^2$ rotating at 500 rpm?

22. A snowball of 0.25 kg is thrown at 10 m/s. How much kinetic energy does it possess? What happens to that kinetic energy after a person is hit with this snowball?

23. A 10 cm long, 5 cm diameter steel rod (density of steel = 8050 kg/m$^3$) is rotating around its axis of symmetry. What is the angular acceleration if a torque of 75 Nm is applied to the rod?

24. How much energy is needed to accelerate a shaft-mounted disk of mass moment of inertia 10,000 kg m$^2$ from 1000 rpm to 30,000 rpm?

25. Shows the physical dimensions of the kinetic energy are consistent with the physical dimension of work.

26. If, during rotation, the mass moment of inertia of a rotating object could be halved, how will its angular velocity change?

27. What is the thermal efficiency of the most efficient heat engine that operates between a cold reservoir at 20°C and a hot reservoir at 250°C?

28. For your home estimate the power installed for lightning and for major appliances (computers, TV sets, fridge, electric stove, washing machine, dryer, etc.). The estimate the electrical energy used in one year.

29. Consider a compound slab, consisting of two materials having thicknesses, $L_1$ and $L_2$, and the thermal conductivities, $k_1$ and $k_2$, respectively. If the outer temperatures are $T_2$ and $T_1$, find the heat transfer rate through the slab in a steady state. Numerical application: $L_1 = 20$ cm, $L_2 = 30$ cm, $k_1 = 50$ W·m$^{-1}$·K$^{-1}$, and $k_2 = 60$ W·m$^{-1}$·K$^{-1}$.

30. A heat engine operates between a temperature of 20°C and 150°C. What is the efficiency of the engine if it works at 75% of the maximum possible Carnot efficiency?

31. A refrigerator operates between a temperature of 21°C and −8°C. What is the COP of the refrigerator if it works at 83% of the maximum possible Carnot COP?

32. A 10 HP electric motor is 85% efficient. How much heat does it generate?

33. What is the coefficient of performance of a refrigerator that uses 150 W to remove 1350 W from a cold chamber? What is the coefficient of performance of the same device if it is used as a heat pump?

34. Determine the temperature of a computer chip package of 1 cm$^2$ surface area that is dissipating 1 W of heat to an airstream which is at 20°C. Assume that all the heat goes into the air and the convective heat transfer coefficient is 100 W/m$^2$·K.

35. Describe the relationship between the temperature of a radiating body and the wavelengths it emits.

36. If the Earth had no atmosphere, its radiation emission would be lost quickly to the outer space, making its temperature about 33 K cooler. Calculate the rate of radiation emitted and the wavelength of the wavelength of maximum radiation emission for the Earth at 255 K.

37. Assuming that the atmospheric pressure at the surface of a fluid is stationary and equal to $10^5$ N/m$^2$, calculate the change in percent due to a wind of 25 m/s, assuming that air density is 1.225 kg/m$^3$.

38. An incompressible fluid flows at a speed of 5 m/s through a pipe of 0.5 m diameter in which a constriction of 0.2 m diameters has been inserted. What is the fluid speed in the constriction?

39. If the flow rate of water that enters and leaves a cooling chamber is 1 kg/s. What is the heat rate that must be extracted from the water to reduce its temperature by 20°C?

40. Calculate the expression of the speed of jet of water emerging from an orifice in a dam at a depth *h* bellow the water surface. Assume the water surface and the jet are at atmospheric pressure, *p*.

41. How much does 1 m$^3$ of water at 4°C weigh (not its mass, but its weight)?

42. An incompressible ideal fluid flows through a pipe with diameter of 0.5 m with a constriction of 0.15 m. If the fluid speed into the pie is 2 m/s, what is the fluid speed inside the constriction?

43. Briefly describes the importance of the international system of units.

44. List the SI fundamental units.

45. Briefly describes the accepted energy units, and what the reasons of using them.

## References and Further Readings

1. F. Bueche, *Introduction to Physics for Scientists and Engineers*, McGraw-Hill, London, UK, 1975.
2. G. Boyle, *Renewable Energy: Power for a Sustainable Future*, Oxford University Press, Oxford, UK, 2012.
3. G. Boyle, B. Everett, and J. Ramage, editors, *Energy Systems and Sustainability*, Oxford University Press, Oxford, UK, 2003.
4. R. E. Putman, *Industrial Energy Systems: Analysis, Optimization and Control*, ASME, New York, 2004.
5. T. D. Eastop, and D. R. Croft, *Energy Efficiency for Engineers and Technologists*, Longman, Harlow, UK, 1990.
6. J. Casazza, and F. Delea, *Understanding Electric Power Systems: An Overview of Technology, the Marketplace, and Government Regulation* (2nd ed.), John Wiley & Sons, Hoboken, NJ, 2010.
7. A. W. Culp, Jr., *Principles of Energy Conversion* (2nd ed.), McGraw-Hill, New York, 1991.
8. V. Quaschning, *Understanding Renewable Energy Systems*, Earthscan, London, UK, 2006.
9. R. A. Ristinen and J. J. Kraushaar, *Energy and Environment*, John Wiley & Sons, Hoboken, NJ, 2006.
10. J. Andrews and N. Jelley, *Energy Science, Principles, Technology and Impacts*, Oxford University Press, Oxford, UK, 2007.
11. E. L. McFarland, J. L. Hunt, and J. L. Campbell, *Energy, Physics and the Environment* (3rd ed.), Cengage Learning, Mason, OH, 2007.
12. F. Kreith and D. Y. Goswami, editors. *Handbook of Energy Efficiency and Renewable Energy*, CRC Press, Boca Raton, FL, 2007.
13. N. Jenkins, J. B. Ekanayake and G. Strbac, *Distributed Generation*, The IET Press, London, UK, 2010.
14. F. M. Vanek and L. D. Albright, *Energy Systems Engineering: Evaluation and Implementation*, McGraw-Hill, New York, 2008.
15. B. K. Hodge, *Alternative Energy Systems and Applications*, John Wiley & Sons, Chichester, UK, 2010.
16. B. Everett and G. Boyle, *Energy Systems and Sustainability: Power for a Sustainable Future* (2nd ed), Oxford University Press, Oxford, UK, 2012.
17. F. R. Spellman, *Environmental Impacts of Renewable Energy*, CRC Press, Boca Raton, FL, 2014.
18. R. A. Dunlap, *Sustainable Energy*, Cengage Learning, Stamford, CT, 2015.
19. V. Nelson and K. Starcher, *Introduction to Renewable Energy (Energy and the Environment)*, CRC Press, Boca Raton, FL, 2015.
20. R. Bansal (Ed.), *Handbook of Distributed Generation: Electric Power Technologies Economics and Environmental Impacts*, Springer, Cham, Switzerland, 2017.
21. R. Belu, *Industrial Power Systems with Distributed and Embedded Generation*, The IET Press, London, UK, 2018.

# 2

## Solar Energy Resources

### 2.1 Introduction, Astronomical Background

Solar energy is considered as one the most potential and techno-economically viable renewable energy sources for electricity generation. In addition, it is clean, plentiful, and sustainable energy sources with almost zero pollution during the system operation. The use of solar energy for energy generation or other applications is growing at a fast rate due to the technological improvements that makes photovoltaic and solar energy systems cheaper, more robust, and efficient resulting in reduction of the overall generation cost. There are still not fully answered questions regarding the nature of the Sun energy generation. However, this is not of critical importance for the solar radiation terrestrial users, more important being the amount of energy, its spectral and temporal distributions, and its daily and yearly variability. Solar energy system design relies on a solar radiation comprehensive assessment and analysis at a particular site or location. Although solar radiation data have been measured and recorded for many locations, they have to be analyzed, processed and interpreted before a sufficient accurate estimate of the available solar radiation for a project can be made. For these reasons, a large emphasis in the current research studies is on the development of the reliable and accurate solar resource assessment methods. Energy decision makers are required to make critical judgments with regard to the energy generation, distribution, demand, storage, and integration. Accurate knowledge of the present and future atmosphere states is vital in making these decisions. A major solar energy drawback represents unpredictable nature and weather dependence of the solar radiation, one of the difficult weather variables to predict. Moreover, the solar radiation variations often do not match with the time energy demand distribution. Topography, ground cover clouds and aerosols all affect solar energy prediction. Research experiments can be costly in terms of resources and computational effort. Site-specific time series solar resource information, along with associated weather data is essential to the successful solar energy system design and deployment. These systems generally must be installed close to the load that they are designed to serve; therefore reliable solar data must be available for the installation site. It is unlikely that any pre-existing measured data will be available for the system optimum sizing or for predicting its performances.

Assessment of renewable energy resources and the setting of targets involve sequential processes in which layers of analysis are applied that progressively reduce the total theoretical opportunity to that which is considered practically achievable by a target date. Earth orbits a star, the Sun, the ultimate source of all energies driving all processes on our planet. The amount of solar radiation that reaches the ground, besides on the daily and yearly apparent motion of the Sun depends on the geographical location (latitude and altitude), the weather and the atmospheric conditions (e.g., cloud cover, turbidity).

Sun's electromagnetic spectrum spans over a large range of wavelengths, from gamma and x-rays, to ultraviolet (UV), visible, infrared, and radio waves. For the purpose of energy applications, the so-called optical solar radiation, spanning from UV to mid-infrared wavelengths that the Earth's atmosphere allows to reach the ground, are of interests.

Solar radiation amount and intensity received by a given surface is controlled, at the global scale, by the geometry of the Earth, atmospheric transmittance and the relative Sun location. Figure 2.1 shows the Earth–Sun geometry and its effects on different seasons. At the local scale, solar radiation is controlled by surface slope, aspect, and elevation. Clear sky solar radiation estimates for sloped surfaces are very important in renewable energy, civil engineering and agricultural applications, which need an accurate estimate of total energy striking a given surface. Figure 2.2 is showing Earth position relative to the Sun's

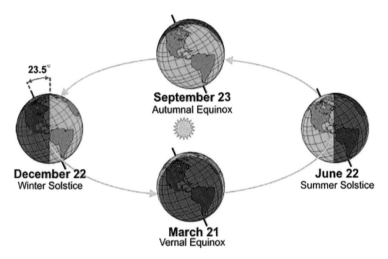

**FIGURE 2.1**
Earth–Sun geometry.

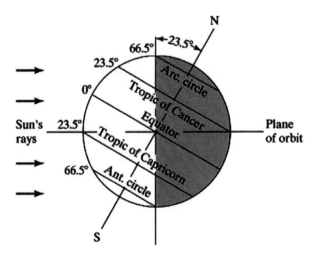

**FIGURE 2.2**
Position of the Earth in relation to the Sun's rays at time of winter solstice.

rays at the winter solstice. At the winter solstice (around December 22), the North Pole is inclined 23.5° away from the Sun.

All points on the Earth's surface north of 66.5° latitude are in total darkness while all regions within 23.5° of the South Pole receive continuous sunlight. At the time of the summer solstice (around June 22), the situation is reversed. At the two equinoxes (March 21 and September 21), both poles are equidistant from the Sun and all points on the Earth's surface have 12 hours of daylight and 12 hours of darkness. During this day a line from the center of the Sun to the center of the Earth passes through the equator and everywhere on Earth, so we have 12 hours of daytime and night, hence the term *equinox* (equal day and night). Table 2.1 provides a list of day numbers for the first day of each month of the year, needed in solar radiation calculations.

Solar radiation, the most important natural energy source, is driving all Earth environmental processes. Sun provides the Earth with an enormous amount of energy, being the primary factor in determining its weather, and thermal environment. The available solar radiation amount for any energy project is the total solar radiation portion that reached the Earth's surface. So knowledge of the Earth–Sun relationships the understanding and the estimate solar radiation intensity, to know how to make, if needed at least simple solar radiation measurements are critical in solar energy. Earth is a nearly spherical body revolving the Sun in a nearly circular path, with it located slightly off the center. It makes one rotation about its axis every 24 hours and completes a revolution about the Sun in approximately 365.25 days. Around January 1, Earth is closest to the Sun while around July 1 it is most remote, about 3.5% far away. The solar intensity incident upon the top of the atmosphere varies inversely with the square of the Earth–Sun distance, so the Earth is receiving about 7% more radiation in January than July. Earth axis of rotation is tilted 23.5° with respect to its orbit about the Sun. Earth tilted position, together with its daily rotation and yearly revolution have profound significance on the surface solar radiation distribution, the length of daylight and darkness hours, and the season changes.

The Sun is the nearest and only star around which the Earth and other planets are rotating. It is a sphere of intensely hot gaseous matter with a diameter of $1.39 \times 10^9$ m, at about $1.5 \times 10^8$ km away from the Earth. As observed from the Earth the Sun disk forms an angle of 32 minutes of a degree, very important in applications, such as solar collectors or concentrator optics. As the Sun is radiating and transmitting radiant energy with a range of wavelengths and intensities, a nuclear fusion-fission reaction is happening inside the Sun core. The Sun nuclear processes create tremendous amounts of energy which are almost endless and the human benefits from are expected to last for generations to come. The Sun energy is transmitted to its surface through convection and radiation, and further is transmitted to its surroundings through solar radiation. The solar energy available at the Earth upper atmosphere

**TABLE 2.1**

Day Number of the First Day of Each Month

| Month | N | Month | N |
|---|---|---|---|
| January | 1 | July | 182 |
| February | 32 | August | 213 |
| March | 60 | September | 244 |
| April | 91 | October | 274 |
| May | 121 | November | 305 |
| June | 152 | December | 335 |

is almost constant and depends on its motions, the Sun–Earth distance and the Sun internal nuclear processes. Passing through the atmosphere, the available solar energy in the upper atmosphere is reduced before it reaches the ground, and depends on the weather and climatic conditions, surface locations and atmospheric conditions. Sun net energy, reaching the ground in the form of radiant energy, has different intensities, depending on the radiant wave spectrum. The spectrum is classified as short wave (0 to 300 nm), infrared (300 to 750 nm) and long wave (750 nm and above). These different radiant spectrum energy intensities are used for different applications, using a variety of techniques and methods. Although the solar energy is inexhaustible and free, it is not quite a convenient energy source, being variable during the day and not readily dispatchable. In contrast, modern lifestyles demand a continuous and reliable energy supply. However, there are ways to overcome the solar energy shortfalls. The chapter discusses the solar energy analysis, the estimates of the energy irradiated from the Sun, Earth–Sun geometrical relationships, solar energy receiver orientations, solar radiation measurements, as well as the importance of acquiring reliable solar data information for design, analysis, operation, and management of solar energy systems.

## 2.2  Sun–Earth Geometry, Extraterrestrial Solar Radiation, Solar Constant

The Sun is the center of our solar system, being in existence for about 5 billion years, and is expected to last at least other 5 billion of years. It consists of 80 percent hydrogen, about 20 percent helium, and about 0.1 percent other elements. Nuclear fusion processes create Sun radiant power. The total mass of particles after the fusion is less than that before, the difference is converted into energy, expressed by the relationship, between mass, $m$ and $c$, the light speed ($\sim 3 \cdot 10^8$ m/s):

$$\Delta E = mc^2 \tag{2.1}$$

Every second, the Sun is losing about 4.3 million metric tons of mass, converted according to Equation (2.1), in a $3.845 \cdot 10^{26}$ W, energy distributed into space as electromagnetic radiation. Sun radiation spectrum varies from very short wavelengths cosmic radiation to very high long wavelength radiation, i.e., from Ångstrom ($10^{-10}$ m) to hundreds of meters. Its spectrum is divided into wavelength bands. Detailed information about solar radiation availability at any location is essential for the design and economic evaluation of solar energy systems. Long-term measured data of solar radiation are available for a large number of locations in the United States and other parts of the world. Where long-term measured data are not available, various models based on climatic data are used to estimate the solar energy availability. Solar energy is in the form of electromagnetic radiation with the wavelengths ranging from approximately 0.3 μm to over 3 μm, which correspond to ultraviolet (less than 0.4 μm), visible (0.4 and 0.7 μm), and infrared (over 0.7 μm). Most of this energy is concentrated in the visible and the near-infrared wavelength range. The Sun generated energy divided by its surface area gives the Sun specific emission, 63.11 MW/m², the radiant power per square meter. One fifth of a square kilometer of the Sun surface emits every year, energy equal with total Earth primary energy demand.

Every object emits radiant energy in an amount that is a function of temperature. The usual way to describe how much radiation an object emits is to compare it to a theoretical body called a blackbody. To a good approximation, the Sun acts as a blackbody (perfect emitter) at about 5800 K temperature. Considering the Sun a blackbody, applying

Stefan-Boltzmann Law ($E_C = \sigma T^4$), with $\sigma = 5.67051 \cdot 10^{-8} \text{W}/(\text{m}^2\text{K}^4)$, its surface temperature is estimate to be ~5800 K. A sphere with the radius equal with the average Sun–Earth distance ($1.5 \cdot 10^8$ km) receives the same total radiant power as the Sun surface. However, the energy density is much lower. This value determines the extraterrestrial radiance at the top of the Earth atmosphere. The extraterrestrial solar radiation also varies due to the Earth's elliptical orbit around the Sun. Another convenient feature of the blackbody radiation curve is given by Wien's displacement rule, which tells us the wavelength at which the spectrum reaches its maximum point:

$$\lambda_{\max}(\mu m) = \frac{2898}{T(K)} \tag{2.2}$$

where the wavelength is in microns ($\mu$m) and the temperature is in kelvins.

**Example 2.1:** Consider the Earth to be a blackbody with average surface temperature 15°C and area equal to 5.1 × 1014 m². Find the rate at which energy is radiated by the Earth and the wavelength at which maximum power is radiated.

**Solution:** Using Stefan-Boltzmann equation, the Earth radiates

$$E = \sigma A T^4 = 5.67051 \times 10^{-8} \cdot 5.1 \times 10^{14} \cdot (273.15 + 15) = 2.0 \times 10^{14} \text{ W}$$

The wavelength at which the maximum power is emitted is given by Equation (2.2).

$$\lambda_{\max}(\text{Earth}) = \frac{2898}{288} = 10.1 \; \mu m$$

The Earth orbits the Sun approximately every 365 days, following an elliptical orbit, with a very small eccentricity, about 0.0167, the Earth path is nearly circular. The Earth's elliptical orbit varies from 14.7 × 107 km in early January, the closest distance to the Sun, the *perihelion* to 15.2 × 107 km in early July, the farthest distance, the *aphelion*. The average Earth–Sun distance is 14.9 × 107 km. However, the Earth is about 4% closer to the Sun at the perihelion than the aphelion. Extraterrestrial radiation increases by about 7% from July 4 to January 3, at which time the Earth reaches the point in its orbit closest to the Sun. Calculation of extraterrestrial solar radiation typically includes an eccentricity correction factor (*eccf*), to account for this variation in the Sun–Earth distance. For many engineering applications, the following hand calculation is used:

$$eccf = 1 + 0.033 \cdot \cos\left(\frac{360n}{365}\right) \tag{2.3}$$

where $n$ is the day of the year counted from January 1 (the day number or Julian day, Table 2.1), such that $1 \leq n \leq 365$. The average amount of solar radiation falling on a surface normal to the rays of the Sun outside the atmosphere of the Earth (extraterrestrial) at mean Earth–Sun distance ($D_0$) is called the solar constant, SC. Measurements by NASA indicated the value of the solar constant to be 1353 W/m² (±1.6%), 429 Btu/h·ft² or 1.94 cal/cm²·min (Langleys/min). This value was revised upward to 1377 W/m² or 437.1 Btu/h·ft² or 1.974 Langleys/min, which was the value used in compiling SOLMET data in the United States. Recently, new measurements have found the value of the solar constant to be 1366.1 W/m². A value of 1367 W/m² is also used by many references. However, the average distance variation causes a variation in radiant power density between 1325 W/m² and 1425 W/m², while the average solar constant is equal to:

$$SC = 1367 \pm 2 \, W/m^2 \tag{2.4}$$

The seasonal variation in the solar radiation availability at the ground can be estimated from the relative movement geometry of the Earth around the Sun. Since the Earth's orbit is elliptical, the Earth–Sun distance varies during the year with about ±1.7% from the average, expressed by the following relationship:

$$I_0 = SC \cdot \left(\frac{D_0}{D}\right)^2 \tag{2.5}$$

Here $D$ is the Earth–Sun distance, and $D_0$ is the annual mean distance, $1.496 \times 10^{11}$ m. The $(D_0/D)^2$ factor may be approximated as:

$$\left(\frac{D}{D_0}\right)^2 = 1.00011 + 0.034221 \cdot \cos(B) + 0.00128 \cdot \sin(B) +$$

$$0.000719 \cdot \cos(2B) + 0.00077 \cdot \sin(2B)$$

where the parameter $B$ is given by:

$$B = \frac{360 \cdot (n - 81)}{365} \tag{2.6}$$

In solar engineering, the following approximate relationship is used with acceptable accuracy:

$$I_0 = SC \cdot \left[1 + 0.034 \cos\left(\frac{360n}{365.25}\right)^o\right] \tag{2.7}$$

The parameter $n$, in both Equations (2.6) and (2.7) is the day number, starting from January 1 as $n = 1$, as given in Table 2.1. The total energy flux incident on the Earth surface is obtained by multiplying $SC$ by $\pi R^2$ (the area of the Earth disk), where $R$ is the Earth radius. The average flux incident on a unit surface area is then obtained by dividing this number to the total surface area of the Earth ($4\pi R^2$), giving:

$$\frac{SC}{4} = 342 \, W/m^2$$

Earth's axis is tilted 23.5° with respect to the plan of its orbit around the Sun. This tilting results in longer days in the northern hemisphere from the spring equinox (approximately March 23) to the autumnal equinox (approximately September 22) and longer days in the southern hemisphere are during the other six months. On the equinoxes, the Sun is directly over the equator both poles are equidistant from the Sun, and the Earth experiences 12 h daylight and 12 h darkness. In the temperate latitude regions ranging from 23.450 to 66.50 north and south, variations in insolation are large. For example, at 40° N latitude, the average daily total extraterrestrial solar radiation varies from 3.94 kWh/m², in December to about 11.68 kWh/m² in June.

To determine the extraterrestrial radiation at a specific location and time, it is necessary to know the Sun location in the sky. Sun position in the sky is a function of time and

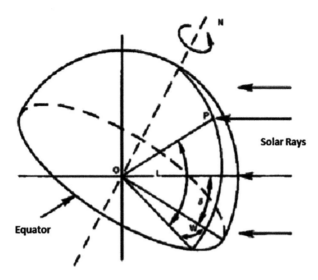

**FIGURE 2.3**
Solar geometry: latitude, hour angle, and the Sun's declination.

latitude, and is defined by its solar altitude and solar azimuth angles. Sun's position relative to a location is determined by the latitude, L, location's hour angle, W, and the Sun's declination angle. These three parameters are shown graphically in Figure 2.3. Latitude is the angular distance north or south of the Earth's equator, measured in degrees along a meridian. The hour angle is measured in the Earth's equatorial plane, and is the angle between the projection of a line drawn from the location to the Earth's center and the projection of a line drawn from the Earth center to the Sun's center. Thus, at solar noon, the hour angle is zero. At a specific location, the hour angle expresses the time of day with respect to solar noon, with one hour of time equal to 15 degrees angle. By convention, the westward direction from solar noon is positive. The Sun's declination is the angle between projection of the line connecting the center of the Earth with the center of the Sun and the Earth's equatorial plane. Declination varies from −23.450 on the winter solstice (December 21), to +23.450 on the summer solstice (June 22), and is expressed by:

$$\delta = 0.006918 - 0.399912\cos\Gamma + 0.070257\sin\Gamma - 0.006758\cos 2\Gamma$$

$$+ 0.000907\sin 2\Gamma - 0.002697\cos 3\Gamma + 0.00148\sin 3\Gamma$$

where $\Gamma$ is the daily angle, in radians given by:

$$\Gamma = 2\pi \frac{n-1}{365}$$

However this relationship is not usually employed in engineering applications. For convenience, Table 2.1 provides a list of day numbers for the first day of each month. Approximate estimates of declination angle, used in practical application are given by the following relationships:

$$\delta = 23.45 \cdot \sin\left[360 \cdot \frac{284 + n}{365}\right] \tag{2.8}$$

and

$$\delta = \arcsin\left[0.4 \cdot \sin\left(\frac{360}{365}(n-81)\right)\right] \qquad (2.9)$$

or

$$\delta = 23.45 \cdot \sin\left[\frac{360}{365}(n-81)\right] \qquad (2.10)$$

**Example 2.2:** Compute solar declination of the twenty-first day of each month of the year.

**Solution:** Using Equations (2.7) through (2.9) the declination angles for twenty-first day of each month of the year are:

| Month | Jan | Feb | Mar | Apr | May | Jun | Jul | Aug | Sep | Oct | Nov | Dec |
|-------|-----|-----|-----|-----|-----|-----|-----|-----|-----|-----|-----|-----|
| $\delta$ | −20.1 | −11.2 | 0.0 | 11.6 | 20.1 | 23.4 | 20.4 | 11.8 | 0.0 | −11.8 | −20.4 | −23.4 |

Solar declination as function of Julian day number is shown in Figure 2.4. Solar altitude angle ($\alpha$) defines the Sun elevation above the location horizon. In the following, the term zenith refers to an axis drawn directly overhead at a site. The solar altitude is related to the solar zenith angle ($\theta_z$), the angle between the Sun rays and the vertical, is given by the following relationships:

$$\theta_z + \alpha = \frac{\pi}{2} = 90°$$

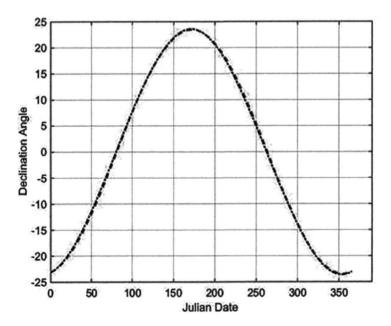

**FIGURE 2.4**
Declination angle vs. Julian date.

$$\cos\left(\theta_z\right) = \sin\left(\alpha_S\right) = \sin(L)\cdot\sin\delta + \cos(L)\cdot\cos\left(H_S\right)\cdot\cos\delta \qquad (2.11)$$

Here $L$ is the local latitude, $\delta$ is the declination angle (computed with Equations (2.7) through (2.9)), and $H_S$ is the solar hour angle (i.e., the angular distance between the Sun and the local meridian line). In other words this is the difference between the local meridian and the Sun meridian, with positive values occurring in the morning before the Sun is crossing the local meridian and negative values in the afternoon. Solar azimuth angle, $\alpha_S$, the angle between the Sun and true north is then given by:

$$\cos\left(180° - \alpha_S\right) = -\frac{\sin(L)\cdot\cos\left(\theta_z\right) - \sin(\delta)}{\cos(L)\cdot\sin\left(\theta_z\right)} \qquad (2.12a)$$

$$\sin\alpha_S = \frac{\cos(\delta)\cdot\sin\left(H_S\right)}{\cos(\alpha)} \qquad (2.12b)$$

For a given day the calculation of declination, altitude and azimuth angles is straightforward for latitude greater than the declination angle. Latitude angle $L$ is the angle between the line from the center of the Earth to the location and the equatorial plane. The solar hour angle, $H_S$ is equal to 15° times the number of hours from local solar noon. The values, east of due south, meaning the morning values are positive, while the values west, the afternoon ones are negative, and the numerical value of 15°/hour is based upon the nominal time, 24-hour required for the Sun to move 360° around the Earth, or 15° per hour. So $H_S$ is calculated by using a simple relationship:

$$H_S = \frac{360}{24}(t-12), \text{ degrees } (t \text{ in h, 24-hour clock}) \qquad (2.13a)$$

and in details,

$$H_S = \begin{cases} \dfrac{15°}{\text{hour}}\cdot(\text{hour before solar noon}) \\[2ex] -\dfrac{15°}{\text{hour}}\cdot(\text{hour after solar noon}) \end{cases} \qquad (2.13b)$$

or

$$H_S = \frac{\text{Minutes from solar noon}}{4\,\text{min/degree}}$$

At sunrise or sunset, the solar zenith angle is 90°, so $\cos(\theta_z) = 0$ and from Equation (2.11), then:

$$\cos\left(H_{SR}\right) = \cos\left(H_{SS}\right) = -\tan(\delta)\cdot\tan(L) \qquad (2.14)$$

where, $H_{SR,\,SS}$ are the hour angle at sunrise or sunset, respectively, related as:

$$H_{SR} = -H_{SS}$$

**Example 2.3:** Compute the solar altitude and solar azimuth angles at 10:00 AM, 12:00 (noon) and 2:00 PM solar time Wilmington, Delaware ($L = 40°$ N) on April 10.

**Solution:** Using one of the Equations (2.8) through (2.10), the declination on April 10 ($n = 101$) is 7.9°. Then, by using Equations (2.11) and (2.12), the following values of the solar altitude and the azimuth angels are computed as:

10:00 AM 48.3°, 48.0°
12:00 PM 58.0°, 63.8°
2:00 PM 48.3°, −48.0°

In the early morning and late afternoon, during the spring or summer, the magnitude of the Sun's azimuth is liable to be more than 90° away from south, which never happens in the fall or winter. Since the inverse of a sine is ambiguous, sin x = sin (180 − x), we need to determine whether to conclude the azimuth is greater than or less than 90° away from south. In other words, Equation (2.12a) is correct, provided that:

$$\cos\left(H_S\right) < \frac{\tan(\delta)}{\tan(L)} \tag{2.15}$$

Otherwise, it means that the Sun is behind the E-W line, and the azimuth angle is:

$$\begin{cases} -\pi + |\alpha_S|, \text{ for morning hours} \\ \pi - \alpha_S, \text{ for afternoon hours} \end{cases} \tag{2.16}$$

In simpler words, if then we have:

$$\cos\left(H_S\right) \geq \frac{\tan(\delta)}{\tan(L)}, \text{ then } |\alpha_S| \leq 90°, \text{ otherwise } |\alpha_S| > 90° \tag{2.17}$$

**Example 2.4:** Find the altitude angle and azimuth angle for the Sun at 3:15 PM solar time in Detroit, Michigan (latitude 42.3° N) on the summer solstice.

**Solution:** The solar declination $\delta$, for the summer solstice is +23.45°. The hour angle is computed as:

$$H = \left(\frac{15°}{h}\right) \cdot (-3.25\text{ h}) = -48.75°$$

Using Equation (2.11), the altitude angle is

$$\sin\beta = \cos(42.3°) \cdot \cos(23.45°)\cos(-48.75°) + \sin(23.45°) \cdot \sin(23.45°)$$

$$= 0.715$$

$$\beta = \sin^{-1}(0.715) = 45.7°$$

The azimuth angle from Equation (2.12) is then:

$$\sin(\alpha_S) = \frac{\cos(23.45°) \cdot \sin(-48.75°)}{\cos(45.7°)} = -0.987$$

$$\alpha_S = -80.8° \text{ and } \alpha_S = 180° - (-80.8°) = 260.8°$$

To resolve this ambiguity, and to decide which of these two options is correct, we apply relationships of (2.16).

$$\cos H = \cos(-48.75°) = 0.660 \quad \text{and} \quad \frac{\tan \delta}{\tan L} = \frac{\tan(23.45°)}{\tan(42.3°)} = 0.478$$

Since, $\cos(H) \geq \tan \delta / \tan(L)$, we are concluding that the azimuth angle is:

$$\alpha_s = -80.8° \text{ (80° west of south)}$$

The solar noon, by definition, the Sun is exactly on the meridian (north-south line), and consequently, the azimuth angle is 0°. Therefore the noon latitude angle (also known as the altitude angle), $\alpha_n$ is then given by:

$$\alpha_n = 90° - L + \delta \qquad (2.18)$$

**Example 2.5:** What are the maximum and minimum noon altitude angles for a location at 45° latitude?

**Solution:** Maximum declination angle is at summer solstice, $\delta = 23.5°$, while the minimum is at winter solstice, when $\delta = -23.5°$. The maximum and minimum noon altitude angles are then:

$$\alpha_n^{\max} = 90° - 45° + 23.5° = 68.5°$$

$$\alpha_n^{\min} = 90° - 45° - 23.5° = 21.5°$$

During an equinox, at solar noon, the Sun is directly over the local meridian, the solar rays are striking a solar collector at the best possible angle, and they are perpendicular to the collector. At other times of the year the Sun is high or low for normal incidence. However, on the average it is possible to find a good tilt angle, maximizing the solar radiation collection. Solar noon is an important reference point for almost all solar calculations. In the Northern Hemisphere, at mid-latitudes, the solar noon occurs when the Sun is due south of the observer. South of the Tropic of Capricorn, the opposite, it is when the Sun is due north, while in the tropics, the Sun may be either due north, due south, or directly overhead at solar noon. On the average, facing a collector toward the equator (in the Northern Hemisphere, meaning facing it south) and tilting it up at an angle equal to the local latitude is a good rule-of-thumb for better annual performances. However, in order to emphasize winter collection, a slightly higher angle in required, and vice versa for increased summer efficiency. The tilt angle, making the Sun's rays perpendicular to the module at noon is given by:

$$\text{Tilt} = 90 - \alpha_n \qquad (2.19)$$

**Example 2.6:** Find the optimum tilt angle for a south-facing photovoltaic module at latitude 32.3° at solar noon on May 1.

**Solution:** From Table 2.1 for May 1st, $n = 121$ and the declination angle (by using Equation (2.10)), is then:

$$\delta = 23.45 \sin\left[\frac{360}{365}(121-81)\right] = 14.9°$$

Using Equations (2.17) and (2.18) the tilt angle of the photovoltaic panel, facing south is:

$$\alpha_N = 90° - 32.3° + 14.9° = 72.6°$$

$$\text{Tilt} = 90° - \alpha_N = 17.4°$$

There is a strong relationship between the solar azimuth and hour angles, the first being the angle on the horizontal plane between the solar radiation projection and the line of the north-south direction, its positive values are indicating that the Sun is west of the south, while the negative values indicate that the Sun is east of the south. The hour angle represents the angular distance between the Sun position at a particular time and its highest position for that day when crossing the local meridian at the solar noon. The length of the day varies for all latitudes during the year, so the solar altitude angle also changes hourly and daily, as expressed in Equation (2.11). To avoid the failures of Equation (2.11) because the arcsine of a negative number does not exist, it is advisable to implement the following equation:

$$\alpha_S = \arctan\left(\frac{\sin(\alpha_S)}{\cos(\alpha_S)}\right) \tag{2.20}$$

When $\delta = -23.45°$, at the winter solstice, the locations in the Northern Hemisphere with latitude above 70° are in darkness, not illuminated at all during the day, and only negative values from Equation (2.20) are obtained, while the South Pole is fully illuminated. For $L = 90°$ S, the solar altitude remains constant at 23.45° during the 24 hours. The locations with latitude lower than 40° S are experiencing the greatest solar altitude during the day, and the solar altitude remains constant at 23.45° during 24 hours. The opposite occurs during the summer solstice (June 22), $\delta = +23.45°$, the Southern Hemisphere locations with latitude lower than 70° S are not illuminated at all during the day, while the North Pole is fully illuminated 24 hours. Locations with latitude lower than 40° N are experiencing the greatest solar altitude during the day, and the solar altitude remains constant at 23.45° during 24 hours. Since the Earth rotates 15° per hour (4 minutes per degree), for every degree of longitude between one location and another, clocks showing solar time would have to differ by 4 minutes. The only time two clocks would show the same time would be if they both were due north/south of each other. To deal with these longitude complications, the Earth is nominally divided into 24 1-hour time zones, with each time zone ideally spanning 15° of longitude. Of course, geopolitical boundaries invariably complicate the boundaries from one zone to another. The intent is for all clocks within the time zone to be set to the same time. Each time zone is defined by a Local Time Meridian located, ideally, in the middle of the zone, with the origin of this time system passing through Greenwich, England, at 0° longitude. Table 2.2 is showing the US time zone meridians. The Sun angles are calculated from the local solar time ($ST$), related to the local standard time ($LST$) through this relationship:

$$ST = LST + ET + \left(MR_{ST} - MR_{local}\right) \cdot 4 \text{ min/degree} \tag{2.21}$$

**TABLE 2.2**

US Time Zone Meridians (West of Greenwich)

| Time Zone | Standard Time Meridian |
|-----------|:---------------------:|
| Eastern Time | 75° |
| Central Time | 90° |
| Mountain Time | 120° |
| Pacific Time | 135° |
| Alaska and Hawaii | 150° |

*ET* is the equation of time. A correction factor accounting for the irregularity of the Earth speed around the Sun, $MR_{ST}$ is the standard time meridian, and $MR_{local}$ is the local longitude. *ET* can be calculated using:

$$ET = 9.87 \cdot \sin 2B - 7.53 \cdot \cos B - 1.5 \cdot \sin B \qquad (2.22)$$

The parameter *B* is given in the Equation (2.5). Together longitude correction and the equation of time give us the final relationship between local standard clock time (*CT*) and solar time (*ST*), expressed as:

$$ST = CT + \frac{4 \text{ min}}{\text{degree}} \left( MR_{ST} - MR_{local} \right)^{\circ} + ET(\text{min}) \qquad (2.23)$$

**Example 2.7:** Find the solar declination on October 5 in Cleveland, Ohio, the equation of time, *ET*, the solar time, *ST* at 2 PM. on this day and location.

**Solution:** For October 5, $n = 279$ and by using Equations (2.8), (2.9) or (2.10) yields to $\delta = -6.02°$. Negative declination makes sense the date is after the Fall Equinox ($n = 265$). On this day of the year, we are using Equation (2.21) to compute *ET*. First, the parameter *B* is computed, using Equation (2.6):

$$B = \frac{360 \cdot (279 - 81)}{365} = 195.2°$$

and

$$ET = 9.87 \times \sin(2 \times 195.82*3.14/180) - 7.53 \times \cos(195.82*3.14/180) -$$

$$-1.5 \times \sin(195.82*3.14/180) = 12.63 \text{ (min)}$$

Cleveland longitude is 81.7° W, and by using Equation (2.22), the solar time is:

$$ST = 2:00 + \frac{4 \text{ min}}{\text{degree}} (75 - 81.7)^{\circ} + 12.63 \text{ (min)}$$

$$= 2:00 - 14.17 \text{ (min)}$$

$$= 1:45.43 \text{ (h)}$$

For L > δ, the solar time is due east (TE) or due west (TW) are calculated by:

$$TE = 12{:}00 \text{ Noon} - \frac{\left( \cos^{-1}\left[ \dfrac{\tan\delta}{\tan L} \right] \text{degrees} \right)}{15°/\text{hour}}$$

$$TW = 12{:}00 \text{ Noon} + \frac{\left( \cos^{-1}\left[ \dfrac{\tan\delta}{\tan L} \right] \text{degrees} \right)}{15°/\text{hour}}$$

(2.24)

Taking in account corrections to solar time (time zone and equation of time), the equation for hour angle in degrees, very useful in engineering and practical applications is given by:

$$H_S = \frac{15°}{\text{hour}}\left(TS - 12 \text{ hour} + ET\right) + \left(MR_{ST} - MR_{local}\right)$$

(2.25)

For solar times earlier than *TE* or later than *TW* the Sun is north (south in the Southern Hemisphere) of east-west line and the absolute value of solar azimuth angle is greater than 90°. In this case, the correct value of solar azimuth angle is $a_s = 180° - |a_s|$. For $L \le \delta$ the Sun remains north, respectively south in the austral hemisphere of the east-west line and the true values of $a_s$ is greater than 90°. For $\alpha = 0$ in Equation (2.8), the **sunrise** ($H_{SR}$) and **sunset** ($H_{SS}$) angles can be determined by:

$$H_{SR(SS)} = \pm\cos^{-1}\left[-\tan L \cdot \tan\delta\right]$$

(2.26)

Since Earth rotates 15°/hour, the hour angle can be converted to time of sunrise and sunset by:

$$\text{Sunrise/Sunset Time (geometric)} = 12{:}00 - \frac{H_{SR(SS)}}{15°/\text{hour}}$$

(2.27)

Equations (2.26) and (2.27) are geometric relationships based on angles measured to the SUN center, hence the designation geometric sunrise in (2.26), being adequate for most of the solar engineering calculations. However, there is difference between weather service sunrise and the geometric sunrise, due to two factors. The first deviation is caused by atmospheric refraction, which bends the Sun's rays, making the Sun appearing to rise about 2.4 min sooner than geometrical calculation and then set 2.4 min later. The second factor is that the weather service definition of sunrise and sunset is the time at which the Sun upper limb (the top) is crossing the horizon, while the geometrical one is based on the center crossing the horizon. This effect is complicated by the fact that at sunrise or sunset the Sun pops up, or sinks, much quicker around the equinoxes when it moves more vertically than at the solstices when its motion includes much more of a sideward component. An adjustment factor Q that accounts for these complications is given by the following, relationship:

$$Q = \frac{3.467}{\cos(L) \cdot \cos(\delta) \cdot \sin\left(H_{SR}\right)} \quad \text{(min)}$$

(2.28)

Since sunrise is earlier when it is based on the Sun top rather than the middle, Q is subtracted from geometric sunrise, while since the upper limb sinks below the horizon later than the Sun middle, $Q$ is added to the geometric sunset. For mid-latitudes, the correction is typically in the range of about 4 to 6 min, which can be included or not depending on the applications.

**Example 2.8:** Determine the sunrise (geometric and conventional) that are occurring in Detroit, Michigan (latitude 42.3°) on July 15 ($n = 197$). Also find conventional sunset.

**Solution:** From Equations (2.8) through (2.10), the solar declination for this location and day is:

$$\delta = 23.45 \cdot \sin\left[\frac{360}{365}(187-81)\right] = 22.7°$$

The hour angle at sunrise, by using Equation (2.25), is then:

$$H_{SR(SS)} = \pm\cos^{-1}\left[-\tan(42.3°)\cdot\tan(22.7°)\right] = \pm112.4°$$

From Equation (2.26), the solar time of the geometrical sunrise is:

$$\text{Sunrise Time (geometric)} = 12:00 - \frac{112.4}{15°/\text{hour}} = 12:00 - 7.493 \text{ h}$$

$$= 4:30.4 \text{ (AM)} - \text{solar time}$$

From Equation (2.27) the adjustment factor is then computed as:

$$Q = \frac{3.467}{\cos(42.3°)\cdot\cos(22.7°)\cdot\sin(112.7°)} = 5.49 \text{ (min)}$$

The upper limb will appear 5.5 min earlier than our original geometric calculation, so:

$$\text{Sunrise} = 4:30.4 - 5.5 \text{ min} = 4:24.9 \text{ (AM)}$$

For Detroit, Michigan, the longitude is 83.04° W in the Eastern Time Zone with local time meridian 75°, and the estimate of the clock time follows.

$$B = \frac{360(187-81)}{365} = 104.2°$$

And the *ET* is the equation of time (Equation (2.21)) gives us:

$$ET = 9.87 \cdot \sin(2 \cdot 104.2°) - 7.53 \cdot \cos(104.2°) - 1.5 \cdot \sin(104.2°)$$

$$= -4.4 \text{ min}$$

The clock time is then computed using Equation (2.23) as:

$$CT = 12:00 - 4(\text{min}/°)(75° - 83.04°) - (-4.4) = 12:00 + 27.76 \text{ min} = 12:27.76$$

The local clock time is 27.76 min later than solar time, so sunrise will be at:

$$\text{Sunrise (upper limb)} = 4:24.9 + 27.76 = 4:52:66 \text{ (AM)}$$

Similarly, the geometric sunset is 7.493 h after solar noon, or 7:30.4 PM (solar time), while the upper limb will drop below the horizon 5.5 minutes later. Then adjusting for the 27.76 minutes difference between Detroit time and solar time gives:

$$\text{Sunset (upper limb)} = 7{:}30.4 + (5.5 + 27.76) = 8{:}03.26 \text{ (PM)}$$

**Example 2.9:** Determine the solar azimuth angle on May 1st for latitude 45° at 11:00 AM

**Solution:** On May 1st, $n = 121$ and using one of the Equations (2.8) through (2.10), $\delta = 14.9°$. At 11:00 AM, the hour angle is $w = -15°$. The solar azimuth angle is computed using Equation (2.10), yielding to:

$$\cos(\theta_z) = \sin(45°)\sin(14.9°) + \cos(45°)\cos(14.9°)\cos(-15°)$$

$$\theta_z = 32.6°$$

## 2.3 Solar Irradiance on the Earth's Surface

The terms *insolation* and *irradiance* are used interchangeably to define solar radiation incident on the Earth per unit of area per unit of time as measured in watts per square meter (W/m²) or watt-hours per square meter per 24-h day (Wh/m²/day). It represents the sunlight power density, an instantaneous quantity and is often identified as the sunlight intensity. Solar Constant is the irradiance received by the Earth at the top of the atmosphere, equal with 1367 W/m². Amount of solar radiation that reached the Earth's surface at any given location and time depends on many factors, such as time of day, season, latitude, surface albedo, the translucence of atmosphere and the weather. Extraterrestrial radiation (*top-of-the-atmosphere radiation*) is the amount of solar energy above the Earth's atmosphere, indicating the solar energy amount that would fall on the ground in the absence of an atmosphere. Sun's energy is nearly constant, varying from year to year by less than 1%. Solar Constant value can be measured outside the Earth's atmosphere on surface perpendicular on the solar radiation direction. It represents the total energy in the solar spectrum. However, this quantity is not sufficient for most of the engineering calculations, being necessary to know the energy distribution with the electromagnetic spectrum. As solar radiation enters the atmosphere of the Earth, some is absorbed, some is scattered, and some passes through unaffected by the atmospheric constituents, being either absorbed or reflected by the ground surface objects. The part of solar radiation that reaches the surface of the Earth with essentially no change in direction is called *direct or beam radiation*. The scattered diffuse radiation reaching the surface from the sky is called the sky *diffuse radiation*. Irradiation, a measure of sunlight energy density, is the time integral of the irradiance measured in kWh/m². Normally, the time frame for integration is one day, meaning the daylight hours. Irradiation is often expressed as *peak sun hours* (psh). The *psh* is simply the time length in hours at an irradiance of 1 kW/m² needed to produce the daily irradiation as obtained from the integration of irradiance over all daylight hours. Irradiance and irradiation both apply to all sunlight components, so at a given time or for a given day, they depend on location, weather and atmosphere conditions, time year, and whether surface of interest is shaded by trees or building or the surface is inclined or horizontal. However, the daily irradiation is numerically equal to the daily psh.

Values of solar radiation measured on the Earth's surface are lower than the Solar Constant. In the space, solar radiation is practically constant. However, on the Earth it varies with the day of the year, day time, the latitude, and the state of atmosphere. Various atmosphere influences, such as reflection, scattering, atmospheric absorption (mainly $O_3$, $H_2O$, $O_2$ and $CO_2$), Rayleigh and Mie scattering reduce the irradiance. The light absorption by different atmospheric gases (water vapor, ozone and $CO_2$) is highly selective, affecting only part of the spectrum. Figure 2.5 shows the spectrum outside of the atmosphere and at the Earth surface. The spectrum describes the light composition and the contribution of different wavelengths to the total irradiance. Water vapor, carbon dioxide, and ozone, for example, absorb significant amount of solar radiation at certain wavelengths. Ozone absorbs a significant amount of radiation in the ultraviolet region of the spectrum, while vapor and $CO_2$ absorb primarily in the visible and infrared spectrum regions. The terrestrial spectrum shows significant reductions at certain wavelengths due to atmospheric constituents' absorption. Molecular air particles with diameters smaller than the light wavelength cause Rayleigh scattering, which tends to increase with frequency. Dust particles and other air pollution particulates with diameter larger than the light wavelength cause Mie scattering, which is highly dependent on the location, usually lower in high mountain regions, and significantly higher in industrial areas. Climatic influences such as clouds, rain, snow or fog can cause additional reductions. In order to determine the amount of irradiation available at a given location and time for solar energy applications it is important to develop appropriate expressions for irradiance on both horizontal and inclined surfaces and to determine the number of hours of sunlight on a given day at a given location.

In extraterrestrial space, solar radiation is practical constant, being determined only by the distance from the Sun and can be estimated with very high accuracy. However, on the Earth it varies with day, time of the day, the latitude, and the state of the atmosphere, so the

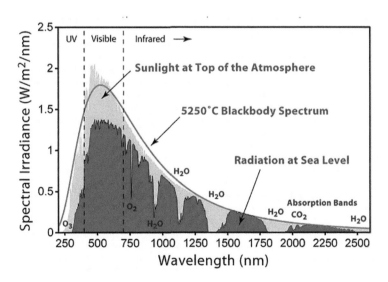

**FIGURE 2.5**
Solar radiation spectrums at the top of the atmosphere and at the ground.

radiation levels on the Earth surface are subject to considerable uncertainties resulting from local climatic and atmospheric effects. In addition to the daily and yearly variations due to the apparent motion of the Sun, irregular variations are caused by the weather conditions and by the atmosphere composition. In solar engineering, the surfaces that capture or redirect solar radiation are known as solar collectors. The solar radiation amount striking a collector depends also on the surface position and landscape. The total solar radiation incident on either *horizontal ($I_H$)* or *tilted ($I_T$) plans* consist of three components: *direct (beam), diffuse,* and *reflected radiation.* The diffuse radiation is the one scattered downward from the atmosphere arriving at the Earth's surface, and the one reflected on the surface from the surroundings. For a horizontal surface, this is expressed as $I_{Hd}$ and for a tilted one as $I_{Td}$. Solar radiation that is reflected from the ground is called albedo radiation, while the sum of all tree sunlight components is called global radiation. The solar radiation reaching the Earth surface without being modified into the atmosphere is called *direct (beam) solar radiation,* ($I_{HB}$ for a horizontal and $I_{TB}$ for a tilted surface). Atmospheric conditions can reduce direct solar radiation by 10% on clear, dry days and up to 10% during dark, cloudy days. The amount of radiation either absorbed or scattered depends on the path length through the atmosphere. A simple concept used to characterize the effect of clear atmosphere on sunlight is the air mass (Figure 2.6), equal to the relative length of the direct radiation beam path through the atmosphere. One of the simplest methods to calculate solar radiation atmospheric absorption for clear sky conditions is through Beer, Lambert, Bouguer law, expressed as:

$$I_B = I_0 \exp(-k \cdot AM) \tag{2.29}$$

where $I_B$ and $I_0$ are the terrestrial and extraterrestrial beam radiation intensities, $k$ is the optical depth or absorption constant for the atmosphere and $AM$ is a dimensionless path length of sunlight through the atmosphere, the air mass ratio. In general, the air mass through which sunlight passes is proportional to the secant of the zenith angle, $\theta_z$, which is the angle measured between the direct beam and the vertical. The air mass coefficient defines the direct optical path length through the Earth's atmosphere, expressed as a ratio relative to the path length vertically upwards, i.e., at the zenith. The air mass coefficient can be used to characterize the solar spectrum after solar radiation has traveled through

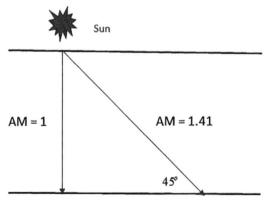

**FIGURE 2.6**
Geometry for air mass definition.

the atmosphere. When the Sun is at zenith corresponds to $AM = 1$, at other times is approximately equal to $1 / \cos\theta_z$, the secant of the zenith angle. For example at a zenith angle of $45°$, $AM = 1.41$, while at a zenith angle of $60°$, the air mass is approximately 2. However, for a path length L through the atmosphere, for solar radiation incident at angle $\theta_z$ relative to the normal to the Earth's surface, the air mass coefficient is:

$$AM = \frac{L}{L_0} \approx \frac{1}{\cos(\theta_z)} = \frac{1}{\sin(\alpha_S)} \tag{2.30}$$

where $L_0$ is the zenith path length (i.e., normal to the Earth's surface) at sea level and $z$ is the zenith angle in degrees.

The air mass number is thus dependent on the Sun's elevation path through the sky and therefore varies with time of day and with the passing seasons of the year, and with the latitude of the observer. However, the Earth is not flat, and, depending on accuracy requirements, this formula is usable for zenith angles up to about $60°$ to $75°$. For zenith angles from $0°$ to $70°$ this is a close approximation to calculate the air mass at sea level, for higher zenith angles, the effect of the Earth's curvature becomes significant and must be taken into account. The above approximation overlooks the curvature of the Earth, and is reasonably accurate for values of $\theta_z$ to around $75°$. The mean atmospheric transmittance is expressed as:

$$\tau_{atm} = \frac{I_B}{I_0} = 0.5 \left( \exp\left(-0.650 AM(z,\alpha)\right) + \exp\left(-0.095 AM(z,\alpha)\right) \right) \tag{2.31}$$

Here $\alpha$ is the solar altitude, defined before and the air mass $AM(z, \alpha)$ at altitude $z$ above sea level, is expressed as:

$$AM(z,\alpha) = AM(0,\alpha) \frac{p(z)}{p(0)}$$

where $p$ is the atmospheric pressure and the air mass $AM(0, \alpha)$ at sea level (0 m altitude) is given by:

$$AM(0,\alpha) = \sqrt{1229 + (614 \sin\alpha)^2} - 614 \sin\alpha \tag{2.32}$$

By obvious extension air mass zero ($AM0$) corresponds to the extraterrestrial radiation. A number of refinements have been proposed to more accurately model the path thickness toward the horizon, such as:

$$AM = \frac{1}{\cos(\theta_z) + 0.50572 \cdot (96.07995 - \theta_z)^{-1.6364}} \tag{2.33}$$

Solar intensity at the collector reduces with increasing air mass coefficient, in a nonlinear fashion due to the complex and variable atmospheric factors involved. For example, almost all high energy radiation is removed in the upper atmosphere (between $AM0$ and $AM1$) and so $AM2$ is not twice as bad as $AM1$. Furthermore, there is great variability in many of the factors contributing to atmospheric attenuation, such as water vapor, aerosols, smog, and the effects of temperature inversions. Depending on level of pollution in the air, overall attenuation can change by up to ±70% toward the horizon, greatly affecting performance particularly toward the horizon, where the effects of the lower layers of atmosphere are amplified by many folds. At $AM = 1$, after absorption has been accounted, the global radiation intensity is reduced from 1367 W/m$^2$ at the top of the atmosphere to just 1000 W/m$^2$ at sea level. Hence, for an $AM = 1$ path length, the sunlight intensity is reduced to 70% of its original $AM = 0$ value. Assuming that the absorption constant depends on the air mass, this observation can be expressed as:

$$I = 1367 \cdot (0.7)^{AM} \tag{2.34}$$

For air masses deferent for the unity, according to Meinel and Meinel, 1976 [1], a better fit to the observed data is given by

$$I = 1367 \cdot (0.7)^{(AM)^{0.678}} \tag{2.35}$$

**Example 2.10:** Calculate the zenith angle needed to produce $AM = 1.8$ at sea level if $AM$ 1.0 occurs at zero degrees, using the air mass definition.

**Solution:** The zenith angle is $\theta_z = \mathrm{acos}(1 / 1.8) = 56.3°$. The surface beam radiation intensity, by using Equation (2.33) is then:

$$I = 1367 \cdot (0.7)^{(1.8)^{0.678}} = 888.56 \text{ W/m}^2$$

Equations (2.33) and (2.34) are fitting comfortably within the mid-range of the expected pollution-based variability. One approximate model for solar radiation intensity versus the air mass is given by:

$$I = 1.1 \times I_0 \times 0.7^{(AM)^{(0.678)}} \tag{2.36}$$

For higher level of the air pollution the following relationship is recommended:

$$I = 1.1 \times I_0 \times 0.56^{(AM^{0.715})} \tag{2.37}$$

where the solar intensity external to the Earth's atmosphere $I_0 = 1.357$ kW/m$^2$, and the factor of 1.1 is derived assuming that the diffuse component of the solar radiation is 10% of the direct component. Table 2.3 summarizes the calculations of the air mass and the solar radiation values for various zenith angles. At greater zenith angles, the accuracy degrades rapidly, with $X = 1/\cos\theta_z$ becoming infinite at the horizon, while the horizon air mass in the more-realistic spherical atmosphere is usually less than 40. Many formulas have been developed to fit tabular values of air mass, and interested readers can found them elsewhere in the literature. Spectrum corresponding to an $AM = 1.5$, zenith angle of 60° is a typical solar spectrum on Earth surface on clear sky with a total of solar irradiance of 1 kW/m$^2$, being used for the solar photovoltaic (PV) cell or PV module calibration. Although the global irradiance can be as high as 1 kW/m$^2$, the value available at the ground is less because of the Earth rotation and the weather conditions. These equations apply only to clear sky situation in standard atmosphere in which there is no air pollution. In this analysis, the air mass is considered as an average quantity applied to solar constant. However, it is generally necessary to know the atmospheric attenuation of solar radiation on spectral basis. Detailed spectral calculations are needed for photovoltaic systems, reflecting-surface

**TABLE 2.3**

Solar Intensity vs. Zenith Angle and Air Mass Coefficient

| $\theta_z$ | Air Mass | Radiation Intensity Eq. (2.33) (W/m²) | Radiation Intensity Eq. (2.34) (W/m²) |
|---|---|---|---|
| — | 0 | 1367 | 1367 |
| 0° | 1 | 840 | 1040 |
| 23° | 1.09 | 800 | 1020 |
| 30° | 1.15 | 780 | 1010 |
| 45° | 1.41 | 710 | 950 |
| 48.2° | 1.5 | 680 | 930 |
| 60° | 2 | 560 | 840 |
| 70° | 2.9 | 430 | 710 |
| 75° | 3.8 | 330 | 620 |
| 80° | 5.6 | 200 | 470 |
| 85° | 10 | 85 | 270 |
| 90° | 38 | — | 20 |

solar calculations or selective surface property calculations. In summary, air mass concept provides a method for solar radiation direct component estimation on Earth for clear sky conditions. Solar irradiance integrated over a period of time is called *solar irradiance*. Of particular importance in the design of solar engineering systems is the irradiance over one day. A simple estimate of the average solar daily radiation on the ground, $G_{av}$ (averaged both over the location and the time of the year) can be obtained using the value of the average extraterrestrial irradiance 342 W/m². The solar radiation observed on the Earth surface is in average 30% lower on account of scattering and reflection, which yields to:

$$G_{av} = 0.7 \times 342 \times 24 \text{ h} = 5.75 \text{ kWh/day} \tag{2.38}$$

Threlkeld and Jordan, 1958 estimated values of $k$ for average atmospheric conditions at sea level with a moderately dusty atmosphere and for monthly average values of the amount of precipitable water vapor for the United States, as shown in Table 2.4. To account for the differences in local conditions from the average sea level conditions, Equation (2.26) is modified by a parameter, called clearness index or number $C_n$ as:

$$I_{BN} = C_n I \exp\left(-\frac{k}{\cos\theta_z}\right) \tag{2.39}$$

**TABLE 2.4**

Average Values of Atmospheric Optical Depth (k), Apparent Extraterrestrial Flux and Sky Diffuse Factor (C) for Twenty-First Day of Each Month, for Sea Level, Mean Atmospheric Conditions and for the United States

| Month | January | February | March | April | May | June | July | August | September | October | November | December |
|---|---|---|---|---|---|---|---|---|---|---|---|---|
| k | 0.142 | 0.144 | 0.156 | 0.180 | 0.196 | 0.205 | 0.207 | 0.201 | 0.177 | 0.160 | 0.149 | 0.142 |
| A (W/m²) | 1230 | 1215 | 1186 | 1136 | 1104 | 1085 | 1085 | 1107 | 1151 | 1192 | 1221 | 1233 |
| C | 0.058 | 0.060 | 0.071 | 0.097 | 0.121 | 0.134 | 0.136 | 0.122 | 0.092 | 0.073 | 0.063 | 0.057 |

As we mentioned before the radiation beam passing through the atmosphere, a good portion of it is absorbed by various gases in the atmosphere, or scattered by air molecules or atmosphere particulates. In fact, over a year's time, less than half of the radiation that hits the top of the atmosphere reaches the Earth's surface as direct beam. On a clear day, with the Sun high in the sky, the solar beam radiation at the surface can exceed 70% of the extraterrestrial solar radiation flux. Attenuation of incoming radiation is a function of the distance that the beam has to travel through the atmosphere, which is easily calculable, as well as factors, such as dust, air pollution, atmospheric water vapor, clouds and turbidity, which are not so easy to account for. A commonly used model is an adaptation of Equation (2.27), an exponential decay function:

$$I_B = A \cdot \exp(-k \cdot AM) \tag{2.40}$$

Table 2.4 gives the values of $A$ and $k$ that are used in the American Society of Heating, Refrigerating, and Air Conditioning Engineers (ASHRAE) Clear Day Solar Flux Model. This model is based on empirical data collected by Threlkeld and Jordan (1958) for a moderately dusty atmosphere with atmospheric water vapor content equal to the average monthly values in the United States. Table 2.4 is also including the diffuse factor, $C$, that is introduced later. For computational purposes, the equation to work with rather than a table of values are developed. Close fits to the values of optical depth $k$ and apparent extraterrestrial ($ET$) flux $A$ given in Table 2.4 are as follows:

$$A = 1160 + 75 \cdot \sin\left(\frac{360(n-275)}{365}\right)$$

$$k = 0.174 + 0.035 \cdot \sin\left(\frac{360(n-100)}{365}\right) \tag{2.41}$$

Here $n$ is again the day number, Table 2.1.

> **Example 2.11:** Find the direct beam solar radiation normal to the Sun's rays at solar noon on a clear day in Detroit, Michigan (latitude 42.3°) on May 21st ($n = 141$).
>
> **Solution:** From Equation (2.40), the apparent extraterrestrial radiation, A and the optical depth, $k$ are:
>
> $$A = 1160 + 275 \cdot \sin\left(\frac{360(141-275)}{365}\right) = 1104 \ \text{W/m}^2$$
>
> $$k = 0.174 + 0.035 \cdot \sin\left(\frac{360(141-100)}{365}\right) = 0.191$$
>
> By using Equation (2.9) the declination angle for May 21st is $\delta = 20.1°$ and the altitude angle at solar noon is:
>
> $$\beta_N = 90^O - L + \delta = 90^O - 42.3^O + 20.1^O = 67.8^O$$
>
> The air mass ratio, $AM$ is
>
> $$AM \simeq \frac{1}{\sin(\beta_N)} = \frac{1}{\sin(67.8°)} = 1.0803$$

And by using Equation (2.39) the predicted value of clear sky beam solar radiation at the Earth' surface is then:

$$I_B = 1104 \cdot \exp(-0.191 \cdot 1.0803) = 898.2 \ \text{W/m}^2$$

## 2.4 Solar Radiation on Horizontal and Tilted Surfaces

Several other angles, as discussed in the previous sections, are used to identify the Sun location in the sky. These angles include the solar azimuth angle ($\gamma$), the solar zenith angle ($\theta_z$), the solar altitude angle ($\alpha$), and the sunset hour angle, $Hs$, as shown in Figure 2.3. If the solar zenith angle is known, the extraterrestrial radiation ($I_0$) can be calculated as follows:

$$I_0 = SC \cdot ecc \cdot \cos(\theta_z) \tag{2.42}$$

Eccentricity may be estimated using Equation (2.3). The units are Joules per second per square meter (Figure 2.7). It is also of interests in solar engineering and solar radiation applications the solar radiation energy received on a horizontal surface outside of the Earth atmosphere. Equation (2.6), taking into the account for the Sun–Earth distance throughout the year, adjusted for the azimuth angle is:

$$I_0 = SC \cdot \left( 1 + 0.034 \cos\left( \frac{360n}{365} \right) \right) \cdot \cos(\theta_z) \tag{2.43}$$

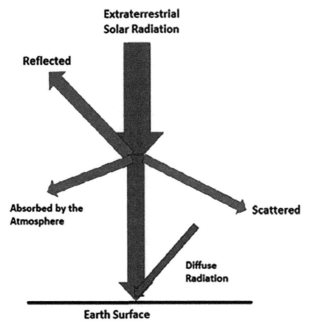

**FIGURE 2.7**
Solar radiation passing through the atmosphere.

This value corresponds to the maximum possible solar energy if there were no atmosphere, and it depends on the solar altitude, Equations (2.11) or (2.12), expressed as:

$$I_{H0} = SC \cdot \left(1 + 0.0333 \cos\left(\frac{360n}{365}\right)\right) \cdot \sin(\alpha_S) \tag{2.44}$$

Here $I_{H0}$ is the extraterrestrial solar radiation on a horizontal surface. By integrating Equation (2.43) from sunshine to sunrise, we can compute the daily extraterrestrial insulation on a horizontal surface at the top of the atmosphere:

$$\bar{I}_{H0} = SC \cdot \frac{24}{\pi} \cdot \left(1 + 0.0333 \cos\left(\frac{360n}{365}\right)\right) \cdot$$

$$\left(\frac{\pi \cdot H_{SS}}{180} \sin(L)\sin(\delta) + \cos(L)\cos(\delta)\cos(H_{SS})\right) \tag{2.45}$$

**Example 2.12:** Determine the daily average solar radiation energy on a horizontal surface outside of the atmosphere at latitude of 40° N on March 5.

**Solution:** The declination angle this day is $\delta = -6.8°$, while the sunset angle (Equation (2.26)) is $H_{SS} = 84.3°$. The daily average for that day is calculated by using Equation (2.45) is:

$$\bar{I}_{H0} = SC \cdot \frac{24}{\pi} \cdot \left(1 + 0.0333 \cos\left(\frac{360 \cdot 64}{365}\right)\right) \cdot$$

$$\left(\frac{\pi \cdot 84.3°}{180} \sin(40)\sin(-6.8°) + \cos(40°)\cos(-6.8°)\cos(84.3°)\right) = 6826.3 \text{ J/m}^2$$

Measuring or calculating the total global solar radiation striking the Earth's surface is far more complicated than determining extraterrestrial radiation. Changing atmospheric conditions, topography, and the changing position of the Sun interact to moderate the solar resource at a given time and location. The problem is further complicated for the solar engineer because the vast majority of solar radiation measurements (and extrapolations based on them) have been made on horizontal surfaces rather than on the tilted surfaces typically employed by efficient solar energy collection. Solar designers must use reference tables that translate horizontal data into directional data for tilted surfaces or make those calculations themselves. The total solar radiation on Earth consists of direct beam radiation that comes on a direct line from the Sun, and diffuse radiation from the sky, created when part of the direct beam radiation is scattered by atmospheric constituents (e.g., clouds and aerosols), and reflected from the surroundings on the surface of interest (see Figure 2.5, for details). Concentrating solar collectors rely almost entirely on direct beam radiation, whereas other collectors, including passive solar buildings, capture both direct beam and diffuse radiation. Radiation reflected from the surface in a collector front may also contribute to total solar radiation. The atmosphere through which solar radiation passes is variable and acts as a dynamic filter, absorbing and scattering solar radiation. Low- and mid-level cumulus and stratus clouds are generally opaque, blocking the direct beam, while high, thin cirrus clouds are usually translucent, scattering the direct beam, but not totally blocking it. On sunny days with clear skies, most of the solar radiation is direct beam radiation, the diffuse radiation only accounts for about

5%–20% of the total. Under overcast skies, diffuse radiation, which is scattered out of the direct beam by gases, aerosols, and clouds, accounts for almost 100% of the solar radiation reaching the surface. Under clear skies, the total instantaneous solar radiation at the planet's surface at midday can exceed 1000 W/m². In contrast, instantaneous midday radiation on a dark, overcast day can be less than 100 W/m².

The impact of clouds on the solar resource shows the inverse relationship between average monthly opaque cloud cover and average daily solar energy, which can significantly affect the insolation. Aerosols, including dust, smoke, pollen and suspended water droplets reduce the transmittance or amount of solar radiation reaching the planet's surface. These factors are influenced by climate and seasonal changes, with significant seasonal variations in aerosol optical depth for various climate regimes. If the annual mean aerosol optical depth and the amplitude of seasonal variations are known for a particular location, the aerosol optical depth for any day of the year can be estimated as follows (Maxwell and Myers 1992):

$$\tau_A = A \cdot \sin\left[\frac{360n}{365} - \varphi\right] + C \tag{2.46}$$

where $\tau_A$ is the aerosol optical depth on day $n$, $A$ is the amplitude of seasonal variations in $\tau_A$, $n$ is the day of the year (from 1 to 365), $\varphi$ is the phase factor that determines the days when the maximum and minimum values occur (for most locations in the northern hemisphere, $\varphi = 90°$) and $C$ is the annual mean aerosol optical depth.

The total solar radiation on Earth consists of direct beam radiation that comes to the surface on a direct line from the Sun, and diffuse radiation from the sky, created when part of the direct beam radiation is scattered by atmospheric constituents (e.g., clouds and aerosols). Concentrating solar collectors rely almost entirely on direct beam radiation, whereas other collectors, including passive solar buildings, capture both direct beam and diffuse radiation. Radiation reflected from the surface in front of a collector may also contribute to total solar radiation. The direct radiation, called also direct beam radiation $I_B$ or direct normal irradiance, is the radiation on a plane normal or perpendicular to the line connecting the observer and the center of the solar disk, within a small solid angle, 5° or smaller, which includes the solar disk. Diffuse radiation, $_{ID}$ is the solar radiation scattered out of the direct beam radiation by the atmosphere into the sky dome hemisphere. The total ("global") hemispherical radiation on a plane is the combination of the direct normal radiation multiplied by the cosine of the incidence angle $\theta$ (between the normal to the plane and the direction from the base of the normal to the center of the solar disk) and the diffuse sky radiation. The atmosphere through which solar radiation passes is variable and acts as a dynamic filter, absorbing and scattering solar radiation. Low- and mid-level cumulus and stratus clouds are generally opaque, blocking the Sun direct beam. In contrast, high, thin cirrus clouds are usually translucent, scattering the direct beam, but not totally blocking it. On sunny days with clear skies, most of the solar radiation is direct beam radiation and diffuse radiation only accounts for about 5%–20% of the total. Under overcast skies, diffuse radiation, scattered out of the direct beam by gases, aerosols, and clouds, accounts for 100% of the solar radiation reaching the Earth's surface. Under clear skies, the total instantaneous solar radiation at the surface at midday can exceed 1000 W/m². In contrast, instantaneous midday radiation on a dark, overcast day can be less than 100 W/m². Total instantaneous solar radiation on a horizontal surface, $I_H$, is compose from direct and sky diffuse radiation (see Figure 2.6), expressed as:

$$I_H = I_{HB} + I_{HD} \tag{2.47}$$

While the daily value of the terrestrial solar radiation is then:

$$\overline{I}_H = \overline{I}_{HB} + \overline{I}_{HD} \tag{2.48}$$

Threlkeld and Jordan (1958), using the ASHRAE clear-day solar flux model, suggested that the sky diffuse radiation on a clear day is proportional to direct beam radiation, by using Equation (2.43) and an empirical sky diffuse factor, $C$ as:

$$I_H = I_{HB} + C \cdot I_{HD} = C_n I \exp\left(-\frac{k}{\sin\alpha}\right)(C + \sin\alpha) \tag{2.49}$$

Values of $k$, the optical depth and $C$, the clearness index are given in Table 2.4.

### 2.4.1 Clear Sky Radiation Models

One of the most used model to estimate the solar radiation, valid for a large number of location has be proposed by Gueymard and Thevenard, 2013. This model was developed to calculate the solar heat gain for fenestration, it was named it the ASHRAE Clear Sky Model. On the basis of the detailed simulations, a simple two-band, two-step solar irradiance model REST2 (Reference evaluation of solar transmittance), that can model clear-sky solar irradiance very accurately was developed. Solar transmittance of clear sky was modeled based on two spectral bands, the first band from 0.29 to 0.7 μm, characterized by absorption by molecules and aerosols, and the second band from 0.7 to 4 μm, characterized by absorption by water vapor and $CO_2$. The two-band clear sky radiation model was used to calculate clear sky solar irradiance for a large number of cases and compared with the data covering a large part of the world. The second step of this model consisted in developing a condensed model depending on only two monthly parameters described later in this section. According to the ASHRAE model, the beam and diffuse components are calculated as:

$$I_{B,N} = I_0 \exp\left(-\tau_B \cdot AM^b\right) \tag{2.50}$$

And

$$I_{D,H} = I_0 \exp\left(-\tau_D \cdot AM^d\right) \tag{2.51}$$

where $I_{B,N}$ is the beam normal irradiance per unit area normal to the Sun rays, $I_{D,H}$ is diffuse horizontal irradiance per unit area on a horizontal surface, $I_0$ extraterrestrial normal irradiance $AM$ is the air mass, $\tau_B$, $\tau_D$ are the beam (direct) and diffuse optical depths ($\tau_B$ and $\tau_D$ are more correctly termed pseudo-optical depths, because optical depth is usually employed when the air mass coefficient is unity), and $b$, $d$ are the beam and diffuse air mass exponents. Values of $\tau_B$ and $\tau_D$ are location specific and vary during the year, embodying the dependence of clear sky solar radiation upon local conditions (elevation, precipitable water content, and aerosols). Their average values are tabulated for the 21st day of each month for all the locations in the tables of climatic design conditions, and are included in Appendix B. The air mass exponents, $b$ and $d$

are correlated to the beam and diffuse optical depths, $\tau_B$ and $\tau_D$ through the following empirical relationships:

$$b = 1.219 - 0.043\tau_B - 0.151\tau_D - 0.204\tau_B\tau_D \tag{2.52}$$

And

$$d = 0.202 + 0.852\tau_B - 0.0071\tau_D - 0.357\tau_B\tau_D \tag{2.53}$$

This radiation model describes a simple parameterization of the solar radiation model, providing accurate predictions of $I_{B,N}$ and $I_{D,H}$, even at sites where the atmosphere is very hazy, polluted or humid most of the time. Then the total solar radiation on a horizontal surface is given by:

$$I_H = I_{B,N}\sin(\alpha_S) + I_{D,H} \tag{2.54}$$

## 2.4.2 Solar Radiation on Tilted Surfaces

Solar radiation on a tilted surface, with a tilt angle of $\beta$ for the horizontal and an azimuth angle, $\theta_z$ is the sum of the beam (direct), sky diffuse and ground reflected solar radiation components, expressed as:

$$I_T = I_{B,C} + I_{D,C} + I_{r,C} \tag{2.55}$$

If $i$ is the incidence angle of the direct (beam) solar radiation on the tilted surface (collector), the instantaneous direct radiation on the surface is given by:

$$I_{B,C} = I_{B,N}\cos i \tag{2.56}$$

The incidence angle is then related to the solar angle, from the geometry of Figure 2.8 by:

$$\cos i = \cos\alpha \cdot \cos(a_S - a_w) \cdot \sin\beta + \sin\alpha \cdot \cos\beta \tag{2.57}$$

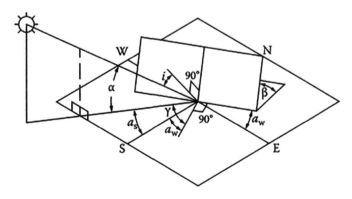

**FIGURE 2.8**
Geometry of the definitions of solar angles for a tilted surface.

The diffuse radiation on the surface ($I_{D,C}$) can be obtained by multiplying the sky diffuse solar radiation on a horizontal surface by the tilt angle, $\beta$ between the sky and the collector surface:

$$I_{D,C} = I_{D,H} \frac{1+\cos(\beta)}{2} = I_{D,H} \cos^2\left(\frac{\beta}{2}\right) \tag{2.58}$$

The last component of the solar radiation striking a solar collector is radiation reflected by the surface in front of the collector. The ground reflectance was estimated to range from 0.1 from dark surfaces to about 0.8 for fresh snow. The amount of reflected radiation can be modeled from the total horizontal solar radiation and the surface reflectivity. The ground reflected solar radiation can be found from the total solar radiation incident on a horizontal surface and the ground reflectance $\rho$ as:

$$I_{r,C} = I_H \cdot \rho = \rho \cdot \left(I_{B,H} + I_{D,H}\right) \tag{2.59}$$

The part of the reflected radiation intercepted by the tilted surface can be found by multiplying the ground reflected radiation by the view factor between the surface and the ground:

$$I_{r,C} = \rho I_H \frac{1-\cos\left(\beta_{tilt}\right)}{2} = \rho\left(I_{B,N} \cdot \sin(\alpha) + I_{D,H}\right) \cdot \sin^2\left(\frac{\beta_{tilt}}{2}\right) \tag{2.60}$$

**Example 2.13:** Find the instantaneous solar radiation at 12:00 noon EST on a solar collector surface ($\beta = 30°$, aw $= +10°$) on February 10th in Tampa, Florida. The values of pseudo optical depths (taken from ASHARE Handbook, 2009) are: $\tau_B = 0.35109$ and $\tau_D = 2.48558$, respectively.

**Solution:** For Tampa International Airport: $L = 27.96$ N (+), Local Time $= 82.54$ W (+) and $ET = 75$ W (+). For February 10, $n = 42$, and the declination angles for this day is $\delta = -12.84°$.

The local time is 12:00 PM. The solar time is given by

$$B = \frac{360}{365}(42-81) = -38.5°$$

$$ET(\text{min}) = -14.5°$$

$$ST = LST + ET + (L_{ST} - L_{Local}) \times 4\frac{\text{min}}{1°} = 11{:}16 \text{ AM}$$

The solar altitude angle is:

$$\sin(\alpha) = \cos(-12.84°)\cos(27.96°)\cos(-11.1°) + \sin(-12.84°)\sin(27.96°)$$

$$\alpha = 47.86°$$

The solar azimuth angle is calculated as

$$\sin(\alpha_S) = \frac{\cos(-12.84°)\sin(-11.1°)}{\cos(47.86°)}$$

$$\alpha_S = -16.24°$$

We need to find out if the absolute value of the azimuth angle is greater than 90°. For $L > \delta$, by using Equation (2.24):

$$TE = 12{:}00 - \left[\cos^{-1}\left(\frac{\tan(-12.84°)}{\tan(27.96°)}\right)\right]\frac{h}{15°} = 4.277 \text{ h}$$

Given that $ST > TE$, the Sun is south, and $\alpha_S = -16.24°$. The air mass, using Equation (2.32) is, $m = 1.3478$, and by using Equations (2.50) and (2.51), the parameters $a$, $b$ for air mass are, $b = 0.6506$, and $d = 0.1722$, respectively. The extraterrestrial solar radiation is given by:

$$I = SC \cdot \left[1 + 0.0333 \cdot \cos\left(\frac{360n}{365}\right)\right] = 1401.1 \text{ W/m}^2$$

Then, the direct radiation and the diffuse solar radiation, on horizontal surface, by using Equations (2.49) and (2.50) are:

$$I_{B,H} = 1401.1 \cdot \exp\left(-0.35109 \times 1.3478^{0.6505}\right) = 914.8 \text{ W/m}^2$$

$$I_{D,H} = 1401.1 \cdot \exp\left(-2.48558 \times 1.3478^{0.1722}\right) = 102.3 \text{ W/m}^2$$

For the geometry of the solar collector, the cosine of the angle of incidence is:

$$\cos(i) = \cos(47.86°)\cos(-16.24° - 10°)\sin(30°) + \sin(47.86°)\cos(30°) = 0.943$$

Then, the direct solar radiation on a collector is, then:

$$I_{B,C} = I_{B,H} \times \cos(i) = 862.6 \text{ W/m}^2$$

While, the diffuse solar radiation on the collector surface is:

$$I_{D,C} = I_{D,H} \cdot \cos^2\left(\frac{\beta}{2}\right) = 102.3 \cdot \cos^2\left(\frac{30°}{2}\right) = 96.3 \text{ W/m}^2$$

Assuming that the solar collector is surrounded by ordinary ground or grass, with $\rho \approx 0.2$, the reflected solar radiation on the collector surface is:

$$I_{r,C} = 0.2 \cdot (862.6 \cdot \sin(47.86°) + 96.3)\sin^2\left(\frac{30°}{2}\right) = 38.1 \text{ W/m}^2$$

Finally, the total solar radiation on a tilted collector surface is

$$I_C = I_{B,C} + I_{D,C} + I_{r,C} = 862.6 + 96.3 + 38.1 = 997 \text{ W/m}^2$$

## 2.5 Daily and Monthly Solar Radiation Estimation Models

The only extant solar radiation data for developing countries is often monthly averages of daily total global horizontal radiation. Given sufficient time and money, an engineer could conduct a multi-year measurement program. Lacking such resources and time, a statistical correlation approach such as the developed by Reddy (1987) can be used. This approach assumes that correlations between monthly averages and daily and hourly distributions at locations having measured data can be used to derive daily and hourly data for locations exhibiting similar clearness indices. One of the earliest methods of estimating solar radiation on a horizontal surface was proposed by the pioneer physicist Angström. It was a simple linear model relating average horizontal radiation to clear-day radiation and to the sunshine level, that is, percent of possible hours of sunshine. First we consider the definition of the monthly clarity index. The monthly average clearness index ($K$) is often used to quantify the relative transmittance of the atmosphere. The clearness index (Iqbal 1983) is defined as:

$$K = \frac{\bar{H}}{\bar{H}_0} \tag{2.61}$$

where $\bar{H}$ is the monthly average daily total of the global horizontal radiation, and $\bar{H}_0$ is the average daily total amount of extraterrestrial radiation incident upon a horizontal surface during that month. Since the definition of a clear day is somewhat nebulous, Page (1966), refined the Angström relationship, based it on extraterrestrial radiation instead of the ill-defined clear day:

$$\bar{H}_H = \bar{H}_{0,H}\left(a + b\,\frac{\bar{n}}{\bar{N}}\right) = \bar{H}_{0,H}\left(a + b\,\frac{\overline{PS}}{100}\right) \tag{2.62}$$

where $\bar{H}_H$ and $\bar{H}_{0,H}$ are the horizontal terrestrial and horizontal extraterrestrial radiation per day averaged for a month, $\overline{PS}$ is the monthly averaged percent of possible sunshine (i.e., hours of sunshine/maximum possible duration of sunshine × 100), $a$ and $b$ are constants for a given site, $\bar{n}$ and $\bar{N}$ are the monthly average numbers of hours of bright sunshine and day length, respectively. The ratio $\bar{n}/\bar{N}$ is also equivalent to the monthly average percent sunshine. $\bar{H}_{0,H}$ parameter can be calculated by finding $\bar{H}_{0,h}$ and averaging it for the number of days in each month, or from measurement data can be used, based on $I_0$ equal to 1367 W/m². The first step in estimating daily and hourly values of solar radiation on horizontal surfaces is the calculation of the clearness index $\bar{K}$ using available monthly average daily total global horizontal data and Equation (2.58), $\bar{H}_0$ can be used to predict the monthly average daily total horizontal diffuse radiation. The monthly average daily total horizontal beam radiation can then be obtained from the global solar radiation data. The correlation between the clearness index and the average daily total horizontal diffuse radiation takes account of the effect of seasonal variation by including the sunset hour angle, $H_{SS}$. The correlation for a clearness index in the range of $0.3 \le \bar{K} \le 0.8$ is given by:

$$\frac{\bar{H}_d}{\bar{H}_0} = 1.391 - 3.561\bar{K} + 4.189\bar{K}^2 - 2.137\bar{K}^3 \text{ for } W_s \leq 81.4°$$

$$\frac{\bar{H}_d}{\bar{H}_0} = 1.311 - 3.022\bar{K} + 3.427\bar{K}^2 - 1.821\bar{K}^3 \text{ for } W_s > 81.4°$$

(2.63)

where $W_S$ is the monthly mean value of the sunset hour angle in degrees on a horizontal surface.

When subjected to statistical analysis, different locations with the same mean monthly clearness index, $\bar{K}$, often display cumulative frequency curves that are more or less identical. This gives rise to the long-term implication that the monthly average distribution of the daily total global horizontal radiation may be independent of location and month. This relationship is most consistent for temperate regions. $K_{max}$ can be calculated as follows:

$$K_{max} = 0.6313 + 0.267\bar{K} - 11(\bar{K} - 0.75)^8$$

(2.64)

The monthly average daily total diffuse radiation, $\bar{H}_d$ can be predicted from the daily total horizontal global radiation and the clearness index using the following relationship, which includes the sunset angle $W_S$ to account for seasonal variation. For $W_S < 81.48°$,

$$\frac{\bar{H}_d}{\bar{H}_0} = 1.0 - 0.2727K + 2.449K^2 + 9.3879K^3 + 9.3879K^4 \text{ for } K < 0.715$$

$$\frac{\bar{H}_d}{\bar{H}_0} = -0.175 \text{ for } K > 0.722$$

(2.65)

For $H_{SS} < 81.48°$,

The hourly horizontal global radiation can also be predicted from daily total horizontal global radiation using a statistical correlation. The correlation is possible if average diurnal solar radiation is assumed to be symmetrical about solar noon. Interested readers can find more on the solar radiation models in the end-of-chapter references or elsewhere in the literature.

## 2.6 Solar Radiation Measurements

### 2.6.1 Solar Design Tools

The National Renewable Energy Laboratory (NREL) completed the National Solar Radiation Database in 1992 to assist designers, planners, scientists and engineers with solar resource assessments and data. The database contains 30 years of hourly values of measured or model solar radiation together with meteorological information from 239 stations across the United States, covering the period from 1961 through 1990. Some of these stations are primary station that measured solar radiation data for at least one year, or secondary stations. Solar radiation data for secondary stations is derived from computer models. In most cases, these stations are located at National Weather Service stations that collected hourly or three-hourly meteorological data throughout the period from 1961 to 1990. An update to the NSRDB includes the data for the period 1991–2005. The solar radiation elements represent the total energy received during the hour preceding the designated times. The Sun position in the sky and the Earth's position in orbit were calculated at the midpoint of the hour proceeding the designated time. Meteorological elements are the values observed at the designated time. The data included present the degree of uncertainty for global horizontal, direct normal, and diffuse horizontal radiation data. They also indicate whether these elements were measured or modeled. Other database products include: (1) hourly, daily, and quality statistics for the solar radiation elements, (2) daily statistics for the meteorological elements and (3) persistence statistics for daily total solar radiation. The statistical summaries include hourly, monthly, annual and 30-year averages and their standard deviations.

Since producing the solar radiation database, NREL has developed new design tools for using the database. A typical meteorological year (TMY) is a dataset of hourly values of solar radiation and meteorological elements for a one-year period. A TMY consists of months selected from different individual years and linked together to form a complete year. For example, if thirty years of data are available, all 30 January would be examined and the one judged most typical would be selected for inclusion in the TMY. Each TMY month is chosen in a similar fashion, with particular attention paid to five statistical elements: global horizontal radiation, direct normal radiation, dry bulb temperature, dew point temperature, and wind speed. Designers and planners often find such a representative dataset useful in predicting the performance and economic viability of a solar energy system. For instance, the TMYs are designed for computer simulations of solar energy conversion and building systems, providing a standard for hourly data for solar radiation and other meteorological elements that supports system performance comparisons and configurations. A TMY is not specific to any particular year and should not be used as an indicator of actual conditions expected over the next year, or even five years. The scarcity of solar radiation measurements in some countries has led to the use of satellite data for estimating solar radiation. An international solar radiation database, update surface radiation budget developed by the NASA Langley Research Center is now available. The dataset uses a model to convert satellite images to daily total or monthly averages of the daily total surface solar radiation estimates with a grid resolution of 100 km.

### 2.6.2 Radiation Measurement Techniques

Solar radiation data are required for resource assessment, model development, system design and collector testing, among other activities in solar engineering. The basic solar radiation measurements are the beam (direct), diffuse and global solar radiation components. The expenses to set and operate of radiometric stations and high maintenance cost make impossible the spatially continuous mapping of solar radiation. Due to the scarcity of real data, the use of representative sites where irradiance data are measured or modeled has been a common practice for solar engineering calculations. However, whereas this practice is acceptable for standard energy calculations, nearby site extrapolation may prove widely inaccurate when site- or time-specific data are needed, being particularly true for concentrating solar power applications where direct normal solar radiation is required. The International Energy Agency Solar Heating and Cooling Program (IEA-SHCP) developed and evaluated techniques for estimating solar radiation at locations between network sites, using both measured and modeled data. In addition to classical statistical techniques, new methods such as satellite-based techniques have been investigated. Although they are less accurate than ground-based measurements, they may be more suitable to generate site- or time-specific data at arbitrary locations and times.

The most commonly used instruments to measure solar radiation today are based on either the thermoelectric or the photoelectric effects. The thermoelectric effect is achieved using a thermopile that comprises collections of thermocouples, consisting of dissimilar metals mechanically joined together, producing a small current proportional to their temperature. When thermopiles are appropriately arranged and coated with a dull black finish, they serve as nearly perfect blackbody detectors that absorb energy across the entire range of the solar spectrum. The hot junction is attached to one side of a thin metallic plate. The other side of the plate is blackened to be highly absorptive when exposed to the Sun radiation. The cold junction is exposed to a cold cavity within the instrument. The output is compensated electrically for the cavity temperature. The amount of insolation is related to the elevated temperature achieved by the hot junction and the electromagnetic force generated. The response is linearized and calibrated so that the output voltage is converted to the radiative flux. The PV sensors are simpler and have instantaneous response and good overall stability. The PV effect occurs when solar radiation strikes a light-sensitive detector, which in the excited state, is producing only by light in a specific range of wavelengths, the atoms release electrons, flowing through a conductor producing an electrical current, proportional to the intensity of the radiation striking the detector. The major disadvantage of these sensors is that their spectral response is not uniform in the solar band. Instruments used to measure the transmission of sunlight through Earth's atmosphere are of two general categories: instruments that measure radiation from the entire sky, and the ones measure only direct solar radiation. Within each of these categories, instruments can be further subdivided into those that measure radiation over a broad range of wavelengths and those that measure only specific wavelengths. The full-sky instruments need an unobstructed 360° view of the horizon, without significant obstacles. Full-sky instruments are called radiometers or, in the case of solar monitors, the pyronameters. The sensor is under one or two hemispherical glass domes. In order to assess the availability of solar energy arriving on the Earth, measurement of solar radiation at some locations is essential. From these measurements, empirical models are developed to predict the availability of solar energy at other locations. Different instruments and their utilities are as follows. A pyrheliometer is an instrument for measurement of the direct solar radiation flux at normal incidence. The instrument is usually attached to an electrically driven equatorial mount which tracks the Sun. A pyranometer is an instrument for measurement of the direct and diffuse irradiance arriving from the whole hemisphere. This hemisphere is usually the complete sky dome. A pyranometer can be used in a tilted position as well in which case it will also receive ground reflected radiation. A pyranometer with a shading device is an instrument that measures diffuse solar irradiance within a solid angle, with the exception of the solid angle subtended by the Sun's disk. The heart of a radiometer (an instrument that measures radiant energy, whether from the Sun or from any other source) is its sensor or detector.

## 2.7 Summary

There are huge resources in solar radiation with the average incident solar power more than 5000 times current world power consumption and demand. The computation of solar radiation data are very important for scientists and engineers involved in creating applications implementing models for solar energy, building design, astronomy, agronomy and agrometeorology. Lamberts cosine law describes the amount of radiation received on the horizontal relative to the Sun's zenith angle. The amount of radiation received on an inclined surface is a function of the angle between the Sun and that of the projection normal to the surface. The axis of rotation of Earth is tilted 23.5 degrees. The amount of radiation received is a function of its latitude and the Sun's declination angle. Knowledge of the solar angles is also critical in the design of passive solar buildings, especially the placement of windows for solar access and the roof overhang for shading the walls and windows at certain times of the year. Seasons arise as the Earth revolves around the Sun because the Earth's tilted rotation axis causes the shaded and sunlit faces of Earth to change. Solar radiation fluxes can be estimated empirically with weather variables like temperature and humidity. The net solar radiation balance is equal to the sum of incident solar radiation minus the fraction reflected (albedo times the incoming solar radiation) plus incoming long-wave energy minus the energy emitted from the surface that is proportional to its temperature to the fourth power. An overall and comprehensive discussion of solar radiation characteristics and parameters is presented and discussed in the view of solar energy resource assessment, as well as presentation of solar radiation models and solar radiation measurements.

## Questions and Problems

1. Which two elements are the main sun components? By which physical process the Sun generates energy?

2. Using appropriate equations and relationships determine the date on which the Sun–Earth distance is maximum, and the date on which this distance will be at a minimum.

3. Estimate the power radiated from a blackbody with a surface of 1 m² and temperature of 1500°C.

4. List the main regions of the solar spectrum.

5. The area of New Mexico is 121,666 square miles. The average annual insolation (hours of equivalent full sunlight on a horizontal surface) is approximately 2000 h. If one-third of the area of New Mexico is covered with solar panels of 10% efficiency, how much electricity can be generated per year? What percentage of US energy needs can be satisfied?

6. What is the difference between the zenith and incidence angles?

7. Calculate the noon Sun angle for your location and the equinoxes and solstices.

8. Calculate the solar declination for spring and fall equinoxes and summer and winter solstices.

9. Calculate the Sun altitude at solar noon on June 21 in Sydney, Australia (latitude 34 S), and San Francisco, California, USA (latitude 34 N).

10. Determine the solar altitude and azimuth angles at 11:00 AM local time at Bucharest, Romania, on July 15.

11. Can the annual variations of the Sun–Earth distance adequately account for seasonal temperature changes?

12. Determine the solar constants for Venus and Mars.

13. Determine the sunset time (in solar time) and the length of daytime of Boston on President Day, Memorial Day, Labor Day, and Thanksgiving Day for 2012 and 2017.

14. Calculate the noon Sun angle for a site located at 50° N latitude on March 5 and August 10.

15. What is the solar time at Boulder, Colorado, on July 15 at 11:00 AM Mountain Standard Time?

16. Verify that the declination reaches the lowest value on the Winter Solstice and the highest value on the Summer Solstice.

17. What is the solar time in El Paso, Texas (latitude: 31.8° N, and longitude: 106.4° W), at 10 AM, and 3 PM, Mountain Standard Time on March 15, July 15, and October 1st?

18. Compute solar elevation angle, azimuth angle at 6, 9, 12, 15 and 18 hours for four locations on Earth at the summer and winter solstice and at the equinox:

    a. Berkeley, CA

    b. Your favorite Northern Hemisphere locale (e.g., Madrid, St Louis, Tokyo, New York, Paris, Rome, Bucharest, San Francisco)

    c. Your favorite Southern Hemisphere locale (e.g., Sydney, Christchurch, Perth, Johannesburg, Sao Paulo, Santiago de Chile, Buenos Aires, Luanda).

    d. The city of origin of one of forefathers.

19. For the chosen locations compute the solar declination angle and day-length for days 50, 180, 270 and 330.

20. Find the altitude angle and azimuth angle of the Sun at the following (solar) times and places:

    a. March 1st at 10:00 AM in New Orleans, latitude 30° N.

    b. August 10 the at 2:00 PM in London, Ontario, Canada, latitude 43° N

    c. July 1st at 5:00 PM in San Francisco, latitude 38° N.

    d. December 21st at 11 AM at latitude 68° N.

21. Determine the June 22 location of the Sun at solar noon at your location.

22. Find the solar altitude and azimuth angles at solar noon in Gainesville, Florida, on February 28 and August 8. Also find the sunrise and sunset times in Gainesville on that day.

23. Repeat the calculations of problem 22 for Sydney, Australia.

24. Calculate the time of sunrise, solar altitude, zenith, solar azimuth, and profile angles for a 60° sloped surface facing 25° west of south at 10:00 AM and 3:00 PM solar time on March 22 at latitude of 43°, north and south. Also calculate the time of sunrise and sunset on the surface.

25. Sunlight of 800 W/m² is incident on a solar panel at 50° t the vertical. What is the solar intensity on a vertical and on a horizontal surface?

26. Calculate the zenith angles needed to produce AM 1.5 and AM 2.0 if AM 1.0 occurs at zero degrees.

27. Calculate the zenith angle at solar noon at a latitude of 45° N on June 1.

28. A house located at 45° latitude north has a roof facing south and elevated at an angle of 30°. A solar panel is attached to the roof. What is the angle of incidence between the Sun and the panel at 10 AM and 2 PM on the fifteenth day of each of the year's month?

29. Calculate the number of hours with the Sun above the horizon on August 8 at Cleveland, Ohio, USA.

30. Calculate the angle of incidence of beam radiation, the surface slope, and the azimuth angle for a surface at latitude 45°, if the declination angle is 21° at 3:00 PM.

31. Find the solar altitude and azimuth angles in San Juan, Porto Rico on (a) June 1 at 7 AM and (b) December 1 at 2 PM Also find the sunrise and sunset times on these days.

32. Determine the location of the Sun at solar noon in Reno, Nevada on March 21. Determine also the local standard time and the sunrise azimuth angle.

33. What should be the tilt angle of a solar collector in Quito, Ecuador (latitude = 0°), in order to capture maximum energy on an annual basis?

34. What is the solar azimuth at sunrise on the summer solstice at 54° north?

35. At what angle should a South-facing collector at 36° latitude be tipped up to in order to have it be normal to the Sun's rays at solar noon on the following dates:
    a. March 21
    b. January 1
    c. April 21
    d. September 10
    e. October 21

36. Find the instantaneous solar radiation at 12:00 noon EST on a solar collector surface ($\beta = 30°$, aw = +10°) on July 10th in Canbera, Australia. The values of pseudo optical depths (taken from ASHARE Handbook, 2009) are: $\tau_B = 0.3548$ and $\tau_D = 2.48584$, respectively.

## References and Further Readings

1. B. Y. Liu and R. C. Jordan, The inter relationship and characteristic distribution of direct, diffuse and total solar radiation, *Solar Energy*, Vol. 4(3), pp. 1–19, 1960.
2. A. B. Meinel, and M. P. Meinel, *Applied Solar Energy—An Introduction*, Addison-Wesley, Reading, MA, 1976.
3. E. Kreith and J. F. Kreider, *Principles of Solar Engineering*, Hemisphere Publishing, Washington, DC, 1978.
4. M. Iqbal, Estimation of the monthly average of the diffuse component of total insolation on a horizontal surface, *Solar Energy*, Vol. 20(1), pp. 101–105, 1978.

5. J. Almorox and C. Hontoria, Global solar estimation using sunshine duration in Spain, *Energy Conversion and Management*, Vol. 11(3), pp. 170–172, 1967.

6. M. Collaress-Pereira and A. Rabl, The average distribution of solar radiation, *Solar Energy*, Vol. 22(2), pp. 155–164, 1979.

7. H. P. Garg, *Treatise on Solar Energy, Vol. 1: Fundamentals of Solar Energy Research*, John Wiley & Sons, Chichester, UK, 1982.

8. J. E. Sherry and C. G. Justus, A simple hourly all-sky solar radiation model based on meteorological parameters, *Solar Energy*, Vol. 32(2), pp. 195–204, 1984.

9. J. A. Davies and D. C. McKay, Estimating solar irradiance and components, *Solar Energy*, Vol. 29(1), pp. 55–64, 1985.

10. T. Muneer and C. S. Saluja, A brief review of models for computing solar radiation on inclined surfaces, *Energy Conversion and Management*, Vol. 25, pp. 443–458, 1985.

11. J. S. Hsieh, *Solar Energy Engineering*, Prentice-Hall, Englewood Cliffs, NJ, 1986.

12. M. Igbal, *An Introduction to Solar Radiation*, Academic Press, Toronto, ON, 1983.

13. G. Zerault, Solar radiation instrumentation, in *Solar Resources* (R. L. Hulstrom ed.), MIT Press, Cambridge, MA, 1989.

14. T. Muneer, *Solar Radiation and Daylight Models* (2nd ed.), Elsevier, Oxford, UK2004.

15. V. Quaschning, *Understanding Renewable Energy Systems*, Earthscan, London, UK, 2006.

16. V. Badescu (ed.), *Modeling Solar Radiation at the Earth's Surface, Recent Advances*, Springer, Berlin, Germany, 2008.

17. *2013 ASHRAE Handbook-Fundamentals*, American Society of Heating, Refrigeration and Air Conditioning Engineers (ASHRAE), Atlanta, GA, 2013.

18. D. Y. Goswami, F. Kreith, and J. F. Krieder, *Principles of Solar Engineering* (3rd ed.), Taylor & Francis Group, Boca Raton, FL, 2007.

19. T. Muneer, S. Younes, and S. Munawwar, Discourses on solar radiation modeling, *Renewable & Sustainable Energy Review*, Vol. 11(4), pp. 551–602, 2007.

20. J. A. Duffie, and W. A. Beckman, *Solar Engineering of Thermal Processes* (3rd ed.) John Wiley & Sons, Hoboken, NJ, 2006.

21. B. K. Hodge, *Alternative Energy Systems and Applications*, Wiley, Chichester, UK, 2010.

22. C. J. Chen, *Physics of Solar Energy*, John Wiley & Sons, Hoboken, NJ, 2011.

23. M. Boxwell, *Solar Energy Handbook*, Greenstream Publishing, London, UK, 2011.

24. M. R. Patel, *Wind and Solar Power Systems: Design, Analysis, and Operation* (2nd ed.), CRC Press, Boca Raton, FL, 2015.

25. V. Nelson and K. Starcher, *Introduction to Renewable Energy (Energy and the Environment)*, CRC Press, Boca Raton, FL, 2015.

26. A. Vieira da Rosa, *Fundamentals of Renewable Energy*, Academic Press, London, UK, 2009.

27. E. E. Michaelidis, *Alternative Energy Sources*, Springer, Berlin, Germany, 2012.

28. N. Enteria and A. Akbarzadeh, *Solar Energy Sciences and Engineering Applications*, CRC Press, Boca Raton, FL, 2013.

29. S. A. Kalogirou, *Solar Energy Engineering: Processes and Systems*, Academic Press, Amsterdam, the Netherlands, 2013.

30. D. Y. Goswami, *Principles of Solar Engineering*, CRC Press, Boca Raton, FL, 2013.

31. G. N. Tiwari, A. Tiwari, and Shyam, *Handbook of Solar Energy Theory, Analysis and Applications*, Springer, Singapore, 2016.

32. M. Martín (Ed.), *Alternative Energy Sources and Technologies*, Springer, Singapore, 2016.

33. R. Belu, *Industrial Power Systems with Distributed and Embedded Generation*, The IET Press, London, UK, 2018.

34. C. Gueymard and D. Thevenard, Revising ASHRAE climatic data for design and standards – Part 2: Clear-sky solar radiation model, *ASHRAE Transactions*, Vol. 119(2), pp.194–209, 2013.

35. A. B. Meinel and M. P. Meinel, *Applied Solar Energy: An Introduction*, Addison-Wesley, Reading, MA, 1976.

36. M. Igbal, *An Introduction to Solar Radiation*, Academic Press, Toronto, Canada, 1983.

37. T. Muneer, *Solar Radiation and Daylight Models* (2nd ed.), Elsevier, Oxford, UK, 2004.

38. J. K. Page, The estimation of monthly mean values of daily total short-wave radiation on vertical and inclined surfaces from sunshine records or latitudes 40°N–40°S, *Proceedings of UN Conference on New Sources of Energy*, Vol. 4, pp. 378–380, 1966.
39. J. L. Threlkeld and R. C. Jordan, Direct solar radiation available on clear days, *Transactions on ASHRAE*, Vol. 64, p. 45–55, 1958.
40. E. L. Maxwell, and D.R. Myers, Daily estimates of aerosol optical depth for solar radiation models. In *Proceedings of the 1992 Annual Conference of the American Solar Energy Society*, pp. 323–327. Pacific Northwest Research Station, Corvallis, OR, 1992.

# 3

## Solar Collectors and Solar Thermal Energy Systems

### 3.1 Introduction to Solar Thermal Energy

Solar energy, generated by the Sun, radiates as electromagnetic radiation and is collected elsewhere, including the Earth. Even, a very small fraction of the total solar radiation is reaching the Earth surface, the world needed energy can easily be supplied by the solar energy, only. However, due to its stochastic nature, two critical components are required for a functional solar energy conversion system—a solar collector and an energy storage unit. The solar collector collects and/or concentrates the solar radiation, converting a part of it into other energy forms (electricity, heat, etc.). The energy storage system is often required because of the solar radiation variability and non-constant nature to hold the energy excess produced during the higher generation periods, and release it when generation drops. Methods of collecting and storing solar energy vary depending on planned uses and applications. There are several solar collector and energy storage systems. The main solar collector types are: *flat-plate solar collectors*, *focusing solar collectors*, and *passive solar collectors*. A *working fluid* (such as air, water, oil or antifreeze fluids) circulates through the solar collectors to transfer the collected energy to the end-user. The collector enclosure is designed to maximize the absorption of solar radiation and the conversion to heat. The solar radiation conversion efficiency into heat, eventually transferred to the working fluid is a complex problem, highly dependent on the solar collector design and type.

The Sun emits radiation in the entire electromagnetic spectrum from gamma rays to radio waves. Its radiant energy is a combination of the energy released by layers with different temperatures, thus different wavelengths, simply known as the solar radiation. The solar radiation spectrum is made of the following components: about 6.4% of the total energy is contained in ultra-violet region (wavelengths $\lambda < 0.38\ \mu m$) of the spectrum, about 48% is contained in the visible region ($0.38\ \mu m < \lambda < 0.78\ \mu m$), and the remaining 45.6% is contained in the infrared region of the spectrum ($\lambda > 0.78\ \mu m$). Beam or direct radiation is that part of the solar radiation, reaching the Earth surface without any direction changes. Diffuse radiation is the radiation where the direction is changed before reaching the surface, through atmosphere scattering processes. The total solar radiation is the sum of the beam, diffuse and reflected solar radiation components. Total solar radiation on a horizontal surface is known as the global solar radiation. The solar irradiance $G$ is the rate at which the radiant energy is incident on a unit surface area, being expressed in $W/m^2$. The incident solar radiation is also known as insolation. Generally, the insolation for a specific time period (commonly one hour) is represented by symbol $I$, while the symbol $H$ is used to specify the daily averaged insolation. The $H$ and $I$ values are indicated by $W\text{-}h/m^2/day$ and $W\text{-}h/m^2/h$, respectively. In the case that both $H$ and $I$ values are measured on an hourly basis, $I$ is numerically equal to $G$. Solar radiation is present on nearly each and every place on the Earth. However, it may or may not be present for the whole year or any period of time. The amount of solar radiation received can change due to the several factors, such as location, time of the day, season, local topography and landscape (mountains, trees or forests), and weather conditions (air temperature, humidity, wind, or atmospheric turbidity). Sun rays are striking the Earth surface at different angles, ranging from 0° (just above the horizon) to 90° (directly overhead).

Solar energy conversion consists of a large technology family, capable of meeting a significant range of energy service needs. Solar technologies can deliver heating, cooling, natural lighting, electricity and fuels for many application hosts. The solar energy use for heating dates back to antiquity. Historically, methods used for collecting and transferring solar heat were passive methods, without active means such as pumps, fans and heat exchangers. Passive solar heating methods utilize natural means such as radiation, natural convection, thermo syphon flow and material thermal properties for solar energy collection and heat transfer. Active solar heating methods, on the other hand, use pumps and/or fans to enhance the fluid flow and heat transfer rates. Passive systems are defined as systems in which the thermal energy flow is by natural means: by conduction, radiation and natural convection. Passive features increase the solar energy use to meet heating and lighting loads and the ambient air use for cooling. For example, window placement enhances (a passive solar use) the solar gains to meet winter heating loads, to provide daylighting. Solar energy conversion to heat is straightforward and any material object placed in the Sun absorbs solar energy. However, maximizing the absorbed energy and reducing losses require specialized techniques and devices such as evacuated spaces, optical coatings and mirrors. Which technique is used depends on the application and temperature range at which the heat needed. The temperatures used can range from 25°C (e.g., swimming pool heating) to 1000°C for a solar power plant, or 3000°C or even higher in the solar furnaces.

Passive solar heating is viewed as a technique for maintaining building comfortable conditions by exploiting the incident solar radiation through the use of glazing, windows, sun-spaces or other transparent materials, while managing the structure heat gains and losses without the dominant use of pumps and fans. Distinctions need to be made between energy conservation techniques and passive solar measures. Energy conservation features are designed to reduce the heating and cooling energy required to the building thermal conditions. Such features include the insulation uses to reduce heating or cooling loads. Similarly, window shading or appropriate window placement is lowering the solar gains, reducing summer cooling loads. Examples of active solar thermal systems are solar collector, thermo-battery and ones of passive solar systems are south-side facing windows and greenhouse. Converting the Sun's radiant energy to heat is the most common and well-developed solar conversion technology. The temperature level and amount of this converted energy are the key parameters that must be known to match a conversion scheme to a specific task effectively. Building solar cooling is achieved by using solar-derived heating to run thermal refrigeration absorption or adsorption cycles. Solar energy for lighting requires no conversion since solar lighting occurs naturally through windows. However, maximizing the effect requires specialized engineering and architectural designs. Solar electricity generation is achieved in two ways, through photovoltaic cells in which the solar energy is converted directly into electricity, and the second one, through a concentrating solar power (CSP) plant, in which the solar energy is converted into high-temperature heat and then via a heat engine-generator unit into electricity. CSP systems use reflectors, lenses, mirrors and usually tracking systems to focus a large area of the incoming sunlight into a small beam focused on the system receiver to achieve very high temperatures. Furthermore, solar-driven systems can deliver process heating and cooling, while other solar technologies are being developed, delivering energy carriers (hydrogen or hydrocarbon fuels, the so-called solar fuels). Each solar technology has differing maturity, and its applicability depends also on the local conditions and supporting government policies. Some technologies are already competitive with market prices in certain locations, and the overall viability of solar technologies is improving. Solar thermal techniques are used for a wide range of applications, such as for domestic hot water, comfort building heating or heat for industrial processes. This is significantly economic, since for many countries heat spending is significant. Notice that solar-based hot water heating systems for buildings, now a mature technology, has annually growing rates of about 15%, being employed in most of the countries.

Solar thermal conversion is based on the fact that any dark surface placed in the sunshine is absorbing solar energy and is heating up. Solar collectors working on this principle consist of Sun-facing surfaces that transfer part of the absorbed energy to a working fluid. To reduce heat losses and to improve the efficiency, glass sheets are usually placed over the absorber surface. However, all solar collectors suffer from heat losses due to radiation and convection, which are increasing as the working fluid temperature increases. Improvements such as the use of selective surfaces, collector evacuation to reduce heat losses, and special glass types improve the efficiency. The simplest thermal conversion devices, the flat-plate collectors are available for operation over a large temperature range, up to 365 K (200°F). These collectors are suitable for providing hot water and space heating services and are also able to operate absorption-type air-conditioning systems. The solar radiation thermal uses for low-temperature heat are technically feasible and economically viable for building hot water and heating aquaculture facilities or swimming pools. The solar thermal power generation involves concentrating solar radiation to increase the radiation flux density (i.e., concentrating of the solar radiation onto a receiver), solar radiation absorption (i.e., radiation energy conversion into heat inside the receiver), transfer of thermal energy to an energy conversion unit, conversion of thermal energy into mechanical energy in a thermal engine (e.g., steam turbine) and conversion into electrical energy.

Solar thermal energy, used for centuries for heating and drying crops is recently applied, in a wide variety of thermal processes, such as power generation, water heating, mechanical crop drying, or water purification, among others. Given the temperature ranges of the solar thermal processes, the most important applications are: (a) less than 100°C for domestic water heating, swimming pools, building heating, distillation and dryers; (b) 150°C or less for air conditioning, cooling and for heating water, oil or air for industrial uses; (c) 200°C–2000°C temperature range for electricity generation and for mechanical power uses; and (d) from 3000°C up to 5000°C in solar furnaces for the material treatments. For processes where temperatures higher than 100°C are required, the solar energy flux is not enough to elevate the working fluid temperature to such levels, and some type of solar radiation concentration (mirrors or lenses) are needed. The ratio of the absorber energy flux to the captured collector energy must be greater than one, while the particular designs can achieve very high concentration ratios. The solar energy spreads out through the space reaching the ground in a diluted form, at an average rate of about 220 W/m². In other words, if one square meter is available for the solar energy conversion into electricity, at 100% ideal efficiency, the energy produced is only 220 W. The challenge of solar energy utilization is to concentrate it. Practical ways to achieve this, discussed below, include the direct and indirect electricity production, while the direct electricity production (solar cells) is discussed in Chapter 4, the indirect method is discussed in this chapter.

Solar thermal power generation is a low-cost application of the complex solar collectors to concentrate the solar radiation, in order to produce high enough temperatures to drive steam turbine generator units.

Solar thermal collectors are classified as low-, medium-, and high-temperature collectors. Low-temperature solar collectors, the flat-plate ones, are used to heat swimming pools or provide domestic hot water. Medium-temperature collectors (often flat-plate types) are used for heating water or air for residential and commercial uses. High-temperature collectors concentrate sunlight through mirrors or lenses are used for heat requirements of 300°C/20 bar pressure or higher in industry or for electricity generation; however, the term used for the last type is the concentrated solar thermal (CST), for industry heat requirements and concentrated solar power when the collected heat is for power generation. CST and CSP are not replaceable in terms of applications. In addition to the solar collector and the working fluid, a complete *active solar system* must have energy storage and/or energy backup unit(s), because the Sun is not shining all the time. Moreover, the passive solar heating system can be used effectively to reduce the heating (and cooling) requirements of houses and buildings. A passive solar system contains no active components, such as solar collectors and pumps, and relies on both building design regular and special features. Walls, ceilings and floors may constitute both the energy collectors and storage systems. Heat can be distributed through natural convection. Building design is optimized to let the solar radiation in and keep it into the winter, and doing the opposite in the summer. The ways used to accomplish these tasks are by using the so-called direct and indirect energy gains. Direct gain refers to systems that admit sunlight directly into the space requiring heat. The maximum sunlight reception is obtained through windows facing south. The sunlight received during the day is absorbed by high-heat-capacity materials of the floor and walls. Indirect-gain passive systems are those that absorb solar radiation in high-heat capacity materials, such as outside concrete walls, and then the accumulated thermal energy is transferred to spaces needing heat. Prior to the examination of specific solar collectors, a brief introduction of the theory and simple models for the prediction of the thermal energy output of various solar collectors is presented. These models are applicable to all collector concepts, separately from any specific type to avoid the impression that these models are useful only to one concept. Usually, the collector performance is determined from experimental data. An important feature of the performance models is that it exploits such data, extending the results to different operating conditions. If experimental data are not available, the designer must rely to the collector analytical performance descriptions.

## 3.2 Passive and Active Solar Thermal Energy Systems

Solar energy conversion systems are often classified into two major systems: *passive and active solar energy conversion systems*. Passive solar energy conversion systems do not involve any moving mechanisms to generate energy. Passive solar heating is a technique for maintaining comfortable conditions in buildings by exploiting the solar irradiance incident on the buildings through the use of windows, sunspaces, conservatories, other transparent materials and managing heat gains and/or losses into the building structure. Solar cooling for buildings is also achieved, for example, by using solar-derived heat to drive thermodynamic refrigeration absorption or adsorption cycles. Solar energy for lighting requires no conversion since solar lighting occurs naturally in buildings through windows. However, maximizing the effect requires specialized engineering and architectural design. *Passive thermal energy systems* are defined as the systems in which the thermal energy flows naturally through processes such as convection, conduction and radiation. Passive solar thermal energy systems utilize non-mechanical techniques to control, process and distribute the captured solar energy into useful energy forms such as heating, lighting, cooling or ventilation. However, if the heat transfer subsystem employs a pump or fan to force the heat transfer fluid flow, the system is referred to as having an active component or subsystem. Passive solar heating is further classified as direct gain, indirect gain or isolated gain. In direct gain solar thermal systems the solar irradiation is directly incident on the structure interior, while indirect gain is characterized as having a thermal mass located between the Sun and structure interior. On the other hand, an isolate gain solar thermal configuration is separated from the structure, but thermal mass is used for thermal energy storage. The passive thermal storage mass is provided by the external or internal walls, floors, or water walls or roofs.

These solar energy conversion techniques include the material with good thermal properties to absorb and retain the incoming solar energy selection, designing spaces that naturally circulate air to transfer the captured energy, and referencing the position of a building to the Sun to enhance energy capture. In some special cases passive solar energy devices can be composed of moving parts, with the distinction that this movement is automatic and powered by the Sun, and are used for heating and lighting purposes. This means that solar energy is used for powering the sun tracking systems and to help transfer of the heating and lighting. Active solar thermal energy conversion systems involve electrical and mechanical components such as solar tracking mechanisms, optical concentrators, pumps, and fans to capture solar energy and process it into usable forms such as heating, lighting, and electricity. The solar panels are oriented to maximize the Sun exposure, concentrate the solar radiation onto collector receiver and efficiently transfer the heat. Active solar energy conversion systems are more expensive and complex than the passive solar thermal energy systems. At this point a distinction must be made between energy conservation methods and passive solar measures. Energy conservation methods are designed to reduce the electricity, heating and cooling energy for the thermal and lighting building requirements, through insulation to reduce the cooling and heating loads, the use of window shading or placement to reduce the solar gains, reducing the summer cooling loads, automated energy use controls with lower energy use thresholds, etc. Passive thermal features are designed to increase the solar energy use to meet the lighting, heating and cooling loads, through the methods briefly discussed here. The major distinction between the energy conservation methods, passive and active solar thermal systems is through a key performance parameter, the net annual energy saved by the passive system installation. This development driving force is the new advanced energy-efficient glazing methods.

### 3.2.1 Direct Use of Passive Solar Energy Systems

Solar energy conversion represents a broad family of technologies in a large range of energy service applications: lighting, comfort heating, hot water for buildings and industry, high-temperature solar heat for electric power and industry, electricity conversion and production of solar fuels. Several solar technologies, such as hot water or pool heating, are competitive and used in areas where they offer the least-cost options and in jurisdictions where governments have taken steps to actively support solar energy. Very large solar electricity installations, approaching 100 MW of power, have been realized, in addition to large numbers of rooftop solar energy installations. Other applications, such as solar fuels, require additional R&D before achieving significant adoption levels. In pursuing any of the solar technologies, there is the need to deal with the variability and the cyclic nature of the solar radiation. One option is to store excess collected energy until it is needed, being particularly effective for handling the lack of night sunshine. For example, a 0.1 m thick slab of concrete in the floor of a home stores much of the solar energy absorbed during the day and release it to the room at night. When collected over a long period of time such as one year, or over a large geographical area, solar energy can offer even greater services. The use of both concepts of time and space, together with energy storage, has enabled designers to produce more effective solar systems. Basic principles are based on the location and material characteristics used in the construction, part of the building structure. One main advantage is durability, because the materials are associated with the building.

The term *passive solar building* represents a qualitative description of buildings that are making significant use of solar gain to reduce heating and cooling energy consumption based on the natural energy flows of radiation, conduction and natural convection. Forced convection based on mechanical means (pumps and fans) is not playing a major role in the heat-transfer processes. The term *passive building* is often employed to emphasize use of passive energy flows into the lighting, heating and cooling, including the redistribution of the absorbed direct solar gains. High-performance energy buildings require higher up-front construction costs, but significantly lower the energy-related costs during their lifetime. However, the building total up-front cost may or may not be higher than for conventional ones, depending on the extent to which heating and cooling systems are downsized, simplified, or even eliminated. Any additional up-front cost tends to be compensated for by the reduced energy costs over the building lifetime. The basic elements of passive solar design are windows, conservatories and glazed spaces, for solar gain and daylighting, thermal mass, sunlight protection elements, and reflectors. Combinations of these basic element combinations lead to different systems, such as direct-gain systems (e.g., windows in combination with storing energy walls, solar chimneys, and wind catchers), indirect-gain systems, mixed-gain systems (direct-gain and indirect-gain systems, conservatories, sun-spaces, and greenhouses) and isolated-gain systems. Passive solar thermal technologies, integrated with the building design and structure, include the following components:

1. Windows with high solar transmittance and high thermal resistance facing toward South as nearly as possible to maximize the direct solar gains into the living space while reducing heat losses through the windows during the cold season and heat gains during the cooling season. Skylights are also used for daylighting in office buildings and in solaria/sun-spaces.

2. Building-integrated thermal storage, the so-called thermal mass, may be sensible thermal storage using concrete or brick materials, or latent thermal storage. The most common type of thermal storage is the direct-gain system in which thermal mass is adequately distributed into the building to directly absorb the solar energy. Energy storage is particularly important through two essential functions: storing much of the absorbed solar energy for slower release, maintaining satisfactory thermal comfort conditions by limiting the increase into the effective room temperatures. Alternatively, a collector-storage wall is used, in which the thermal mass is placed directly next to the glazing, with air circulation between the wall cavity and the room. However, this system has not gained much acceptance because it limits the outdoor views. Hybrid thermal storage with active charging and passive heat release is also employed in part of a solar building while direct-gain mass is also used for solar-heated air from a building-integrated

photovoltaic-thermal system to heat a ventilated concrete slab. Isolated thermal storage passively coupled to a fenestration system or solarium/sunspace is another option in passive design.

3. Well-insulated opaque envelope appropriate for the climatic conditions can be used to reduce heat transfer to and from the outdoor environment. In most climates, this energy efficiency aspect is integrated with the passive design. A solar technology with opaque envelopes and transparent insulation is combined with thermal mass to store solar energy.

4. Daylighting technologies and advanced solar control systems, such as controlled shading (internal or external) and fixed shading devices, are suited for workplace daylighting applications. These technologies include electro-chromic and thermo-chromic coatings, or transparent photovoltaics, which, in addition to a passive daylight transmission function, also generate electricity. Daylighting is a combination of energy conservation and passive solar design, aiming to use most of the available natural daylight. These techniques include: shallow-plan design, allowing daylight to penetrate all rooms and corridors, building light wells, taller windows, allowing light to penetrate deep inside rooms, task lighting directly over the workplace, rather than lighting the whole building interior, and deep windows, revealing, lighting room surfaces and reducing the glare.

5. Sun-spaces are a particular case of the direct gain passive solar system, but with most surfaces transparent, that is, made up of fenestration. Solariums are becoming increasingly attractive both as a retrofit option for existing houses and as an integral part of new buildings.

Some basic rules for optimizing the use of building passive solar heating are the following: well insulated buildings to reduce overall heat losses, having a responsive, efficient heating system, as possible South facing, the glazing being concentrated on the southern side, as the main living rooms, with rooms such as bathrooms on the opposite side, avoiding shading by other buildings to benefit from the mid-winter Sun, and thermally massive to avoid summer overheating or on certain winter sunny days. Clearly, passive solar technologies cannot be separated from the building technologies. Thus, when estimating the passive solar gain contributions, the following must be distinguished: (1) buildings specifically designed to harness direct solar gains using passive systems, defined here as solar buildings, and (2) buildings that harness solar gains through near-south facing windows with this orientation by chance rather by design. Few reliable statistics are available on the adoption of passive design in residential buildings. Furthermore, the contribution of passive solar gains is missing in existing national statistics. Passive solar is reducing the energy demand but is not part of the energy supply chain, considered into the energy statistics. In most climates, unless effective solar gain control is employed, there is a need to cool the space during the summer. However, the need for mechanical cooling may often be eliminated by designing for passive cooling, based on the use of heat and solar protection techniques, heat storage in thermal mass and heat dissipation techniques. The specific contribution of the passive solar and energy conservation techniques depends strongly on the climate. Solar-gain control is particularly important during periods when only some heating is required. In adopting larger window areas, enabled by their high thermal resistance, active solar gain control becomes important in solar buildings for both thermal and visual considerations. Passive solar system applications are mainly of the direct-gain type, but they can be further subdivided into the following categories: multi-story residential buildings, detached or semi-detached solar homes, designed to have a large south-facing façade to provide a large solar capture area. Perimeter zones and their fenestration systems in office buildings are designed primarily based on daylighting performance, to reduce lighting and cooling loads, but passive heat gains may be desirable during the heating season. In addition, residential or commercial buildings may be designed to use natural or hybrid ventilation systems for cooling or fresh air supply, in conjunction with designs for using daylight throughout the year and direct solar gains during the heating season. These buildings may profit from low summer night temperatures by using night hybrid ventilations that utilize both mechanical and natural ventilation processes.

Other important passive solar application, viewed as direct solar application is natural drying of agricultural products, vegetables and grains. Grains and other agricultural products have to be dried before being stored so that insects and fungi do not render them unusable. Examples include wheat, rice, coffee, copra, certain fruits, and timber. Solar energy dryers vary mainly as to the solar energy use and the system configurations and components. Solar dryers constructed from wood, metal and glass sheets have been evaluated extensively and used widely to dry a full range of crops. Solar drying of crops and timber is commonly used, either by using natural processes or by concentrating the heat in specially designed structures. Solar drying techniques are particularly suited to the warmer climates since there is abundant sunlight the air temperature is high and relatively constant over the whole year. A high and stable air temperature is actually just as important as the sunshine itself, since it limits loss of generated heat, allowing to simpler solar driers to maintain the temperature of the drying crop during the day around 40°C. The temperature within the solar drier is higher than that outside it. Consequently water on and in the product evaporates. The air takes up more and more of this moisture until a certain equilibrium is reached. Ventilation ensures that the saturated air is replaced with less saturated air, and the product eventually dries out. Drying is intended to evaporate and dispel the water in a product, to make it unavailable to micro-organisms. Good ventilation is also of crucial importance. It determines the exchange of warmth from the absorbent surface to the air next to it and the water evaporation on and in the product. Stronger ventilation can lead to a lower average temperature but also to a more efficient overall heat transfer, leading to a reduction in the relative humidity and improved drying. In choosing a specific dryer, the following six criteria must be considered: the use of locally available construction materials and skills, the investment and maintenance costs, drying capacity, holding capacity, adaptability to different products, drying times and the end product quality. Solar driers are often constructed of locally available materials, making them well suited for domestic manufacture. Solar driers are divided into two categories:

1. Driers directly using sunlight, in which heat absorption occurs primarily by the product itself. These are further divided into: traditional open air drying racks, covered racks (protecting against dust and insects) and drying boxes provided with insulation and absorptive material.
2. Driers in which the sunlight is employed indirectly, the drying air is heated in a space other than that where the product is stacked. The products are not exposed to direct sunlight. Various constructions are possible, including ones using fans in order to optimize the air circulation.

A drier that operates optimally is usually the result of a number of adjustments whose value is established by trial and error and drying tests. It is therefore important that if a solar drier is bought or made, these adjustments are made. With the regard to the temperature regulation the important technical factor is that the available sunlight is dependent on the season and the location and usually limited to 4–7 kW.h/day/m². The absorbent area can be effectively increased by directing extra sunlight through reflectors. The absorber angle is specified by the latitude, the solar collector facing the Sun, being out of shadow as far as possible. It is a simple matter to insulate the drier better to raise the degree of heat absorption (air warmth uptake). The wall of a covered drier, through which the sunlight cannot pass, is better replaced by insulating material which lines the box and is painted black. The heat collector of an indirect drier can be improved by enlarging the absorptive capacity and reducing heat loss, by means of insulation and keeping hot-air-glass contact to a minimum. This is usually only worthwhile if the airflow has been artificially increased. In the absence of forced ventilation, the chimney-effect is crucial. The difference in height between the air intake and outlet determines the draught and therefore the natural ventilation. A chimney will help provided that: it is wide enough; if it is too small it will obstruct the draught, and it is warm enough. Advanced direct solar thermal energy applications, such as solar cooling and air conditioning, industrial solar heat applications and desalination/water treatment are still in the early stages of development and commercialization, with only a few first-generation systems in operation or construction. Considerable cost reductions are expected if the research and development efforts are increased.

### 3.2.2 Active Solar Heating and Cooling

Active solar heating and cooling technologies use the Sun and mechanical elements to provide either heating or cooling, often using thermal storage. In a solar heating system, the collector transforms solar irradiance into heat and uses a carrier fluid to transfer that heat to an energy storage tank, where it can be used when needed. The two most important factors in choosing the correct type of collector are the following: (1) the service to be provided by the solar collector, and (2) the related desired range of temperature of the heat-carrier fluid. An uncovered absorber, also known as an unglazed collector, is likely to be limited to low-temperature heat production. A solar collector can incorporate many different materials and be manufactured using a variety of techniques. Its design is influenced by the system in which it will operate and by the climatic conditions of the installation location. Active solar water heaters rely on electric pumps and controllers to circulate the carrier fluid through the collectors. Three types of active solar water-heating systems are available. Direct circulation systems use pumps to circulate pressurized water directly through the collectors. These systems are appropriate in areas that do not freeze for longer periods and do not have hard or acidic water. Antifreeze indirect-circulation systems pump heat-transfer fluid, usually a glycol-water mixture, through the collectors. Heat exchangers transfer the heat from the fluid to the water for use. Drain-back indirect-circulation systems use pumps to circulate water through the collectors. The water in the collector and the piping system drains into a reservoir tank when the pumps stop, eliminating the risk of freezing. Such systems must be carefully designed and installed to ensure that the piping always slopes downward to the reservoir tank, and the stratification is carefully considered in the water tank design. A solar combined-system provides both solar space heating and cooling as well as hot water from a common array of solar thermal collectors, usually backed up by an auxiliary non-solar heat source. Solar combined-systems may range in size from those installed in individual properties to those serving several in a block heating scheme. A large number of different types of solar combined-systems are produced. The systems on the market in a particular country may be more restricted, however, because different systems have tended to evolve in different countries.

Solar cooling can be broadly categorized into solar electric refrigeration, solar thermal refrigeration, and solar thermal air-conditioning. In the first category, the solar electric compression refrigeration uses PV panels to power a conventional refrigeration machine. In the second category, the refrigeration effect can be produced through solar thermal gain; solar mechanical compression refrigeration, solar absorption refrigeration and solar adsorption refrigeration are the three common options. In the third category, the conditioned air can be directly provided through the solar thermal gain by means of desiccant cooling. Both solid and liquid sorbents are available, such as silica gel and lithium chloride, respectively. Solar electrical air-conditioning, powered by PV panels, is of minor interest from a systems perspective, unless there is an off-grid application. This is because in industrialized countries, which have a well-developed electricity grid, the maximum use of photovoltaics is achieved by feeding the produced electricity into the public grid. Solar thermal air-conditioning consists of solar heat powering an absorption chiller, and it can be used in buildings. Deploying such a technology depends heavily on the industrial deployment of low-cost small-power absorption chillers. This technology is being studied within several research entities and centers.

## 3.3  Collecting Solar Energy, Tracking Sunlight

The Sun radiation spectrum is very similar to that of a blackbody at 5780°K, as was discussed in Chapter 2. The radiation is emitted in all directions from the Sun, and the radiation that falls on the outside Earth atmosphere is 1367 W/m$^2$, the solar constant, a yearly average, as the irradiance varies slightly over the year, due to the Earth elliptic orbit around the Sun and to the changes in the solar activity. Different atmospheric gases (e.g., $CO_2$ and water vapor) absorb and scatter solar radiation of different wave lengths the irradiance reaching the ground after passing through the atmosphere is reduced to approximately 1000 W/m$^2$, the so-called one sun. The solar spectrum extends from 0.3 to 3.0 μm and includes near infrared radiation, visible light and ultraviolet radiation, while the highest intensity is in the area of visible light, at approximately 0.5 μm. At higher latitudes, the angle of incidence of the solar irradiance on the Earth surface is higher, resulting in lower ground solar irradiance. The distance through which solar radiation has to travel in the atmosphere is also longer at higher latitudes, which means that more energy is absorbed and reflected before reaching the ground. Because of these two effects, the average irradiance and annual irradiation are lower at high latitudes. The irradiance is also reduced by cloudiness, humidity and the concentration of particles in the air. As the total annual global irradiation on a horizontal surface varies from a value of 600 kWh/m$^2$ to over 2300 kWh/m$^2$ depending on the location, there are different opportunities of using solar energy. For example in the Mediterranean regions with similar climatic conditions, the global annual irradiation is raging usually from 1400 to 1800 kWh/m$^2$ and even over 2300 kWh/m$^2$ in some areas.

Chapter 2 presented and discussed the methods to determine the rate and amount of the solar energy at a specific site, as well as the concepts of the solar radiation, irradiance, solar constant, or air mass, the difference between the amount of energy falling on a surface pointing at the Sun, and a surface parallel to the Earth surface. The Sun incidence angle, altitude angle and azimuth angle are calculated according to the equations are also discussed in previous chapter. In order to collect solar energy here on the Earth, it is important to know the angle between the Sun rays and a collector surface (*aperture*). When a solar collector is not pointing (or more exactly, when the collector *aperture normal* is not pointing) directly at the Sun, a part of the energy that could be collected is being lost. In this section, the equations used to calculate the angle between the collector aperture normal and the Sun central ray are developed, for both fixed and tracking collectors. These equations are used to provide insight into the collector tracking and orientation design by predicting the integrated solar radiation energy that is incident on the collector aperture and could be collected by the solar collector. Sun is changing its position during the day and from one season to another. A sun tracker, is an equipment that is used for orienting a solar collector or panel, toward the Sun. Solar panels require a high degree of accuracy to ensure that the concentrated sunlight is directed precisely to the PV module or solar collector. Solar tracking systems can substantially improve the amount of power produced by a system by enhancing morning and afternoon performances. However, for low-temperature solar thermal applications, sun trackers are not usually used, because such systems are relatively expensive, compared to the cost of adding more solar collectors to the system. In addition, some restricted solar angles are required for winter operation, influencing the average year-round system capacity. For solar electric applications, sun tracker costs are acceptable when compared to the cost of a PV or solar thermal electrical system, making them effective such applications. From the maintenance point of view, the solar trackers need to be inspected and lubricated on a regular basis. There are several techniques to track the Sun position and to adjust the panel or collector positions. One of the most common techniques of the sun tracking is to use the relationship between the angles of the light source and the differential current generated in two photodiodes due to the shadow produced by a cover over them. With the knowledge of the position of the Sun with respect to a particular location on Earth, learned in the previous chapter the solar collector orientation may be determined. The collector angle ($\theta_C$) between the Sun and collector normal is:

$$\cos(\theta_{Col}) = \sin(\alpha)\cos(\beta_{Tilt}) + \cos(\alpha)\sin(\beta_{Tilt})\cos(\alpha_S - \alpha_2) \tag{3.1}$$

where $\alpha_S$ is the azimuth angle normal to the collector surface, computed with Equation (2.12), $\beta_{Tilt}$ represents the tilt angle computed through Equation (2.18), $\alpha$ is the solar altitude angle, as given by the Equation (2.11), and $\alpha_2$ is the collector tilt angle from the ground. These angles are shown in the schematic diagram in Figure 3.1. If $\theta_C$ is greater than 90°, then the Sun is behind the collector, and no energy is collected.

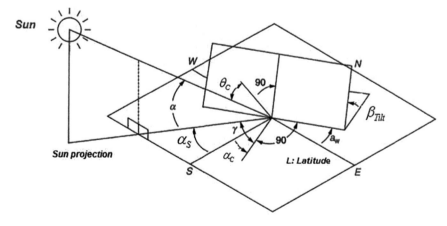

**FIGURE 3.1**
Solar angles of a tilted surface.

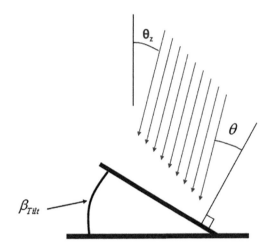

**FIGURE 3.2**
Incidence angle on fixed solar panel.

As was briefly discussed in the previous chapter, there is a set of useful relationships among these angles. Equations relating the angle of incidence of the beam radiation on a surface (the angle between the beam radiation on a surface and the normal to that surface, as shown in Figure 3.2), $\theta$, to the other important critical solar angles are:

$$\cos(\theta) = \sin(\delta)\sin(L)\cos(\beta) - \sin(\delta)\cos(K)\sin(\beta)\cos(\gamma)$$

$$+\cos(\delta)\cos(L)\cos(\beta)\cos(H_R) + \cos(\delta)\sin(L)\sin(\beta)\cos(\gamma)\cos(H_R) \quad (3.2a)$$

$$+\cos(\delta)\sin(\beta)\sin(\gamma)\sin(H_R)$$

and

$$\cos(\theta) = \cos(\theta_Z)\cos(\alpha_2) + \sin(\theta_Z)\sin(\alpha_2)\cos(\gamma - \gamma_S) \quad (3.2b)$$

Here, $\theta_Z$ is the zenith angle, $\gamma$ is the surface azimuth angle, $H_R$ is the hour angle, and $\gamma_S$ is the solar azimuth angle. The angle $\theta$ may exceed 90°, which means that the Sun is behind the surface. When using Equation 3.2a, it is necessary to ensure that the Earth is not blocking the Sun (i.e., that the hour angle is between sunrise and sunset). For vertical surfaces, $\beta = 90°$, Equation 3.2a becomes:

$$\cos(\theta) = -\sin(\delta)\cos(L)\cos(\gamma) + \cos(\delta)\sin(L)\cos(\gamma)\cos(H_R)$$

$$+\cos(\delta)\sin(\beta)\sin(\gamma)\sin(H_R) \quad (3.3a)$$

For horizontal surfaces, the incidence angle is the Sun zenith angle, $\theta_z$. Its value is the range 0°–90° when the Sun is above the horizon. For this situation, $\beta = 0$, and Equation 3.2a becomes:

$$\cos(\theta_Z) = \cos(\delta)\cos(L)\cos(H_R) + \sin(L)\sin(\delta) \quad (3.3b)$$

**Example 3.1:** Calculate the angle of incidence of the beam solar radiation on a surface located at Cleveland, Ohio, at 2:30 (solar time) on March 16, if the surface is tilted 45° from the horizontal and is pointing 15° west of the south.

**Solution:** The surface azimuth angle is 15°, as given in this example, and by using the relationships of Chapter 2, for the declination angle and hour angle, the values in this case are:

$$\delta = 23.45 \cdot \sin\left(360\,\frac{284 + (59 + 16)}{365}\right) = -2.4°$$

$$H_R = +37.5° \left(+15° \text{ per hour times } 1.5 \text{ h for after noon}\right)$$

Using a slope $\beta = 45°$, the latitude, L of Cleveland, Ohio of 41.5° N, and Equation (3.2), the angle of incidence is:

$$\cos(\theta) = \sin(-2.4°)\sin(41.5°)\cos(45°) - \sin(-2.4°)\cos(41.5°)\sin(45°)\cos(15°)$$

$$+ \cos(-2.4°)\cos(41.5°)\cos(45°)\cos(37.5°)$$

$$+ \cos(-2.4°)\sin(41.5°)\sin(45°)\cos(15°)\cos(37.5°)$$

$$+ \cos(-2.4°)\sin(45°)\sin(15°)\sin(37.5°) = 0.628 \text{ or } \theta = 51.1°$$

Given the collector angle, the total daily radiation to a surface is determined from integrating the solar irradiance, $I_0\cos(\theta_C)$, from sunrise to sunset. For a horizontal surface, the collector angle and the zenith angle are identical, so $\theta_C = \theta_Z$. If the solar extraterrestrial irradiance reaching a horizontal panel is known, then the total radiation received is obtained by integrating the between the sunrise and sunset hours or for a needed time interval. The total irradiance on any plane is the sum of the beam (direct), reflected and diffused components, as discussed in Chapter 2. Cumulative irradiance is the amount of energy that is hitting an area over a certain period of time. The daily insolation $H$ is the total energy per unit area received in one day from the Sun is computed by:

$$H = \int_{0}^{24\,h} G(t) \cdot dt \tag{3.4}$$

As discussed in Chapter 2, the daily insolation varies with latitude, season and over the years. The seasonal variation at high latitudes is more significant than at lower latitudes. The clear sky solar radiation on a horizontal surface varies due to three factors: (a) daily variation, (b) receiving surface orientation and (c) atmospheric absorption variations. In practice *clear day radiation* is a notional quantity, because actual weather and site conditions vary widely from those assumed in its calculation. $H_0$ is the computed by:

$$H_0 = I_0 \frac{24}{\pi}\left[\cos(L) \cdot \cos(\delta) \cdot \sin(H_S) + H_S \sin(L) \cdot \sin(\delta)\right] \tag{3.5}$$

In similar fashion, the total extraterrestrial radiation to a south facing tilted surface (Northern Hemisphere) at a tilted angle, $\beta_C$ can be determined as:

$$H_{0(Tilted)} = I_0 \frac{24}{\pi} \left[ \cos(L - \beta_C) \cdot \cos(\delta) \cdot \sin(\bar{H}_S) + \bar{H}_S \sin(L - \beta_C) \cdot \sin(\delta) \right] \qquad (3.6)$$

For the whole day by integrating from the sunrise to the sunset the extraterrestrial radiation on a horizontal plane for the day $n$ (Julian day) is given by relation:

$$H_0 = \frac{24 \times 3600 \times SC}{\pi} \left[ 1 + 0.033 \cos \frac{360 \cdot n}{365} \right]$$

$$\times \left( \cos(\delta) \cos(L) \sin(H_{SS}) + \left( \frac{\pi \bar{H}_{SS}}{180} \right) \sin(\delta) \sin(L) \right) \qquad (3.7)$$

Here, $L$ is the location latitude, $\delta$ is the declination (computed with one of the relationships of Chapter 2), $SC$ is the solar constant (1367 W/m²), $H_{SS}$ and $\bar{H}_{SS}$ are the sunset time on the horizontal and tilted surfaces, respectively, computed with the relationships of Equation (3.8).

$$H_{SS,SR} = \pm \arccos \left( -\tan(L - \beta_C) \sin(\delta) \right) \qquad (3.8a)$$

and

$$\bar{H}_{SS} = \min \left\{ H_S, \arccos \left( -\tan(L - \beta_C) \sin(\delta) \right) \right\} \qquad (3.8b)$$

Equation (3.8a) gives the hour angle for sunrise ($H_{SR}$) and sunset ($H_{SS}$), respectively. This equation is based on assumption that the Sun center, being at the horizon. In practice, sunrise and sunset are defined as the times when the upper limb of the Sun is on the horizon. The surface azimuth angle, the deviation of the projection on a horizontal plane of the normal to the surface from the local meridian, with zero due south, east negative, and west positive, $-180° \le \gamma \le 180°$. The solar azimuth angle, $\gamma_S$ the angular displacement from south of the beam radiation projection on the horizontal plane, while the displacements east of south are negative and west of south are positive, with the range from 180° to $-180°$. For north or south latitudes between 23.45° and 66.45°, $\gamma_S$ ranges between 90° and $-90°$ for days, having duration less than 12 hours, and for days longer than 12 hours, between sunrise and sunset, $\gamma_S$ is greater than 90° and less than $-90°$ early and late in the day when the Sun is north of the east-west line in the northern hemisphere or south of the east-west line in the southern hemisphere. For tropical latitudes, $\gamma_S$ can have any value when $\delta^-L$ is positive in the northern hemisphere or negative in the southern. Thus $\gamma_S$ is negative when the hour angle is negative and positive when the hour angle is positive, being computed by:

$$\gamma_S = sign(H) \left| \arccos \left( \frac{\cos(\theta_z) \cdot \sin(L) - \sin(\delta)}{\sin(\theta_z) \cdot \cos(L)} \right) \right| \qquad (3.9)$$

**Example 3.2:** Determine the daily solar radiation in J/m² on a horizontal surface, $H_0$ at latitude 41.5° N on April 10, 2017.

**Solution:** By using the relationships of Chapter 2, for the declination and the sunset angles in this case are:

$$\delta = 23.45 \cdot \sin \left( 360 \frac{284 + (90 + 15)}{365} \right) = 9.36°$$

and

$$H_{SS} = \cos^{-1}\left(-\tan\left(41.5°\right)\cdot\tan\left(9.38°\right)\right) = 98.4°$$

Now, the daily solar radiation on a horizontal surface in the extraterrestrial region can be obtained from Equation (3.6) as follows:

$$H_0 = \frac{24\times3600\times1367}{\pi}\left[1+0.033\cos\left(\frac{360\cdot105}{365}\right)\right]$$

$$\times\left(\cos\left(9.36°\right)\cos\left(41.5°\right)\sin\left(98.4°\right)+\left(\frac{\pi\cdot98.4°}{180}\right)\sin\left(9.36°\right)\sin\left(41.5°\right)\right)$$

$$= 34.159 \text{ MJ}/\text{m}^2$$

Small variations in the tilt angles from the horizontal (±10°) and the orientation from the South (±20°) are not reducing the collector performance significantly. Whereas maximum energy production occurs for the solar energy conversion systems, having the solar collectors or PV panels oriented directly South at a fixed angle near latitude, the maximum economic return depends on the electricity rate structure, which usually is including different rates due to seasonal and time-of-use energy demand. In such situations, fixed solar panels may be more optimally oriented to maximize power production during on-peak periods rather than overall maximum generation. The amount of direct solar radiation is dependent on both the solar angle and the panel tilt angle. Solar thermal or photovoltaic panels need to be installed at an angle relative to the Earth surface to maximize the amount of solar energy collected at specified times of the year. There are types of the solar collectors that are tracking the Sun by moving in prescribed ways to minimize the angle of incidence of beam radiation on their surfaces, maximizing in this way the incident beam solar radiation. The angles of incidence and the surface azimuth angles are needed for the operation of such solar collectors. The relationships in this section are useful in radiation calculations for the moving surfaces. Tracking systems are classified by their motions. Rotation can be about a single axis, which could have any orientation but which in practice is usually horizontal east-west, horizontal north-south, vertical, or parallel to the Earth's axis or it can be about two axes. For purposes of solar process design and performance calculations, it is often necessary to calculate the hourly radiation on a tilted surface of a collector from measurements or estimates of solar radiation on a horizontal surface. The most commonly available data are total radiation for hours or days on the horizontal surface, whereas the need is for beam and diffuse radiation on the plane of a collector. The geometric factor $R_b$, the ratio of beam radiation on the tilted surface to that on a horizontal surface at any time, can be calculated exactly by appropriate use of Equation 3.9, while Figure 3.2 indicates the angle of incidence of beam radiation on the horizontal and tilted surfaces. The ratio $G_{b,tilt}/G_b$, the solar radiation on horizontal to the solar radiation on tilted surface, is given by:

$$R_b = \frac{\text{Total Radiation on Tilted Surface}}{\text{Total Radiation on Horizontal Surface}} = \frac{G_{b,tilt}}{G_b} = \frac{\cos\left(\theta\right)}{\cos\left(\theta_z\right)} \tag{3.10}$$

Here, $\cos(\theta)$ and $\cos(\theta_z)$ are both determined from Equation (3.2) (or from similar equations, derived from this equation). Notice that the optimum azimuth angle for solar collectors or panels is usually 0° in the northern hemisphere (or 180° in the southern hemisphere). Several models have been developed, with various degree of complexity, as the basis for calculating the total irradiations. The differences are largely in the way that is treating the diffuse and reflected terms. The simplest model is based on the assumptions that the beam radiation predominates and that the diffuse and reflected radiations are effectively concentrated, all radiation being treated as beams. This leads to substantial overestimation of the irradiation, and the procedure is not recommended. More accurate methods are based on various assumptions about the directional distribution of the diffuse radiation incident on the tilted surface. However, problems can arise in calculating radiation on a tilted surface at times near sunrise and sunset. For example, solar radiation data may be recorded before sunrise or after sunset due to reflection from clouds and/or by refraction of the atmosphere. The usual practice is to either discard such measurements or treat the radiation as all diffuse as the impact on solar system performance is negligible.

**Example 3.3:** For the case of Example 3.1 determine the ratio, $R_b$.

**Solution:** In Example 3.1, the declination angle in this case is −2.4°, and the hour angle is computed by using: $HRA = 15° \times (LST−12) = 15° \times (13.5−12) = 22.5°$, where LST is local standard time in fractional hours. The incidence angle on the tilted surface is computed by using Equation (3.2b) is 51.1°, while $\cos(\theta) = 0.6283$. For the horizontal surface, from Equation (3.3b), the incidence angle is:

$$\cos(\theta_Z) = \cos(-2.4°)\cos(41.5°)\cos(22.5°) + \sin(41.5°)\sin(-2.4°) = 0.5018$$

By using Equation (3.10), the ratio, $R_b$ is then:

$$R_b = \frac{\cos(\theta)}{\cos(\theta_z)} = \frac{0.6283}{0.5018} = 1.383$$

For diffuse radiation, the conversion factor for diffuse radiation ($R_d$) is defined as the ratio of diffuse radiation incident on an inclined surface in $W/m^2$ to that on a horizontal surface in $W/m^2$. Unfortunately there are not established methods for finding the distribution of diffuse radiation over the sky, making the conversion factor estimation difficult. However, diffuse radiation can be estimated by considering the sky as the isotropic source of diffuse radiation. For reflected solar radiations are radiations reflected from the ground and other objects near the surface of interest. For a tilted surface at an angle $\beta_{tilt}$ from the horizontal surface, the conversion factors for diffuse radiation, $R_d$ and for the reflected radiation (assuming that the reflected radiations are diffuse and isotropic), $R_r$ are given by these relationships:

$$R_d = \frac{1 + \cos(\beta_{tilt})}{2} \tag{3.11a}$$

and

$$R_r = \frac{1 - \cos(\beta_{tilt})}{2} \tag{3.11b}$$

It may be useful to be mentioned that both the beam and diffuse components of solar radiation are undergoing reflection from the ground and the surroundings of the panel or collector.

**Example 3.4:** Determine the conversion factors for beam, diffuse and reflected radiations for an inclined surface at 45° from the horizontal with orientation of 30° west of south and located at latitude of 28.8° N at 1:30 PM on February 16, 2013.

**Solution:** The declination angles, n = 47 is −13°, while the hour angle is 22.5°. The solar panel/collector azimuthal angle and the incidence angles are computed, by using Equations (3.2) and (3.3) as:

$$\cos(\theta_z) = \sin(-13°)\sin(28.8°) + \cos(-13°)\cos(28.8°)\cos(22.5°) = 0.683$$

and

$$\cos(\theta_i) = 0.999$$

The correction factors are then:

$$R_b = \frac{\cos(\theta_i)}{\cos(\theta_z)} = \frac{0.999}{0.683} = 1.463$$

$$R_d = \frac{1 + \cos(45°)}{2} = \frac{1 + 0.707}{2} = 0.854$$

and

$$R_r = \frac{1 - \cos(45°)}{2} = \frac{1 - 0.707}{2} = 0.147$$

## 3.3.1 Thermal Solar Collector Capture, Heat Transfer, and Loss Mechanisms

The thermal and performance analysis of the solar collectors are presented and discussed here, with reference to the major solar collector types. The basic parameter to consider is the collector thermal efficiency, as well as the input and output energy and the collector losses. The solar collector thermal efficiency is defined as the ratio of the useful energy delivered to the energy incident on the solar collector aperture. The incident solar flux consists of direct and diffuse radiation. While flat-plate collectors can collect all types of radiation (direct, diffuse and even reflected components), the concentrating solar collectors can utilize the direct radiation only if their concentration ratio, as defined later is greater than 10. The prediction of collector performance requires information on the solar energy absorbed by the collector absorber plate. The solar energy incident on a tilted surface can be found by using the methods presented before. As can been seen earlier, the incident solar radiation has three components: beam, diffuse and ground-reflected radiation. The calculation is strongly dependent on the radiation model employed.

## 3.3.2 Radiative Properties and Characteristics of Materials

When electromagnetic radiation, so the solar radiation is striking the interface between the two medium, a part of it is reflected, a part is absorbed and, and if the material is transparent, a part of the radiation is transmitted, as shown in Figure 3.3. The incident solar radiation fraction reflected is defined by the reflectance $\rho$, the fraction absorbed is defined

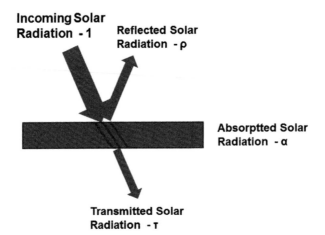

**FIGURE 3.3**
Schematic representation of the electromagnetic radiation transmittance, $\tau$, absorptance, $\alpha$, and reflectance, $\rho$.

by the absorptance $\alpha$, and the fraction transmitted through the medium as the transmittance $\tau$. Notice that the opaque bodies do not transmit any radiation, $\tau$ being 0. According to the first law of thermodynamics (the energy conservation), these three electromagnetic radiation fractions or components, transmission, reflection and absorption must add up to unity or to incoming energy flux (as shown in the diagram of Figure 3.3), as expressed by these two relationship:

$$G = G_\rho + G_\alpha + G_\tau \tag{3.12a}$$

$$1 = \frac{G_\alpha}{G} + \frac{G_\rho}{G} + \frac{G_\tau}{G} = \alpha + \rho + \tau \tag{3.12b}$$

The radiation reflection can be specular or diffuse. If the angle of incidence is equal to the angle of reflection, the reflection is specular, and when the reflected radiations are distributed uniformly into all directions, the reflection is diffuse. An actual surface is neither specular, nor diffuse, but a highly polished surface approaches and can approximate the specular reflection, whereas rough surfaces reflect diffusely. Another important radiative property is called emittance $\varepsilon$, the ratio of the radiative emissive power of an actual surface to that of an *ideal blackbody* surface. All of the radiative properties of materials, $\alpha$, $\tau$, $\rho$, and $\varepsilon$, are functions of the wavelength and electromagnetic radiation direction, such dependence being used in the design of solar energy devices and systems. For example, selective absorbers are used for solar collectors and passive heating systems, and glazing materials are used for daylighting and solar collectors. Reflectance $\rho$ is a property of any reflecting surface, as ones associated with the solar collectors. Properly designed concentrators capture radiation fractions of 0.95 or higher, and silver and glass mirrors can have a reflectance of 0.94 and new aluminum reflecting surfaces have a reflectance of about 0.86. The transmittance $\tau$ is the solar radiation fraction passing through transparent cover materials, such as the ones that sunlight passing through the absorber. Cover sheets of glass or plastic are used on flat-plate collectors, above the absorber to reduce convective heat loss. The transmittance in Equation (3.11) is the average overall transmittance, representing the total reduction in the transmitted energy by all the system covers. For example, flat-plate collectors may have two or more cover sheets, often of different materials, with the collector transmittance, $\tau$ being the product of each individual cover transmittance. Cover transmittance depends on the passing light wavelength. Glass for example transmits most electromagnetic radiation in the visible spectrum range, being almost opaque in the infrared region. Therefore, an absorber covered with glass receives most of the incoming, short wavelength solar radiation, but is not transmitting most of the long wavelength radiation coming from the absorber, while plastic covers have higher transmittance values at longer wavelengths. Solar collectors using plastic covers are used for nighttime cooling since radiation losses to the nighttime sky are significant. This is also why greenhouses in warm climates use plastic rather than glass as a cover.

The absorption term, $\alpha$ represents the fraction of solar energy incident upon the surface, absorbed by the object, the remainder being reflected. Good blackbody surfaces have an absorption greater than 0.98, however, as surfaces degrade, this value is decreasing. This is important to point out that this radiation property in the solar *visible* spectrum. For most real surfaces, the absorption varies as a function of the wavelength of the incident electromagnetic radiation. There is a class of surfaces used in solar collectors, called *selective surfaces* that have a higher absorptance in the visible spectrum than at longer wavelengths, thereby reducing the thermal radiation losses. In summary, when performing an energy balance on the solar collector absorber plate or receiver, there are four important mechanisms that reduce the solar energy amount that is incident on the collector aperture: imperfect reflection, imperfect geometry, imperfect transmission and imperfect absorption. This is a general statement, and different collectors may include or not these losses depending on the design and materials used. Concentrating solar collectors require the use of reflecting surfaces with high specular reflectance in the solar spectrum or refracting devices with high transmittance in the solar spectrum. Reflecting surfaces are usually highly polished metals or metal coatings on suitable substrates. With opaque substrates, the reflective coatings must always be front surfaced, for example, chrome plating on copper or polished aluminum. If a transparent substrate is used, however, the coating may be front or back surfaced. In any back-surfaced reflector, the radiation must pass through the substrate twice and the transmittance of the material becomes very important. The optical transmission behavior can be characterized by two wavelength-dependent physical properties, the index of refraction n and the extinction coefficient $k$. The index of refraction, which determines the speed of light in the material, also determines the amount of light reflected from a single surface, while the extinction coefficient determines the amount of light absorbed in a substance in a single pass of radiation.

The generation of higher working temperatures as needed to operate a conventional steam engine or to provide heat to an industrial process requires the use of focusing devices in connection with a basic absorber–receiver system. Operating temperatures up to 4000 K (6740°F) or even higher have been achieved in pilot experiments, or steam-based engines in industrial processes have also proved technologically feasible. Today, a number of focusing devices and equipment for the steam generation to produce electricity are under construction and research, and preliminary cost estimates are suggesting that the solar power in favorable locations is similar to that of conventional thermal powers. The first problem encountered in the design of equipment for solar energy utilization is the low flux density, making necessary large surfaces to collect solar energy for large-scale utilization, and the larger the surfaces, the more expensive the energy becomes. When the Sun is directly overhead on a cloudless day, 10 $m^2$ of surface could theoretically provide energy at 70% collection efficiency and 30% conversion efficiency at a rate of approximately 2 kW. However, several factors are reducing this amount in practice.

The total solar energy reaching the Earth is made up of two parts: the energy in the direct beam and the diffuse energy from the sky. The amount of direct energy depends on the cloudiness and the position of the Sun and is obviously greatest on clear days. Some solar radiation falling on clouds is diffused by scattering, but clouds do not absorb all of the energy. The effect of clouds is mainly to increase the percentage of diffuse energy in the total energy reaching the surface, and diffuse irradiance in summer months with high and broken clouds can be as high as 400 W/m². Thick clouds let less energy pass than thin clouds and scatter proportionally more of the total energy back into space. The second practical limitation that is not apparent from the macroscopic energy view is that most of the solar energy falls on remote areas and would therefore require some means of transmission to be useful to the industrialized nations. The mean amount of energy available on a horizontal plane is greatest in the continental desert areas around latitudes 25°N and 25°S of the equator and falls off toward both the equator and the poles. The highest annual mean irradiance is 300 W/m² in the Red Sea area. Clouds reduce the mean global irradiance considerably in equatorial regions. However, whereas in northern climates the mean horizontal surface global irradiance varies from season to season, it remains relatively constant in equatorial regions. Typical values of mean annual horizontal surface irradiance are for Australia, approximately 200 W/m², for U.S. about 185 W/m², and for United Kingdom about 105 W/m².

### 3.3.3  Heat Transfer and Loss Mechanisms in Solar Collectors

Energy balance on a solar thermal collector, the surface that absorbs the incoming radiation, assumed insulated is the balance of the energy inflow and outflow. In a flat-plate collector, this is called the *absorber plate* and for a concentrating collector, it is often called the *receiver*. In subsequent sections, details of construction, structure, surface types for major solar collectors are presented. However, at hare such details are not discussed, only a brief description of the heat transfer and losses. The energy balance on a solar collector absorber or receiver, calculated as the rate of useful energy leaving the absorber, equal to the difference between the rates of the radiation incident on absorber and the absorber thermal energy losses is written as:

$$\frac{dQ_{useful}}{dt} = \frac{dE_{solar}}{dt} - \frac{dQ_{losses}}{dt} \quad (W) \tag{3.13a}$$

Or in shorter notation, the above equation is expressed as:

$$\dot{Q}_{out} = \dot{E}_{rad} + \dot{Q}_{Losses} \quad (W) \tag{3.13b}$$

Then the solar thermal collector useful energy is the rate of thermal energy leaving the collector, described in terms of the rate of energy being added to a heat transfer fluid passing through the collector receiver or absorber, is given by:

$$\dot{Q}_{useful} = \dot{m}C_P \left( T_{out} - T_{in} \right) \tag{3.14}$$

Here in Equations (3.12) and (3.13), $\dot{Q}_{out}$ is the rate of *useful (output) energy* leaving the collector absorber (W), $\dot{E}_{rad}$ is the rate of the solar radiation incident on absorber (W), basically on the solar collector absorbing surface, $\dot{Q}_{losses}$ is the rate of absorber thermal energy losses (W), $\dot{m}$ is the mass flow rate (kg/s), $C_p$ is the specific heat of the collector heat transfer fluid (J/kg. K), $T_{out}$ is the heat transfer fluid temperature leaving the absorber (K), and $T_{in}$ is the temperature of heat transfer fluid entering the absorber (K). The rate of optical (short wavelength) radiation incident on absorber or receiver is the solar irradiance for that collector type and its tracking (global/total solar irradiance for a flat-plate collector and direct beam solar irradiance for a concentrating collector). Since the collector capture area may not be aimed directly at the Sun, the irradiance must be reduced to account for the angle of incidence. The collector area on which the solar irradiance falls is called the collector aperture area. The incident solar radiation is then:

$$\dot{E}_{IN} = I_{Col} \times A_{Col} \text{ (W)} \tag{3.15}$$

Here, $I_{Col}$ is the solar irradiance entering the collector aperture (global (total) or direct (beam) radiation, in W/m$^2$), and $A_{Col}$ is the collector aperture area, in m$^2$. This input solar energy is reduced by the losses as it passes from the collector aperture to the absorber. These processes depend on the type and design of the specific collector, but here are included the important optical loss mechanisms, and neglecting the unimportant terms in the chapter sections, when specific types of collectors are discussed. The rate of optical (short wavelength) energy reaching the absorber or receiver is the product of the incoming solar radiation multiplied by a number of factors, all less than 1.0, describing the collector energy reduction:

$$\dot{E}_{Opt} = \Gamma \cdot \rho \cdot \alpha \cdot \tau \cdot I_{Col} \cdot A_{Col} \text{ (W)} \tag{3.16}$$

where, $\Gamma$ represents the capture fraction (fraction of reflected energy entering or impinging on the collector receiver), $\rho$ is the reflectance of any intermediate reflecting surfaces, $\tau$ is the transmittance of any glass or plastic cover sheets or windows of the collector and $\alpha$ is the absorptance of absorber or receiver surface. Notice, that the first two terms above apply only to concentrating collectors. The capture fraction $\Gamma$ is a measure of the quality of the reflecting surface shape and the receiver size. Often this coefficient is described in terms of the reflected energy fraction not impinging on or entering the collector receiver. A poorly shaped concentrator or a collector receiver too small is making this number considerably less than 1.0. Transmittance of the cover, as we specified earlier also depends on the wavelength of light passing through it. In summary then, when performing an energy balance on the solar collector absorber plate or receiver, there are four important mechanisms that reduce the amount of solar energy that is incident on the collector aperture, such as the imperfect reflection, imperfect geometry, imperfect transmission and imperfect absorption. This is a general statement, and different collector types either are including or not these losses depend on the collector specific design. In order to design a solar energy conversion system, the impinging solar radiation on the collecting (active) panel surface and the system efficiency, $\eta_{sys}$ are required. However, as mentioned before the solar irradiation depends on several factors, such as the latitude and the altitude of the area, the season, the day, the hour, the degree of nebulosity and the dust content, the water vapors and the greenhouse gases in the atmosphere. In order to evaluate the performance level, the multiannual average value of the horizontal daily solar irradiation, $H$ is needed, being defined as the solar irradiation energy, reaching a horizontal surface of 1 m$^2$ daily. Remember that the value of H is determined from statistical data, depending on the solar collector location. A simple thermal energy (heat) output of the solar thermal system is then computed by:

$$Q_{ST} = \eta_{ST} \times H(t) \tag{3.17}$$

**Example 3.5:** Estimate heat generated by a STE, having a system efficiency of 0.75, located at mid-latitude in Central Europe, where H is equal to 3.95 kWh/m²/day. If this specific energy is delivered to a boiler, heating the water, what is the mass $m$ of the heated water by a 2.5 m² of the STE system, with a difference in temperature of $\Delta\theta$, from 10°C to 90°C?

**Solution:** The STE generated heat is computed by using Equation (3.17), as:

$$Q_{ST} = 0.75 \times 3.95 \text{ kWh/m}^2/\text{day} = 2.97 \text{ kWh/m}^2/\text{day}$$

The daily water quantity heated by this system is estimated by Equation (3.14) as:

$$m = \frac{Q_{ST(daily)}}{C_{PT} \cdot \Delta\theta} = \frac{2.97 \text{ kWh/m}^2/\text{day} \times 2.5 \text{ m}^2}{4.187 \text{ kJ}/(\text{kg}) \cdot (90-10)} \approx 62.5 \text{ kg/day}$$

Once the solar energy (short wavelength) radiation has interacted with the collector absorber or receiver surface, the absorber or receiver temperature rises above ambient temperature. This in turn starts a process of heat loss from the absorber as with any surface heated above the temperature of the surroundings. The loss mechanisms are convection, radiation and conduction, all being dependent on, among other things, the temperature difference between the absorber and the surroundings. Because solar thermal collectors are designed to transfer heat to a fluid, there is a balance between the heat transfer rate of the transfer fluid and the heat losses by radiation, convection and conduction, as defined by Equation (3.16). Since heat losses increase with temperature, the balance between heat removal and heat losses sets the collector operating temperature. If the fluid, used for heat transfer removes too much heat, the absorber temperature decreases, reducing the heat losses. If not enough heat from the absorber is removed, temperature increases, increasing the rate of heat losses, posing a major problem for concentrating collectors, because when not enough heat is being removed (e.g., if fluid flow is interrupted), the absorber temperature can increase to its melting temperature. The collector heat losses are expressed as:

$$\dot{Q}_{Losses} = \dot{Q}_{Loss,convection} + \dot{Q}_{Loss,radiation} + \dot{Q}_{Loss,conduction} \text{ (W)} \tag{3.18}$$

Convective heat loss of a solar collector receiver is proportional to the absorber or receiver surface area and the temperature difference between the absorber surface and the surrounding air. As with the other heat loss equations, this is a simplified model. Usually several convective processes exist, causing an absorber or receiver to lose heat. For example, a flat-plate collector often has a glass cover sheet between the absorber plate and outside ambient air. There is one convection process between the hot absorber and the cover sheet, and a second between the cover sheet and outside air. Wind is increasing the heat transfer coefficient on the cover sheet and must be included in any analysis of convective heat loss. For parabolic dish concentrators, the absorbing surface is typically placed inside of a cavity. This protects it from wind, and naturally driven air currents. Little is known about convective heat loss from an open cavity, but it is quite clear that the position of the cavity and its internal temperature, along with wind speed and direction all affect the rate of heat loss from a cavity. Finally, the average absorber or receiver temperature, $T_r$ is not a fixed or measurable quantity. The temperature of the absorber or receiver near the heat transfer fluid inlet is lower than near the outlet, and both are less than intermediate surfaces not in contact with the heat transfer fluid. Further, since convection is a surface phenomenon, also is driven by the surface temperature, this temperature may be that of a paint or coating rather than the metal below. Even with all of these imperfections, it is instructive to consider the convective heat loss as being proportional to surface area and the difference between some average temperature, and ambient temperature. Since convective heat loss is the major heat loss term for most solar collectors, inventors and designers have incorporated many features to collector design to reduce this term. Examples that are discussed under the different collector designs are: multiple transparent cover sheets for flat-plate collectors, glass tubes surrounding linear absorbers with a vacuum drawn in the intervening space, concentration of solar energy so that the absorber area is small relative to the capture area, absorbers within cavities incorporating glass windows. Radiation heat loss is important for collectors operating at temperatures only slightly above ambient, and becomes dominant for collectors operating at higher temperatures. The rate of radiation heat loss is proportional to the emittance of the surface and the difference in temperature to the fourth power. Terms in this equation over which the collector designer has some control, are the surface emittance and receiver. Surfaces that have a low emittance often have a low absorptance as well, are reducing the absorbed solar energy. However there is a class of surface coatings, the *selective coatings* that have low values of emittance when the surface is at relatively low temperatures, but high values of absorptance for solar energy. The other term, which may be minimized, is the receiver surface area. As with convection loss, concentration of solar energy is the main design tool for reducing radiation heat loss by reducing receiver surface area. In addition, cavity receivers can be used since they have small openings through which concentrated solar energy passes, onto larger absorbing surfaces.

Solar collectors are operating outdoors, usually facing the open-sky, capturing the sky solar radiation. The equivalent radiation temperature of the sky depends on the air density and its moisture content. When the relative humidity is high and at sea level, the sky temperature can be assumed to be the same as ambient air temperature. However, for low relative humidity or at high altitudes, the sky radiation temperature can be 6°C–8°C less than ambient temperature. The final mode of heat loss to consider in collector design is heat conduction. This is generally described in terms of a material constant, the thickness of the material and its cross-section area. Conduction loss is usually small compared to convection and radiation losses and therefore is combined with the convection loss term in most analyses. However, it is discussed here for completeness, and to emphasize its importance to ensure that this mode of heat loss is minimized in any collector design. In flat-plate collectors, the sides and back surface of the absorber plate should incorporate good insulation and the insulation should be thick enough to render this heat loss insignificant. Another important mode of conduction loss is the way the high-temperature absorber is attached to the frame and support structure. Use of low conductance materials such as stainless steel can reduce conduction loss into the frame or support casing. However, since most design issues around conduction can be handled without reducing the solar input, the term is generally combined with the convective heat loss term. The relationships describing the three heat loss mechanisms, convection, conduction and radiation are:

$$\dot{Q}_{Loss,convection} = \overline{h}_{Col} \cdot A_r \left( T_r - T_{amb} \right) \ \ (\text{W}) \tag{3.19a}$$

$$\dot{Q}_{Loss,conduction} = \overline{k} \cdot \Delta \overline{x} \cdot A_r \left( T_r - T_{amb} \right) \ \ (\text{W}) \tag{3.19b}$$

and

$$\dot{Q}_{Loss,radiation} = \varepsilon \cdot \sigma \cdot A_r \left(T_r^4 - T_{sky}^4\right) \quad (\text{W}) \tag{3.19c}$$

Here, $\bar{h}_{Col}$ is the average overall convective heat transfer coefficient (W/m²K), $A_r$ is the surface area of receiver or absorber (m²), $T_r$ is receiver average temperature (K), $T_{amb}$ is the ambient air temperature (K), $\bar{k}$ is the equivalent average collector conductance (W/m. K), $\Delta \bar{x}$ is the average thickness of insulating material (m), $\varepsilon$ is the emittance of the absorber surface (or cavity in the case of a cavity receiver), $\sigma$ is the Stefan-Boltzmann constant (5.670 × 10⁻⁸ W/m² K⁴), and $T_{sky}$ is the equivalent black body temperature of the sky (K). In order to provide a single expression for the useful energy produced from a solar collector based on an energy balance of the receiver or absorber, we can combine Equations (3.12), (3.14) and (3.16) into a single equation for the solar collector energy balance. This equation is often used to develop an understanding of how and why specific collector types are designed.

$$\dot{Q}_{useful} = \dot{m}C_P\left(T_{out} - T_{in}\right) = \bar{h}_{col}' \cdot A_{col()r}\left(T_r - T_{amb}\right)$$
$$+ \bar{k} \cdot \Delta \bar{x} \cdot A_{col(r)}\left(T_r - T_{amb}\right) + \varepsilon \cdot \sigma \cdot A_{col(r)}\left(T_r^4 - T_{sky}^4\right) \quad (\text{W}) \tag{3.20}$$

where, $\bar{h}_{col}'$ is the combined convection and conduction coefficient (W/m²K), and $A_{col(r)}$ is the area of collector aperture or receiver (m²). The equation states that the rate of useful energy produced by a solar collector equals the optical (short wavelength) energy absorbed on the absorber surface, minus the rate of the absorber heat losses, while the convection heat loss term is combined for simplicity with the convection term.

> **Example 3.6:** The temperature on the house inside wall is 80°F and the outside temperature is 10°F. The house wall is 13.5 ft high and 360 ft wide. Assuming for this house is exposed to ambient air, what is the rate of heat transfer by radiation? The wall emissivity, $\varepsilon$ is 0.5.
>
> **Solution:** The area of the house wall, expressed in square meters is:
>
> $$A = 13.5 \times 360 \times \left(\frac{m}{3.281 \text{ ft}}\right)^2 = 83.605 \text{ m}^2$$
>
> The wall total heat transfer, by using Equation (12.18) is then:
>
> $$\dot{Q} = \sigma\varepsilon \cdot A \cdot \left(T_{Indoor}^4 - T_{Outdoor}^4\right)$$
>
> $$= 5.67 \times 10^{-8} \frac{W}{m^2 K^4} \times 0.5 \times 83.605 \times \left(\left(80 + 460\right)^4 - \left(10 + 460\right)^4\right)\right)\left(\frac{1K}{1.8R}\right)^4$$
>
> $$= 8181 \text{ W}$$

In considering the heat transfer mechanisms, it is worth to notice that since the conduction and convection increase linearly with temperature, *radiation increases with temperature to the fourth power, thus at higher temperatures, heat transfer by radiation always exceeds that due to conduction and convection.* The solar collection system type and characteristics are determining the temperature at which the storage materials and transport thermal fluid are charged and the charging maximum rates. Thermo-physical properties of the energy storage material at any temperature are important in determining the suitability of any material for specific energy storage applications. For example, some of the flat-plate liquid-type collectors may use water as the storage material, while the air-type flat plate collectors for space heating may use rock or pebble beds as the thermal storage medium. If the storage material can be used as the collector heat-exchanger fluid, the need of a collector-to-storage heat exchanger is avoided. Such criteria favor the use of the liquid storage materials. Water and glycol-water mixtures are the most common storage materials for flat-plate and moderately concentrating solar collectors. For parabolic trough concentrators, high-temperature oils are the choice. For higher-concentration and higher-temperature collectors, e.g., central receiver tower, molten salts is often used. An important parameter characterizing any energy conversion system, including solar energy systems is the efficiency. The solar energy collection efficiency, $\eta_{col}$ of the solar thermal collectors or photovoltaic collectors is defined as the ratio of the rate of useful thermal energy leaving the collector, to the useable solar irradiance falling on the aperture area. Simply stated, collector efficiency is:

$$\eta_{Col} = \frac{\dot{Q}_{useful}}{A_a \cdot I_a} \tag{3.21}$$

where $\dot{Q}_{useful}$ is the rate of (useful) energy output (W), $A_a$ is the collector aperture area (m$^2$), and $I_a$ is the solar irradiance falling on the collector aperture (W/m$^2$). The definition of collector efficiency may differ depending on the type of collector. The rate of useful energy output from thermal collectors is the heat addition to a heat transfer fluid as defined by Equation (3.13). The incoming solar irradiance falling on the collector aperture, $I_a$, multiplied by the collector aperture area represents the maximum amount of solar energy that could be captured by that collector. Another important concept used in solar energy conversion is the optical efficiency. The optical efficiency of a solar collector is defined as the rate of optical (short wavelength) energy reaching the absorber or receiver, divided by the appropriate solar resource, expressed as:

$$\eta_{opt} = \Gamma \cdot \rho \cdot \tau \cdot \alpha \tag{3.22}$$

Here, $\Gamma$ is the capture fraction (fraction of reflected energy entering or impinging on receiver), while the other parameters were defined before. The optical efficiency $\eta_{opt}$ depends on the solar collector type and configuration and varies with the angle of incidence as well as with the relative values of diffuse and beam solar radiation. Optical efficiency also referred as the zero-loss efficiency (i.e., the efficiency in the ideal case, when there is zero temperature difference between the absorber and the ambient air, so there are no thermal losses) is used in separating out the non-thermal solar collector performances. The solar collector glass cover is reducing the irradiance on the absorber through reflection and absorption, while the reflectors, having the reflectance weakly dependent of incident angle and tending to degrade over the time. Optical efficiency is setting the solar collection efficiency maximum limit, very useful in solar energy system analysis and modeling. Basically the collector optical efficiency describes the collector irradiance losses due to absorption and reflection in the glass cover and the absorber, while for the concentrating collectors there also are losses in the reflector. Models of the solar collector performance, predicting the collector output under varying solar irradiance, operating temperature and weather conditions, are very important to the system design, analysis and development. Prevalent solar collector performance models permit the useful energy output prediction, under varying solar irradiance, ambient temperature and operating temperature.

**Example 3.7:** The cover of a solar collector has 0.96, 0.97, and 0.93 reflectance, transmittance, and absorptance, respectively. If the solar collector capture fraction is 0.93 what is the collector optical efficiency.

**Solution:** Applying Equation (3.22), the collector optical efficiency is:

$$\eta_{opt} = \Gamma \cdot \rho \cdot \tau \cdot \alpha = 0.90 \times 0.96 \times 0.97 \times 0.93 = 0.805 \text{ or } 80.5\%$$

This is a reasonable value for the optical efficiency of many solar collectors.

## 3.4 Solar Collectors, Types, and Characteristics

Solar collectors are collecting and converting the incident solar radiation into thermal energy by absorbing them. The major component of many solar energy systems is the solar collector, a device absorbing the incoming solar radiation, converting it into heat, and transferring the heat to a working fluid (usually air, water, oil or anti-freezing fluid). This converted heat is extracted by flowing fluid in the tube of the collector for further utilization in different applications. In essence a solar collector consists of a receiver that absorbs the solar radiation and transfers the thermal energy to a working fluid. The solar energy collected can be carried from the circulating fluid either directly to an electricity generating units, to a hot water and space heating equipment or to thermal energy storage tank from for the use at night or on cloudy days. A solar collector is an efficient device in converting the solar energy into heat. A special collector part is the coating, the absorber surface coating, spluttered or fused at a very high temperature, with metal sheet inside the collector, making it efficient and effective. The solar collector aperture area or absorber surface collects the solar radiation, transforming it into heat. All efficient collector panels are often made of a strong non-corrosive aluminum body. The energy absorbed by the solar collector is kept inside it by a toughened transparent glass, covering the entire solar panel, preventing the collected heat to be carried out through convection. Together with the frame, the glass also protects the absorber from adverse weather conditions. Typical frame materials are aluminum, galvanized steel or fiberglass-reinforced plastic. Because of the radiant energy nature (spectral characteristics, variability, or changes in diffuse to global fractions), the solar thermal energy system application, the collector analysis and design are presenting unique and unconventional problems in heat transfer, optics, control and material sciences.

Classifications of solar collectors are made according to the working fluid type (water, air, oils or antifreeze fluids) or the solar receiver type used (non-tracking or tracking types). Solar collectors are also distinguished by their motion, i.e., stationary, single axis tracking and two axes tracking and the operating temperature. A third classification criterion is to distinguish between non-concentrating and concentrating solar collectors. The main reason for using concentrating collectors is not the amount of the collected energy, but its temperature levels. This is done by decreasing the area from which heat losses occur (the receiver area) with respect to the aperture area (i.e., the area that intercepts the solar radiation). The ratio of the aperture to receiver area is called the concentration ratio or geometric concentrating factor. In the solar thermal collector systems, the heat-transporting fluid receives the heat from the solar collector and delivers it to the thermal (energy) storage system, boiler of a steam turbine-generator or to an equipment heat exchanger. Thermal storage systems can store heat for a few hours or longer, releasing during cloudy hours and at night, when the solar collectors are not collected solar energy. Thermal electric conversion system receives thermal energy, transfers it to a boiler and drives turbine-generator units, generating the electricity, supplied to the electrical load or to the power grid. Applications of solar thermal energy systems range from simple solar cooker of 1 kW rating to complex solar central receiver of a thermal power plant of 200 MW rating. Solar collectors are also classified as: (a) non-concentrating collectors, and (b) concentrating (focusing) collectors. Solar collectors are also distinguished as low-, medium-, or high-temperature heat exchangers. Solar collectors and concentrators may be classified as the following main types: (a) flat plate type without focusing, (b) parabolic trough type with line focusing, (c) parabolic dish with central focusing, (d) Fresnel lens with a center focus and (e) heliostats with center receiver focusing.

There are several STE types and configurations that are used to convert the solar energy into a useful form of energy, thermal energy or electricity. A block diagram showing three of the most basic system types is presented in Figure 3.4. In upper configuration, the solar energy is captured and converted into heat which is then supplied to a thermal load such as house heating, hot water heating or to industrial processes. This system may or may not include thermal energy storage, and often include auxiliary energy sources so that the energy demands are met during periods with no sunshine. The solar collectors concentrate sunlight, converted to heat and transferred to a working fluid to higher temperatures. The heat transfer fluid is eventually used to generate steam that drives the power conversion units, to generate electricity. Thermal energy storage provides backup for operations during periods without adequate sunshine. Usually, if the solar thermal power systems (lower configuration of Figure 3.4) are connected to a power grid, no energy storage or

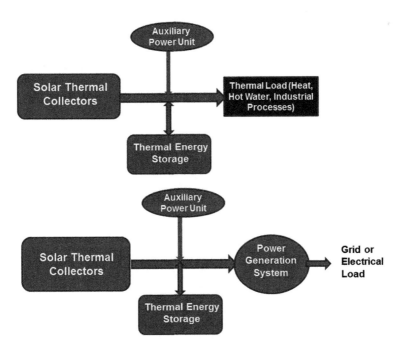

**FIGURE 3.4**
Block diagram of a solar energy conversion system with auxiliary power unit (thermal or electrical energy) and energy storage.

auxiliary energy units are strictly needed. However, if the solar energy conversion system is the main electricity source in any application, then the energy storage and auxiliary energy supply are usually incorporated. If the thermal route is chosen, the heat storage is used to extend the operating periods. Auxiliary energy may be supplied as heat before the power conversion or as electricity. The solar collectors are the key component in most of the solar energy conversion system. The solar collector technology and practice requires good theoretical understandings in order to fully operate, design and use them.

The solar collector function is quite simple: the incoming irradiance is intercepted and changed into a useable energy form to meet specific demands. Among the solar collectors, the flat-plate thermal solar collectors are one of the most commonly used, having quite simple construction and operation. A large plate of blackened material is oriented in such a manner that the falling solar energy on the plate is absorbed and converted into thermal energy. Tubes or ducting are used to remove heat from the plate and to transfer it to a fluid, eventually carried to the load(s). One or more transparent (glass or plastic) plates are often placed in front of the absorber plate to reduce heat losses and for protection. Likewise, opaque insulation is placed around the backside of the absorber plate for the same purpose. Operating temperatures up to 100°C are typical for non-concentrating solar collectors, while much higher in thousands of degrees range are common for concentrating solar collectors. Flat-plate collectors have the advantage of absorbing both the beam normal insolation and the solar energy that has been diffused into the sky and/or reflected from the ground. Flat-plate solar collectors are seldom tracked to follow the Sun daily path. However their fixed mounting usually provides a South tilt to minimize the angle between the Sun rays and the surface at noon. Tilting flat-plate collectors toward the South provides a higher rate of energy at noon and higher total daily energy. In general, flat-plate panels are mounted in a fixed position. However, it is not quite uncommon flat-plate panels mounted on mechanisms that track the Sun about one tilted axis, to increase the daily output of the panels.

**Example 3.8:** A residence requires 72 kWh of heat on a winter day to maintain a constant indoor temperature of 21°C. (a) How much solar collector surface area is needed for an all-solar heating system that has 20% efficiency? (b) How large does the storage tank have to be to provide this much energy? Assume the average solar energy per square meter and per day for the area is 6.0 kWh/m²/day.

**Solution:**

1. Daily thermal energy per unit of area converted into thermal energy is:

$$\text{Thermal Energy} = \frac{6.0 \times 0.20}{1.0} = 1.20 \text{ kWh/m}^2/\text{day}$$

The minimum converter area is then:

$$A_{Collector} = \frac{72}{1.2} = 60 \text{ m}^2$$

2. If we are assuming the storage medium, water, the most common storage medium in residential applications, heat capacity of water is 1 kcal/kg/°C, and the temperature difference is that between the hot fluid and the cold water going into the storage tank, about 40°C). Therefore, the required mass of water for a day's worth of heat is:

$$\text{Tank Mass} = \frac{72}{1.116 \times 10^{-3} \times 40} = 1612.9 \text{ or } \approx 1613 \text{ kg}$$

There are basically three types of thermal solar collectors: *flat-plate, evacuated tube* and *concentrating systems,* the later ones designed to achieve very high operating temperatures. Although there are great geometric differences, their purpose remains the same: *to convert the solar radiation into heat to satisfy some specific energy needs.* The heat produced by solar collectors can supply energy demand directly or be stored. To match demand and production of energy, the thermal performance of the collector must be evaluated. To evaluate the energy produced by a solar collector properly, it is necessary to consider its material physical properties. Solar radiation, mostly short wavelength, passes through a translucent cover and strikes the receiver. Low-iron glass is commonly used as a glazing cover due to its high transmissivity. The cover also greatly reduces the heat losses. The receiver optical characteristics are very similar to those of a blackbody, especially regarding high absorbitivity. Higher thermal conductivity is improved by adding selective coatings. Together with the radiation absorption, an increase of the receiver temperature is achieved. The glazing material essentially becomes opaque at the long-wavelengths favoring the greenhouse effect. A combination of the cover high transmissivity and the receiver high absorbitivity brings higher performances for well-designed solar collectors. Solar collectors range from unglazed flat-plate solar collectors operating at about 10°C above the ambient temperature to central receiver concentrating solar collectors operating above 1000°C.

### 3.4.1 Flat-Plate Solar Collectors

Flat-plate solar collectors are the most widely used kind of collectors in the world for domestic solar water heating and solar space heating applications. Flat-plate collectors are used typically for temperature requirements up to 75°C either higher temperatures can be obtained from high efficiency collectors. In this case, the water is changed to other heat transfer liquid because of its boiling temperature of 100°C. A flat-plate collector consists of an absorber surface (a dark, thermally conducting surface), a trap for re-radiation losses from the absorber surface (such as glass, which transmits shorter-wavelength solar radiation but blocks the longer-wavelength radiation from the absorber), a heat-transfer medium such as air, water, oil or antifreeze fluid and a thermal insulation behind the absorber surface. These collectors are of two basic types based on the heat-transfer fluid: (a) liquid type, where heat-transfer fluid may be water, mixture of water, oil and antifreeze fluid, and (b) air type, where heat-transfer medium is air (used mainly for drying and space heating requirements). A flat-plate solar collector consists of a waterproof, metal or fiberglass insulated box containing a dark-colored absorber plate, the energy receiver, with one or more translucent glazing. Flat-plate collectors can absorb incident solar energy from direct beam, diffuse and ground reflected albedo radiation components. The inclusion of the latest two insolation components means that in most of the climate types there are only few instances where any viable long-term performance advantages are gained by tracking a flat-plate collector to follow the Sun daily path across the sky. Most flat-plate collectors have south-facing fixed mountings that usually provide a static inclination that is cognoscenti of the maximum average diurnal solar energy collection period, the annual durations of utilizable solar energy, the prevalence of diffuse conditions and any diurnal or annual bias of the predominant times of hot water withdrawal. A flat-plate solar collector usually consists principally of tubes through which a heat transfer fluid is conveyed connected with good thermal contact with an absorber plate that, often solar selectively coated, absorbs incoming solar radiation, an aperture cover plate that inhibits outgoing long-wave thermal radiation losses and traps an insulating air layer so inhibiting convective heat losses, and thermal insulation to maximize heat loss and a casing to provide weather protection together with mechanical integrity to the sides and back. The tubes are usually integrated fully in the absorber plate by a variety of manufacturing techniques to ensure good thermal contact between the tubes and the plate. Absorber plates are typically made out of metal due to its high thermal conductivity and painted with special selective surface coatings in order to absorb and transfer heat better than regular black paint can. The glazing covers reduce the convection and radiation heat losses to the environment.

Figure 3.5 shows the typical components of a flat-plate collector. These systems are always mounted in a fixed position optimizing the energy gain for the specific application and particular location. Flat-plate collectors can be mounted on a roof, in the roof itself, or be freestanding. Figure 3.5 shows a schematic representation of a typical flat-plate collector used for domestic heating. A *working fluid* (air, water, oil or antifreeze fluid) circulates through its tubes. The enclosure, with its black metal surface between the tubes and insulation at the bottom, is designed to maximize the solar radiation absorption and its conversion to heat: the glass cover provides the greenhouse effect. The efficiency of conversion of solar radiation to heat stored in the working fluid is a complex issue, being dependent on collector design. Man-made collectors are much less efficient than the *natural collectors*, the animal furs. The fur of polar bears, for example, has been reported to have an efficiency of about 95%. The most sophisticated and most expensive, solar collectors have maximum efficiencies of about 70% or less. Typical values on a cold winter day, when they are most used, are about 20%. Flat-plate collectors are an important part of many solar thermal energy systems. It is one of the simplest in design and both direct and diffuse radiations are absorbed by collector and converted into useful heat. These collectors are suitable for heating to temperature below 100°C. The main advantages of flat-plate collectors are: (a) it utilizes the both the beam as well as diffuse radiation for heating, and (b) are requires less maintenance than any other collectors. Their major disadvantages include: (a) large heat losses by conduction and radiation, (b) no sun-tracking and (c) low water temperature is achieved.

Anti-reflective coatings and surface texture can also improve transmission significantly. The effect of dirt and dust on collector glazing may be quite small, and the cleansing effect of an occasional rainfall is usually adequate to maintain the transmittance within 2%–4%

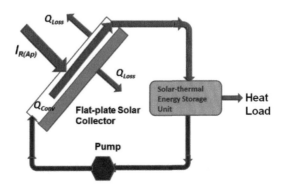

**FIGURE 3.5**
Typical solar energy collection systems with a flat-pale collector.

of its maximum value. The glazing should admit as much solar irradiation as possible and reduce the upward loss of heat as much as possible. Although glass is virtually opaque to the long-wave radiation emitted by collector plates, absorption of that radiation causes an increase in the glass' temperature and a heat loss to the surrounding atmosphere by radiation and convection.

Since flat-plate solar collectors are capable of absorbing both direct (beam) and diffuse solar irradiance, the appropriate aperture irradiance is the total irradiance falling on the solar collector aperture ($Ia = I\rho t$). Equations for calculating total irradiance on an aperture area (surface) were discussed in Chapter 2, and the Equations (3.4) through (3.6) and (3.11) are used for general cases. Various schemes of orientation and tracking are often used and are applicable to determining the angle of incidence. Equation (3.11) includes both direct and scattered and reflected energy. Adding the appropriate useful energy term to Equation (3.21), for solar thermal flat-plate collectors, the following definition of the flat-plate collector efficiency is obtained:

$$\eta_{flat-plate} = \frac{\dot{m}C_P\left(T_{out} - T_{in}\right)}{I_{t\rho} \cdot A_a} \tag{3.23}$$

Figure 3.6a is showing the schematic diagram of the energy and heat flows through a flat-plate collector. The question is how to measure its thermal performance, i.e., the useful energy gain or the collector efficiency. Thus it is necessary to define step by step the singular heat flow equations in order to find the governing equations of the collector system. Considering the schematic diagram of a solar thermal system, employing a flat-plate solar collector, the system energy flows and performances can be determined. The principal energy gain and loss mechanisms for a solar collector are shown in Figure 3.6b. Its power losses can be divided into optical losses, occurring until absorption of the radiation, and thermal losses, occurring after absorption of the solar radiation. The power absorbed (useful thermal power) is either removed by the heat transfer fluid as useable heat or is lost to the environment. The radiation ($G$ or $I_{R(ap)}$) and the thermal power, with thermal energy, in power units and describe the collector energy flow, which means energy per time unit. To reduce thermal losses, most of the non-concentrating solar thermal collectors

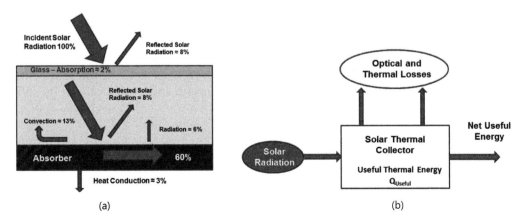

(a)                                        (b)

**FIGURE 3.6**
(a) Flat-plate collector energy fluxes and losses and (b) collector energy flows.

are insulated to the back and the side and covered with a transparent cover at the front. The transparent cover reduces the amount of solar radiation that reaches the absorber due to reflections of both sides of the cover as well as due to absorption of a small part of the radiation by the cover. The transmitted radiation reaches the absorber, where a small part is reflected and the remaining is absorbed and converted into thermal power. If $I_{R(ap)}$ is the solar radiation intensity (W/m²), incident on the solar collector aperture plane, having a collector surface area of $A_{col}$ (m²), then the total amount of solar radiation received by the collector is the product of the irradiance and area. However, a part of this radiation is reflected back to the sky, another part is absorbed by the glazing and the rest is transmitted through the glazing and reaches the absorber plate as short wave radiation, as shown Figure 3.6a. Therefore the conversion factor indicates the solar radiation percentage penetrating the solar collector transparent cover (the transmission, determined by the collector cover transmittance, $\tau$) and the percentage being absorbed (absorption coefficient, $\alpha$). Basically, it is the product of the rate of transmission of the cover and the absorption rate of the absorber, given by:

$$Q_{in} = I_{R(ap)} \cdot A_{Col} \cdot (\tau \cdot \alpha) \tag{3.24}$$

As the solar collector absorbs solar radiation, converting it into heat its temperature becomes higher than the surrounding temperature, $T_{amb}$ and the heat is lost to the atmosphere mainly by convection and radiation. The heat loss rate, $Q_{Loss}$, depends on the solar collector overall heat transfer coefficient ($U_L$) and the collector temperature, $T_{Col}$, expressed as:

$$Q_{Loss} = U_L \cdot A_{Col} \cdot (T_{Col} - T_{amb}) \tag{3.25}$$

Notice that the useful energy gain depends strongly on the energy losses from the top surface of the collector both due to convective and radiative heat transfer processes. The losses from the bottom and from the edges of the collector do always exist. Their contribution, however, is not as significant as the losses from the top. Thus, the rate of useful energy extracted by the solar collector, $Q_{Conv}$ expressed as a rate of extraction under steady state conditions, is proportional to the rate of useful energy absorbed by the solar collector, less the amount of the energy lost by the solar collector to its surroundings (Equation 3.25), expressed as:

$$Q_{Conv} = I_{R(ap)} \cdot A_{Col} \cdot (\tau \cdot \alpha) - U_L \cdot A_{col} \cdot (T_{Col} - T_{amb}) = mC_P \cdot (T_{out} - T_{in}) \tag{3.26}$$

The heat extraction rate from the solar collector can also be measured by the amount of heat carried away in the working fluid passing through the collector, as expressed by the Equation (3.14). The collector efficiency, defined as the useful energy output over the total insulation over the collector area ($I_{R(ap)} \cdot A_{col}$) is:

$$\eta = \tau \cdot \alpha - U_L \cdot \frac{T_{Col} - T_{amb}}{I_{R(ap)}} \tag{3.27}$$

However, Equations (3.26) and (3.27) are somewhat inconvenient in practical applications due to the difficulty in defining the collector average temperatures, being also unlikely to be uniform over the absorber area, ASHARE (American Society for Heating, Refrigeration, and Air-conditioning Engineers) proposed an empirical adjusting factor

$F_{HRF}$ (*collector heat removal factor*) to permit the use of the inlet transport fluid temperature in place of the absorber plate temperature. This adjustment factor is justified from
empirical data and is viewed as the ratio of the net heat actually collected to the net
heat that can be collected if the collector absorber plate is entirely at the transport fluid
entering temperature, $T_{in}$. In this way it is convenient to relate the actual useful solar
collector energy gain to the useful gain if the whole collector surface were at the transport fluid inlet temperature, so the Equations (3.26) and (3.27) are rewritten as:

$$Q_{Conv} = F_{HRF} \cdot \left[ I_{R(ap)} \cdot A_{Col} \cdot (\tau \cdot \alpha) - U_L \cdot A_{Col} \cdot (T_{in} - T_{amb}) \right] \tag{3.28a}$$

and

$$\eta = F_{HRF} \cdot \left[ (\tau \cdot \alpha) - U_L \cdot \frac{A_{Col} \cdot (T_{in} - T_{amb})}{I_{R(ap)}} \right] \tag{3.28b}$$

Since the thermal losses are not exactly proportional to the operational temperature
difference $(T_{in} - T_{amb})$, often, the more precise quadratic solar collector efficiency curve
is used:

$$\eta = F_{HRF} \cdot \tau \cdot \alpha - a_1 \frac{T_m - T_{amb}}{I_{R(ap)}} - a_2 \frac{(T_m - T_{amb})^2}{I_{R(ap)}} = \eta_0 - a_1 \frac{T_m - T_{amb}}{I_{R(ap)}} - a_2 \frac{(T_m - T_{amb})^2}{I_{R(ap)}} \tag{3.29}$$

Here, the mean temperature of the collector $(T_m)$, defined, in operating terms, as the
average value of the inlet and outlet temperatures of the collector (fluid) medium:
$T_m = (T_{out} + T_{in})/2$ is rather used. Therefore, the collector is typically characterized with
the three parameters: the maximum efficiency, $\eta_0$, the linear, $a_1$, and the quadratic loss
factor, $a_2$. The efficiency of a collector at a specific operating point is also called the
*instantaneous efficiency*. To compare the performance of solar collectors, their efficiency
curves are usually presented as a function of the main influencing factors: the operational temperature difference $(T_m - T_{amb})$ and the solar radiation $(I_R$ or $G)$ as shown in
Figure 3.7. Notice that a value of the heat removal factor $(F_{HRF})$ is not obtained by itself
and is not needed. If further it is assumed that $F_{HRF}$, $\tau$, $\alpha$, $U_L$ are constants for a given
collector and flow rate, then the efficiency is a linear function of the three parameters
defining the operating condition: the solar irradiance, the fluid inlet temperature and
the ambient air temperature. These are straight lines on the diagram of Figure 3.7. The
line slope represents the collector heat loss rate. For example, collectors with cover
sheets have less of a slope (lower (bottom) line of Figure 3.7), than those without cover
sheets (upper (top) line of Figure 3.7). The line intercept on the efficiency axis is the optical efficiency. Most low-temperature solar collector performance data are presented in
terms Equation (3.28b). It should be remembered that both of these terms are multiplied
by the heat removal factor, $F_{HRF}$ making them a function of flow rate. Therefore the flow
rate for any $\Delta T/I_{R(ap)}$ curve should be specified. The maximum possible useful energy
gain in a solar collector occurs when the whole collector is at the inlet fluid temperature.
Heat removal factor can be considered as the ratio of the heat actually delivered to that
delivered if the collector plate were at uniform temperature equal to that of the entering fluid. The actual useful energy gain, $Q_{Col}$ is found by multiplying the collector heat
removal factor $(F_{HRF})$ with the maximum possible useful energy gain. Equation (3.28a),

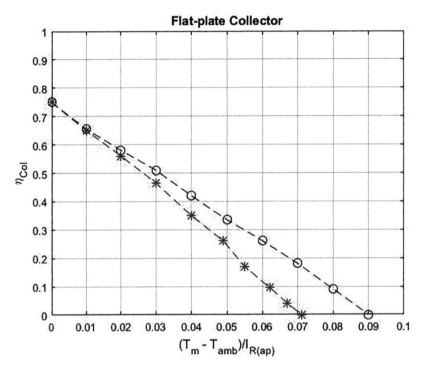

**FIGURE 3.7**
Performance of a typical flat-plate thermal collector (one glass cover, black-painted absorber, water transfer fluid and ambient temperature 25°C). The lower (bottom) line shows the performance as a function of the $\Delta T/I_{R(ap)}$ of a collector with cover sheets, while the upper (top) line for the ones without coversheets.

known as the *Hottel-Whillier-Bliss equation* is a widely used relationship for measuring the solar collector energy gain and performance.

The performance of a flat-plate solar collector can be approximated by measuring these three parameters in test experiments. Notice the in practice, $U_L$ is not a constant as heat losses are increasing as the temperature of the collector rises further above ambient temperature (thermal conductivity of materials varies with temperature). If the solar collector efficiency $\eta_{Col}$ is plotted against $(T_{in}-T_{amb})/I_{R(ap)}$, then the slope of this line $(-F_{HRF}\cdot U_L)$ is the heat loss rate of the collector. Collectors with cover sheets have less of a slope than those without cover sheets, meaning less heat losses. There are two important operating points on this linear characteristic. The first point is the maximum collection efficiency, the so-called the optical efficiency, occurring when the fluid inlet temperature equals the ambient temperature $(T_{in} = T_{amb})$. For this condition, the $\Delta T/I_{R(ap)}$ value is zero and the intercept is $F_{HRF}(\tau\alpha)$. The other point of interest is the intercept with the $\Delta T/I_{R(ap)}$ axis, reached when the useful energy is no longer removed from the solar collector, which can happen if fluid flow through the collector stops, power failure. In this case, the incoming optical energy must equal the heat loss, requiring that the temperature of the absorber increase until this balance occurs. This maximum temperature difference (the stagnation temperature) is defined by this point. For well-insulated collectors or concentrating collectors the stagnation temperature is reaching very high levels causing fluid boiling and, in the case of concentrating collectors, the absorber surface can melt. A measure of a flat-plate collector performance is the collector efficiency,

$\eta_{Col}$ defined as the ratio of the useful energy gain, $Q_{Conv}$ to the incident solar energy over a specific time period:

$$\eta_{Col} = \frac{\int_{t}^{t+\Delta t} Q_{Conv} \cdot dt}{A_{Col} \int_{t}^{t+\Delta t} I_{R(Ap)} \cdot dt} \tag{3.30}$$

**Example 3.9:** Assuming a unity $F_{HRF}$, calculate the efficiency of two flat-pale solar collectors with the following characteristics, transmittance of collector glazing, 0.90 and 0.90, absorptance of collector absorber, 0.90 and 0.90, front $U_L$ values are 6 and 3.5 W/m²K, back $U_L$ values are 1 and 1.5 W/m²K, $T_{amb}$ equal to 283 K, irradiance 800 W/m², $T_{abs}$ equal to 303 K, respectively.

**Solution:** Collector efficiencies are computed by using Equation (3.30), for each set of data, as:

$$\eta_{Col-1} = 1 \cdot \left[ 0.9 \cdot 0.9 - \frac{(6+1) \cdot (303 - 283)}{800} \right] = 0.635 \text{ or } 63.5\%$$

$$\eta_{Col-2} = 1 \cdot \left[ 0.9 \cdot 0.9 - \frac{(3.5+1.5) \cdot (303 - 283)}{800} \right] = 0.685 \text{ or } 68.5\%$$

The FPC types used for water heating applications are as follows: (a) Full pipe and fin types are used with a fin made of highly conducting materials, such as copper or aluminum, are used for domestic water heating; (b) Full water sandwich type, in which both the wetted area and the water capacity are high, and because the thermal conduction is only across the skin thickness (short distance) of the materials, low-conductivity materials are used (plastic and steel), and such FPCs are commonly used for heating swimming pools with plastic panels); and (c) Semi-sandwich type, in which a medium-conductivity material, such as steel, is commonly used. In some case aluminum may also be used. The constructional details and main components of the flat plate collectors are given here:

1. *Insulated Box:* The rectangular box is made of thin glass insulated sheet and is insulated from sides and bottom using glass or mineral wool of thickness 5–8 cm to reduce losses from conduction to back and side wall. The box is tilted at due south and a tilt angle depends on the latitude of location. The face area of the collector box is kept between 1 and 2 m².

2. *Transparent Cover:* This allows solar energy to pass through and reduces the convective heat losses from the absorber plate through air space. The transparent tampered glass cover is placed on the top of the rectangular box to trap the solar energy and sealed by rubber gaskets to prevent the leakage of hot air. It is made of plastic or glass but glass is most favorable because of its transmittance and low surface degradation. However with development of improved quality of plastics, the degradation quality has been improved. The plastics are available at low cost, lighter weight and can be used to make tubes, plates and cover, but are suitable only for low temperature application 70°C–120°C, with single cover plate or up to 150°C using double cover plates. The glass cover thickens of 3–4 mm is

used and 1–2 covers with spacing 1.5–3 cm are generally used between plates. The temperature of glass cover is lower than the absorber plate and is a good absorber of thermal energy and reduces convective and radiative losses of sky.

3. *Absorber Plate:* It intercepts and absorbs the solar energy. The absorber plate made of copper, aluminum or steel with thickness of 1–2 mm is the most important collector part, along with the tubes containing the working fluid to be heated. The plate absorbs the incident solar radiation through cover plate and transfers the heat to the tubes, with minimum heat losses. The plate is black painted, provided with selective material coating to increase its absorption and to reduce the emission. The absorber plate has high absorption (80%–95%), low transmission and reflection.

4. *Tubes:* The plate is attached to a series of parallel tubes or a serpentine tube through which working fluid passes. The tubes are made of copper, aluminum or steel in the diameter 1–1.5 cm, brazed, soldered on top/bottom of the absorber with water equally distributed in all the tubes. The header pipe, of larger diameter is made of same material as tube. Today the tubes are made of plastic with low thermal conductivity and higher coefficient of expansion than the metals. Copper and aluminum are corroded with saline liquids and steel tubes are used at such places.

5. *Removal of Heat:* These systems are best suited to applications that require low temperatures. Once the heat is absorbed on the absorber plate it must be removed fast and delivered to the place of storage for further use. As the liquid circulates through the tubes, it absorbs the heat from absorber plate of the collectors. The heated liquid moves slowly and the losses from collector increases because of rise of high temperature of collector, lowering the efficiency. Flat-plate solar collectors are less efficient in cold weather than in warm weather. The collector plate absorbs as much of the irradiation as possible through the glazing, while losing as little heat as possible upward to the atmosphere and downward through the casing back. The collector plates transfer the heat to the transport fluid. The collector surface absorptance for shortwave radiation depends on the nature and color of the coating and on the incident angle. Usually black or dark colors are used, however various color coatings have been proposed mainly for aesthetic reasons.

Factors affecting the performance of flat-plate collectors are listed here. (a) The efficiency of collector is directly related with the falling solar radiation and increases with rise in temperature. (b) The increased number of cover plate reduces the internal convective heat losses but also prevents the transmission of radiation inside the collector, usually more than two cover plates are not used to optimize the system. (c) The more space between the absorber and cover plate the less internal heat losses, so the collector efficiency increases. However on the other hand, the space increase between them provides the shading by side wall in the morning and evening, reducing the absorbed solar flux by 3%. The spacing between absorber and cover plate is kept 2–3 cm to balance the problem. (d) The flat-plate collectors do not track the Sun and should be tilted at angle of latitude of the location for an average better performance. However with changing declination angle with seasons the optimum tilt angle is kept L ± 15°. The collector is placed with South facing at northern hemisphere to receive maximum radiation throughout the day. (e) Some materials like nickel black ($\alpha = 0.89$, $\varepsilon = 0.15$) and black chrome ($\alpha = 0.87$, $\varepsilon = 0.088$), copper oxide ($\alpha = 0.89$, $\varepsilon = 0.17$) etc. are applied chemically on the surface of absorber in a thin layer of thickness 0.1 µm. These chemicals have high degree of absorption ($\alpha$) to short wave radiation (<4 µm) and low emission ($\varepsilon$)

of long wave radiations (>4 μm). The higher absorption of solar energy increases the temperature of absorber plate and working fluid. The top losses are reduced and the collector efficiency increases. The selective surfaces are able to withstand temperatures up to 400°C, cost less, and need to be corrosive resistant, and the material property should not change with time. (f) With increase in the inlet working fluid temperature the losses increase. The high temperature fluid absorbed the less heat from absorber plate because of low temperature difference and increases the top loss coefficient. Therefore the efficiency of collector is reduced with rise in inlet temperature. (g) The collector efficiency decreases with dust particles accumulate on the cover plate because the transmission radiation decreases. Frequent cleaning is required to get the collector maximum efficiency. Because of their high heat loss coefficient, ordinary flat-plate collectors are not practical for elevated temperatures above 100°C. When higher temperatures are required, the heat loss coefficient must be reduced, which is accomplished by two methods: evacuation and concentration, either one or in combination. While several attempts have been made to build evacuated tube flat-plates, they do not seem to hold any promise of commercial success. FPC performance are determined from: (a) controlled indoor tests by using a solar simulator; however, non-availability of a solar simulator matched to the solar spectrum limits the accuracy of indoor tests, or (b) carefully performed outdoor tests under steady-state conditions; in this case weather variability limits that the accuracy of outdoor tests. These tests are the ultimate check on FPC performance under field conditions.

### 3.4.2 Tubular Solar Energy and Evacuated-Tube Collectors

While flat-plate collectors are all essentially made the same way and perform the way from one brand to other, evacuated tube collectors vary widely in their construction and operation. One method of obtaining temperatures between 100°C and 200°C is to use evacuated tubular collectors. The advantage in creating and being able to maintain a vacuum is that convection losses between glazing and absorber can be eliminated. Notice that the evacuated tube solar collectors perform better in comparison to flat plate solar collectors, in particular for high temperature operations. However, they are no real competition for flat-plate solar collectors, because of difficulties in manufacturing and maintenance of the metal-to-glass vacuum seal. One of the most significant developments is the use of double-glass evacuated tubular solar water heaters. The mechanism of this type of solar water heater is driven by natural circulation of the fluid in the collector and the storage tank. It consists of all-glass vacuum tubes, inserted directly into a storage tank, with water in direct contact with the absorber surface. There are different configurations of the evacuated tubular solar collectors. These solar collectors are constructed of a number of glass tubes. Each tube is made of annealed glass and has an absorber plate within the tube, because tube is the natural configuration of an evacuated collector. During the manufacturing process in order to reduce heat losses through conduction and convection, a vacuum is created inside the glass tube. The only heat loss mechanism remaining is radiation. The absence of air in the tube creates excellent insulation, allowing higher temperatures to be achieved at the absorber plates. In order to improve an efficiency of the evacuated tube collector there are several types of concentrators depending on its concave radius established.

There are many possible designs of evacuated collectors, but in all of them selective coating as an absorber is used because with a nonselective absorber, radiation losses would dominate at high temperatures, and eliminating convection alone would not be very effective. When higher temperatures are desired, the heat loss coefficient needs to be reduced, being usually accomplished by two methods: evacuation and concentration, either single or in combination. Convection and conduction heat losses through the cover can very well be reduced without significantly affecting the transmission by using vacuum as insulation. For flat-plate collectors, the air pressure differences cause large forces on the cover and casing, and to guarantee the tightness over two decades is a huge challenge. However, circular glass tubes are resistant to the pressure due to their edgeless shape, and for glass to glass-sealed tubes, the vacuum tightness is very high and reliable over time. The evacuated tube solar collector is therefore made-up of many parallel glass tubes each covering an absorber strip. While several attempts have been made to build evacuated flat plates, they do not seem to hold significant promise of commercial success. Two constructions of evacuated tube collectors are on the market. In single-walled tubes, the vacuum exists within the entire tube, and the absorber fin sits in the vacuum space; this construction requires a glass-metal seal to be able to effectively remove the heat; here, it is a challenge to maintain the vacuum reliable under varying operating conditions. In double-walled tubes, a second glass tube is located within the outer one and the vacuum exists only between these double walls; here, a glass-glass seal can be used since the heat removal construction is not placed in the vacuum space. In the so-called Sidney tube, the inner glass tube is selectively coated and works as the absorber. Within the inner glass tube, a metal heat removal construction is placed. There are different methods to transfer heat from the absorber in a single glass tube to the header of the collector. Either the heat transfer fluid flows through the pipe in the tube or a *heat pipe* is used. In summary, two general methods exist for significantly improving the performance of solar collectors above the minimum flat-plate collector level. The first method increases solar flux incident on the receiver. The second method involves the reduction of parasitic heat loss from the receiver surface. Tubular collectors, with their inherently high compressive strength and resistance to implosion, afford the only practical means for completely eliminating convection losses by surrounding the receiver with a vacuum on the order of $10^{-4}$ mm Hg. Tubular collectors have a second application. They may be used to achieve a small level of concentration (1.5–2.0) by forming a mirror from a part of the internal concave surface of a glass tube. This reflector can focus radiation on a receiver inside this tube. Since such a receiver is fully illuminated, it has no parasitic "back" losses. The performance of a non-evacuated tubular collector is slightly improved by filling the envelope with high-molecular-weight noble gases. External concentrators of radiation may also be coupled to an evacuated receiver for performance improvement over the simple evacuated tube.

From operation point of view, a heat pipe provides an elegant way of extracting heat from an evacuated solar collector. A heat pipe is a sealed pipe filled with a liquid (e.g., water or ethanol) under low pressure designed for heat transfer. The absorber heat evaporates the liquid, which rises to the header (upper end bulb or canula), where it condenses and transfers the heat to the heat exchanger, where the heat pipe header is in contact with heat exchanger working liquid. By using the latent heat of this evaporating/condensing mechanism, a large heat transfer capacity in a narrow connection is possible. In the header the vapor condenses back to the volatile liquid, descending back down to the heat pipe completing the cycle. The heat pumps are transferring heat more rapidly than other type of heat exchange devices. For the dry connected heat pipes, the condenser is connected conductively with the heat transfer circuit metal, while for the wet connections the condensing end is immersed into the heat transfer circuit fluid. The dry connection allows for easier replacement, while at the top end of the collector, the header, and the heat removal fluid flows through the collector loop. Heat pipe is hermetically sealed tube that contains a small amount of heat transfer liquid. When one portion of tube is heated the liquid evaporates and condenses at the cold portion, transferring heat with great effectiveness because of the latent heat of condensation. The heat pipe contains a wick or is tilted (or both) to ensure that the liquid follows back to the heated portion to repeat the cycle. It is easy to design a heat pipe (e.g., by giving it the proper tilt) so that it functions only in one direction. This so-called *thermal diode effect* is very useful for the design of solar collectors, because it automatically shuts the collector off and prevents heat loss when there is insufficient solar radiation. Heat pipes have lower heat capacity than liquid-filled absorber tubes, minimizing warm-up and cool-down losses. Heat pipe provides the method of transferring larger heat amounts from the focal area of a high-concentration solar collector to a fluid with only small temperature difference. It consists of a circular pipe with an annular wick layer situated adjacent to the pipe wall. The circular pipe is perfectly insulated from outside to avoid thermal losses from the circular pipe. Solar energy falls on evaporator and the fluid inside evaporator boils. The vapor migrates to the condenser where heat of vapor is transferred to a circulation fluid loop. The heat available with circulating fluid is further carried away to the end use point. The circulation fluid after releasing its heat is transferred to the boiler by capillary action in the wick or by gravity and cycle repeats. Gravity return heat pipes can operate without wick but cannot be operated horizontally as a result.

Evacuated-tube devices have been proposed as efficient solar energy collectors since the early twentieth century. With the recent advances in vacuum technology, evacuated-tube collectors can be reliably mass produced. Their high-temperature effectiveness is essential for the efficient operation of solar air-conditioning systems and process heat systems. Evacuated-tube collectors are fabricated from either concentric glass tubes or a metal tube end-sealed to and within a glass tube. An enclosed evacuated annular space and a selective absorber surface provide a very low overall heat loss particularly when operated at higher temperatures. An evacuated-tube solar collector consists of rows of parallel glass tubes connected to a header pipe. The air within each tube is removed reaching vacuum pressures around 10–3 mbar. This creates high insulation conditions to eliminate heat loss through convection and radiation, for which higher temperatures than those for flat-plate collectors can be attained. A variant to the vacuum is that the tube can use a low thermal conductivity gas such as xenon. Each evacuated tube has an absorber surface inside. Depending on the mechanism for extracting heat from the absorber, evacuated-tube solar collectors fall into either a direct-flow or heat-pipe classifications. In direct-flow tubes, the working fluid flows through the absorber. These collectors are classified according to their connecting-material joints as glass–metal or glass–glass and, further, by the arrangement of the tubes (such as concentric or U-pipe). Inside each evacuated tube, a flat or curved metallic fin is attached to a copper or glass absorber pipe. The fin is coated with a selective thin film whose optical properties allows high absorbance of solar radiation and impedes radiative heat loss. The glass–metal collector type is very efficient, although it can experience loss of vacuum due to the junction of materials with very different heat expansion coefficients. Within this type, the fluid can follow either a concentric or a U-shape path; for both, the working fluid flows in and out at the same end (the header pipe). The concentric configuration could incorporate a mechanism to rotate each single-pair fin pipe up to the optimum tilt incidence angle, even if the collector is mounted horizontally. However, the U-pipe configuration is the most typical direct-flow/ evacuated-tube solar collector. The evacuated space between the glazing and absorber eliminates convective loss and long-wave thermal radiative heat loss is inhibited by the deposition of the spectrally selective absorber coating on the absorber surface. Evacuated-tube solar collectors usually have low thermal mass. The ability to heat-up rapidly (often from higher maintained overnight temperatures than a flat plate collector) gives low utilizable insolation thresholds providing good low insolation performance. Heat removal in evacuated tube collectors can be indirect often using a volatile fluid in the absorber via a closed heat pipe, as discussed before. More frequently water, as a heat transfer fluid, moving in a thermosyphon through the collector is employed. Evacuated tube collectors can have copper absorbers or use a selectively coated glass inner glass tube as the absorber. A heat pipe provides the most elegant way of extracting heat from an evacuated collector.

### 3.4.3 Solar Collector Components

*Absorber:* The purpose of the absorber is to absorb as much as possible of the incident solar radiation, reemit as little as possible, and allow an efficient transfer of the heat to the working fluid. The most common forms of absorber plates in use are shown in Figure 3.6a. The materials used for absorber plates include copper, aluminum, stainless steel, galvanized steel, plastics, and rubbers. Copper is the most common material used for absorber plates and tubes because of its high thermal conductivity and high corrosion resistance, but is quite expensive. For low-temperature applications (up to 50°C or 120°F), plastic materials (e.g., ethylene propylene polymer) is used to provide inexpensive absorber material. To compensate for the low thermal conductivity, larger surface area is used for heat transfer. The plate and tubes of both a flat-plate and metal-in-glass evacuated tube solar energy collector absorber are usually made of metal with high thermal conductivity such as copper or aluminum. Good heat transport is provided through the plate to the heat transfer fluid. An ideal absorber plate has a high solar absorbance surface to selectively absorb as much as possible of the incident insolation, together with low emittance to long-wave thermal radiation so long-wave radiative heat losses are low. In order to increase the solar radiation absorption and to reduce the absorber emission, the metallic absorber surfaces are painted or coated with flat black paint or selective coating. A selective coating has high absorptivity in the solar wavelength range (0.3–3.0 μm). Such solar selective absorbers often consist of two layers with appropriately optical properties, often a thin upper layer exhibiting high absorptance to the solar radiation whilst being relatively transparent to thermal radiation is deposited on an underlying surface whose high reflectance providing low emittance to thermal long-wave radiation. Alternatively a heat selective mirror that has high solar transmittance with high infrared reflectance can be placed on top of a non-selective high absorptance material. An example of this latter of selective surface combination is "black chrome" that consists of microscopic chromium particles deposited on a metal substrate; the chromium particles reflect long-wave thermal radiation, but shorter wavelength insolation passes between the chromium particles. In very low cost and low-temperature solar energy collector applications, such as swimming pool heating, these guidelines for absorber materials and fabrication are inappropriate as the minimum initial cost usually dominates design choices. Black paint which has a high absorptivity but not being selective is equally high emittance is used on absorbers or is the dark pigment in unglazed often ground-mounted plastic coils through which water is solar heated.

*Aperture cover:* Most glasses are almost completely transparent to the shortwave radiation the associated with insolation, but nearly opaque to long-wave thermal radiation. When employed as an aperture cover, glass inhibits successfully radiative heat loss from an absorber plate to ambient or to the lower temperature radiative heat sink that the sky often constitutes. Glass is thus the aperture cover material employed most commonly for solar energy collectors. It is desirable that a large part of the incoming direct, diffuse and ground reflected solar radiation is transmitted through the cover and used efficiently to heat the transfer fluid. This means that the transmittance of an aperture cover must be high which requires both low reflectance and absorptance. The reflectance of a material depends on its refractive index and the angle of insolation incidence. It can vary for different wavelengths of insolation transmittance and decreases with increasing angle of insolation incidence. An aperture cover glazing with a smaller refractive index exhibits a lower reflectance and higher transmittance. The overall solar transmittance of an aperture cover glazing is the normalized sum of each quantized wavelength of the solar radiation spectrum transmitted by the aperture cover. Since iron absorbs light in the visible part of the spectrum, the transmittance of glass for the solar spectrum decreases with increasing iron concentrations. Thus low-iron glass typically with iron content below 0.06% is preferred for flat plate collector's aperture cover glazing. A simple and inexpensive collector consists of a black painted corrugated metal absorber on which water flows down open, rather than enclosed in tubes. This type of collector is called a trickle collector and is usually built on-site. Although such a collector is simple and inexpensive, it has the disadvantages of condensation on the glazing and a higher pumping power requirement.

*Selective surfaces:* Two types of special surfaces are very important in solar collectors, selective absorbers and reflecting surfaces. Essentially, typical selective surfaces consist of a thin upper layer, which is highly absorbent to shortwave solar radiation but relatively transparent to long-wave thermal radiation, deposited on a surface that has a high reflectance and a low emittance for long-wave radiation. Selective surfaces are particularly important when the collector surface temperature is higher than the ambient air temperature. Lately, a low-cost mechanically manufactured selective solar absorber surface method has been proposed. Two types of special surfaces in solar collector systems are selective absorbers and reflecting surfaces. Selective absorber surfaces combine a high absorptance for solar radiation (short solar radiation wavelengths) with a low emittance for the surface normal operating temperature. This combination of surface characteristics is possible because 98% of the energy in incoming solar radiation is within wavelengths below 3 μm, whereas 99% of the radiation emitted by black or gray surfaces at 400 K is at wavelengths longer than 3 μm. However, the actual (real) surfaces do not approach this performance. Performance of selective absorber coatings degrades as the temperatures are above 400°C. Since the interest in concentrating solar thermal power has increased and plants as large as 300 MWe are under construction, there is a need for selective solar absorbers capable of maintaining high performance at temperatures of 400°C and above. An energy efficient solar collector should absorb incident solar radiation, convert it to thermal energy and deliver the thermal energy to a heat transfer medium with minimum losses at each step. It is possible to use several different design principles and physical mechanisms in order to create a selective solar absorbing surface. Solar absorbers are based on two layers with different optical properties, which are referred as tandem absorbers.

*Reflecting surfaces:* Concentrating solar collectors are requiring the use of reflecting surfaces with high specular reflectance in the solar spectrum or refracting devices with high transmittance in the solar spectrum. Reflecting surfaces are highly polished metals or metal coatings on suitable substrates. With opaque substrates, the reflective coatings must always be front surfaced, for example, chrome plating on copper or polished aluminum. If a transparent substrate is used, the coating may be front or back surfaced. In any back-surfaced reflector, the solar radiation must pass through the substrate twice and the transmittance of the material becomes very important. Table 3.1 presents typical values for the normal specular reflectance for the direct (beam) solar radiation of new (not-used) surfaces.

*Transparent materials and glazing:* the purpose of glazing or transparent cover is to transmit the shorter-wavelength solar radiation and to block the longer-wavelength re-radiation from the absorber plates and to reduce the heat losses by convection for the absorber plate top. Glass and special plastic materials are the most common used glazing materials. The optical transmission behavior can be characterized by two wavelength-dependent physical properties—the index of refraction $n$ and the extinction coefficient $k$. The index of refraction, which determines the speed of light in the material, also determines the amount of light reflected from a single surface, while the extinction coefficient determines the amount of light absorbed in a substance in a single pass of radiation. The angle $i$ is called the angle of incidence, being equal to the angle at which a beam is reflected from the surface, separating the two media. Angle $\theta_r$ is the angle of refraction, which is defined as shown in the Figure 3.8. The incidence and refraction angles are related by the Snell's law, as:

$$\frac{\sin(i)}{\sin(\theta_r)} = \frac{n_i}{n_r} = n \tag{3.31}$$

where $n_i$ and $n_r$ are the two refractive indices and $n$ is the index ratio for the two substances forming the interface. Typical values of refractive indices for of the some common materials that are used for solar collectors are given in Table 3.2. For most materials of interest in solar applications, the values range from 1.3 to 1.6, a fairly narrow range. By having a gradual change in index of refraction, reflectance losses are reduced significantly. The reflectance of a glass–air interface common in solar collectors may be reduced by a factor of 4 by an etching process. If glass is immersed in a silica-supersaturated fluo-silicic

**TABLE 3.1**

Secular Reflectances of the Common Used Solar Reflector Materials

| Material | P |
|---|---|
| Silver | $0.94 \pm 0.02$ |
| Gold | $0.76 \pm 0.03$ |
| Aluminized acrylic (second surface) | 0.86 |
| Anodized aluminum | $0.82 \pm 0.05$ |
| Aluminum surface | 0.82–0.92 |
| Back-silvered plateglass | 0.88 |
| Copper | 0.75 |
| Aluminized C-mylar | 0.76 |

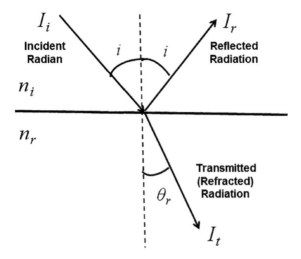

**FIGURE 3.8**
Diagram showing incident, reflected, and refracted beams of light and incidence and refraction angles for a transparent medium.

**TABLE 3.2**

Refractive Index for Common Materials in the Visible Range against Air

| Material | Refractive Index |
| --- | --- |
| Air | 1.00 |
| Clean polycarbonate | 1.59 |
| Diamond | 2.42 |
| Glass (solar collector type) | 1.50–1.52 |
| Plexiglass (polymethyl methacrylate) | 1.49 |
| Mylar (polyethylene terephthalate) | 1.64 |
| Quartz | 1.55 |
| Tedlar (polyvinyl fluoride, PVF) | 1.45 |
| Teflon (polyfluoroethylenepropylene) | 1.34 |
| Water—liquid | 1.33 |
| Water—solid | 1.31 |

acid solution, the acid inter-acts with the glass, leaving a porous silica surface layer, which has an index of refraction intermediate between glass and air.

*Headers (manifolds), insulation, containers or castings:* There are two headers, namely, the lower header, which allows fluid to pass through, and upper header, which is used to discharge hot water after heating. Headers are used to facilitate the flow of heat transfer fluid, and together with the insulation are minimizing the heat losses. The insulation is a non-conducting material at the bottom of tube and fins to minimize heat loss from the back and sides of the FPC. Container or casing surrounds the various components and protects them from dust and moisture, etc. It is important to mention that area of the glass cover and the absorber is equal.

*Heat transfer fluid:* In any STE the HTF, especially those used in solar thermal power plants must satisfy several technological requirements in terms of stability (e.g., at high temperatures), low freezing points, cost, corrosivity and flammability. Commonly used the common HTS for thermal power plants include synthetic oil and molten salts, while for residential STEs the common used are water, air and anti-freezing liquids. Synthetic oil, typically a diphenyl/biphenyl oxide is by far the most used HTFS in solar thermal power plants. Molten slats are typically a mixture of $NaNO_3$ and $KNO_3$ in various proportions, the most common used being the so-called solar salt consisting of 60% $NaNO_3$ and 40% of $KNO_3$. The use of molten salt IN solar thermal power plants has several advantages, such as non-flammable, environmentally friendly, low cost, up to 600°C operation temperatures, with the main drawbacks being the requirement for expensive anti-freezing systems, due to the high solidification temperatures, 120°C–200°C.

*Energy storage:* Energy storage is necessary whenever there is a mismatch between the energy available and the energy demand of any solar energy application. Energy storage is especially important in solar energy applications because of the seasonal, diurnal, and intermittent solar energy nature. Nature provides solar energy storage in a number of ways, such as plant matter (biomass), ocean thermal energy, hydro-potential at high elevation by evaporation from water bodies and subsequent condensation. In fact, even the fossil fuels are a stored form of solar energy. Natural solar energy storage provides long-term buffers between supply and demand.

*Power conditioning, control and protection system:* There are types of solar energy conversion systems, including solar collectors that may require power electronic and control equipment in order to operate properly. Load requirements of electrical energy usually vary with time. The energy supply has certain specifications like voltage, current, frequency, power etc. The power conditioning unit performs several functions such as control, regulation, conditioning, protection, automation, etc. For example, in residential STE mechanical and/or electronically operated controls are used for supplying stored thermal energy for hot water and/or into space for heating.

## 3.5 Concentrating Solar Energy

When higher temperatures are required, concentrating solar collectors are used. Solar energy falling on a large reflective surface is reflected onto a smaller area, and then converted into heat. The surface absorbing the concentrated energy is smaller compared to the surface capturing the solar energy and therefore can attain higher temperatures before heat loss due to radiation and convection. Most concentrating collectors can only concentrate the parallel insolation coming directly from the Sun (beam or direct normal insolation), and must follow (track) the Sun's path across the sky, by using sun tracking subsystems. The solar concentrators are broadly classified as follows: (a) tracking type, further classified based on the tracking system used (continuous, intermittent, one-axis tracking and two-axis tracking), and (b) non-tracking type, having the concentrator axis fixed. Solar concentrators are also classified on the basis of optical components as: reflecting or refracting types, imagining or non-imagining types, and line-focusing and point-focusing types. The refracting or reflecting surface can be a single piece or a composite surface, as well as symmetric or asymmetric types. Four types of solar concentrators are in common use, parabolic troughs, parabolic dishes, central receivers and Fresnel lenses. Figure 3.9 shows the schematics of the main solar concentrating collector concepts. A parabolic trough concentrates incoming solar radiation onto a line running the length of the trough. A tube (receiver) carrying heat transfer fluid is placed along this line, absorbing concentrated solar radiation and heating the fluid inside. The trough must be tracked about one axis. Because the surface area of the receiver tube is small compared to the trough capture area (aperture), temperatures up to 400°C can be reached without major heat loss. Figure 3.9c shows one of the parabolic trough concentrator. A parabolic dish concentrates the incoming solar radiation to a point. An insulated cavity containing tubes or some other heat transfer device is placed at this point absorbing the concentrated radiation and transferring it to a gas. Parabolic dishes must be tracked about two axes. A central receiver system consists of a large field of independently movable flat mirrors (heliostats) and a receiver located at the top of a tower. Each heliostat moves about two axes, throughout the day, to keep the Sun's image reflected onto the receiver at the top of the tower. The receiver, typically a vertical bundle of tubes, is heated by the reflected insolation, thereby heating the heat transfer fluid passing through the tubes. A Fresnel lens concentrator, such as shown in Figure 3.9, uses the refraction rather than the reflection to concentrate the solar energy incident on the lens surface to a point. Usually molded out of inexpensive plastic, these lenses are used especially in photovoltaic concentrators. Their use is not to increase the temperature, but to enable the use of smaller, higher efficiency photovoltaic cells. As with parabolic dishes, point-focus Fresnel lenses must track the Sun about two axes.

Concentrating solar technologies are focusing the solar energy from a very large aperture area onto a small receiver area by means of systems of lenses, mirrors or reflectors. When the concentrated sunlight is converted to heat, very high temperatures are achieved; the higher the concentration ratio, the higher the temperature is. High-concentrating, sun-tracking solar systems are used to produce high-temperature heat for industrial processes or to drive steam turbines to produce electricity, with new CSP power plants with

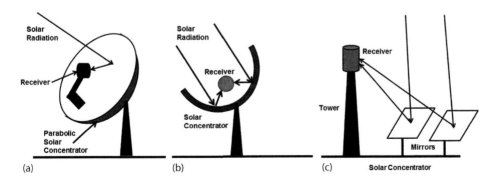

(a)  (b)  (c)

**FIGURE 3.9**
Three commonly used reflectors for concentrating solar energy to attain high temperatures: (a) parabolic trough, (b) central receiver, and (c) parabolic dish concentrators.

in development and research. However, these high temperature systems are also used in heat applications, in which heat is utilized in combined heat and power installations. Concentrating collectors present certain advantages as compared with the conventional flat-plate solar collectors, such as:

1. The working fluid can achieve higher temperatures in a concentrator system for the same solar energy collecting surface. This means that a higher thermodynamic efficiency can be achieved.
2. It is possible with a concentrator system, to achieve a thermodynamic match between temperature level and task, operating higher temperature devices.
3. The thermal efficiency is greater because of the small heat losses relative to the receiver area.
4. Reflecting surfaces require less material and are structurally simpler than FPC. For a concentrating collector the unit area cost of the solar collecting surface is less than that of a FPC.
5. Owing to the relatively small receiver area per unit of collected solar energy, selective surface treatment and vacuum insulation to reduce heat losses and improve the collector efficiency are economically viable.

Their major disadvantages are:

1. Concentrator systems collect little diffuse radiation depending on the concentration ratio.
2. Some form of tracking system is required so as to enable the collector to follow the Sun.
3. Solar reflecting surfaces may lose their reflectance with time and are requiring periodic cleaning and refurbishing.

Concentrating solar thermal collectors generally need to track the Sun in one or two axes in order to present high efficiency and concentration. Concentrators are special collector devices that are directing radiation onto the solar collector receiver, devices designed to absorb and convert the solar energy to some other energy forms (e.g., heat). The aperture of the concentrator is the opening through which the solar radiation enters the concentrator. Concentrator is the solar collector component that directs radiation onto the receiver. The concentrator is characterized by a geometric concentration factor or ratio $C_g$, i.e., the ratio between the incident power density and the power density at the concentrator output. The aperture of the concentrator is the opening through which the solar radiation enters the concentrator. In this way, the solar energy is concentrated to produce higher output temperatures and/or reduce collector cost by replacing an expensive absorber area with less expensive reflector. Concentration ratio is the factor by which an intervening concentrating mirror, lens or luminescent system increases the insolation flux on a solar energy absorbing surface. The geometric concentrator ratio is the collector aperture area ($A_a$) is divided by the absorber surface area ($A_R$), expressed as:

$$C_g = \frac{A_a}{A_R}$$

(3.32)

**Example 3.10:** Estimated the average power density (per unit of area) that is concentrated by a parabolic dish, having a diameter of 13.5 m on a receiver with diameter of 6 cm. Assume that the midday solar radiation is 630 W/m².

**Solution:** The concentrator ratio, Equation (3.31) is:

$$C_g = \frac{A_a}{A_R} \frac{\pi(13.5/2)^2}{\pi(6\times10^{-2}/2)^2} = 50,625$$

The power focused on the receiver is:

$$P_{rcvr} = 630\times50,625 = 31.89375 \text{ MW}$$

Thermal losses are dependent largely on the heat loss characteristics of, and proportional to, absorber area. The $\varepsilon$ is the angular size of the Sun disc, and $n$ is the refractive index of the last material insolation traversed for the refractive index of air, respectively. Equation (3.32), can be expressed in the case of the solar collectors concentrating into two dimensions (2-D), such as a parabolic dish and the ones concentrating in one dimension, such as a parabolic trough as:

$$C_{g(2-D)} = \left(\frac{d_a}{d_R}\right)^2 \tag{3.33a}$$

and, respectively

$$C_{g(2-D)} = \left(\frac{d_a}{d_R}\right) \tag{3.33b}$$

Insolation can be concentrated by an one-dimensional or a two-dimensional system to theoretical upper limits of each system, given by:

$$C_{max,1-D} = \frac{n}{\sin\left(\dfrac{\varepsilon}{2}\right)} \tag{3.34a}$$

and

$$C_{max,2-D} = \frac{n^2}{\sin^2\left(\dfrac{\varepsilon}{2}\right)} \tag{3.34b}$$

Typically values for $C_{max,1-D}$ is 216, while for $C_{max,2-D}$ is 46,000 for concentrators operating in the refractive index of air. In order to maintain high concentration high concentration ratio 1-D concentrators need accurate solar tracking systems as concentration decreases sharply at off-normal insolation incidence. However, for larger maximum concentration ratios 2-D systems need less-precise tracking accuracy to yield acceptable optical performance. These highest concentration values, related to the highest temperature that can be achieved can be estimated from the solar radiation inverse square law. By the time that solar radiation, emitted by the Sun reaches the Earth, it spreads over a sphere whose area is larger by the factor $1/\sin^2(\theta_S) \approx 1/(\theta_S)^2$, where $\theta_S$ is half of the Sun subtended angle in the sky, or approximately 1/213 radians. It is easy to show that the maximum possible concentration ratios for 1-D and 2-D solar concentrators are:

$$C_{g(1-D)} = \frac{1}{\sin(\theta_S)} = 213$$

and

$$C_{g(2-D)} = \frac{1}{\sin^2(\theta_S)} = 45,300$$

In practice, the actual 1-D and 2-D concentration ratios of the concentrating solar collectors are only half or less of these values. Solar collectors fall short of the above limits due to several factors, such as imperfect optics, tracking errors, reflection coefficients lower to 100%, and indirect radiation of some components of the solar radiation sue to the atmospheric scattering or clouds. However, the main solar concentrating collector system measure is not the concentrating geometric ratio, which determines the maximum temperature, but the overall system efficiency. In order to determine the concentrator efficiency, let us quantify the power inputs $P_{absorbed}$ and the power losses $P_{losses}$ at the absorber, per unit of surface of the absorber.

$$P_{absorbed} = \alpha \cdot \tau \cdot \rho \cdot C_g \times P_{solar} \tag{3.35}$$

where $P_{solar}$ (W/m²) is the incident sunlight per unit of surface area of the collector, $\alpha$ is the absorption coefficient of the absorber surface, $\tau$ is the transmission coefficient of the absorber, and $\rho$ is the reflection coefficient of the mirrors of the collector. The total losses $P_{losses}$ are the combination of the losses by conduction-convection and radiation losses by radiation. The quantity of heat received by the heat transfer fluid depends on the efficiency factor $F_R$ of the absorber-fluid transfer, similar as previously discussed. Therefore, the efficiency of the solar concentration system is given by:

$$\eta_{SC} = F_R \cdot \alpha \cdot \tau \cdot \rho - \frac{F_R \cdot \left[ U_L \left( T_{absorber} - T_{amb} \right) + \varepsilon \cdot \sigma \left( T_{absorber}^4 - T_{amb}^4 \right) \right]}{C_g \cdot P_{solar}} \tag{3.36}$$

where $U_L$ is the overall heat transfer coefficient, $T_{absorber}$ and $T_{amb}$ are the absorber and ambient temperatures (in $K$), respectively, $\varepsilon$ is the absorber emissivity coefficient and $\sigma$ is the Stefan-Boltzmann constant. The last term in Equation (3.36) is the solar concentrating system losses. The solar concentrating efficiency is decreasing with the absorber temperature and increases with the geometric concentration factor or ratio. However, the conclusions based only on concentration factors, without system thermal analysis are likely incorrect. The geometric concentration ratio (factor), $C_g$ in Equation (3.36) is higher often much than 1, meaning that the loss term (e.g., the last term in the equation) is thus significantly smaller for concentrating solar collectors, compared with the flat-plate collectors, where the geometric concentration factor is ~1, resulting in higher efficiencies for the concentrating solar collectors.

**Example 3.11:** Let consider a solar concentrating systems with the following numerical characteristics: $P_{solar}$ = 800 W/m², $F_R U_L$ = 21 W/m², $C_g$ = 75, and $F_R \alpha \tau \rho = F_R \varepsilon$ = 0.81, estimates the solar concentrating system for two ambient temperature, 100°C and 500°C.

**Solution:** Plunging the system numerical values into Equation (3.34) the required efficiencies are:

$$\eta_{SC} = 0.81 - \frac{20 \cdot (100 - 25) + 0.81 \cdot 5.67 \times 10^{-8} \left(100^4 - 25^4\right)}{75 \cdot 800} = 0.785 \text{ or } 78.5\%$$

$$\eta_{SC} = 0.81 - \frac{20 \cdot (500 - 25) + 0.81 \cdot 5.67 \times 10^{-8} \left(500^4 - 25^4\right)}{75 \cdot 800} = 0.643 \text{ or } 64.3\%$$

Both Earth diurnal rotation about its axis and the annual rotation about the Sun must be followed continually by a two-axis tracking solar energy collector to maintain the direct insolation component at normal incidence to the aperture plane. Two-axis solar tracking is essential to achieve concentration ratios from 100 to 1000. Such solar concentrators can elevate the working fluids to temperatures at values required to generate electricity using steam turbines or Stirling engines or to provide high-grade thermal energy for industrial processes. Single axis trackers are usually oriented either horizontal east-west, or inclined north-south achieving normal incidence once a day or twice a year respectively. Stationary concentrating systems with a concentration ratio in the range from 1 to 3 can be integrated into buildings, having no moving parts, complex mounting and associated mechanical systems. Most systems are based on 2-D non-imaging compound parabolic concentrators (CPCs). Building integration of high temperature solar thermal applications is attractive as a means of lowering installation cost. The initial investment cost is reduced if lower cost reflector materials are replacing the more expensive evacuated tube collectors. The 2-D CPC is termed an ideal concentrator, with a concentration ratio of $1/\sin\theta_{max}$, since all the light incident at angles less than the angle of acceptance arrives at the absorber.

*Fresnel lens* are optical devices for concentrating light that is made of concentric rings that are faced at different angles so that light falling on any ring is focused to the same point. *Parabolic trough collectors* are high-temperature (above 360 K) solar thermal concentrators with the capacity for tracking the Sun using one axis of rotation. It uses a trough covered with a highly reflective surface to focus sunlight onto a linear absorber containing a working fluid that can be used for medium temperature space or process heat or to operate a steam turbine for power or electricity generation. Central receivers, also known as *power towers* are solar power facility that uses a field of two-axis tracking mirrors known as *heliostat*. The heliostat is a device that tracks the movement of the Sun). Each heliostat is individually positioned by a computer control system to reflect the Sun's rays to a tower-mounted thermal receiver. The effect of many heliostats reflecting to a common point creates the combined energy of thousands of suns, which produces high-temperature thermal energy. In the receiver, molten nitrate salts absorb the heat energy. The hot salt is then used to boil water to steam, which is sent to a conventional steam turbine-generator unit to produce electricity. Planar and non-concentrating types are providing concentration ratios of up to four times of that of the flat-plate solar collectors. Line focusing type produces a high density of radiation on a line at the focus. Cylindrical parabolic concentrators are of this type and they could produce concentration ratios of up to ten times higher. Point focusing types generally produce much higher density of radiation in the vicinity of a point. Concentrators of paraboloid-like shapes are examples of point focus concentrators. For example, the geometric concentration for the cylindro-parabolic concentrator shape is:

$$C_g = \frac{L - d_{abs}}{\pi \cdot d_{abs}} = \frac{1 - \left(\sin(\theta_{Sun}) / \cos(\phi_R)\right)}{\left(\sin(\theta_{Sun}) / \cos(\phi_R)\right) \cdot \left(\phi_R - \theta_{Sun} + \pi / 2\right)} \tag{3.37}$$

Here, $L$ is concentrator parabola opening width, $d_{abs}$ is the absorber diameter, $\theta_{Sun}$ is the solid angle under the Sun is seen from the collector, and $\phi_R$ is the half opening angle of the parabola. While for a linear Fresnel concentrator the concentration ratio or factor can be expressed as:

$$C_{g(LFR)} = \frac{n \cdot W_{LFR}}{d_{abs}} \tag{3.38}$$

where $n$ is the number of primary mirrors, $d_{abs}$ is the absorber diameter, and $W_{LFR}$ is mirror aperture width all in SI unit. The overall aperture with in this case is the aperture with of each mirror tomes the number of mirrors. Notice LFRs may have a second reflector collecting solar radiation from the primary mirrors to the absorber tube. Another compound-curvature collector types are using spherical geometry or parabolic shapes in order to improve the solar energy collection. However, the concentration ratios of the spherical ones are much smaller than the parabolic solar concentrators. The concentration ratios for paraboloid and spherical concentrators are determined from their basic geometry and ray tracing models are:

$$C_{g(\mathrm{Pr}bC)} = \frac{\sin^2(\phi)}{4 \cdot \sin(\theta_{max})} \tag{3.39a}$$

and, respectively

$$C_{g(SC)} = \frac{\sin^2(\phi)}{\sin(\theta_{max})} \tag{3.39b}$$

For a compound parabolic concentrator, a non-imaging type of solar concentrating system, basically formed from two distinct parabolic segments, the geometric concentrating ratio is the same as for one-dimensional maximum geometric concentration factor:

$$C_{g(CPC)} = CR_{1-D,max} = \frac{1}{\sin(\theta_{max})} \tag{3.40}$$

Here, $\theta_{max}$ is the shape acceptance half-angle, and $\phi$ is the paraboloid or sphere the rim half-angle, as shown in Figure 3.10. Notice that the receiver radius for perfect optics is sized to collect all Sun rays with the acceptance half-angle.

There three most used types of concentrating solar power systems in use are *linear concentrator, dish/engine* and *power tower systems*. Linear concentrator systems collect the solar energy using long rectangular, curved (U-shaped) mirrors, tilted toward the Sun, focusing sunlight on tubes (or receivers) that run the mirror length. The reflected sunlight heats a fluid flowing through the tubes. The hot fluid then is used to boil water in a boiler of a conventional steam-turbine generator to produce electricity. There are two major types of linear concentrator systems: *parabolic trough systems*, where receiver tubes are positioned along the focal line of each parabolic mirror; and *linear Fresnel reflector systems*, where one receiver tube is positioned above several mirrors to allow the mirrors greater mobility in tracking the Sun. A dish/engine system uses a mirrored dish similar to a very large satellite dish. The dish-shaped surface directs and concentrates sunlight onto a thermal receiver, which absorbs and collects the heat and transfers it to the engine generator. The most common type

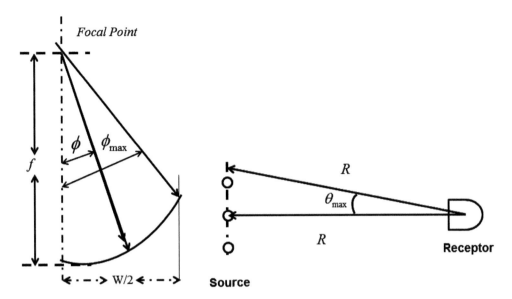

**FIGURE 3.10**
Solar concentrating collector geometry and parameters.

of heat engine used in dish/engine systems is the Stirling engine. This system uses the fluid heated by the receiver to create mechanical power, then used to run a generator or alternator to produce electricity. A power tower system uses a large field of flat, sun-tracking mirrors known as heliostats to focus and concentrate sunlight onto a receiver on some power towers use water/steam as the heat-transfer fluid. Other advanced designs are experimenting with molten nitrate salt because of its superior heat-transfer and energy-storage capabilities. The energy-storage capability, or thermal storage, allows the system to continue to dispatch electricity during cloudy weather or at night. In this technology large fields of parabolic systems that are secured on a single-axis solar tracking support are installed in a modular parallel-row configuration aligned in a north-south horizontal direction. Each of the solar parabolic collectors track the movement of the Sun from east to west during daytime hours and focus the Sun's rays to linear receiver tubing that circulates a heat transfer fluid (HTF), that in turn passes through a series of heat exchangers where the heat is transferred as superheated vapor that drives steam turbines. After propelling the turbine, the spent steam is condensed and returned to the heat exchanger via condensate pumps. At present the technology has been successfully applied in thermal electric power generation.

### 3.5.1 Shadowing and Shadow Types

One of the important steps before designing an STE configuration is to investigate the solar location characteristics. In order to the harvest the maximum possible solar energy all solar panel and collectors, in addition to be mounted at the optimum solar tilt angles must be totally exposed to the Sun rays without any shading that may be cast by surrounding buildings, trees, vegetation, or even the other solar panels or collectors, located in the proximity. To achieve this objective the solar power system site must be analyzed for year-round shading, taking into account the seasonal rise and fall of solar angle and its impact on the shadow direction and area, through measurements and calculations. The shaded area is not only reducing the output proportionally to the size of the shaded area, but often the losses are much higher. Within a chain of modules the output is that of the weakest (shaded) module. If shade cannot be avoided at certain times, specific measures are required in order to leave the sunny modules to fully function. Ideally, solar collectors and PV arrays are installed in a shade-free location. However, grid-connected systems or residential solar thermal systems are often found in urban and suburban areas and the collectors or panels are usually installed on roofs, where some shading is sometimes inevitable. Shading can reduce the solar energy system output considerably and ideally should be avoided. As far as PV arrays or STE systems for stand-alone applications are concerned, they are usually located in rural areas, having sufficient available land around the building, can be installed where there is no shade. A shadow cast on a PV system has a much greater effect on the solar yield than, for example, in the case of solar thermal systems. Shading can be classified as temporary, resulting from the location or from a building, or caused by the system itself (self-shading). This goes for seasonal shading due to trees or vines, and even the shade from deciduous trees in the winter when they are bare. Watch for plant growth over time that can shade the panels. Typical temporary shading includes factors such as snow, leaves, bird droppings and other types of soiling. Snow is a significant factor, especially in mountainous areas. Dust and soot soiling in industrial areas or fallen leaves in forested areas are also significant factors. Snow, soot and leaves collecting on the PV array cause shading, the effect of which will be lower if the array self-cleans, for example if it is washed away by flowing rainwater. A till angle of about $10^\circ$ or more is usually sufficient to achieve this. Greater tilt angles increase the flow speed of the rainwater and, hence, help to carry away dirt particles. This type of shading can be reduced by increasing the tilt of the PV array. Snow on a PV array melts faster than the surrounding snow, so that, generally, shading only occurs on a few days. In snowy areas, arranging the standard modules horizontally enables losses caused by snow to be reduced by half. In this way, shading caused by the snow generally affects only two and not four rows of cells on each module, as is the case in the vertical arrangement.

Shading caused by leaves, bird droppings, air pollution, and other dirt has a stronger and longer-lasting impact. If a system is heavily affected by this, regular cleaning of the PV modules will noticeably increase the solar yield. In a normal location and with sufficient tilt, it can be assumed that the loss due to soiling amounts from 2% to 5%. Generally, this loss is considered acceptable. If heavy soiling is present, the modules may be cleaned using water and a gentle cleaning implement (as a sponge) without using detergents. Shading resulting from the location covers all shading produced from the photovoltaic system's surroundings. Neighboring buildings, trees, and even distant tall buildings can shade the system. It must be taken into account that, due to the growth of trees and shrubs, vegetation may shade the system only after years. Overhead cables running over the building also have a negative effect, casting a small but effective moving shadow. Shading resulting from the building involves direct shadows, which should therefore be viewed as particularly critical. Special attention should be paid to chimneys, antennae, lightning conductors, satellite dishes, roof and facade protrusions as well as roof superstructures. Some shading can be avoided by moving the photovoltaic modules or the object causing the shading (e.g., an antenna). If this is not possible, the impact of the shading can be minimized by taking it into account when selecting how the cells and modules are wired up and during the system's concept. Self-shading of the modules may be caused by the row of modules in front. Space requirements and shading losses can be minimized through optimization of the tilt angles and distances between the module rows. A poorly designed or installed mounting system may also cause micro-shading in sloping roof installations.

In design of many of the solar energy systems or building windows, the estimates of the possibility and the extent of shading are often required. In order to determine the shading, it is needed to know the shade cast as a function of the time for every day of the year. There are quite comprehensive mathematical models that can be used for this purpose; however, simpler graphical methods are often employed, being suitable for quick, practical applications. Such methods are usually sufficient, since often the objectives are usually to determine whether selected solar collector or panel positions are suitable or not, rather to exactly estimate the shading amount. A knowledge of the angle of solar elevation, $\alpha_s$, and the azimuth, $\theta_z$, allows us to calculate the length and the location of a simple shadow. For example to describe the shadow due to an electric pole or communication tower that is $h$ meters or ft high at the specific time of day when the solar elevation is $\alpha_0$ and the azimuth angle is $\theta_{z0}$, as shown in the diagrams of Figure 3.11. In this figure, we observe that the shadow has length $|OP|$ where in the right triangle, $\triangle OPQ$ is found that the relationship between the pole height and shading length is:

$$\tan(\alpha_0) = \frac{h}{|OP|} \Rightarrow |OP| = \frac{h}{\tan(\alpha_0)} \quad \text{(m or ft, depending of the h unit)} \qquad (3.41)$$

Note that the tip of the shadow is located at the point $P = (x_0, y_0)$ that is $x_0$ ft East and $y_0$ ft North of the base of the post. By analyzing the triangle, $\triangle ORP$, and after a little bit of calculation the shadow tip coordinates are found as given by:

$$x_0 = -\frac{h \cdot \sin(\theta_0)}{\tan(\alpha_0)}, \text{ in m or ft}$$

$$y_0 = -\frac{h \cdot \cos(\theta_0)}{\tan(\alpha_0)}, \text{ in m or ft}$$

$$(3.42)$$

The shadow determination is facilitated by the estimation of a surface-oriented solar angle, the so-called *solar profile angle, p*, as shown in Figure 3.11. The solar profile angle

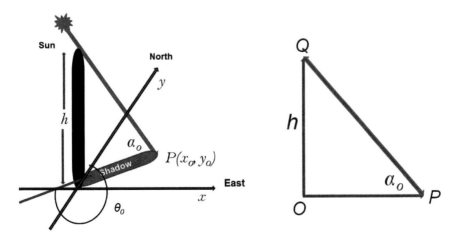

**FIGURE 3.11**
Pole shade and triangle diagrams.

is the angle between the normal to the surface (vertical plane), and in terms of the solar altitude angle, solar azimuth angle and the surface azimuth angle is expressed by the equation:

$$\tan(p) = \frac{\tan(\alpha_S)}{\cos(\theta_Z - \gamma)} \tag{3.43a}$$

When the surface due to South, i.e., the surface azimuth angle is 0°, this equation is simplified to:

$$\tan(p) = \frac{\tan(\alpha_S)}{\cos(\theta_Z)} \tag{3.43b}$$

The projection of the Sun's path on the horizontal plane is called the *sun path diagram*, as the diagrams shown in Figure 3.12. Solar altitude and azimuth angles for given latitude can be conveniently represented in such graphical format, that can be used to find the position of the Sun in the sky at any time of the year. Sun path diagrams, besides of helping us to build the intuition of the Sun position during the day, have also a very practical design and operation applications in the field when trying to predict shading patterns at a specific site or location, a very important consideration for the solar energy conversion systems, which are sensitive to the shadow. The concept is simple, the solar altitude and the solar azimuth

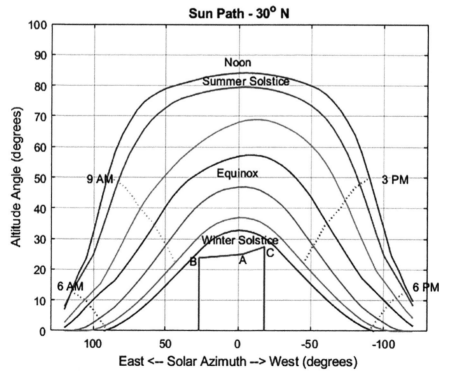

**FIGURE 3.12**
Sun path diagram for 30°N latitude.

angles are function of latitude, hour angle and declination. Using only two parameters, in a two-dimensional plot, the other two parameters can be correlated. For given latitude and a given day, the solar altitude angle and the azimuth angle are function of the hour angle, therefore the solar time. By entering in a sun path diagram for given latitude with appropriate values of the declination and hour angle, the point of intersection of the corresponding lines represents the instantaneous location of the Sun. The solar altitude can then be read from the diagram circles. Usually different Sun path diagrams are plotted for different latitudes, and are showing the complete variations of the hour angles and declination for the full year. Lines of constant declination are labeled by the values of the angles, while points of constant hour angles are clearly marked. The azimuth angles are the normal compass coordinates where the South is 180°, so the azimuth angle coordinates run from 0° to 360°, and the elevation angles coordinates from 0° to 90°. Since the Sun's path is symmetric about the summer solstice, the lines for the other year half are not included, being simply the duplicate of those of the first half of the year. The plot of sun path diagram may be shown in polar coordinates. Notice that the diagrams of Figure 3.11 are referring to the Northern Hemisphere, while for the Southern Hemisphere the declination angles' signs are reversed.

In order to determine the shading, apart from the Sun position also valid descriptions of the surroundings are needed, being needed a sketch of the azimuth and altitude angles for trees, buildings, and other obstructions along the southerly horizon that can be drawn on top of a Sun path diagram. Sections of the sun path diagram that are covered by the obstructions indicate periods of time when the Sun will be behind the obstruction and the site will be shaded. The Sun path diagrams are used in practical applications to determine the periods of the year and hours of the day when shading is taking place at the site of interest. Solar collector, often, especially in power generation applications are installed in multi-row configurations, facing South. These arrangements are requiring having an accurate estimate of the shading possibility by the front rows of the second and the subsequent rows. The maximum shading in such cases occurs at the local solar noon, which is easily estimated by computing the noon solar altitude and checking whether the shadow formed is affecting the subsequent solar collector rows.

**Example 3.12:** A south-facing solar collector is installed on a building, located at 30° N, sited in front of an existing taller building, with the solar collector side located 15° East of South. Plan and elevation views are shown below. Determine the time interval when the collector is shaded.

**Solution:** Profile upper angle limit for shading point $S$ is 30°, and the solar profile angle is 15° of true South, the point A on the sun path diagram (Figure 3.12), the solar profile angle being the solar altitude angle in this case. The distances S–B and S–C are then:

$$\text{S} - \text{B}: \sqrt{8^2 + 12^2} = 14.4 \text{ m}$$

$$\text{S} - \text{C}: \sqrt{4^2 + 12^2} = 12.7 \text{ m}$$

and the altitudes angles for these points are:

$$\text{Point B: } \alpha_B = \tan^{-1}\left(\frac{10-2}{14.4}\right) = 29.1°$$

$$\text{Point C: } \alpha_C = \tan^{-1}\left(\frac{10-2}{12.7}\right) = 32.3°$$

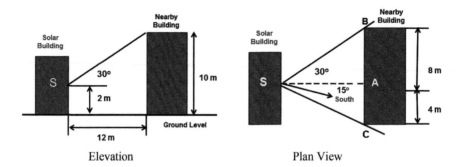

Elevation                    Plan View

The point of interest is shaded during the period indicated by the area under contour A-B-C of Figure 3.12 being straightforward to determine the hours when the shading occurs, whereas the time of the year is estimated from the declination angles vales. Solar collectors for large power or heat applications are often installed in multi-row settings, facing the true South, therefore there is a need to estimate the shading possibility by the front rows, the maximum shading, in such cases occurring at the local solar noon, which is easily to estimate by finding the noon altitude angles, $\alpha_{noon}$, and checking whether the shadow formed are shading the subsequent row.

### 3.5.2  Optimum Solar Collector, Panel Orientation, and Solar Collection Estimates

The insulations over specific periods, such as daily are the total energy density (energy per unit of area, W/m²) received from the Sun in that specific time interval, such as 24 hours for daily estimates, computed by:

$$H_{24-hr} = \int_{t=0\ hr}^{t=24\ hr} I(t) \cdot dt \tag{3.44}$$

However, during one day the Sun is illuminating the solar system only between the Sunrise and Sunset and the number of hours, the so-called *the day length* is computed with:

$$N = \frac{2}{15} \cos^{-1}\left(-\tan(L) \cdot \tan(\delta)\right) \tag{3.45}$$

The daily insulation varies with latitude and season, and the seasonal variation at high latitudes is the most significant. Its seasonal variations are due to three factors: the day length variations (estimated by Equation (3.45)), the receiving surface orientation, and the variations into the atmospheric absorption. The daily hour number of insolation, N has larger variations at higher latitudes, for example it varies from 16 hours in mid-summer to 8 hours in mid-winter at latitudes around 45°. Moreover, the horizontal plane at any location is much more oriented toward the Sun in summer than winter, the overall direct radiation on horizontal surface is reduced in winter by a factor of $\cos(\theta_z)$, as well as the daily or hourly insolation. The solar radiation atmospheric attenuation also increases with solar azimuth, $\theta_z$, so the direct (beam) solar radiation decreases in winter and the seasonal insolation variations are increasing in winter beyond the geometrical effect variations. Notice that the *clear day* solar radiation diagrams of Figure 3.13 are less than the extraterrestrial radiation due to the atmospheric attenuation. In practice, the so-called *clear day radiation* is a notional parameter, because the actual weather and site conditions quite far from those assumed in calculations. However, the graphs of

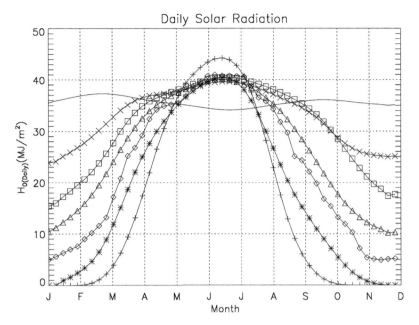

**FIGURE 3.13**
The seasonal and latitude variations of the daily insolation on horizontal plane with the clear skies.

Figure 3.13 are indicating that the changes in the average daily insulation on horizontal surface are function of latitude and season. A common used approximation of the hourly irradiance, as function of the sunrise time, $t_{SR}$ after sunrise, often used in practical applications for clear days is:

$$I_{hr} = I_{hr}^{\max} \cdot \sin\left(\frac{\pi \cdot t_{SR}}{N}\right) \tag{3.46}$$

Integrating this equation over daylight interval (i.e., from Sunrise to Sunset) for a clear day the total daily irradiance can be approximated by this relationship:

$$H_{h,(24-hr)} \approx \left(\frac{2N}{\pi}\right) \cdot I_{h,(hr)}^{\max} \tag{3.47}$$

The extraterrestrial radiation on a horizontal surface is a useful parameter in solar energy conversion system design and analysis, used to estimate the irradiance, in this case the hourly values, on surface of interest as:

$$I_{h,hr}^{\max} = I_{0,(hr)}^{\max} \cdot \cos(\theta_z) \tag{3.48}$$

However, the irradiance on horizontal surface, by using one of the common irradiance notation $G_0$ is expressed as:

$$G_{0,(h)} = I_{0n} \cdot \cos(\theta_z) \tag{3.49}$$

Here, $I_{0n}$ is given by the Equation (2.7) and the cos ($\theta_z$) by the Equation (3.3b). Integration of this equation from $H_{SR}$ (sunrise) to $H_{SS}$ (sunset) gives the daily solar energy, expressed as:

$$H_{0,(daily)} = \frac{24 \times 3600 \times I_{0n}}{\pi}\left[\cos(L)\cos(\delta)\cos(H_S) + \frac{\pi \cdot H_S}{180}\sin(L)\sin(\delta)\right] \qquad (3.50)$$

Here the sunset angles for horizontal surfaces (and zenith angle of 90°), and for tilted surfaces, with $\beta$, the tilt angle (for an incident angle, $\theta$ equal to 90°) are given by:

$$\cos(H_S) = -\tan(\delta)\cdot\tan(L)$$

and

$$\cos(H_S) = -\tan(\delta)\cdot\tan(L-\beta)$$

The monthly averaged values, $\bar{H}_0$ of the irradiance equal to $H_{0,(daily)}$ (measured in MJ/day) for the month average day and $M_0$ equals $\bar{H}_0$ multiplied by the number of days in that month, converted from MJ/day to kWh/month. The intensity of the direct (beam) radiation, as mentioned before varies with weather, air quality, altitude and the solar zenith angle, $\theta_z$. It is therefore almost impossible to estimate the solar radiation intensity or irradiance without measuring it at the location of interest. However, there are empirical relationships that are giving and approximate estimate, quite accurate for clear weather conditions, moderate elevations and dry air, as:

$$I_{bn} \simeq I_{0n}\cdot\exp\left(-0.347\left(\frac{1}{\cos(\theta_z)}\right)^{0.678}\right) \qquad (3.51)$$

The intensity of the diffuse sky solar radiation of the same conditions is about 10% of the direct radiation (it can be sometimes much higher), while the total solar radiation intensity is very often about $1.1I_{bn}$ or even lower. The hourly irradiance, used in Equation (3.46) at latitude $\pm 50°$, and if the $I_h^{max} \approx 900$ W/m$^2$ and the day number, $N \approx 16$ hours, then $H_{h,(daily)} \approx 33.0$ MJ·m$^{-2}$·day$^{-1}$, while in midwinter at the same latitude the $I_h^{max} \approx 200$ W/m$^2$ and the day number, $N \approx 8$ hours, then $H_{h,(daily)} \approx 3.70$ MJ·m$^{-2}$·day$^{-1}$. In the tropical regions, the, $I_h^{max} \approx 950$ W/m$^2$ while the daylight periods are not varying to much from, $N \approx 12$ hours throughout the year, thus $H_{h,(daily)} \approx 26.0$ MJ·m$^{-2}$·day$^{-1}$, for all months. These calculations are not taking into account cloud presence, dust in the atmosphere or other weather affecting conditions so the average measured values are always less than those computed and mentioned here, about 50% to 70% of the clear sky values, with only the desert areas having larger average values.

## 3.6 Solar Thermal Systems for Electricity Generation

Concentrating solar power technologies concentrate solar energy to produce high-temperature heat that is then converted into electricity. Electrical power is produced when the concentrated light is converted to heat which drives a heat engine, usually a steam turbine connected to an electrical power generator. The collectors concentrate the sunlight, collect it as heat energy and store it. Then, the heat energy is used generate steam that runs heat engines to produce electricity, which is transferred to the grid. The three most advanced CSP technologies currently in use are parabolic troughs (PT), central receivers (CR), and dish engines (DE). Planar and non-concentrating types systems are providing concentration ratios of up to four and are of the flat-plate geometrical shape types. Line focusing type produces a high density of radiation on a line at the focus. Cylindrical parabolic concentrators are of this type and they could produce concentration ratios of up to ten. Point focusing type generally produces much higher density of radiation in the vicinity of a point. Parabolic type concentrators are examples of point focus concentrators. CSP technologies are considered one of today's most efficient power plants; they can readily substitute solar heat for fossil fuels, fully or partially, to reduce emissions and provide additional power at peak times. Dish engines are better suited for distributed power, from 10 kW to 10 MW, while parabolic troughs and central receivers are suited for larger central power plants, 30–200 MW and higher. The diagrams of Figure 3.9 show graphical depiction of the common solar concentrators used in solar thermal power plants. The solar resource for power generation from concentrating systems is very plentiful, which can provide sufficient electric power for the entire country if it could be arranged to cover only about 9% of the state of Nevada, about 100 miles square. The amount of power generated by a concentrating solar power plant depends on the direct sunlight. These technologies use only direct (beam) solar radiation to concentrate the solar energy. The solar thermal energy is used in a CSP plant to produce high-temperature heat, which is then converted to electricity via a heat engine and generator units. Furthermore, solar-driven systems can deliver process heat and cooling, and other solar technologies are being developed that will deliver energy carriers such as hydrogen or hydrocarbon fuels, known as solar fuels. The leading contenders for exploiting solar energy for electricity generation are the PV and CSP facilities. Concentrating solar power systems are from 2 to 10 kW or larger enough, up to 200 Mw or even higher to be grid-connected. Some existing systems installed use thermal storage during cloudy periods and are combined with natural gas resulting in hybrid power plants that provide grid-connected dispatchable power. Other schemes for converting sunlight to power include solar salt ponds and solar chimneys.

Solar thermal electricity generation is defined as the process by which directly collected solar energy is converted to electricity through the use of some sort of heat to electricity conversion device. Mostly this is a heat engine, but there are other options such as a thermoelectric pile converter or a fan converter as in solar chimneys. Solar thermal electricity on grid was not achieved until the 1980s, although the basic technology for the production of mechanical energy, which could be eventually converted into electricity using a conventional electrical generator, had been under development for about 150 years, in France, U.S. and other countries. Solar thermal power has likely the greatest potential of any renewable energy sources, but has been delayed in market development since the 1980s because of market resistance to large plant sizes and poor political and financial support. However, there is rapid development occurring both in the basic technology and the market strategy and prospects for rapid growth appear now to be very bright for newer approaches. In the near-term, CSP appears to be leading PV for utility-scale power generation due in part to its maturity and relative cost. In addition, there are two other advantages of solar thermal over PV. First, most CSP units can be constructed with an integral thermal energy storage system thereby providing the capability of generating electricity into the evening hours. Second, solar thermal plants can be equipped with auxiliary burners in order to generate electricity when sunlight is unavailable (at night time and during inclement weather). Most of the CSP plants employ a thermal cycle that is similar to those of coal and nuclear power plants except that the heat source is from sunlight. The thermal-to-mechanical energy conversion efficiency of a solar thermal electric generating station is Carnot cycle limited, that is, the maximum theoretical thermal efficiency is:

$$\eta_{Carnot} = 1 - \frac{T_L}{T_H} \tag{3.52}$$

Here, $T_L$ and $T_H$ are the system sink (cold) and source (hot) temperatures, respectively. Since $T_L$ is approximately equal to the ambient environmental temperature at the power plant site, this relation motivates the design of a thermal cycle that adds heat at the highest possible temperature $(T_H)$, and hence, it promotes the utilization of sunlight concentrators to achieve high temperature operation. However, the Carnot efficiency is the maximum achievable in a thermal energy system. In order to compute the overall actual efficiency of a solar thermal energy conversion system, the all losses occurring during each conversion phases must be included. In any system where there are multiple energy conversion processes occurring, the efficiencies of each subsequent conversion result in an ever decreasing net energy output. An efficiency of interest is the *overall thermal efficiency* of the power plant, defined as the net electrical produced by the solar thermal power divided by the insulation at the plat location, being equal with product of the solar plat collection efficiency, $\eta_{col}$ and the thermal efficiency of the cycle:

$$\eta_{overall} = \frac{W_{net}}{E_{rad}} = \eta_{col} \times \eta_{thermal} \tag{3.53}$$

Typical solar thermal power plants have solar collection efficiencies in the range of 50%–70% and their Rankine cycles have efficiencies about 40%. Therefore the solar plant overall thermal efficiencies calculated for peak power of well-designed solar thermal power plants are in the range of 20%–30%. However, the solar thermal power plants are usually not operating continuously from sunrise to sunset the actual overall thermal efficiencies are even lower in the range 14%–20%. One of the important solar thermal power plant limitations is non-continuously operation. Their actual operation is limited by the available solar energy, which is also periodic and somewhat unpredictable. Moreover, a solar thermal power plant is commencing the operation at sunrise and do not stop operating at sunset, due to the solar collector geometry, limiting the plant operation hours and the total plant generated power. The plant operation hours and the collected solar energy are highly dependent on the power plant location latitude. To improve the solar plant overall efficiencies it is critical to improve its thermal cycle efficiency, for example through the use of well-designed and well-functioning Sterling cycles and/or to use thermal energy storage. In order to characterize the operation of a solar thermal power plant the so-called *average efficiency* may be used. The average efficiency is based on the average energy that a solar thermal power plant is producing over a specific peroid of time, $T$, typically one day or one year, defined by:

$$\eta_{avrg} = \frac{\dfrac{1}{T}\displaystyle\int_0^T W_{net} \cdot dt}{\dfrac{1}{T}\displaystyle\int_0^T E_{rad} \cdot dt} = \frac{1}{T}\int_0^T \eta_{Col} \cdot \eta_{th} \cdot dt \tag{3.54}$$

**Example 3.13:** A solar thermal power plant, peak power 18 MW (peak) is built in a region with annually averaged insulation is 900 W/m², and due to the local weather conditions the heliostat area is limited to 15 m². If the overall power plant efficiency is 25%, determine the number of the heliostats needed for this power plant.

**Solution:** The collected and transferred power from a single heliostat is:

$$P_{Heliostat} = \eta_{plant} \cdot I_{h,0} \cdot A_{heliostat} = 0.25 \times 900 \times 15 = 3375 \text{ W}$$

The number of heliostat is then:

$$Nr \text{ heliostats} = \frac{18 \times 10^6}{3375} = 5333.3 \text{ or } 5334 \text{ heliostats}$$

Another use of solar concentrator technology that generates electric power from the Sun is a construction that focuses concentrated solar radiation on a tower-mounted heat exchanger, usually employing Sterling engines as thermo-mechanical convertor. The system basically is configured from thousands of sun-tracking mirrors, commonly referred to as heliostats that reflect the Sun's rays onto the tower. The receiver contains a fluid that once heated, by a similar method to that of the parabolic system, transfers the absorbed heat in the heat exchanger to produce steam that then drives a turbine to produce electricity. Power generated from this technology produces up to 400 MW of electricity. The heat transfer fluid, usually a molten liquid salt, can be raised to 550°F. The HTF is stored in an insulated storage tank and used in the absence of solar ray harvesting. Recently a solar pilot plant located in southern California called Solar Two, which uses nitrate salt technology, has been producing 10 MW of grid-connected electricity with a sufficient thermal storage tank to maintain power production for 3 hours, which has rendered the technology as viable for commercial use.

### 3.6.1 Systems with Parabolic Troughs or Linear Fresnel Collectors

There are four major categories of concentration collectors, each of which is discussed below:

1. Parabolic trough collectors (PTC)
2. Linear Fresnel collectors (LFR)
3. Solar power towers (SPT), with heliostat field collectors
4. Parabolic dish reflectors (PDR)

The main purpose of the concentrating solar energy on the smaller receiver area is to produce high temperatures, therefore high thermodynamic efficiencies and to be able to operate thermal engines of the electricity generation units.

**Parabolic trough collectors** are the most-used CSP technology for solar thermal power applications. A parabolic trough consists of a linear parabolic mirror concentrating the received and collected solar energy onto a receiver (tube) positioned along its focal line. They are made by bending a sheet of reflective material into a parabolic shape. A metal black tube, covered with a glass tube to reduce heat losses, is placed along the focal line of the receiver. The heat transfer fluid (HTF), usually water or synthetic oil is pumped through the receiver tube and capture the solar energy through the tube walls. It is sufficient to use a single axis tracking of the Sun thus long collector modules are built. The collected heat is routed to a heat exchanger to produce superheated and pressurized steam, feed to an industrial process or a steam turbine-generator to generate electricity. The collector can be orientated in an East–West direction, and tracking the Sun, along a single axis from North to South. Over one year period, a horizontal North–South through field usually collects more energy than a horizontal East–West oriented collector. The North–South field is collecting a lot of energy during the summer and much less during winter. The East–West field collects more energy in the winter than a North–South field and less in summer, providing a more constant annual output. Therefore, the choice of orientation depends on the application and whether more energy is needed during summer or during winter. PTCs can effectively produce heat at temperatures up to 400°C, being the most mature solar thermal technology to generate heat for electricity generation or process heat applications. However, use of oil-based heat transfer media restricts operating temperatures today to 400°C, resulting in only moderate steam qualities.

In order to provide viable heat production PTCs must achieved their tasks, regardless the input energy fluctuations, i.e., solar radiation, so an effective control scheme is needed to provide the operating requirements for such solar thermal system. Most of the currently operational solar thermal plants provide the thermal energy in the form of super-heated oil by the PTCs. Such PTC systems (fields) are usually connected to electricity generating units or to sea-water desalination facilities, both are more efficient if are operating continuously. In order to do these, both must be provided with constant hot oil at pre-specified temperatures, despite the ambient temperature and solar irradiance variations. Such constrain requires the use of a thermal storage tank as a buffer between the solar collectors and the industrial processes or steam turbine. A new approach is to operate a gas fired boiler in parallel to the solar PTC field in order to compensate for any shortfalls in the solar-produced steam. However, the maintaining a constant supply of solar produced thermal energy is not the main task of the control scheme because is not cost effective strategy, because theoretically by oversizing the solar collector field that can be with system parts only, during low radiation intervals. The control aim is rather to regulate the outlet temperature of the collector field suitably adjusting the oil flow rate, so it is supplying the thermal energy in a usable form at desired operating temperature, improving the overall system efficiency, the solar field is maintained in the readiness state for full-scale operation resumption, while avoiding unnecessary shutdowns and startup procedures which are both wasteful and time consuming. Moreover, if the control strategy is optimum the plant can operate close to design limits, significantly improving its productivity.

**Linear Fresnel reflector** technology relies on an array of linear mirror strips that concentrate light on to a fixed receiver mounted on a linear configuration. The receiver consists of tubes where heat transferring fluid circulates. The LFR field can be imagined as a broken-up parabolic trough reflector, while higher concentration can be achieved. The concentrated solar energy is then used as a heat source for a conventional power plant or industrial processes. The main advantage of this type of system is that it uses flat or elastically curved reflectors which are cheaper compared to parabolic glass reflectors. Additionally, these are mounted close to the ground, thus minimizing structural and mechanical requirements. However, even higher concentrations are obtained, LFR systems are cheaper than parabolic mirrors, but more complex tracking mechanisms are needed. Absorbers are located at the focal point of the mirrors and are consisting of an inverted air cavity with a glass cover enclosing insulated tubes. Notice that quite a wide range of LFR concentration technologies exists or are under developments.

**Tower power plants with central receiver systems** have an array of reflectors, the heliostats that are able to tack the Sun movement and to concentrate the solar radiation onto the central receiver, located at the tower top, in this way dealing with one of the main characteristics of the solar radiation is that it is a diffuse form of energy, meaning that the energy density incident on a given area is quite low and typically less than 1 kW/m², often significantly less. Let us consider a small solar thermal power plant that produces maximum power out of 15 MW with an overall thermal efficiency 25%. We would need approximately 60 MW of heat power for the operation of such power plant. A simple calculation shows that even at the maximum rate of incident radiation of 1 kW/m², at least 60,000 m² of reflecting area are required. This area corresponds to a large number of solar reflectors, placed in a predefined pattern with all of them reflecting the Sun energy to the central receiver. The receiver working fluid is heated and then used for power generation. A SPT consist of the heliostat field, receiver unit, steam and electricity generation units and an integrated control subsystem. Usually, each system unit or component has its specific control device. The integrated control subsystem communicates with different subsystems in order to coordinate the different units in such a way that the SPT operates in a safe and most efficient ways. Central receiver systems are employing large mirror fields, the heliostats, individually oriented to redirect sunlight toward the power tower top. The heat transfer media of the thermal power cycle may be pumped to the top of the central receiver or a mirror can be located at the top of the tower to reflect the light to a ground-level receiver. Configurations using multiple towers have also been proposed and are under research. The HTF, typically a molten salt mixture, is circulated to transport its received heat to a secondary fluid such as water. The fact that the HTF can also be used as a thermal energy storage medium is an advantage, but the high freezing point of a salt requires deployment of heat tracing and its concomitant parasitic energy use.

**Solar dish-engine electricity systems** are usually small energy conversion units compared to other CSP systems. Typical system powers are usually in the range of 5–25 kWe. A parabolic dish engine system consists of a parabolic collector dish, deigned to concentrate only the direct solar radiation that enters the system parallel to its optical axis onto the receiver located at the reflector focal point. The solar dish has to be oriented always toward the Sun through a proper sun-tracking system, and being a point-concentrating system requires a two-axes tracking. Like other point-concentrating systems it reaches very high concentration ratios, which are much higher than the ones of line-concentrating systems. These high concentration ratios, about 2000 are resulting in high receiver temperatures. Realized systems have reached temperatures above 800°C. Such high operating temperatures allow a high thermal-mechanical energy conversion efficiency and, consequently, high solar-to-electric efficiencies. Dish-Stirling systems have reached the highest solar-to-grid peak efficiency among CSP systems, 30% or higher. Nevertheless, under typical conditions they have an average solar-to-electric efficiency between 16% and 25%. The heat conversion to mechanical energy in such systems is done by compressing a working cold fluid, heating the compressed fluid and finally expanding it in a piston-type engine or a turbine to produce work to drive an electrical generator for electricity production. Solar dish-engine systems are modular and have comparatively high conversion efficiencies and can be deployed either individually or in groups. As solar dishes are equipped with Stirling engines or micro gas turbines, an important additional advantage in comparison to Rankine-cycle systems is that the water consumption is minimal. However, there are also even smaller systems for the domestic use. In general, the solar dish-engine systems are equipped with Stirling engines as thermal engines, as prime-mover for electric generator. However, also gas microturbines are used for all power ranges. Because of the reduced size of solar dish-engine systems they are in particular suitable for decentralized, standalone or off-grid applications. Dish technology is the oldest of the solar technologies, dating back to the nineteenth century when a number of companies developed solar powered steam and Stirling-based systems. The geometrical properties of paraboloid mirrors, especially their very good concentrating and collimating properties are well established for a very long period of time, being described by the Greek mathematician and geometer Diocles around 200 BC. Nowadays the dish geometry has a lot of applications: satellite dishes, reflecting telescopes, radio telescopes, parabolic microphones, solar heater and many lighting devices such as spotlights, car headlights, par cans and LED housings.

The Stirling engine was invented in 1816 by the Scottish clergyman and "part-time engineer" Robert Stirling. It is the second oldest thermal engine after the steam engine, developed during eighteenth century. The first solar application of the Stirling engine is attributed to the Swedish-American inventor and mechanical engineer John Ericsson, who built in 1872 the dish-Stirling device. The modern dish-Stirling technology was developed in the late 1970s and in the early 1980s, mainly by the U.S. and European companies and organizations, and such systems are in operation in many countries around the world. Solar dish systems are power conversion units that, as any CSP system, use direct radiation to provide electricity. Their distinctive features are the paraboloid collector and a heat engine (Stirling engine or micro gas turbine) that is connected directly to a receiver, which is located in the focal point of the paraboloid mirror. The main components of a solar dish/engine system are a parabolic collector, a receiver, a heat engine (Stirling engine, micro gas turbine) and a generator. The parabolic collector, which concentrates the direct solar radiation in its focal point, is also referred to as the parabolic dish. The receiver in (or closed to) the focal point of the parabolic dish receives the concentrated radiation and converts it into thermal energy. The heat is subsequently transferred to the thermal engine, where it is converted into mechanical energy. The mechanical energy drives the electric generator that generates the electric energy. Receiver, heat engine and generator are assembled in many actual systems in one integrated and compact unit, the power conversion unit. The power conversion unit is attached to the collector because it always remains in the same position in relation to it (in the focus or slightly behind it). Additional components of a solar dish/engine system are a bearing structure, a tracking system and a control unit. Geometrically, the collector, the solar dish, is a rotationally symmetric section of a rotational paraboloid or some kind of approximation to that. A paraboloid mirror has a focal point in which the direct radiation is concentrated that reaches the mirror parallel to its optical axis. Direct solar radiation, which has a certain beam spread (and is not exactly parallel to the optical axis), is concentrated in a more or less extended focal spot in the focal plane. Like in any other concentrating system, the concentration ratio is one of the central parameters of the collector. It is decisive for the possible operating temperatures of the Stirling engine. However, the geometrical concentration ratio is only an approximation of the real mean radiation concentration. It does not take into consideration the limited reflectivity of the mirror, and the geometrical mirror imperfections that may scatter a part of the incident light away from the receiver aperture. It also does not take into account shading effects on the collector, caused by the energy conversion unit and its bearing structure. A solar dish has the shape of a paraboloid section or an approximation to it. While the geometrical figure of a paraboloid is infinite in its dimensions, a paraboloid mirror covers just a section of it. Thus, in order to define the shape and the size of a paraboloid mirror we need, first, a description of the paraboloid and, second, a description of the section the paraboloid mirror covers.

### 3.6.2 Salt Solar Pond, Solar Chimney Systems, and Solar Furnaces

A *solar pond* is a reservoir or pool of saltwater which collects and stores or transfer solar energy as heat. In a clear natural pond about 30% of the solar radiation reaches a depth of 2 m or so. This solar radiation is absorbed at the bottom of the pond. The hotter water at the bottom becomes lighter and hence rises to the surface, where the heat is lost to the ambient air and, hence, a natural pond does not attain temperatures much above the ambient. If some mechanism can be devised to prevent the mixing between the upper and lower layers of a pond, then the temperatures of the lower layers will be higher than of the upper layers. The simplest method is to make the lower layer denser than the upper layer by adding salt in the lower layers. The salt used is generally sodium chloride or magnesium chloride because of their low cost. Ponds using salts to stabilize the lower layers are called *salinity gradient ponds*. There are other ways to prevent mixing between the upper and lower layers. One of them is the use of a transparent honeycomb structure which traps stagnant air and hence provides good transparency to solar radiation while cutting down heat loss from the pond. The honeycomb structure is made of transparent plastic material. The saltwater naturally forms a vertical salinity gradient also known as a *halocline*, in which low-salinity water floats on top of high-salinity water. The layers of salt solutions increase in concentration (and therefore density) with depth. Below a certain depth, the solution has a uniformly high salt concentration. When the Sun' rays contact the bottom of a shallow pool, they heat the water adjacent to the bottom. When water at the bottom of the pool is heated, it becomes less dense than the cooler water above it, and convection begins. In the case of a solar salt pond, a shallow pool with depth of 2–3 m with salt forms the solar collector for the energy systems, as shown in Figure 3.14.

Solar ponds heat water by impeding this convection to the pond surface. Salt is added to the water until the lower layers of water become completely saturated. High-salinity water at the pond bottom does not mix readily with the low-salinity water above it, so when the bottom layer of water is heated, convection occurs separately in the bottom and top layers, with only mild mixing between the two. This greatly reduces heat loss, and allows for the high-salinity water to get up to 90°C while maintaining 30°C low-salinity water. In a solar pond has three zones with the following salinity with depth:

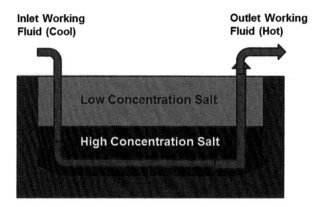

**FIGURE 3.14**
A salt solar pond schematic diagram.

(1) the surface convective zone (0.3–0.5 m), with salinity less than 5% concentration; (2) the non-convective zone (1–1.5 m), where the salinity increasing with depth; and (3) the storage zone (1.5–2 m, and 20% salinity). This hot, salty water then is pumped for use in electricity generation, through a turbine or as a thermal energy. Extraction of thermal energy in the lower layers of the pond can be easily accomplished without disturbing the non-convicting salt gradient zone above. Hot water can be extracted from a solar pond without disturbing the concentration gradient. This is achieved by installing the water outlet at the same height as the water inlet. Hot brine can be withdrawn and cold brine returned in a laminar flow pattern because of presence of density gradient. There are two different solar salt pond implementations: (1) the saturated salt pond and (2) the non-convective gradient pond. In either case, a pipe system, carrying a low boiling point working fluid (e.g., ammonia) is placed near the bottom of the pond to extract the heat (~100°C) for use in a thermal power conversion cycle. It is noteworthy that the solar salt pond itself forms an integral TES mechanism. Solar salt ponds have been demonstrated on a small scale; the largest one was located near the Dead Sea in Israel, initially sized for a 5 MWe output using a 1 km² pond, and operated with 1% solar-to-electric efficiency, until 1989. There are several issues with solar salt ponds, such as water loss due to evaporation, the need to maintain a critical water-salt balance to form the required salt concentration gradient, inefficiency due to the small temperature difference (Carnot efficiency), meaning that the thermal efficiency is very low, and the surface area of the pond needs to be large for a reasonable power output, leading to high capital cost. For small or model ponds because of presence of density gradient, heat exchangers consisting of pipes can be placed in hot lower layers, but this entails not only the initial installation cost but the continued pumping loses associated with the heat transfer fluid. To estimate the performance of the initial temperature of the solar pond, density, conductivity, specific heat capacity, thickness, depth, incoming solar radiation and the time increment for each node are required as input boundary conditions. Since the solar ponds are horizontal solar collectors, the best sites are at low or moderate latitudes, i.e., ±40°. Each potential site has to be evaluated for its geological characteristics and structure, free of fissures and stresses, free moisture, the soil conductivity increases with greatly with moisture, source of cheap slat and water, and in fairly flat area to avoid large construction cost. The solar radiation falling on a solar pond surface is in part reflected at the water surface and in part absorbed at the bottom. Water is a selective absorber of the electromagnetic radiation, so the solar radiation, only shorter wavelengths are reaching the pond bottom.

The solar radiation, penetrating the solar pond surface and the transmittance of the solar radiation through the salt water layers of the pond are calculated through empirical relationships, such as ones given here:

$$h(x) = H_h \cdot \left[0, 36 - 0.08 \cdot \ln(x)\right] \tag{3.55}$$

and

$$\tau(x) = 0.237 \cdot \exp(-0.032 \cdot x) + 0.193 \cdot \exp(-0.45 \cdot x)$$

$$+ 0.167 \cdot \exp(-3 \cdot x) \tag{3.56}$$

$$+ 0.179 \cdot \exp(-35 \cdot x)$$

Here, $H_h$ is the solar radiation reaching the top surface of the solar pond after less the reflective losses, $h(x)$ is the solar radiation that penetrates the solar pond surface, reaching the depth of $x$, and $\tau(x)$ is the transmittance of water at depth $x$.

**Example 3.14:** Find a solar pond transmittance for a depth of 1.05 m.

**Solution:** Applying Equation (3.51b) the transmittance is:

$$\tau(1.05) = 0.237 \cdot \exp(-0.032 \cdot 1.05) + 0.193 \cdot \exp(-0.45 \cdot 1.05)$$

$$+ 0.167 \cdot \exp(-3 \cdot 1.05)$$

$$+ 0.179 \cdot \exp(-35 \cdot 1.05) = 0.357$$

A *solar chimney power* plant is a hybrid plant that utilizes sunlight-induced temperature differences to create a natural draft within a tall tower. The solar chimney's three essential elements—glass roof collector, chimney and wind turbines–have thus been familiar from time immemorial. Air is heated by solar radiation under a low circular glass roof open at the periphery; this and the natural ground below it form a hot air collector. Continuous 24-hour operation is guaranteed by placing tight water-filled tubes under the roof. The water heats up during the daytime and emits its heat at night. These tubes are filled only once, no further water is needed. In the middle of the roof is a vertical chimney with large air inlets at its base. The joint between the roof and the chimney base is airtight. As hot air is lighter then cold air it rises up the chimney. Suction from the chimney then draws in more hot air from the collector, and cold air comes in from the outer perimeter. Thus solar radiation causes a constant up-draught in the chimney. The energy this contains is converted into mechanical energy by pressure-staged wind turbines at the base of the chimney, and into electrical energy by conventional generators. The draft created within the tower is then harnessed using wind turbines, thereby making the solar chimney a hybrid solar–wind system. Thermal mass elements can be placed within the collector region of the facility so that heat storage occurs during daylight, and then during nighttime or inclement conditions the heat is released and the draft continues at a reduced level for some limited time period. A pilot solar chimney plant was constructed at Manzanares, Spain and operated from 1982 to 1988, having 50 kWe power, a tower of 194.6 m height and 10.16 m diameter, with collector area covering 46,800 m² at average height of 1.85 m. The plant produced a temperature rise of 20°C yielding an upward air flow of 9 m/s under load conditions, with overall efficiency less than 1%.

A single solar chimney with a suitably large roof glazed area and a taller chimney can be designed to generate up to 200 MW, 24 h a day, so even with few solar chimneys can replace a large power plant. Solar chimneys operate simply and have a number of advantages:

1. The collector uses all solar radiation, both direct and diffused. This is crucial for tropical countries where the sky is frequently overcast. Other large scale solar thermal power plants, from parabolic through and central receiver systems are using only direct radiation.

2. Due to the heat storage system, the solar chimney operates 24 hours on pure solar energy. The water tubes under the glass roof absorb part of the radiated energy during the day and release it into the collector at night, producing electricity at night as well.

3. Solar chimneys are particularly reliable and not liable to break down, in comparison with other solar generating plants. Turbines, transmission, and generator subjected to a steady flow of air are the plant's only moving parts. This simple and robust structure guarantees operation that needs little maintenance and no combustible fuel.

4. Unlike conventional power stations (or other solar thermal power systems), solar chimneys do not need cooling water, a key advantage in the many areas with major water supply problems.

5. The materials needed for solar chimneys, concrete, and glass, are available everywhere in sufficient quantities. In fact, with the energy taken from the solar chimney itself and the stone and sand available in the desert, they can be reproduced on site.

6. Solar chimneys can be built even in less industrially developed countries. The industry already available in most countries is entirely adequate for their requirements. No investment in high-tech manufacturing plant is needed. Even in poor countries, it is possible to build a large plant without high expenditure by using local resources and work-force, creating large numbers of jobs and reduces the capital investment requirement and the cost of generating electricity.

Solar chimneys convert only a small percentage of the solar heat collected into electricity, and thus have a *poor efficiency*, but this disadvantage is compensated by their cheap, robust construction and low maintenance costs. Solar chimneys need large collector areas, but they are confined to the regions with high solar radiation, which usually have enormous deserts and unutilized areas, so *the land use* is not a particularly significant factor, although of course deserts are complex biotopes that have to be protected. Solar furnaces are made of high concentration and thus high-temperature collectors of the parabolic dish and heliostat type. These solar thermal systems are primarily used for material processing. Solar material processing involves affecting the chemical conversion of materials by their direct exposure to concentrated solar energy. A diverse range of approaches are being researched for applications related to high added-value products such as fullerenes, large carbon molecules with major potential commercial applications in semiconductors and superconductors, to commodity products such as cement. None of these processes, however, have achieved large-scale commercial adoption. Some pilot systems are shortly described here. A solar thermo-chemical process has been developed, which combines the reduction of zinc oxide with reforming of natural gas leading to the co-production of zinc, hydrogen and carbon monoxide. At the equilibrium chemical composition in a black-body solar reactor operated at a temperature of 1250 K at atmospheric pressure with a solar concentration of

2000, efficiencies between 0.4 and 0.65 have been found, depending on product heat recovery. A 5 kW solar chemical reactor has been employed to demonstrate this technology in a high-flux solar furnace. A 2 kW concentrating solar furnace is used to study the thermal decomposition of titanium dioxide at temperatures of 2300–2800 K in an argon atmosphere.

## 3.7 Solar Thermal Energy Storage

In conventional power plants, the fuel (or the uranium) in the storage facilities of the power plant simultaneously represents the energy storage. Additional energy storage is generally not required. The use of an energy storage system in a solar thermal power plant has many advantages: higher annual solar contribution, reduction of part-load operation, power management and buffer storage. An energy storage system has the function of a buffer during the day as the solar irradiation onto the Earth's surface varies with time (e.g., night and day, seasonal changes, diurnal variation and weather). With the aid of the storage the security of the energy supply is increased. In some solar thermal power plants the energy storage system delivers enough energy to operate the plant for several hours after sunset. A solar thermal power plant without an energy storage system can therefore only operate between sunrise and sunset, unless it is hybridized. Hence either an energy storage system and/or fossil co-firing should be integrated into the plant, as presented in Figure 3.15. A key advantage of solar thermal compared to photovoltaics is the capability of integrating thermal energy storage (TES) within a CSP plant. Not only TES systems can allow the plant to continue generating electricity during periods of sunlight loss (i.e., cloudiness), more importantly, electricity production can be continued after sunset and into the evening, which usually corresponds to a peak utility load period. Similarly, solar thermal plants can incorporate auxiliary burners to produce heat at night and during inclement weather. In the near term, TES is expected to provide capacity factors approaching 40%, and in the long term up to 70%. Solar thermal power plant comprises *power plants* which first convert *solar radiation* into *heat*. The resulting *thermal energy* is subsequently

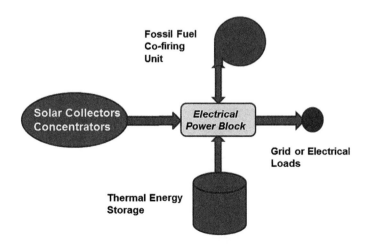

**FIGURE 3.15**
A solar energy system with thermal energy storage and fossil fuel co-firing unit.

transformed into *mechanical energy* by a thermal engine, and then converted into *electricity*. Processes in a solar thermal power generation system consists of concentrating solar radiation through the collector systems, increasing radiation flux density (concentrating of the solar radiation onto a receiver), absorption of the solar radiation (conversion of the solar radiation energy into thermal energy, heat inside the receiver), and the transfer of thermal energy to an energy conversion unit, conversion of thermal energy into mechanical energy using a thermal engine (e.g., steam turbine) and then conversion of mechanical energy into electrical energy using a generator.

When using a thermal energy storage and fossil back-up system, CSP plants can be operated as base load power plants, which is an advantage, being for example important for seawater desalination since such facilities are operating in base load modes. However, without thermal energy storage and fossil back-up system, CSP has no any advantages against PV, because CSP technologies are more complicated than PV, lacking also the PV modular expandability. With such technologies, CST systems can cover a range of scale, capacity, applications and costs, while each combination having its own advantages and disadvantage. Since solar energy availability varies by the time of the day, month or year (seasonal), and often has intermittency during one day due to the weather conditions, the energy storage systems can be of short-term or long-term type depending on the application:

1. Buffering with the objective of overcoming cloudiness for short periods from a few minutes to a couple of hours;
2. Delivery period displacement, such as energy collection during the day for delivery during peak load period of evening hours or during night;
3. Delivery period extension by storing enough energy during the day to have stored energy available for hours before sunrise and after sunset or even for longer times of non-sunshine hours of a day to a number of days; and
4. Seasonal storage, which would store heat during summer for usage during winter.

There two major types of the thermal energy storage systems: *sensible heat* and *latent heat* types. The term sensible heat describes the heat which is absorbed or released by a material as a result of a change in temperature, whereupon the material does not undergo a change of aggregate state. When storing latent heat, the property of materials to absorb or release heat energy during a phase change is used. Latent heat storage systems therefore use a phase change material (PCM) as storage medium. In principle there are three possible phase changes: solid—solid, solid—liquid, and liquid—vapor thermal energy systems. The phase change solid—solid and liquid—vapor types are seldom used in practical applications. In the present days, the latent heat storage systems, the phase change solid—liquid plays an important role in the TES applications. Both, sensible heat and latent heat are discussed in a later chapter. The current thermal energy storage types can be divided into four main groups: (1) thermal energy storage systems for sensible heat, further divided into: (a) *indirect storage systems*, and (b) *direct storage systems*; (2) latent heat thermal energy storage; (3) steam accumulator; and (4) thermo-chemical energy storage system. There are four indirect storage concepts, introduced here, that are often employed in solar thermal applications: 2-tank molten salt indirect thermal energy storage, packed-bed thermal energy storage (regenerator); sand thermal energy storage, and concrete heat storage.

The 2-tank molten salt indirect storage system, a commercially available technology, is based on nitrate salts, is mainly in parabolic trough power plants, being one the most widely used technology in solar thermal power generation. The receivers of the collectors carry temperature-stable synthetic oil, the heat transfer fluid, which is heated to 400°C. In a boiler the heat from the heat transfer fluid is passed to the steam cycle. The generated steam drives a steam turbine, which in turn drives a generator for electricity production. As the storage medium is different than the HTF, a heat exchanger is used so that the heat from the heat transfer fluid is passed for storage. The indirect storage system comprises two tanks: a hot-salts and cold-salts tank, which are filled with altogether 28,500 tons of molten salt, allowing peak load operation for up to 8 hours after sunset. In summer, it is possible to run the turbine nearly 24 hours a day. The composition of the salt is 60% sodium nitrate ($NaNO_3$) and 40% potassium nitrate ($KNO_3$). During the day the hot synthetic oil from the collectors is not only passed through the boiler but also through a heat exchanger for passing the heat to the (separate) storage cycle. When charging, the salt from the cold-salts tank is pumped through the heat exchanger, upon which the salt is heated to approximately 386°C and is pumped into the hot-salts tank. During the night or in periods with low solar irradiance, the flow direction is reversed. The salt is pumped from the hot-salts tank through the heat exchanger into the cold-salts tank. In the heat exchanger the salt passes the heat to the synthetic oil. The heated synthetic oil is then passed through the boiler for producing steam. The power plant control system ensures that the salt never falls below a temperature of 292°C, if necessary by means of electrical heating, to prevent the salt from solidifying. Solidification would take place at approximately 220°C. Packed-bed thermal energy storage (pebble bed or rock pile storage) consists of a container filled with a bed of loosely packed particulate material with a high heat capacity e.g., pebbles, gravel or rocks. This storage type was designed to utilize air as the heat transfer fluid. However, it would also be possible to use liquid media. At both the top and bottom of the storage there is a duct through which the air is forced. When charging, the hot air enters the storage through the top duct, passes through the pebble bed transferring the heat to the storage material, and leaves the storage through the bottom duct. In the discharge process, the direction of the air circulation is reversed. Cold air enters the storage through the bottom duct and is heated up as it travels upwards to the top where it exits through the top duct. The hot air can then be passed through a steam generator and eventually the electricity is generated. The solar tower technology, using air as heat transfer fluid in its primary cycle is often built with a similar packed-bed thermal energy storage system. The sand storage is particularly important in the future when solar tower power plants are being constructed near to or in sand-desert regions. The advantages of sand thermal energy storage systems are: low cost, high storage capacity, suitable for high temperatures, and environmentally friendly. Similar to the packed-bed thermal energy storage, the sand thermal storage is also utilizing the air from the primary cycle of a solar tower power plant with air receiver as the thermal transfer fluid. As well as using regenerators as thermal energy storage using ceramics or concrete as storage material, it is also possible to utilize sand as a cheaper alternative. Sensible heat, $Q_{SH}$, is stored in a material of mass $m$ and specific heat $C_P$ by raising the temperature of the storage material from $T_1$ to $T_2$. Thermal energy stored as latent heat in a material that undergoes phase transformation at a temperature that is useful for the application. If a material with phase change temperature $T_m$ is heated from $T_1$ to $T_2$, such that $T_1 < T_m < T_2$, then the thermal energy $Q_{LH}$ is stored in a mass $m$ of the material. The types of thermal energy storage are expressed by these two equations:

$$Q_{SH} = \int_{T_1}^{T_2} m \cdot C_P \cdot dt = \int_{T_1}^{T_2} \rho \cdot V \cdot C_P \cdot dt \tag{3.57}$$

and

$$Q_{LH} = \int_{T_1}^{T_m} m \cdot C_{P(1)} \cdot dt + m \cdot \lambda + \int_{T_m}^{T_2} m \cdot C_{P(2)} \cdot dt \qquad (3.58)$$

Here $\rho$ and $V$ are density and volume of the sensible heat storage material, and $\lambda$ is the heat of phase transformation. For moderate temperature changes, such as for solar space and water heating systems, the density and specific heat may be considered constants, therefore, $Q = \rho \cdot V \cdot C_P \cdot \Delta T$. Most common sensible heat storage materials are water, organic oils, rocks, ceramics, and molten salts. Water has the highest specific heat value of 4190 J/kg°C. The most common medium for storing sensible heat for use with low- and medium-temperature solar systems is water. Water is cheap and abundant and has a number of particularly desirable properties. The most common PCMs used for solar energy storage undergo solid–liquid transformation. For such materials, the thermal energy stored is written from Equation (3.38) as:

$$Q_{LH} = m \cdot \left[ C_{P(s)} \cdot (T_m - T_1) + \lambda + C_{P(l)} \cdot (T_2 - T_m) \right] \qquad (3.59)$$

where $C_{P(s)}$ and $C_{P(l)}$ are the average specific heats in the solid and liquid phases, respectively.

> **Example 3.15:** A company is interested in 2 hours of buffer storage for a 1.5 $MW_{th}$ solar thermal power plant that operates between 285°C and 395°C. Estimate the amount of material that is needed if sodium nitrate, $NaNO_3$, is used as the latent heat storage medium.
>
> **Solution:** The melting temperature, average solid and liquid specific heats, and the latent heat of sodium nitrate can be found from Table A2 are: 307°C, 1.27 kJ/kg°C, 1.64 kJ/kg°C, 177 kJ/kg. The energy that is required for 3 hours of storage is:
>
> $$Q_{Th(3-h)} = 3 \text{ h} \times 1500 \text{ kW} = 4500 \text{ kWh} = 4500 \text{ kWh} \times \frac{3600}{h} = 16.2 \times 10^6 \text{ kJ}$$
>
> The mass of material is then computed by using Equation (3.39) as:
>
> $$m = \frac{16.2 \times 10^6}{1.27 \cdot (307 - 285) + 177 + 1.64 \cdot (395 - 307)} = 46,618.7 \text{ kg or } 46.619 \text{ tons}$$

## 3.8 Summary

Solar energy technology embraces a family of technologies capable of being integrated amongst themselves, as well as with other renewable energy technologies. The solar energy technologies can deliver heat, cooling, electricity, lighting, and fuels for a host of applications. Solar thermal energy systems are increasingly popular due to the characteristics, performances, low cost, easy operation and maintenance and volatility of the energy costs. The most common types of solar collectors are presented and discussed in this chapter. Solar thermal energy systems are used to heat and cool buildings, by using both active and passive methods, to provide hot water for domestic use, industrial processes, or swimming pools, to operate engines and pumps, for the desalinate water for drinking purposes, to generate electricity, for chemistry applications and many more applications. However, at the present moment two major methods exist by which sunlight can be converted into directly usable energy: conversion to thermal energy, then to electricity or directly conversion to electricity, by photovoltaic effect. The various types of solar collectors described here include flat-plate, evacuated tube, parabolic trough, Fresnel lens, parabolic dish and heliostat field collector. Solar thermal collector technologies in the first phase are relatively mature. Well-performing, affordable, standard collectors for hot water heating are on the market with efficiencies close to the physical limits and which are durable and of high quality. However, since research and development in the solar thermal technology was rather small over the last decades, there is large room for further improvements for specific application areas and new concepts. The optical and thermal analysis of the solar collectors is briefly presented as well as methods to evaluate their performances. Solar energy conversion to heat (thermal conversion) is straightforward, because any material object placed in the Sun absorbs solar energy. However, maximizing and maintaining that absorbed energy can take specialized techniques and devices such as vacuums, phase-change materials, optical coatings, and mirrors. Which technique is used depends on the application and temperature range at which the heat is to be delivered, and this can range from 25°C (e.g., for swimming pool heating) to 1000°C (e.g., for dish/Stirling solar thermal electrical power), and even over 3000°C in solar furnaces. Generation of electricity can be achieved in either of two ways. Solar energy con be converted directly into electricity in a solid-state semiconductor device, via photovoltaic cell, or the solar energy is used in a concentrating solar power plant to produce high-temperature heat, then converted to electricity via a heat engine and generator. Both approaches are currently in use. Power plants of 200 MWe power or even larger are feasible to design and build based on the current demonstrated technology to date. Solar-fossil hybrids are likely to be the next step in development of this technology. It is also noteworthy that besides electricity generation, CSP facilities have been proposed for other uses including desalination, coal gasification, water splitting to yield hydrogen and destruction of hazardous chemicals. The solar energy use for lighting requires no conversion per se; sunlight occurs naturally in buildings through windows, but maximizing the effect requires careful engineering and architectural design. In addition, the passive solar heating is a technique for maintaining buildings at comfortable conditions by exploiting the solar rays that are incident on the buildings' exterior, without using pumps and fans. Solar cooling for buildings can also be achieved, for example, by using solar-derived heat to drive a special thermodynamic cycle called absorption refrigeration. Furthermore, solar devices can deliver process heat and cooling, and other solar technologies are being developed that will deliver fuels such as hydrogen or hydrocarbons, The types of solar collectors are often classified by the following criteria: (a) location, i.e., space-based solar power or surface units, (b) conversion method, i.e., photovoltaic or thermal, (c) fixed or tracking, i.e., stationary or following the Sun and (d) if tracking then number of axes of rotation of the device, i.e., single axis or two axes. Refractive optics is used to concentrate the Sun's irradiance onto a receiver. A square Fresnel lens, incorporating circular facets, is used to turn the Sun rays to a central focal point. A receiver that is mounted at this focal point is receiving a much larger amount of solar energy and very high temperatures can be obtained. The tracking of a collector or panel can be full, tilted N-S axis with tilt adjusted daily, polar N-S axis with E-W tracking, horizontal E-W axis with N-S tracking or horizontal N-S axis with E-W tracking. The variation of solar flux for each case depends on a local Sun path diagram, and the method of tracking used. To this local parameters like shading and shadows, due for example to overhangs or local build-ings, have to be factored in. One of the beautiful characteristics of solar equipment and solar thermal applications is that it can be made in varying degrees of perfection and in a wide range of sizes and costs.

## 3.9 Questions and Problems

1. What types of the energy storage options are available for solar energy applications?
2. What are the modes of heat transfer? Briefly explain the physical mechanism of each mode.
3. List the major types of solar collectors.
4. Why the energy storage units are needed in many of the solar energy conversion systems?
5. Briefly describe the properties and characteristics of each of the main types of the solar collectors.
6. What are the differences between the active and passive solar energy systems?
7. What are the components of a flat-plate solar collector and the ones of an evacu-ated-tube solar collector?
8. Briefly explain how a solar thermal collector works.
9. List the advantages and disadvantages of the active and passive solar energy systems.
10. The transmittance of a transparent slab, such as the window glass, depends on what: (a) reflectivity, (b) the incidence angle, or (c) both of them.
11. What is a heliostat and central receiving tower?
12. What are the reasons for using energy storage in solar energy conversion systems?
13. What are the modes of heat transfer? Briefly explain the physical mechanism of each mode.
14. List the main advantages of the concentrating solar collectors compared to the conventional solar collectors.
15. Why are energy storage units needed in many passive or active solar thermal energy systems?
16. The solar collectors of a weather station, located at latitude of 30° N, facing south have a surface of 45 m². What is the solar energy absorbed in 6 hours?
17. Calculate the zenith and solar azimuth angles for latitudes of 28.5°, 36° and 45° at (a) 10:30 AM on February 21 and (b) 6:00 PM on July 1.
18. Determine for Phoenix, Arizona, which is located at 112° W longitude and latitude of 33.43° N, the solar altitude and azimuth angles.
19. For the conditions of the previous problem, find the collector angle for a wall that faces east-southeast and is tilted at an angle equal to the location (site) latitude.
20. What are the components of a flat-plate solar collector?

21. Determine the daily solar radiation in $J/m^2$ on a horizontal surface, $H_0$ in the extra-terrestrial region at latitude 36° N on August 15, 2017.

22. Calculate the sunrise time, solar altitude, surface zenith, solar azimuth, and profile angles for a 45° inclined surface facing 30° West of South at 2:00 PM solar time on April 21st, located at latitude of 36°. Also calculate the surface sunrise and sunset times.

23. Determine the total solar radiation on a horizontal surface for 12.30 PM of the monthly average daily hours of bright sunshine, observed at tamp, Florida, USA in May 2013.

24. A 2.5 $m^2$ south-facing solar panel tiled at 20° is located at 35° N latitude and 90° W longitude. Determine the incidence angle and the total daily solar radiation for June 2nd at 9:30 AM, 11:30 AM, 1:30 PM and 3:30 PM.

25. What is the ratio of beam radiation to that on a horizontal surface for a solar panel located at Madison, Wisconsin, at 11:30 (solar time) on February 22 if the surface is tilted 45° from the horizontal and is pointing 15° west of south?

26. Calculate $R_b$ for a surface at latitude 43° N at a tilt 36° toward the south for the hour 9–11 solar time on March 12.

27. Calculate the beam, diffuse, reflected, and total solar radiation on a south-oriented tilted surface inclined at 40° at a location having latitude of 40°N for 9:30 AM 11:30 AM 1:30 PM and 3:30 PM on May 21, June 21, and August 21, 2015.

28. Calculate the angle of incidence of beam radiation on a surface located at Cleveland, Ohio 41° N, 81° W, on February 10, at 10 AM and on July 10, at 2 PM if the surface is oriented 20° east of south and tilted at 40° to the horizontal.

29. The solar collector efficiency depends on: (a) thermal conductivity, (b) working fluid density (c) specific heat, and (d) none of these.

30. A solar water heater is equipped with an effective collector area of 1.8 $m^2$, and the daily cumulative solar insolation onto the collector is 4 $kWh/m^2$-day in February. If the average efficiency of the solar water heater is 60%, how many kilo-calories (kcal) of heat can be collected by this solar water heater during a day? (Note: 1 cal = 4.186 J = 4.186 W·s).

31. If the minimum heat demand is 8100 kcal/day, and there is a certain solar thermal system which can offer a heat supply of 1720 $kcal/m^2$ in a day. With the absence of auxiliary heating device, calculate the required installation area of the solar panel. If the effective area of this solar panel is 0.8 $m^2$/panel, how many solar panels should be installed to satisfy this heat demand?

32. A flat-plate collector, with an area of 6.3 $m^2$ is tested during the night to measure the overall heat loss coefficient. Water at 63°C circulates through the collector at a flow rate of 0.062 l/s. The ambient temperature is 7.5°C and the exit temperature is 45°C. Determine the overall heat loss coefficient.

33. Determine the daily solar radiation in $J/m^2$ on a horizontal surface, located at 30° N and 40° N latitudes on April 30, July 30 and October 30, respectively.

34. For the FPC collector have the performance characteristics of Figure 3.7 (upper (top) line in the diagram), having glass cover transmissivity, 0.91 and the surface absorptivity 0.92, find: the collector heat removal factor, the overall conductance, $U_L$, and the rate at which the collector can deliver useful energy when the incident solar irradiation is 250 $Btu/ft^2 \cdot h$.

35. Calculate and compare the efficiency of the two flat-place collectors, operating in an area with 220 Btu/ft²·h and ambient temperature of 40°F, having the following characteristics:
    1. Collector A: $\tau\cdot\alpha = 0.95$, $U_L = 6.10$ Btu/ft²·°F, and $T_{avg} = 70°F$.
    2. Collector B: $\tau\cdot\alpha = 0.91$, $U_L = 5.00$ Btu/ft²·°F, and $T_{avg} = 65°F$.

    Hint: Assume an ideal heat removal factor and use Equation (3.27).

36. A flat-plate solar collector, with an aperture are of 4.5 m², tested for direct solar radiation normal to the collector plane with the following test data:

| $Q_U$ (MJ/h) | $G_T$ (W/m²) | $T_{in}$ (°C) | $T_{amb}$ (°C) |
|---|---|---|---|
| 9.05 | 880 | 18.5 | 10.0 |
| 1.13 | 880 | 30.0 | 10.0 |
| 1.34 | 880 | 50.5 | 10.0 |
| 1.59 | 885 | 66.0 | 10.0 |
| 1.98 | 890 | 84.0 | 10.0 |

    Determine the $F_{HRF}(\tau\alpha)$ and $F_{HRF}\bullet U_L$ for this solar collector from the line of best fit of the test data, based upon the aperture area.

37. Compute and plot for a absorber temperature range from 0°C to 1500°C, in a 10°C increment, the efficiency of a solar concentrating system, characterized by the numerical values of the Example 3.11 and for the following geometric concentrating factors (aperture ratio), 10, 50, 75, 100, 500, 1000, and 2000.

38. Two collectors are used to provide hot water, one a flat-plate solar collector with parameters (Equation 3.29): $\eta_0 = 0.65$, $a_1 = -4.28$, $a_2 = -0.031$, and an evacuate-tube solar collector with parameters: $\eta_0 = 0.33$, $a_1 = -1.15$, $a_2 = -0.011$, all SI units. Determine the temperatures above which the evacuate-tube solar collector performs better, for irradiations 900, 600, and 300 W/m².

39. The average insolation on a flat-plate solar collector at a specific location is 7500 kJ/(m²·day). This solar collector is used to continuously provide 39,500 Btu/h or heat to an industrial process. If the collector efficiency is 43.5%, what is the area of this solar collector?

40. A central receiver thermal power plant is located in a region with an annual average peak insulation of 750 W/m² is consisting of 1500 heliostats, each with area of 24 m². If the plant solar collection efficiency is 59.5% and the power cycle efficiency is 37.5%, what is the overall annually averaged solar plant efficiency?

41. A 12 MW (peak) solar thermal power plant is proposed for Baton Rouge, Louisiana (30.5° N, 91.2° W). The overall power plant efficiency is estimated to be 26.5%, determine the minimum and the optimum area of the heliostats that must be used for this development.

42. Calculate the solar radiation intensity, penetrating a solar pond at level 0.85 m, if the surface intensity is 670 W/m² and the water surface reflection losses count for 7.6%.

## References and Further Readings

1. F. Bueche, *Introduction to Physics for Scientists and Engineers*, McGraw-Hill, New York, 1975.
2. K. L. Coulson, *Solar and Terrestrial Radiation*, Academic Press, New York, 1975.
3. B. Anderson, *Solar Energy: Fundamentals in Building Design*, McGraw-Hill, New York, 1977.
4. J. P. Lunde, *Solar Thermal Engineering*, John Wiley & Sons, New York, 1980.
5. J. C. Zimmerman, Sun Pointing Programs and Their Accuracy, Sandia National Laboratories Report SAND81-0761, September 1981.
6. E. E. Anderson, *Fundamentals of Solar Energy Conversion*, Addison-Wesley, Reading, MA, 1982.
7. A. Rabl, *Active Solar Collectors and their Applications*, Oxford University Press, New York, 1985.
8. M. Iqbal, *An Introduction to Solar Radiation*, Academic Press, Toronto, ON, 1983.
9. J. A. Duffie and G. A. Backman, *Solar Engineering of Thermal Processes*, John Wiley & Sons, New York, 1991.
10. B. Norton, *Solar Energy Thermal Technology*, Springer-Verlag, London, UK, 1992.
11. J. Gordon (Ed.), *Solar Energy: The State of the Art*, James and James, London, UK, 2001.
12. I. Dincer and M. Rosen, *Thermal Energy Storage: Systems and Applications*, John Wiley & Sons, New York, 2002.
13. F. P. Incropera and D. P. DeWitt, *Introduction to Heat Transfer*, John Wiley & Sons, New York, 2002.
14. R. Siegel and J. R. Howell, *Thermal Radiation Heat Transfer* (4th ed.), Taylor & Francis Group, New York, 2002.
15. M. Norm, *Methods of Testing to Determine Thermal Performance of Solar Collectors*, ANSI/ASHRAE Standard 1993–2003, Washington, DC, 2003.
16. F. P. Incropera, D. P. Dewitt, T. L. Bergman, and A. S. Lavine, *Introduction to Heat Transfer*, (5th ed.) John Wiley & Sons, New York, 2007.
17. F. Kreith, and D. Y. Goswami, *Handbook of Energy Efficiency and Renewable Energy*, CRC Press, Boca Raton, FL, 2007.
18. D. Y. Goswami and F. Kreith, *Energy Conversion*. CRC Press, Boca Raton, FL, 2008.
19. M. R. Patel, *Wind and Solar Power Systems* (2nd ed.), CRC Press, Boca Raton, FL, 2006.
20. G. Boyle, *Renewable Energy—Power for a Sustainable Future*, Oxford University Press, Oxford, UK, 2012.
21. V. Quaschning, *Understanding Renewable Energy Systems*, Earthscan, London, UK, 2006.
22. R. A. Ristinen and J. J. Kraushaar, *Energy and Environment*, John Wiley & Sons, Hoboken, NJ, 2006.
23. J. Andrews and N. Jelley, *Energy Science, Principles, Technology and Impacts*, Oxford University Press, Oxford, UK, 2007.
24. A. Vieira da Rosa, *Fundamentals of Renewable Energy Processes* (2nd ed.), Academic Press, Waltham, MA, 2009.
25. B. K. Hodge, *Alternative Energy Systems and Applications*, John Wiley & Sons, Hoboken, NJ, 2010.
26. C. J. Chen, *Physics of Solar Energy*, John Wiley & Sons, Hoboken, NJ, 2011.
27. B. Everett and G. Boyle, *Energy Systems and Sustainability: Power for a Sustainable Future* (2nd ed), Oxford University Press, Oxford, UK, 2012.
28. F. Kreith and J. F. Kreider, *Principles of Sustainable Energy*, CRC Press, Boca Raton, FL, 2012.
29. J. A. Duffie and W. A. Beckman, *Solar Engineering of Thermal Processes* (4th ed.), John Wiley & Sons, Hoboken, NJ, 2013.
30. N. Enteria and A. Akbarzadeh, *Solar Energy Sciences and Engineering Applications*, CRC Press, Boca Raton, FL, 2013.
31. S. A. Kalogirou, *Solar Energy Engineering: Processes and Systems*, Academic Press, Amsterdam, the Netherlands, 2013.

32. D. Y. Goswami, *Principles of Solar Engineering*, CRC Press, Boca Raton, FL, 2013.
33. R. A. Dunlap, *Sustainable Energy*, Cengage Learning, Stamford, CT, 2015.
34. V. Nelson and K. Starcher, *Introduction to Renewable Energy (Energy and the Environment)*, CRC Press, Boca Raton, FL, 2015.
35. G. N. Tiwari, A. Tiwari, and Shyam, *Handbook of Solar Energy Theory, Analysis and Applications*, Springer, Singapore, 2016.
36. M. Martín (ed.), *Alternative Energy Sources and Technologies*, Springer International Publishing Switzerland, Switzerland, 2016.

# 4

## Photovoltaic Systems and Applications

### 4.1 Introduction

PV systems are an empowering technology allowing new and old applications in a better and in more sustainable way, by generating direct current (DC) electrical power from semiconductors or other materials when they are illuminated by light. As long as light is shining on the solar PV cell (the individual element of a PV system) generates electricity. A solar or PV cell converts solar radiation directly into electricity, by absorbing light the electrons of some materials (semiconductors or special organic materials) acquire enough kinetic energy to move from valence band to conduction band within the material. Semiconductors are the materials that have shown the best performance of photovoltaic effect. The PV panels and modules consist mainly of semiconductor materials, with Silicon being the most commonly used. One of the major tasks in controlling PV cells for power generation is improving cell efficiency and maximizing energy extraction. The solar panels are only a part of a complete PV solar system, and are the heart of the PV systems. One must have also mounting structures to which PV modules are fixed and directed toward the Sun. First practical use of solar cells was the generation of electricity on the orbiting satellite Vanguard 1 in 1958. The solar cells were made from single crystal silicon wafers, with an efficiency of 6%. For quite some time, the PV system space applications were the only ones. However, the energy crisis in the seventies of the twentieth century accelerated the search for new energy sources for terrestrial applications, resulting in a growing interest for PV solar energy and other renewable energy sources. The major obstacle of using solar cells for terrestrial electricity generation has been a much higher solar electricity generation cost, when compared to the price of conventional electricity generation. Therefore, there has been much effort to reduce the cost of solar electricity to levels comparable to the conventional electricity.

Large-scale use of PV solar energy, which is considered an environmentally friendly energy source, can lead to a substantial decrease in the pollutant emissions due to the burning of the fossil fuels. However, when we closely look at the contribution of the PV solar energy to the total energy production in the world we see that the PV solar energy contribution is only a tiny part of the total energy production. At present, the total energy production is estimated to be $1.6 \times 10^{10}$ kW compared to $1.0 \times 10^6$ kWp that can be delivered by all solar cells installed worldwide. By Wp (Watt peak) we understand a unit of power that is delivered by a solar cell under a standard illumination. Some of the advantages and disadvantages of photovoltaics are listed here. Note, that they include both technical and nontechnical issues. Often, the advantages and disadvantages of photovoltaics are almost completely opposite of conventional fossil-fuel power plants. Major PV system advantages include: fuel source, the Sun is vast and essentially infinite, so no use of fuels or water, there are virtually no pollutant emissions, no combustion processes or no radioactive fuel for disposal, low operating costs (no fuel involved into the PV system operation), no noise, no moving parts (making a PV system very robust and resilient), PV are operating at ambient temperature (no high temperature corrosion or safety issues), high reliability of PV modules (average lifetime about 20 years or more), modularity (small or large increments), quick and easy installation, can be integrated into new or existing building structures, can also be installed at nearly any point-of-use, daily PV output peak may match local demand and last a high public acceptance. PV electricity is generated wherever there is light, solar or artificial, PV system can operate even in cloudy weather conditions. When PV starts to make a substantial contribution to the energy production and consequently to the decrease in the gas emissions depends on the growth rate of the PV solar energy production and market. However, the solar cells and PV solar panels are already on the market. One of the major advantages of the PV solar systems is that the solar panels are modular and can be combined, wired and connected together to deliver exactly the required power, in a "custom-made" or "application-configuration" topology. The reliability and small operations and maintenance costs, modularity and expandability, are major advantages of PV solar energy systems in many rural or in remote energy generation applications. Among the major disadvantages are: PV systems need light to operate, unlike a solar thermal panel which can tolerate some shading, PV modules are sensitive to shading (PV modules can be affected considerably even by shading of the tree branch, and if enough cells are hard shaded, a module is not converting energy and become a sink of energy on the entire system), high initial installation costs, overshadowing the low maintenance costs and no fuel costs, large area needed for large scale applications, PV systems generate direct current (DC), so DC appliances or inverters are needed in off-grid applications, and energy storage is needed to offset the generation variability and intermittency.

The PV module output depends on sunlight intensity and cell temperature; therefore components that condition the DC output and deliver it to batteries, energy storage units, grid, and/or loads are required for a smooth operation of the PV systems. These PV system components are referred to as charge regulators or power conditioning units. For applications requiring AC power, DC-AC inverters are needed. All these additional components of a PV system are called balance of system (BOS). Finally, equipment, electric motors, the household appliances, such as radio or TV set, lights, etc., being powered by the PV solar system are called electrical loads.

## 4.2 Photovoltaic Basics, Operation, and PV Cell Materials

The solar energy conversion into electricity takes place in a semiconductor device that is called a solar cell. A solar cell is a unit that delivers a certain amount of electrical power, characterized by an output voltage and current. In order to use solar electricity for practical devices, which require a certain voltage or current for their operation, several solar cells are connected together to form a solar PV module and panel. For large-scale generation of solar electricity the solar panels are connected together into a solar array. PV panels are part of a complete PV solar system, which, depending on the application, may comprise electricity storage unit(s), DC-AC inverters that connect a PV solar system to the electrical grid, power condition and control unit(s), Sun tracking systems, and other miscellaneous electrical components or mounting elements. PV cell energy conversion consists of two essential steps: absorption of light generating an electron-hole pair, while the electrons and holes are then separated by the device structure, with electrons to the negative terminal and holes to the positive terminal, thus generating electricity. The basic processes due to the photovoltaic effect are: *the generation of the charge carriers due to the absorption of photons in the materials that form the junction, subsequent separation of these charge carriers in the junction, and finally the collection of the photo-generated charge carriers at the junction terminals.* PV cell structure consists of an absorber layer, in which the incident radiation photons are efficiently absorbed resulting in the creation of electron-hole pairs. In order to separate the photo-generated electrons and holes from each other, the so-called "semipermeable membranes" are attached to the both sides of the cell absorber. The important requirement for the semi-permeable membranes is the selectively, allowing only one charge carrier type to pass through. An important issue for and efficient PV cell design is that the electrons and the holes generated in the absorber layer are reaching the membranes, requiring that the absorber layer thickness is smaller than the charge carrier diffusion lengths. A membrane that let electrons go through and blocks holes is a material with large electron conductivity and small hole conductivity. An example of such material is an n-type semiconductor, having much larger electron conductivity than the conductivity of the holes, caused by a large difference in electron and hole concentrations. Electrons can easily flow through the *n*-type semiconductor while the hole transport, the minority charge carriers in such material, due to the recombination processes is very limited, while the opposite holds in a *p*-type semiconductor. The total current density flowing through the *p-n* junction in the steady state is constant across the junction therefore we can determine the total current density as the sum of the electron and hole current densities, $J_n$ and $J_p$, at the edges of the depletion region, expressed as:

$$J = J_n + J_p \qquad (4.1)$$

Solar PV cells consist of a *p-n* junction fabricated in a thin semiconductor layer as shown in Figure 4.1. Electrons can be located in either the valence band or the conduction band, depending on the structure absolute temperature and/or external excitations like radiation photons. Semiconductors, characterized as being perfect insulators at absolute zero temperature, become increasingly conductive as temperature is increased. As temperature

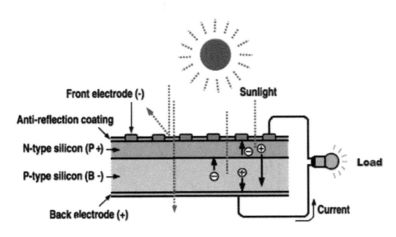

**FIGURE 4.1**
PV cell diagram.

becomes greater, sufficient energy is transferred to a fraction of electrons, causing them to move from the valence band to the conduction band and holes to be generated into the valence band. The increase in temperature responsible for this entire process is a direct result of external energy. In PV systems, the incident photons due to illumination a source of energy are the external forcing. Initially, the electrons in the semiconductor fill up the valence band but when light hits the semiconductor, some electrons acquire enough energy to move to the conduction band, where they can move freely, and may create an electric current. The electron leaving the valence band leaves a positively charged hole behind and the valence band is no longer full, aiding the current flow. Usually solar cell materials are doped in a controlled way to reduce the energy required for the electrons to move from the valence band to the conduction band. The main parameters that are used to characterize the performance of solar PV cells are the peak power, $P_{max}$, the short-circuit current density, $J_{SC}$, the open-circuit voltage, $V_{OC}$, and the fill factor, $FF$. These parameters are determined from the illuminated $J$-$V$ characteristic diagram as one illustrated in Figure 4.2. The cell conversion efficiency, $\eta$, can be determined from these parameters.

The *short-circuit current*, $I_{SC}$, is the current that flows through the external circuit when the electrodes of the solar cell are short circuited. The short-circuit current of a solar cell depends on the photon flux density incident on the solar cell, determined by the spectrum of the incident light. For the standard solar cell measurements, the AM1.5 spectrum is used. Short-circuit current depends on the area of the solar cell, so the short-circuit current density is often used to describe the solar cell maximum current. The maximum current that the solar cell can deliver strongly depends on the optical properties (absorption in the absorber layer and total reflection) of the solar cell. In the ideal case, the $J_{SC}$ is equal to the $J_{ph}$, considering the diffusion length and the charge carrier lifetimes, no surface recombination, and uniform generation. Crystalline silicon solar cells can deliver under an AM1.5 spectrum a maximum current density of about 46 mA/cm². In laboratory test conditions for the type cSi solar cells the measured $Jsc$ is about 42 mA/cm², while commercial solar cells can have $J_{SC}$ over 35 mA/cm². The *open-circuit* voltage is the voltage at which no current flows through the external circuit. It is the maximum voltage that a solar

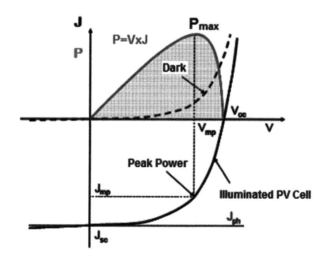

**FIGURE 4.2**
J-V characteristics of a *p-n* junction under dark and illumination conditions.

cell can deliver. The $V_{OC}$ corresponds to the forward bias voltage, at which the dark current compensates the photo-current. The $V_{OC}$ depends on the photo-generated current density and can be calculated, by assuming that the PV cell net current is zero as:

$$V_{OC} = \frac{kT}{q} \ln\left(\frac{J_{ph}}{J_{SC}} + 1\right) \qquad (4.2)$$

Regardless the PV cell types, developed using different semiconductor materials, the operating principle is the same. The most commonly PV cell structure is configured as a large-area silicon *p-n* junction. When a piece of *p*-type silicon is placed in contact with an *n*-type silicon one, electron diffusion occurs from the high electron concentration (*n*-type side) region into the low electron concentration region (*p*-type side), while holes flow in the opposite direction by diffusion processes. These charge carriers flow from the diffusion current $J_D$. The electrons diffusing across the *p-n* junction, recombines with holes on the *p*-type side. This process does not happen indefinitely due to the electric field created by the charge imbalance on either side of the junction. This electric field established across the *p-n* junction generates the drift current $J_S$, that opposes and eventually balances out the diffusion current $J_D$. The region where electrons and holes have diffused across the junction is called the *depletion zone*. The solar cell junction gives the device its diode behavior, enabling charge separation. As a point of reference the ideal diode current density in the dark, $J_{dark}$ (no illumination) is given by:

$$J_{dark} = J_S\left(e^{\frac{V}{kT}} - 1\right) \qquad (4.3)$$

where $J_S$ is the (dark) saturation current density, $V$ is the applied voltage bias, $k$ (=1.380658 · $10^{-23}$ J/K) is the Boltzmann constant, $q$ (= 1.602 · $10^{-19}$ C) is the elementary (electron) charge, and $T$ is the junction temperature (K). A positive applied voltage results in an exponential increase in current, while a negative one yields a very small current. In a *p-n* junction with zero applied voltage and in the dark, is still possible that majority charge carriers to diffuse across the junction even though the established electric field at the junction that is opposing to this process. Once majority carriers diffuse across the boundary became minority charge carriers and eventually recombine in a diffusion length. Minority carriers, thermally generated within a diffusion length from the junction have a good probability of crossing the junction, becoming majority carriers. At equilibrium these two currents (diffusion and drift) are equal and the net current is zero. When a forward voltage is applied to the junction (a positive bias at the *p*-type material and a negative one at the *n*-type) resulting in an electric field that is opposing to the electric field created by the space charge junction region. Since the space charge region is depleted of charge carriers, its resistivity is much higher than the rest of the device and the applied voltage bias is dropped almost entirely across this region. The net junction electric field is reduced since the applied electric field is opposing to the built-in field, lowering the barrier height and increases the diffusion current across the junction. There is a very little effects are on the drift current, which depends only on the thermally generated minority carrier concentrations, within a diffusion length of the depletion region, which doesn't changes as the diffusion current. The increased diffusion current enhances the junction minority carrier injection at the junction and therefore increased recombination with majority carriers which are supplied by the external circuit. The removal of minority charge carriers by recombination allows more majority charge carriers to diffuse across the junction to become minority carriers, sustaining the external current. The diffusion current can be viewed as a recombination current where a greater recombination results in a larger current across the junction. The parameter $J_S$ from Equation (4.3) is a measure of the diode recombination process, a larger $J_S$ corresponds to larger recombination. It depends on the material quality and temperature, increasing as $T$ increases and decreasing as the material quality increases. There are two other important characterization parameters beside $Jsc$ and $Voc$ for solar PV cells, the maximum power and the fill factor, $FF$, which is defined as:

$$FF = \frac{J_{mp}V_{mp}}{J_{sc}V_{OC}}$$

(4.4)

where $J_{mp}$ and $V_{mp}$ are the current density and voltage at the maximum power point. The fill factor is a measure of the real *I-V* characteristic, and its value is greater than 0.7 for good PV cells. The fill factor decreases as the cell temperature increases. Thus, by illuminating and loading a PV cell so that the voltage equals the PV cell's $V_{mp}$, the output power is maximized. The PV cells can be loaded using resistive loads, electronic loads, or batteries.

> **Example 4.1:** Calculate the fill factor of a PV cell having an open-circuit voltage of 0.21 V, short-circuit current 5.6 mA, a maximum voltage of 0.125 V and maximum current of 3.1 mA.
>
> **Solution:** Using Equation (4.4) the fill factor is:
>
> $$FF = \frac{0.125 \times 3.1}{0.21 \times 5.6} = 0.33$$

Different resistive loads across the terminals produce different currents and voltages at the terminals. $J_{mp}$ and $V_{mp}$ correspond to the load where the power ($P = JV$) is maximized. We have to notice that in engineering applications we are interested in currents, rather than current densities. Figure 4.2 shows the current and voltage relationship, the *I-V* characteristics, with most important parameters marked. The figure is also showing how the PV cell maximum achievable power relates to a "non-real" power determined by $J_{sc}$ and $V_{oc}$. The fill factor then provides a quantification of the shape of the J-V curve. If the *p-n* junction is *forward-biased* by a positive potential, $V$, connected to the p-side, the so-called forward current is flowing:

$$I_D = I_S\left[\exp\left(\frac{qV}{kT}\right) - 1\right]$$

(4.5)

Here, $I_S$, is the saturation current, computed by multiplying $J_S$ with cell area. The reverse saturation current increases with increasing temperature. The typical order of magnitude of $J_S$ (saturation current density) is $10^{-8}$ A/m². This is also known as the "leakage current" or "diffusion current." Temperature affects solar cell characteristics primarily in the following two ways: directly via $T$ in the exponential term in (4.5) and indirectly via its effect on the reverse-diode saturation current $IS$ and the photo-generated current $IL$. The saturation current density dependence on the energy gap and cell temperature is expressed as:

$$J_S = nT^3\exp\left(-E_g/kT\right)$$

(4.6)

Here $n$ is the non-ideality factor (or diode quality factor), ranging from 1 to 2 ($n = 1$ is for an ideal case), $E_g$ is the semiconductor energy gap, about 1.12 eV for Si. Saturation current is simple computed by multiplying the current density by the diode cross-sectional area.

**Example 4.2:** Determine the value of the saturation current for Silicon at 60°C.

**Solution:** Substituting the known values in the above equation, assuming ideal PV cell case (A = 1), we get for the saturation current density:

$$J_s = 1 \cdot (273.15 + 60)^3 \exp\left(\frac{1.12 \times 1.6 \times 10^{-19}}{1.38 \times 10^{-23} \times (273.15 + 60)}\right) = 4.365 \times 10^{-10} \, \text{A/m}^2$$

PV cells are usually modeled as a current source in parallel with a diode. If there is no light present to generate photocurrent, the PV cell behaves like a diode. A PV cell, consisting of *p-n junction*, when no light presented behaves as a conventional diode. Therefore, a simple diode can describe the equivalent circuit (as shown in Figure 4.3). When solar radiation (light) falls on a *p-n* junction, the photons with enough energy can create electron-hole pairs, through photoelectric effect. In an ideal PV cell, the total current $I_C$ is equal to the current $I_L$ generated by the photoelectric effect minus the diode current $I_D$, according to the equation below. The field across the depletion zone forces the electrons to the *n*-side and the holes to the *p*-side of the junction, producing a reverse current $I_L$, and a photocell current $I_C$, given by:

$$I_C = I_L - I_D = I_L - I_S\left[\exp\left(\frac{qV}{kT}\right) - 1\right] \tag{4.7}$$

Here $I_D$ is the diode current, see Figure 4.3 for details. Another fundamental parameters obtained from the PV cell I-V characteristics are the short-circuit current and the open circuit voltage. The short-circuit current, $I_{SC}$, is the higher value of the current generated by the cell and is obtained under short-circuit conditions, i.e., $V = 0$, and is equal to $I_L$, the photocurrent. The photocurrent $I_L$ is also influenced by the temperature too. The open circuit voltage corresponds to the voltage drop across the diode when it is traversed by the photocurrent, $I_L$, which is equal to $I_D$ when the generated current is $I_C = 0$. The calculation of the open-circuit voltage is shown in Example 4.3.

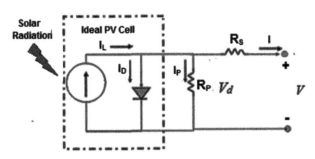

**FIGURE 4.3**
PV cell equivalent circuit.

**Example 4.3:** Determine the voltage for a zero overall solar cell current.

**Solution:** Substituting $I = 0$, in Equation (4.7) and solving for voltage, we are getting:

$$0 = I_L - I_S\left[\exp\left(\frac{qV}{kT}\right) - 1\right]$$

and

$$\exp\left(\frac{qV}{kT}\right) = \frac{I_L}{I_S} + 1$$

Or solving the above equation for the open-circuit voltage, expressed as:

$$V_{OC} = \frac{kT}{q}\ln\left(\frac{I_L}{I_S} + 1\right) = V_{Th}\ln\left(\frac{I_L}{I_S} + 1\right) \tag{4.8}$$

As $I_L \gg I_S$, open-circuit voltage is less than the semiconductor energy band gap, Equation (4.8) become:

$$V_{OC} \simeq V_{Th}\ln\left(\frac{I_L}{I_S}\right) \tag{4.9}$$

This equation is the open circuit voltage. Here the thermal voltage, having a value of 0.0259 V for 300 K is given by:

$$V_{Th} = \frac{kT}{q} \tag{4.10}$$

Notice that Equations (4.7–4.9) are in terms of currents rather than in terms of current densities. The diode saturation current $I_S$, is related to the semiconductor band gap energy and the cell area, $A_{cell}$ (m$^2$), through the following reasonable estimate:

$$I_S = 1.5 \cdot 10^9 A_{cell} \cdot \exp\left(-\frac{E_g}{kT}\right) \text{ (A)} \tag{4.11}$$

In PV cells, the photon absorption, taking place in semiconductor materials, results in the charge carrier generation and the subsequent photo-generated charge carries separation. Therefore, semiconductor layers are the most important PV cell parts. PV cells are depicted in two ways. A diagram is showing the device physical structure and the electron transport processes, contributing to the energy-conversion process (Figure 4.1). The same processes are also shown on the semiconductor band diagram (energy levels). An ideal PV cell is represented as a current source in parallel with a diode, as shown in the equivalent circuit of Figure 4.3 (inside the dash-dot square). The currents owing through the illuminated PV cell are shown in Figure 4.3. In the dark, PV cell act as a regular diode, corresponding to the electric circuit element in Figure 4.3. The current under these conditions depends on the bias voltage $V$, where positive bias decreases the barrier height, increases the recombination currents. A negative bias (negative voltage) has an opposite effect. Electron-hole pairs created by the photons result in (1) minority carriers swept through the junction from both sides of the junction; (2) additional majority carriers, some of which are collected by the electrodes, with the remainder undergoing recombination.

**Example 4.4:** A PV cell has a saturation current of $12.5 \cdot 10^{-13}$ A and light-induced current at 20°C of 0.85 A. Find the open-circuit voltage.

**Solution:** The thermal voltage at 20°C is 0.0255 V by using Equation (4.9) the open-circuit voltage is:

$$V_{OC} \simeq 0.0255 \times \ln\left(\frac{0.85}{12.5 \times 10^{-13}}\right) = 0.695 \text{ V}$$

When a solar cell is connected to an external circuit, the photo-generated current then travels from the *p*-type semiconductor-metal contact, through the wire, powers the load and continues through the wire until it reaches the *n*-type semiconductor-metal contact. Under a certain sunlight illumination, the current passed to the load from a solar cell depends on the external voltage applied to the solar cell normally through a power electronic converter for a grid-connected PV system. If the applied external voltage is low, only a low photo-generated voltage is needed to make the current flow from the solar cell to the external system. Nevertheless, if the external voltage is high, a high photo-generated voltage must be built up to push the current flowing from the solar cell to the external system. This high voltage also increases the diffusion current so that the net output current of the solar cell is reduced. To analyze the behavior of a solar cell, it is useful to create a model which is electrically equivalent. An ideal solar cell can be modeled by a current source, representing the photo-generated current $I_L$, in parallel with a diode, representing the *p-n* junction of a solar cell (see Figure 4.3). In a real solar cell, there exist other effects, not accounted for by the ideal model. Those effects influence the external behavior of a solar cell, which is particularly critical for integrated solar array study. Two of these extrinsic effects include: (1) current leaks proportional to the terminal voltage of a solar cell and (2) losses of semiconductor itself and of the metal contacts with the semiconductor. The first is characterized by a parallel resistance $RP$ accounting for current leakage through the cell, around the edge of the device, and between contacts of different polarity (Figure 4.3). The second one is characterized by a series resistance $RS$, which causes an extra voltage drop between the junction voltage and the terminal voltage of the solar cell for the same flow of current. PV cell series resistance results from the inherent resistance of the cell itself to the charge carrier flow, from the resistance between cell and metal contact and the resistance of the metal contact themselves. The shunt (parallel) resistance of the cell is primarily the effect of the manufacturing defects, allowing the current to flow through PV cell and its edges. Expanding Equation (4.7), through the simplified circuit model (Figure 4.3), and the following associated equation, including *n* is the diode ideality factor (typically between 1 and 2), the output current is expressed as:

$$I_C = I_L - I_S\left(e^{\frac{q(V+I \cdot R_S)}{nkT}} - 1\right) - \frac{V + I \cdot R_S}{R_P} \tag{4.12}$$

Similar, the open circuit voltage, considering the diode ideality factor is given by:

$$V_{OC} = \frac{nkT}{q}\ln\left(\frac{I_L}{I_S} + 1\right) \tag{4.13}$$

The mathematical model of a solar cell is described by the Equations (4.7–4.10, 4.12) and (4.13), while the last one gives the voltage at the cell external terminals, as shown in Figure 4.3.

$$V = V_d - I \cdot R_S \tag{4.14}$$

**Example 4.5:** An ideal diode has a reverse saturation current of 1.12 nA, and a photocurrent of 1.05 A, when is operating at 30°C. A 5 Ω load is connected to this cell. Compute the PV cell output power.

**Solution:** From Equation (4.10), the thermal voltage at 30°C is:

$$V_{Th} = \frac{1.38 \times 10^{-23}(273.15 + 30)}{1.602 \times 10^{-19}} = 0.0261 \text{ V}$$

The output cell current, calculate from Equation (4.7) (ideal PV cell) is:

$$I_C = I_L - I_S\left[\exp\left(\frac{V}{V_{Th}}\right) - 1\right]$$

The load voltage (Figure 4.3) is then:

$$V = RI_C = RI_L - RI_S\left[\exp\left(\frac{V}{V_{Th}}\right) - 1\right] = 5 \times 1.05 - 5 \times 1.12 \times 10^{-9}\left(\exp\left(\frac{V}{0.0261}\right) - 1\right)$$

The above equation is a nonlinear one equation with unknown $V$, load voltage. Solving this equation iteratively, the load voltage is:

$$V \approx 0.539 \text{ V}$$

The cell output power is then:

$$P = \frac{V^2}{R} = \frac{0.539^2}{5} = 0.0581 \text{ W or 58.1 mW}$$

## 4.2.1 Double-Diode and Simplified Single-Diode PV Cell Models

An extended model of a single diode PV cell model derived from the minority carrier diffusion equation, considering the minority carrier current densities, leads to a model of PV cell with an ideal current source in parallel with two diodes, the so-called two-diode cell model. Equation (4.7) is re-formulated in this case as:

$$I_C = I_L - I_{S1}\left[\exp\left(\frac{qV}{nkT}\right) - 1\right] - I_{S2}\left[\exp\left(\frac{qV}{2kT}\right) - 1\right] \tag{4.15}$$

Here, $I_{S1}$ and $I_{S2}$ are the dark saturation current due to the recombination in the quasi-neutral region (diffusion), and the dark saturation current due to the recombination in the space charge region, respectively. The PV circuit models are based on the linearity assumption, meaning that the current flowing to a cell is the superposition of the junction bias and illumination currents. From this assumption the I-V cell characteristics can be obtained, from which a practical PV cell can be modeled as a hybrid voltage-current source depending on the operating point. Once the PV cell electrical characteristics are derived, the open-circuit voltage, short-circuit current and the maximum power point can be estimated. The PV cell double-diode or double exponential model is accepted as describing the real cell electric behavior, especially for the one made of polycrystalline silicon, being regarded as more accurate at low illuminations. However, when the studies are focusing on PV module or arrays, the implications of the low-illumination effects are less significant, PV power generation occurring at high illumination values.

The PV cell single-diode model described in the previous chapter subsection is quite general and can be used for different solar cells where the parameters are representing the source physical phenomenon. For example the cell shunt resistance is due to the recombination effects far from the dissociation site, while the series resistance considers each charge carrier conductivity, being affected by the space charge and traps. The presence of the shunt resistance in the cell circuit model is mainly due to the leakage current of the *p-n* junction, which depends on the PV cell fabrication process. The shunt resistance has stronger effect at constant current region and low illuminations, having not a significant effect when a PV cell is part of PV generation system, operating at high illumination values and near the MPP, therefore neglecting it has less impact on the model validity. The shunt resistance can be neglected, and the Equation (4.11) is written as (Figure 4.4):

$$I_C = I_L - I_S \left( e^{\frac{q(V + I \cdot R_S)}{nkT}} - 1 \right) \tag{4.16}$$

In common practice, the term −1 in Equation (4.16) is neglected, being much smaller than the exponential. The simplified PV cell model is considered as the reference model for PV system simulation analysis and for PV source emulation studies. From Equation (4.16) the PV cell model inversion, i.e., the PV cell representation as $V = f(I)$, very useful in source emulation can be determined in straightforward manner:

$$V = \frac{nkT}{q} \ln \left( \frac{I_L - I}{I_S} \right) - I R_S \tag{4.17}$$

We should note that the logarithm is null for $I = I_L - I_S$, and become negative for $I_L - I_S < I < I_L$. Since the saturation current is much smaller than the other currents, an additional care must be taken during the experiments, for values near the short-circuit conditions.

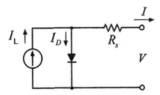

**FIGURE 4.4**
Simplified single-diode model of practical PV cells.

**Example 4.6:** A PV cell has a short-circuit current, 1.25 mA and a saturation current, $2.0 \cdot 10^{-14}$ A and is operating at room temperature, 25°C. If the applied voltage increases from 0.2 V to open-circuit voltage, in steps 0.1 V calculate the photo-cell currents, the output power, the open-circuit voltage, the maximum voltage and current and the fill factor.

**Solution:** Appling a slightly modified Equation (4.7), the photo-cell currents are calculated and listed in the table below. At room temperature 25°C or 298 K, the thermal voltage is approximately 0.026 V. At $V = 0$, the short-circuit current equal $I_L$, and the output power is computed by $P = V \cdot I_C$. The open-circuit voltage is

$$V_{OC} \approx 0.026 \times \ln\left(\frac{1.25 \times 10^{-3}}{2 \times 10^{-14}}\right) = 0.65 \text{ V}$$

$$I_C = I_L - I_S\left[\exp\left(\frac{V}{V_{Th}}\right) - 1\right] \approx I_L - I_S \exp\left(\frac{V}{V_{Th}}\right)$$

| $V(V)$ | 0.2 0.3 0.4 0.5 0.6 0.65 |
|---|---|
| $I_C$ (mA) | 1.250 1.250 1.250 1.246 1.040 0.00 |
| $P$(mW) | 0.250 0.375 0.500 0.623 0.624 0.00 |

The maximum power is 0.624 mW, and $V_{mp} = 0.6$ and $I_{mp} = 1.04$ mA. We can also estimate the fill factor:

$$FF = \frac{0.624}{0.65 \times 1.25} = 0.77$$

Notice that important PV cell characteristics are also consisting of the output current *IC* and output power *PC* versus output voltage *VC* characteristics. Figure 4.5 shows typical *I-V* and *P-V* characteristics of a solar cell under different illuminations, with the consideration of parallel and series resistance. From this diagram, we noticed that if the external voltage applied to the solar cell is low, the net output PV cell current, depending primarily on the photo-generated current, is almost constant. Therefore, as the external voltage increases, more power is out-putted from the solar cell. If the external voltage is around the forward conduction voltage of the p-n junction diode, the net output current drops significantly and the output power reduces.

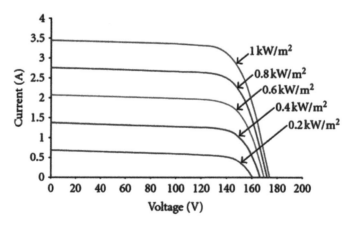

**FIGURE 4.5**
PV cell I-V characteristics.

### 4.2.2 Non-Ideal PV Cell Characteristics and Behavior

Expanding the Equations (4.7) and (4.10) that are obtained from the simplified circuit model shown above and the following associated equation, where A, mentioned above is the diode ideality factor (typically between 1 and 2), and $R_S$ and $R_P$ represents the series and shunt cell resistances. PV cells can be operated over a wide range of voltages and currents. By varying the load resistance from zero (a short circuit) to infinity (an open circuit), the *MPP* of the cell can be determined. On the *I–V* curve, the maximum power point (*Pm*) occurs when the product of current and voltage is at maximum. No power is produced at the short-circuit current with no voltage, or at the open-circuit voltage with no current. Therefore, *MPP* is somewhere between these two points. Maximum power is generated at about the "knee" of the curve. This point represents the maximum efficiency of the solar device in converting sunlight into electricity. The fill factor, defined before is the ratio between the maximum power ($P_{max} = J_{mp} \cdot V_{mp}$) generated by a solar cell and the product of $V_{oc}$ with $J_{sc}$ (Equation 4.4). Assuming that the solar cell behaves as an ideal diode the fill factor can be expressed as a function of open-circuit voltage, $V_{OC}$, as:

$$FF = \frac{v_{OC} - \ln(v_{OC} + 0.72)}{v_{OC} + 1} \tag{4.18}$$

where $v_{oc} = V_{oc} \cdot q/(kT)$ is the normalized open-circuit voltage. Equation (4.16) is a good approximation of the ideal value of *FF* for normalized open-circuit voltages higher than 10.

**Example 4.7:** Calculate the fill factor of a PV cell, assuming that the PV cell is operating at 300 K, the diode saturation current density is $1.95 \cdot 10^{-9}$ A/m², and photo-current density is 35 mA/cm².

**Solution:** The open-circuit voltage is computed using Equation (4.8) in current densities

$$V_{OC} = \frac{kT}{q} \ln\left(\frac{J_{ph}}{J_S} + 1\right) = 0.0259 \ln\left(\frac{350}{1.95 \times 10^{-9}} + 1\right) = 0.671 \text{ V}$$

The normalized open-circuit voltage is then:

$$v_{OC} = 0.671 / 0.0259 = 25.913 \text{ V}$$

The fill factor is given by the Equation (4.17), as:

$$FF = \frac{25.913 - \ln(25.913 + 0.72)}{25.913 + 1} = 0.84$$

The open-circuit voltage, is not affected significantly by solar irradiation levels, varies significantly with the operating temperature module, the higher the temperature, the lower the open-circuit voltage, reducing the maximum power that can be generated by the PV cell. However, the fill factor is not changing drastically with a change in $V_{oc}$. For PV cells with particular absorber, large $V_{OC}$ variations. For example, at standard illumination conditions, the difference between the maximum open-circuit voltage measured in a laboratory and a typical commercial solar cell is about 120 mV, giving a maximal *FF* of 0.85 and 0.83, respectively.

However, the variation in maximum *FF* can be significant for solar cells made from different materials. For example, a *GaAs* PV cell may have a *FF* approaching 0.89. The output power, *P*, from a photovoltaic cell, depends also on the load resistance, *R*, and is given by:

$$P = VI = RI^2 \tag{4.19}$$

**Example 4.8:** An ideal PV cell has the saturation current 1.05 nA and is operation gar 35°C and the photocurrent is 0.95 A. Compute the output voltage and the output power when the load connected to the cell draws 0.5 A.

**Solution:** For the given operating temperature he thermal voltage of the PV cell is:

$$V_{Th} = \frac{kT}{q} = \frac{1.38\times10^{-23}(273.15+35)}{1.602\times10^{-19}} = 0.0265 \text{ V}$$

From Equation (4.7) we can calculate the output voltage

$$V = \ln\left[(I_c - I)\times10^9 + 1\right]V_{Th} = \ln\left[(0.95-0.5)\times10^9 + 1\right]\cdot0.0265 = 0.53 \text{ V}$$

The output power is then

$$P = V\cdot I = 0.53\times0.5 = 0.265 \text{ W}$$

Substituting Equation (4.7) in the power Equation (4.17) yields to:

$$P = V\left[I_L - I_S\left[\exp\left(\frac{qV}{kT}\right) - 1\right]\right] \tag{4.20}$$

Equation (4.19) can be differentiated with respect to *V*, and by setting the derivative equal to 0, the external voltage, $V_{mp}$, that gives the maximum PV cell output power can be obtained:

$$\exp\left(\frac{qV_{mp}}{kT_{cell}}\right)\cdot\left[1+\frac{qV_{mp}}{kT_{cell}}\right] = 1+\frac{I_{SC}}{I_S} \tag{4.21}$$

Equation (4.21) is an explicit equation of the maximum output voltage, which maximizing the output power in terms of short-circuit current, the saturation current and the absolute value of the cell temperature, $T_{cell}$. If these three parameters are known, then $V_{mp}$ can be obtained from Equation (4.21), via iterative methods. Wenham et al. (2007) found the following expression for the maximum voltage:

$$V_{mp} = V_{OC} - \frac{kT_{cell}A}{q}\ln\left[1+V_{mp}\frac{kT_{cell}A}{q}\right] \tag{4.22}$$

This equation is showing that the $V_{mp} < V_{OC}$. For cell temperature of 300 K with an initial guess, $V_{OC}$, an iterative solution of the maximum voltage is expressed as:

$$V_{mp}^k = V_{OC} - \frac{kT_{cell}A}{q}\ln\left[1+V_{mp}^{k-1}\frac{kT_{cell}A}{q}\right], \quad \text{for } k = 0,1,2,... \tag{4.23}$$

Substituting Equation (4.21) into Equation (4.7), the load current, $I_{mp}$, which maximizes the output power can be found:

$$I_{mp} = I_{SC} - I_S \left[ \exp\left( \frac{qV}{kT_{cell}} \right) - 1 \right] = \frac{qV_{mp}}{kT_{cell} + qV_{mp}} \left( I_{SC} + I_S \right) \tag{4.24}$$

And the maximum PV cell output power is computed as:

$$P_{max} = \frac{qV_{mp}^2}{kT_{cell} + qV_{mp}} \left( I_{SC} + I_S \right) \tag{4.25}$$

The conversion efficiency is computed as the ratio between maximum generated power and the incident power, being another measure of the performances of the PV cells. Efficiency is commonly reported for a PV cell temperature of 25°C and incident light at an irradiance of 1000 W/m² with a spectrum close to that of sunlight at solar noon. Any improvements in the cell efficiency are directly connected to cost reduction of the photovoltaic systems. The irradiance value $P_{in}$ of 1000 W/m² from the AM1.5 spectrum is the standard for measuring the PV cell conversion efficiency, while $I_{max}$ and $V_{max}$ correspond to a given SR(t) solar radiation intensity. The maximum solar PV cell conversion efficiency can be expressed as:

$$\eta_{max} = \frac{P_{max}}{P_{in}} = \frac{V_{max} \times I_{max}}{\text{Incident Solar Radiation} \times \text{Cell Area}} = \frac{V_{OC} \times I_{SC} \times FF}{SR(t) \times A_{Cell}} \tag{4.26}$$

The maximum efficiency ($\eta_{max}$) found from a light test is not only an indication of the performance of the device under test, but, like all of the *I-V* parameters, can also be affected by ambient conditions such as temperature and the intensity and spectrum of the incident light. For this reason, it is recommended to test and compare PV cells using similar lighting and temperature conditions. Efficiency of an actual PV cell is the ratio of the electrical power output $P_{out}$, compared to the solar power input, $P_{in}$, into the PV cell:

$$\eta = \frac{P_{Out}}{P_{IN}} \tag{4.27}$$

**Example 4.9:** Calculate the power output from a solar cell under standard test conditions (SR(t) = 1000 W/m² and $T_c$ = 25°C), when the efficiency is 16.5%, and FF = 0.82, and the PV aperture area is 4.02 · 10⁻⁴ m².

**Solution:** From Equations (4.26) to (4.27), the PV cell power output is

$$P_{Out} = 0.165 \times 1000 \times 4.02 \times 10^{-4} \times 0.82 = 0.054 \, \text{W}$$

During operation, the efficiency of PV cells is further reduced by the power dissipation across internal resistances, the shunt resistance ($R_P$) and series resistance ($R_S$). For an ideal cell, $R_P$ would be infinite and would not provide an alternative path for current to flow, while $R_S$ would be zero, resulting in no further voltage drop before the load. Decreasing $R_P$ and increasing $R_S$ will decrease the fill factor (*FF*) and $P_{max}$ as shown in Figure 4.6. If $R_P$ is decreased too much, $V_{OC}$ will drop, while increasing $R_S$ excessively can cause $I_{SC}$ to drop

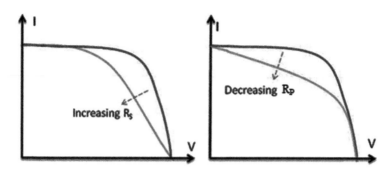

**FIGURE 4.6**
Effect of the diverging of $R_S$ and $R_P$ from ideality.

instead. It is possible to approximate the series and shunt resistances, $R_S$ and $R_P$, from the slopes of the I-V curve at $V_{OC}$ and $I_{SC}$, respectively. The resistance at $V_{OC}$, however, is at best proportional to the series resistance but it is larger than the series resistance. $R_P$ is represented by the slope at $I_{SC}$.

Other parameters that are affecting the efficiency of PV cells are the semiconductor band gap ($E_g$), operating temperature, incident light, type and purity of the material, and parasitic resistances. Construction of the I-V (current-voltage) is critical to the estimate of the PV cell efficiency. To have higher values of the open-circuit voltage, the saturation current must be small. A well accepted minimum saturation current density estimate (from Equation 4.6) is:

$$J_S = 1.5 \times 10^5 \exp\left(-E_g / kT\right) \tag{4.28}$$

This equation is stating that if *band gap* increases, *the saturation current* decreases, and $V_{OC}$ became larger, the maximum value of $V_{OC}$ increases with increasing band gap. Subsequently, $V_{OC}$ has an effect on $V_{mp}$, which in turn affects the efficiency. This is one of the reasons why GaAs cells with a band gap of 1.4 eV are more efficient than Si-based PV cells with a band gap of 1.12 eV. Another reason that GaAs cells are more efficient is due to their characteristics related to light absorption. In direct band-gap materials, for example GaAs, the light is absorbed quickly, with less probability of passing straight out the back. Most PV cells are constructed of silicon, an indirect band-gap material, which means that the emission or absorption of a photon is also required for photon energies near that of the band gap. When more energy is converted in the PV cell material, less energy is available for external use.

**Example 4.10:** A 1.0 cm² silicon solar cell has a saturation current of 1 nA and is illuminated with sunlight yielding a short-circuit photocurrent of 75 mA. Calculate the solar cell efficiency and the fill factor, if the PV cell is operating at 27°C and is illuminated with 1000 W/m².

**Solution:** For the operating conditions the thermal voltage is 0.0259 V and the open-circuit voltage, by using Equation (4.9) is:

$$V_{OC} \simeq V_{Th}\ln\left(\frac{I_L}{I_S}\right) = 0.0259 \times \ln\left(\frac{75 \times 10^{-3}}{10^{-9}}\right) = 0.530 \text{ V}$$

The maximum power is again estimated by making the derivative of power versus voltage equal to zero (the condition for maximum), yielding to:

$$\frac{dP}{dV} = 0 = I_S\left(e^{V_{mp}/V_{Th}} - 1\right) - I_L + \frac{V_{mp}}{V_{Th}}I_S \cdot e^{V_{mp}/V_{Th}}$$

where $V_{mp}$, is the voltage corresponding to the maximum power point. The maxim power point voltage is obtained by solving the following transcendental equation:

$$V_{mp} = V_{Th} \ln \frac{1 + I_L / I_S}{1 + V_{mp} / V_{Th}} \tag{4.29}$$

Using iteration with a starting value of 0.5 V, the following successive values for $V_{mp}$ are obtained:

$$V_{mp} = 0.405, 0.406, 0.395 \text{ V}$$

The current, $I_{mp}$, corresponding to the voltage, $V_{mp} = 0.405$ V, is calculated using Equation (4.7) was found 0.050 A. The PV efficiency, for one Sun (1000 W/m²) is then:

$$\eta = \frac{V_{mp}I_{mp}}{P_{IN}} = \frac{0.406 \times 0.05}{1000 \times \left(10^{-4} \text{cm}^2 / 1 \text{ m}^2\right)} = 0.203 \text{ or } 20.3\%$$

To compute the fill factor, we need the compute open-circuit voltage, by using Equations (4.8) or (4.9), $V_{OC} = 0.625$ V, then the fill factor is:

$$FF = \frac{V_{mp}I_{mp}}{V_{OC}I_{SC}} = \frac{0.406 \times 0.050}{0.53 \times 0.075} = 0.51 \text{ or } 51\%$$

The actual PV cell efficiency is quite low (10%–20% for most of the marketed PV cells). However, we have to keep in mind that during the exploitation of renewable resources the conversion efficiency direction is different from the efficiency, corresponding to the conversion of non-renewable resources, which are irreparably lost after use. A low PV system efficiency consists of the need of a larger capturing surface, but not necessarily of a low eco-balance. The efficiency depends partly on illumination, usually not specified by manufacturers, and on the temperature (a few tenths of a %/°C). Temperature is an important parameter, since the PV cells are exposed to solar radiation, which tends to overheat them, part of the absorbed radiation, not converted into electric energy, is dissipated as heat. So the cell temperature is always higher than the ambient temperature. The colder it is, the more efficient the PV cell is. The maximum efficiency ($\eta_{MAX}$), found from a lab test is only an indication of the performance of the device, but, like all of the I-V parameters, is also affected by ambient conditions such as temperature, the incident light intensity and spectrum. For this reason, it is recommended to test and compare PV cells using similar ambient conditions. The standard approach for defining PV cell efficiency depends on the cell temperature, $T_{cell}$, which is calculated using the ambient temperature and the reference value of the cell temperature, the nominal operating cell temperature (NOCT). The nominal operating cell temperature is defined as the temperature reached by the open circuit cells in a photovoltaic module under test conditions, 800 W/m² irradiance on the PV cell surface, 20°C air temperature, and 1 m/s wind velocity. However, these conditions may vary depending on the climate zone nature. Most PV manufacturers provide temperature elements for their crystalline PV modules based on the NOCT as the cell temperature ($T_{cell}$), which has a standard equation of:

$$T_{cell} = T_a + \frac{SR}{800}(NOCT - 20) \tag{4.30}$$

where $T_{cell}$ is the cell temperature, $T_a$ is the ambient temperature, $SR$ is the average illumination (W/m²), and NOCT is the nominal operating cell temperature. Several correlations have been developed over the years for the cell temperature ($T_{cell}$) as a function of climatic parameters (solar radiation, ambient air temperature, wind speed, etc.). There are also several correlations available to calculate the effect of the cell temperature on the efficiency of a PV cell ($\eta_c$). However, in most of the practical applications the following linear relation for the cell efficiency is used without incurring significant loss in accuracy:

$$\eta_c = \eta_{ref}\left(1 + \beta_{ref}\left(T_{cell} - T_{ref}\right)\right) \tag{4.31}$$

where $\eta_{ref}$ is the efficiency of the PV cell at temperature $T_{ref}$. The temperature coefficient $\beta_{ref}$ is determined by the cell material, which usually is provided by the manufacturer, and on the $T_{ref}$, and is expressed as:

$$\beta_{ref} = \frac{1}{T_0 - T_{ref}} \tag{4.32}$$

Here $T_0$ is the maximum temperature at which the efficiency of the PV cells decreases to zero. For a crystalline Si cell this temperature is about 270°C. The values of $\eta_{ref}$ for silicon based PV technologies ranges from 0.002 to 0.005. Open-circuit voltage drops about 0.0037/°C. When the NOCT is not given, another approach to estimating cell temperature (Masters 2013), based on the following assumption is used:

$$T_{cell} = T_a + \gamma\left(\frac{\text{Insulation}}{1000 \text{ W/m}_2}\right) \tag{4.33}$$

where $\gamma$ is a proportionality factor that has some dependence on the wind speed and how well ventilated the PV modules are when installed. Typical values of $\gamma$ range between 25°C and 35°C, meaning that in 1 sun (1000 W/m²) of insolation, PV cells tend to be 25°C–35°C hotter than their environment.

It worth to notice that the short-circuit current is direct proportional to incident solar radiation, while the open-circuit (no-load) voltage varies little with solar radiation, depending logarithmically on solar radiation ($I_L$ in Equation 4.8 is proportional to the solar radiation), and quite often in practical applications this variations is ignored. Short-circuit current for different solar radiation values is approximated by this relationship:

$$I_{SC} = I_{SC-STC} \cdot \frac{SR}{SR_{STC}} \tag{4.34}$$

Here, $I_{SC-STC}$ is the short-circuit current corresponding to the standard test conditions, $SR$ is the actual radiation on the device surface (W/m²), while $SR_{STC}$, is the standard radiation equal to 1000 W/m². The photo-generated current $I_L$ is also influenced, how we pointed out in a section above, by the temperature too, as well as by the irradiance level, as expressed in the following:

$$I_L = \left(I_{L(STC)} + k_1 \cdot \Delta T\right)\frac{SR}{SR_{STC}} \tag{4.35}$$

where $I_{L(STC)}$ is the photo-generated current at the nominal condition, $\Delta T$ is the difference between actual temperature $T$ and nominal temperature $STC$ (25°C), respectively. On the other, hand the open-circuit voltage decreases with temperature. For Si PV cells the temperature open-circuit voltage variation coefficient, $K_T$ has values between 2.3 and 3.7 mV/°C. The open-circuit voltage for temperatures others than standard is approximated by the following relationship:

$$V_{OC} = V_{OC-STC} - K_T(t - 25) \tag{4.36}$$

Where $V_{OC-STC}$ is the PV cell open-circuit voltage at standard temperature (25°C), $t$ is the PV cell temperature, in °C. However, in most of the PV system design calculations, the variations of the short-circuit current the open-circuit voltage and the fill factor are neglected, the most important design parameter being the solar radiation. As the irradiance level increases, the maximum power increases greatly and the MPP voltage drops slightly due to the temperature rise according to the above relationships.

> **Example 4.11:** Estimate cell temperature, open-circuit voltage, and maximum power output for a 4.75-W PV cell, having the open circuit voltage 0.671 V, under conditions of 1-sun insolation and ambient temperature 30°C. The module has a NOCT of 50°C.
>
> **Solution:** The cell temperature is computed using Equation (4.23):
>
> $$T_{cell} = T_a + \frac{SR}{800}(\text{NOCT} - 20) = 30 + \frac{1}{0.8}(50 - 20) = 67.5°C$$
>
> The new open-circuit voltage and output power are:
>
> $$V_{OC} = 0.671\left(1 - 0.0037 \cdot (67.5 - 25)\right) = 0.565 \text{ V}$$
>
> $$P_{Out} = 0.475\left(1 - 0.005 \cdot (67.5 - 25)\right) = 3.761 \text{ W}$$

## 4.2.3 PV Cell Construction, Technology, and Materials

When sunlight strikes a PV cell, the photons are dislodging the electrons from the cell material atoms. These free electrons can move through the cell, creating and filling in holes in the cell, generating electricity. The physical process in which a PV cell converts sunlight into electricity, as we discussed in previous chapter subsections is known as the photovoltaic effect. A single PV cell produces up to 2 W of power, too small even for powering any low-power electronics. To increase power output, several PV cells are connected together to form modules, which are further assembled into larger units called panels and arrays. PV system modular nature enables designers to build PV systems with various power output for any application type. A complete PV system consists of PV modules, and the so-called "balance of system," the support structures, wiring, energy storage, power conversion devices, etc. i.e., everything else in a PV system except the PV modules. Two major types of PV systems are available in the marketplace today: flat plate and concentrators. As the most prevalent type of PV systems, flat plate systems build the PV modules on a rigid and flat surface to capture sunlight. Concentrator systems use lenses to concentrate sunlight on the PV cells, increasing the cell power output. Flat plate systems are typically less complicated, employ larger number of PV cells, while the concentrator systems use smaller areas, requiring a more sophisticated and expensive tracking systems. Unable to focus diffuse sunlight, concentrator systems do not work well under cloudy conditions.

PV cells are made of semiconductor or composite materials. Most of these materials are crystalline or thin films, varying from each other in terms of light absorption efficiency, energy conversion efficiency, manufacturing technology and cost. Crystalline materials, the most common used in the PV module manufacturing, with single-crystal silicon cells as the most common used in the PV cell manufacturing. High-purity polycrystalline, melted in a quartz crucible, in which a single-crystal silicon seed is dipped into the molten polycrystalline mass. The seed is pulled slowly from the melt, so a single-crystal ingot is formed. The ingots are sawed into thin wafers about 200–400 µm thick. The thin wafers are then polished, doped, coated, interconnected and assembled into PV modules and eventually into PV panels or arrays. A single-crystal silicon structure has a uniform molecular structure, and this high uniformity is resulting in higher energy conversion efficiency (the ratio of electric power produced by the PV cell to the amount of available sunlight power). The higher a PV cell's conversion efficiency, the more electricity it generates for a given area of exposure to the sunlight. The conversion efficiency for single-silicon commercial modules ranges between 15% and 20%. Not only are they energy efficient, single-silicon PV modules are also highly reliable for outdoor power applications. About half of the PV module manufacturing cost is the cost of the wafer manufacturing, a time-consuming and expensive batch process in which ingots are cut into thin wafers with a thickness no less than 200 micrometers thick. If the wafers are too thin, the entire wafer can break in manufacturing process and the subsequent PV module process. Due to this strict thickness requirement, PV cell manufacturing requires a significant amount of raw silicon and half of this expensive material is lost as sawdust during the wafer cutting and polishing.

Polycrystalline silicon PV cells, consisting of small single-crystal silicon grain are less efficient than single-crystalline silicon PV cells. The grain boundaries in poly-crystalline silicon hinder the flow of electrons and reduce the power output of the cell. The energy conversion efficiency for a commercial PV module made of polycrystalline silicon ranges from 10% to 14%. Common approaches to produce polycrystalline silicon PV cells are to slice thin wafers from blocks of cast polycrystalline silicon. A more advanced approach is the "ribbon growth" method in which silicon is grown directly as thin ribbons or sheets with the needed thickness, so no sawing is needed, lowering the manufacturing cost compared to the single crystal process. The most commercially developed ribbon growth approach is the edge-defined film-fed growth. Compared to single-crystalline silicon, polycrystalline silicon material is stronger and can be cut into one-third the thickness of single-crystal material. It also has slightly lower wafer cost and less strict growth requirements. However, their lower manufacturing cost is offset by the lower PV cell efficiency. The average price for a polycrystalline module made from cast and ribbon is slightly lower than that of a single-crystal PV module.

Gallium Arsenide (GaAs) is a compound semiconductor made of gallium (Ga) and arsenic (As), having a crystal structure similar to that of silicon, and a high level of light absorptivity. To absorb the same amount of sunlight, GaAs requires only a layer of few micrometers thick while crystalline silicon requires a wafer of about 200–300 µm thick. GaAs has much higher energy conversion efficiency than crystal silicon, reaching about up to 30%. Having high resistance to heat it is the ideal choice for concentrator systems in which cell temperatures are higher. GaAs PV cells are also better for space applications where strong resistance to radiation damage and higher solar cell efficiency are required.

The most important drawback of the GaAs PV cells is their higher cost of the single-crystal substrate on which the GaAs is grown on. Therefore GaAs materials are used in concentrator systems where only a small area of GaAs cells is needed.

Crystalline silicon has been the workhorse of the PV cells for the past three decades. However, recent progress in the thin-films technology has led many industry experts to believe that thin-films PV cells will eventually dominate the marketplace and to realize the goals of PV, a low price and reliable source of energy. Thin-film materials are one of the newest cell technologies. In a thin-film PV cell, a thin semiconductor layer is deposited on low-cost supporting layer such as glass, metal or plastic foil. Since thin-film materials have higher light absorptivity than crystalline materials, the deposited layer of PV materials is extremely thin (a few micrometers or even less than a micrometer, e.g., a single amorphous cell can be as thin as 0.3 μm). Thinner material layers yield to significant cost reductions. The deposition techniques in which PV materials are sprayed directly onto substrate are also cheaper one. Thin-film manufacturing process is faster, uses less energy and mass production is easier than the crystalline SI ingot-growth process. However, thin-film cells suffer from poor conversion efficiency due to non-single crystal structure, requiring larger array areas, and area-related costs and mountings. Constituting about 4% of total PV modules used in US, the industry sees great potentials of thin-film technology to achieve low-cost electricity. Materials used for thin film PV modules include:

1. Amorphous Silicon (a-Si), used mostly in consumer electronic products which require lower power output and cost of production, amorphous silicon has been the dominant thin-film PV material since it was first discovered in 1974. Amorphous silicon is a non-crystalline form of silicon. A significant advantage of a-Si is its high light absorptivity, about 40 times higher than that of single-crystal silicon. Therefore only a thin layer of a-Si is sufficient for making PV cells (about 1 μm thick as compared to 200 μm or more for crystalline silicon). It can be deposited on various low-cost substrates, including steel, glass and plastic, and the manufacturing process requires lower temperatures and thus less energy. So the total material costs and manufacturing costs are lower per unit area as compared to those of crystalline silicon cells. Despite the promising economic advantages, a-Si still has two major drawbacks to overcome, the low cell energy conversion efficiency (5%–9%), and is the outdoor reliability problem in which the efficiency degrades within a few months of exposure to sunlight, losing about 10%–15%.

2. Cadmium Telluride (CdTe) is a polycrystalline semiconductor compound made of cadmium and tellurium to manufacture by processes such as high-rate evaporation, spraying or screen printing. The conversion efficiency for a CdTe commercial module is about 7%, similar to that of a-Si. The instability of cell and module performance is one of the major drawbacks of using CdTe for PV cells. Another disadvantage is its toxicity. Although very little cadmium is used in CdTe modules, extra precautions have to be taken in manufacturing process.

3. Copper Indium Diselenide (CuInSe2, or CIS) is a polycrystalline semiconductor compound of copper, indium and selenium, CIS has been one of the major research areas in the thin film industry, because it has the highest "research" energy conversion efficiency of 17.7%, not only the best among all the existing thin film materials, but very close to the 18% research efficiency of the polycrystalline silicon PV cells. A prototype CIS power module has a conversion efficiency of 10%. Being able to deliver such high energy conversion efficiency without suffering from the outdoor

degradation problem, CIS has demonstrated that thin film PV cells are a viable and competitive choice for the future. CIS is also one of the most light-absorbent semiconductors, 0.5 μm can absorb 90% of the solar spectrum. CIS is an efficient but complex material, making it difficult to manufacture. Safety issues might be another concern in the manufacturing process as it involves hydrogen selenide, an extremely toxic gas. CIS is not commercially available yet although there are plans to commercialize CIS thin-film PV modules.

Multi-junction PV cells and organic PV cells are two new technologies still under development.

Multi-junction cells are made up of various layers, which are sensitive to various solar spectrum wavelengths (at the research stage, except for space industry). This characteristic is very useful for increasing the PV cell efficiency of up to 40%. Organic cells are made up of polymers, which are semiconductor plastic materials (currently at the research stage still), having also the property to absorb photons and generate a current. They are quite cheap but, for now, have very low efficiencies, lower values, less than 5%. Their lifespan in external environments still needs to be improved, in order for mass applications to be considered. Organic solar cells are thin-film polymer solar cells. The polymer solar cell is made using organic semiconducting materials such as copper phthalocyanine, polyphenylene vinylene and carbon fullerenes. These solar cells are less costly, have a high optical absorption coefficient, and the energy band gap can be tailored by changing the chain length of polymer. The energy-conversion efficiency of organic solar cells is low compared with inorganic solar cells. Lower stability, smaller life, and degradation are the major limitations of organic solar cells. The advantage of converting the production of crystalline solar cells from mono-silicon to multi-silicon is to decrease the flaws in metal contamination and crystal structure. Multi-crystalline cell manufacturing is initiated by melting silicon and solidifying it to orient crystals in a fixed direction producing rectangular ingot of multi-crystalline silicon to be sliced into blocks and finally into thin wafers. However, this final step can be abolished by cultivating wafer thin ribbons of multi-crystalline silicon.

Organic and polymer solar cells are built from thin-films (typically 100 nm) of organic semiconductors such as polymers or small-molecule compounds like pentacene, multi-phenylene vinylene, copper phthalocyanine and carbon fullerenes. About 5% is the highest efficiency currently achieved using conductive polymers. Interest in such materials lies in their mechanical flexibility and disposability, being largely made from plastic opposed to traditional silicon, the manufacturing process is cost effective (lower-cost material, high throughput manufacturing) with limited technical challenges (not require high-temperature or high vacuum conditions). Donor-acceptor pair forms the basis of organic PV cell operation where light excites the donor causing the electron to transfer to the acceptor molecule, generating a hole for the cycle to continue. The photo-generated charges are transported and collated at the opposite electrodes to be utilized, before they recombine. Typically the cell has a glass front, a transparent indium tin oxide (ITO) contact layer, a conducting polymer, a photoactive polymer and finally the back contact layer (Al, Ag, etc.). Since ITO is expensive researchers are looking into using carbon nanotube films as the transparent contact layer. In 2007 the US PV thin film technology reached a PV market share of about 65%. However, researches for better efficiency and lower cost are still in place. Nanotechnology seems to support sustainable economic growth by offering lower costs, however, at low PV cell efficiency, although not ideal offers additional alternatives. Organic photovoltaic (OPV) devices are increasingly pursued due to their low fabrication costs and fairly easy processing. Their light weight, mechanical flexibility and large-scale roll-to-roll production capability are additional advantages compared to traditional Si-based photovoltaics. In a typical OPV device, a blend of conjugated polymer (electron donor) and a fullerene derivative (electron acceptor) is used as an active layer sandwiched between the cathode and the anode. The interpenetrating network of donor and acceptor components forms the bulk hetero-junction system for the charge carrier separation upon illumination and subsequently transports the opposite charges toward the electrodes. Among the active layers, multi (3-hexylthiophene)-phenyl-C61-butyric acid methyl ester combination remains the promising system researched till date and shows greater than 5% power conversion efficiency. The device performance depends critically on the nano-scale morphology and phase separation of the blend components.

In addition to the above types, a number of other promising materials, such as cadmium telluride (CdTe) and copper indium diselenide (CuInSe2), are used today for PV cells. The main trends today concern the use of polymer and organic solar cells. The attraction of these technologies is that they potentially offer fast production at low cost in comparison to crystalline silicon technologies, yet they typically have lower efficiencies (around 4%), and despite the demonstration of operational lifetimes and dark stabilities under inert conditions for thousands of hours, they suffer from stability and degradation problems. Organic materials are attractive, primarily due to the prospect of high-output manufacture using reel-to-reel or spray deposition. Other attractive features are the possibilities for ultra-thin, flexible devices that can be integrated into appliances or building materials, and tuning of color through its chemical structure.

Another device type is the nano-PV, considered the third-generation PV; with the first generation is the crystalline silicon cells, and the second generation is amorphous silicon thin-films. Instead of the conductive materials and a glass substrate, the nano-PV technologies rely on coating or mixing "printable" and flexible polymer substrates with electrically conductive nanomaterials. This type of photovoltaics is expected to be commercially available on larger scale, reducing tremendously the high costs of PV systems. The result of this process is a stack of crystalline layers with different band gaps, tailored to absorb most of the solar radiation. Also compound semiconductor cells have been shown to be more robust when expose to outer space radiation. Since each type of semiconductor has different characteristic band gap energy allows ta more efficient light absorption, at a certain wavelength, hence higher absorption of radiation over a portion of the electromagnetic spectrum. These hetero-junction devices layer various cells with different band-gaps which are tuned utilizing the full spectrum. Initially, light strikes a wide band-gap layer producing a high voltage therefore using high energy photons efficiently enabling lower energy photons transfer to narrow band-gap sub-devices which absorb the transmitted infrared photons. Gallium arsenide (GaAs) and indium gallium phosphide (InGaP) multi-junction devices have reached the highest efficiency of 39% with a record 40.8% from a metamorphic triple-junction OV cell. These cells were originally fabricated on GaAs substrates however, in order to reduce the cost and increase robustness, and because the lattice-matched to GaAs (often with Ge substrates). The first cells had a mono junction much like the Si $p$-$n$ junction, however because of the ability to introduce ternary and quaternary materials such as InGaP and aluminum indium gallium phosphide, dual and triple junction devices were grown in order to capture a larger solar spectrum band, therefore increasing the PV cell efficiency.

Limitations found in other PV technologies are lessened by the nanoscale components due to their ability to control the energy band-gap will provide flexibility and interchangeability, in addition to enhancing the probability of charge recombination. Carbon nanotubes (CNT) are constructed of a hexagonal lattice carbon with excellent mechanical and electronic properties. The nanotube structure is a vector consisting of number lines and columns defining how the grapheme (individual graphite layer) sheet is rolled up.

Nano-tubes can be either metallic or semiconducting. Carbon nanotubes can be used as reasonably efficient photosensitive materials as well as other PV materials. PV nanometer-scale tubes when coated by special $p$ and $n$ type semiconductor materials to form a $p$-$n$ junction can generate electrical current. Methodology that is enhancing and increasing the surface that is available to produce electricity. In the last years researchers fabricated and tested photodiodes, formed from an individual carbon nanotube, converting light to electricity in an extremely efficient process, that can important for next-generation high-efficiency solar cells. Currently nanotubes are used as the transparent electrode for efficient, flexible polymer solar cells. Naphthalocyanine (NaPc) dye-sensitized nanotubes have been developed and resulted in higher short circuit current however the open circuit voltage is reduced. There are also several researcher working on inorganic based nano-particle solar cells, based on nanoparticles of CdSe, CdTe, CNTs and nano-rods made out of the same material, trying to get rid of the complications of using a polymer based solar cells. The efficiencies are still in the 3%–4% range but much research is being conducted in this field.

Today, the solar cell technologies are categorized into three generations, divided according to the time of evolution of technology advancement. Research and development activities, being conducted for efficiency improvement and reduction in production costs for each PV cell generation. Large part of the solar cell market is still covered by first-generation solar cells. First-generation of solar cells is based on Si wafer technology, which includes monocrystalline and polycrystalline silicon solar cells. The processing technology involved for the manufacturing of first-generation solar cell requires higher energy and labor. These solar cells are single-junction PV cells with a theoretical efficiency of about 33%, while the actual conversion efficiency of is in the range of 15%–20%. These solar cells are widely used amongst all generations of PV cells. Second-generation solar cells include the amorphous solar cells. The efficiency of these PV cells is lower compared to that of first-generation solar cells, but the manufacturing cost is lower. This solar cell technology does not require high-temperature processing unlike first-generation solar cells, being manufactured by depositing the thin film of the above materials on the substrates (Si, glass or ceramics) using chemical vapor deposition, molecular beam epitaxy, or spin-coating technique. Third-generation technologies mainly focus on the improvement of second-generation the energy-conversion efficiency and light-absorption coefficient, while keeping the production cost close to that of second-generation PV cells. The efficiency enhancement is achieved by manufacturing multi-junction PV cells, improving the light-absorption coefficients, and using techniques to increase the carrier collection.

## 4.3  Photovoltaic Modules, Arrays, and Systems

The solar cell is the basic unit of a PV system. An individual solar cell produces direct current and power typically between 1 and 2 W, hardly enough to power most applications. For example, in case of crystalline silicon solar cells with a typical area of $10 \times 10$ cm$^2$ an output power is typically around 1.5 W$_p$, with $V_{OC} \approx 0.6$ V and $I_{SC} \approx 3.5$ A. Most of the commercial crystalline silicon solar PV cells have open-circuit voltage $V_{OC}$ ranging from around 0.55–0.72 V at a cell temperature of 25°C. With a cell area of $A_{cell} = 100$ cm$^2$ (4 inch wafer) such cells have short-circuit current ISC from 3 to 3.8 A; with $A_{cell} = 155$ cm$^2$ (5 inch wafer) the figure ranges from 4.5 to 6 A; with $A_{cell} = 225$ cm$^2$ (6 inch wafer) the figure ranges from around 6.8 to 8.5 A; and with 400 cm$^2$ (8 inch wafer) the figure ranges from around 13–15 A. For optimal power yield, the voltage is around $V_{MPP}$, from 0.45 to 0.58 V.

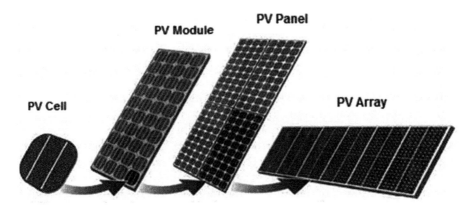

**FIGURE 4.7**
PV Solar cell, module, panel and array configurations.

Hardly any appliance can be operated at such low voltages. Hence voltages that are usable for PV system operation can only be generated by using several PV cells wired in series. If higher currents are needed for a specific application, a series of solar cells must be wired in parallel. For practical applications, the PV cells are arranged in series configuration to form a module, and modules can then be connected in parallel-series configurations to form panels and arrays, as shown in Figure 4.7. A number of PV panels connected in series and/or parallel forms a PV array. These series and parallel connections of PV cells allow for an unlimited interconnection of PV cells to form large PV solar generator fields having several megawatts of power. In some cases it is useful to wire in series anywhere from 32 to 72 PV cells, housed in a single enclosure. The entities thus created are referred as PV modules, solar modules or simply modules. PV modules with 36 cells, operating voltages between 15 and 20 V, and output ranging from 50 to 200 Wp are widely used, because a viable 12 V power source is formed from one module and battery. On the other hand, modules comprising 72 PV cells for operating voltages between 30 and 40 V are suitable for stand-alone 24 V systems, with output up to about 200 Wp, with an area of 1.5 m² and 18 kg weight. The output of the largest mass-produced polycrystalline module currently on the market is 300 Wp (80 cells wired in series). The largest monocrystalline module with an output of around 315 Wp (96 cells wired in series) is manufactured by the Sun power. Larger PV modules for building integration are custom designed.

The PV module manufacturing process for crystalline solar cells is relatively labor intensive. The packing method is extremely cost intensive in that the requisite materials, especially for modules with aluminum frames, consuming considerable energy amounts for manufacturing. Laminates with thicker (e.g., 6–10 mm) glass that can be set into facades or roofs are made with plate glass, being often integrated into buildings. Manufacturers are also selling special plastic or glass solar roof tiles for rooftop PV systems. These tiles are usually larger than normal tiles. Framed PV modules are superior to their laminate counterparts in terms of handling, mechanical stability and lightning protection. However, in framed modules, particularly those flush-mounted on the roof, over time a permanent grit layer forms between the outermost cells and the frames, reducing the energy yield. Hence the frames on the outside of such modules (i.e., the Sun-facing side) should be as thin as possible, with 5–15 mm clearance left on all sides between the PV cells and frames. When connecting PV cells or modules in series, they must have the same current rating to produce an additive voltage output, and similarly, modules must have the same voltage rating when connected in parallel to produce larger currents. The voltage in solar cells wired in series is cumulative, which means that the cumulative voltage of all cells wired in series of $N_S$ units, is $N_S$ times as high as the voltage in one cell. Owing to the power source characteristics of solar cells, the level of current in series-wired configurations is determined by the weakest cell.

The life span of PV modules is largely determined by how well they are protected against the ambient environment. Some vendors indicate a 30-year life span and grant 2–5-year full warranties and in some cases limited performance guarantees for 10–26 years. The fronts of most such products are well protected against hail, by tempered highly transparent low-iron glass. In addition, solar cells are hermetically packed in a transparent plastic material such as ethyl vinyl acetate. The rear protection elements are made of plastic or glass, depending on the manufacturer. A classic PV module integrates relatively thin (e.g., 3–4 mm) glass and has a robust metal frame (usually made of aluminum) that provides the requisite mechanical stability and good edge protection. Modules with thin-film silicon solar cells or the like also have plastic frames. A PV array is a collection series, parallel, or both series and parallel, connected PV modules. The size of a PV array depends on the requirement of electrical power. Often, t he DC power produced from a PV array is converted into AC power using an inverter and fed to the different electrical loads. PV modules are connected in series to achieve the desired voltage; then such series-connected strings are connected in parallel to enhance the current and hence power output from the PV panel or PV array. The size of the PV array decides its capacity, which may be in the range of kilowatts, or even megawatts, depending on the application specifics.

Voltage generated by the PV modules and arrays depends primarily on the design and the cell materials, where the electric generated current is primarily determined by the incident solar radiation and the cell area. The amount of current generated by photon excitation in a PV cell at a given temperature is affected by incident light in two ways:

1. By the intensity of the incident light.
2. By the wavelength of the incident rays.

The materials used in PV cells have different spectral responses to incident light, and exhibit a varying sensitivity with respect to the absorption of photons at given wavelengths. Each semiconductor material will have an incident radiation threshold frequency, below which no electrons will be subjected to the photovoltaic effect. Above the threshold frequency, the kinetic energy of the emitted photoelectron varies according to the wavelength of the incident radiation, but has no relation to the light intensity. Increasing light intensity will proportionally increase the rate of photoelectron emission in the photovoltaic material. In actual applications, the light absorbed by a solar cell will be a combination of direct solar radiation, as well as diffuse light bounced off of surrounding surfaces. Solar cells are usually coated with anti-reflective material so that they absorb the maximum amount of radiation possible.

In the outdoor environment the PV module output current magnitude depends on the solar irradiance and can be increased by connecting solar cells in parallel. The voltage of a solar cell does not depend strongly on the solar irradiance, depending mainly on the cell temperature. PV modules can be designed to operate at different voltages by connecting solar cells in series. It is usually assumed that all the PV cells and modules making up a PV generator are identical and work under the same conditions. However, in reality, the PV cell or module characteristics are subject to variations, dye to uneven sunlight on the solar cells, unclean PV cells, variation and inconsistence of the cell parameters, from manufacturing process or other factors. I–V characteristics for various insulations, at a constant cell temperature, under laboratory test conditions, provide insufficient information concerning practical applications, since insolation can make a PV module quite hot. Depending on mounting type, module design and wind conditions, cell temperature $T_{cell}$ at 1 kW/m² insolation normally ranges from 20°C to 40°C above ambient temperature $T_a$. If the PV module's available electrical power is drawn off by the outer circuit, $T_a$ is somewhat lower than at open or short-circuit, where insolation is converted into heat. This phenomenon can be used for purposes such as thermo-graphic searches for inactive modules in large solar generator fields. The cell temperature increase relative to ambient temperature $T_{cell}$ due to the PV module design is determined using nominal operating cell temperature (NOCT), defined as the fixed temperature, the module would exhibit in the AM1.5 spectrum at open circuit at an ambient temperature of 20°C, 1 m/s wind speed and $SR_{NOCT} = 800$ W/m² irradiance. The cell temperature in a PV module is computed by using Equation (4.23).

PV cell models, developed for a single PV cell, can be easily extended to series and parallel connections of PV cells inside a PV module, consisting of several cells connected in series and in parallel to provide some desired output voltage and current. Equation (4.16) can be easily adapted for a PV module, and then the current-voltage-characteristics of a PV module could be described as:

$$I_C = I_L - I_S \left( \exp\left( \frac{q\sum(V + I \cdot R_S)}{nkT} \right) - 1 \right) - I_{R_P} \tag{4.37}$$

Usually, the cell parallel resistance is considered to stay constant. When PV cells are wired in series, they all carry the same current, and at any given current their voltages add, and overall PV module voltage $N_S$ PV cells connected in series is:

$$V_{module} = V_1 + V_2 + V_3 + \ldots + V_{N_S} = \sum_{k=1}^{N_S} V_k \tag{4.38}$$

The module current, for similar PV cells is:

$$I_{module} = I_1 = I_2 = \ldots = I_{N_S} \tag{4.39a}$$

While in the case of dissimilar solar cells, a common practice is to set it to:

$$I_{module} = I_{Min} \tag{4.39b}$$

The PV module current is equal to the individual cell current. In the case of $N_S$ identical PV cells, having each an output voltage $V_{cell}$, connected in series, then the PV module voltage is:

$$V_{module} = N_S \cdot V_{cell} \tag{4.40}$$

**Example 4.12:** A PV module consists of 36 PV cells, each having a short-circuit current of 3 A, and the open-circuit voltage 0.6 V. If the cells are connected in series, then:

    a. Compute the module short-circuit current and open-circuit voltage.
    b. If one of the PV cell has, due to some technical reasons a short-circuit current of 2 A and an open-circuit voltage of 0.3 V, what are the module new short-circuit current and open circuit voltage.

**Solution:**

    a. From Equations (4.39b) and (4.40), the PV module open-circuit voltage and short-circuit current are:

$$V_{\text{module}} = 36 \times 0.6 = 21.6 \text{ V}$$

$$I_{\text{module}} = I_1 = I_2 = \ldots = I_{N_S} = 3 \text{ A}$$

    b. In the case of one dissimilar PV cell, the module parameters are:

$$V_{\text{module}} = 35 \times 0.6 + 0.3 = 21.3 \text{ V}$$

$$I_{\text{module}} = I_{Min} = 2 \text{ A}$$

If a number, $N_P$ of PV cells are connected in parallel, the voltage for identical PV cells is:

$$V_{\text{module}} = V_1 = V_2 = \cdots = V_{N_P} \qquad (4.41a)$$

While in the case of dissimilar solar cells, a common practice is to set it to:

$$V_{\text{module}} = \frac{1}{N_P} \left( V_1 + V_2 + \cdots + V_{N_P} \right) \qquad (4.41b)$$

The PV module current is equal to the sum of all NP individual solar cell currents:

$$I_{\text{module}} = I_1 + I_2 + \cdots + I_{N_P} = \sum_{k=1}^{N_P} I_k \qquad (4.42)$$

If $N_P$ identical PV cells are wired in parallel, then the module voltage equals a PV cell voltage, while the PV module current is given by:

$$I_{\text{module}} = N_P I_{cell} \qquad (4.43)$$

**Example 4.13:** A PV module is made up of 28 cells in series, each having an open-circuit voltage of 0.5 V and short-circuit current of 2 A. A PV panel is made up of two such PV modules, connected in parallel. If the maximum voltage and maximum current of this PV panel are 12.5 V and 3.85 A, what is the fill factor of the panel.

**Solution:** The voltage of a module is thus the sum of the cell voltages and the drain current is the maximum current delivered by a cell. We thus have an equivalent generator of 14 V–2 A. The voltage of a panel is the voltage of an individual PV module and the current is the sum of the maximum module currents: 14 V and 4 A, respectively. The fill factor of the PV panel is then:

$$FF = \frac{12.5 \times 3.85}{14 \times 4} \approx 0.86 \text{ or } 86\%$$

It is important to realize that when PV cells with a given efficiency are incorporated into PV modules, the module efficiency is less than the individual PV cell efficiency; even if the cells are exactly identical electrically, there additional PV module losses, affecting the overall module efficiency. When cells are operated at or closer to their maximum power point, located on the cell *I–V* curve at the point where the cell undergoes a transition from a nearly ideal current source to a nearly ideal voltage source. If the cell *I–V* curves are not identical, since the current in a series combination of cells is the same in each cell, each cell of the combination is not necessarily operate at its maximum power point. Instead, the PV cells operate at a current consistent with the rest of the cells in the module, which may not be the maximum power current of each cell. For the same reasons that the PV module efficiencies are less than the PV cell efficiencies in the module, the PV array efficiency is less than the individual module efficiency. However, since a large array are built with sub-arrays, operating essentially independently of each other, in spite of the decrease in efficiency at the array level, PV arrays that produce in excess of 1 MW are in operation at acceptable efficiency levels. The bottom line is that most efficient operation is achieved if the PV modules are made of identical cells and if arrays consist of identical modules. In the above relationships, the cell resistances where ignored. However, a more accurate estimate of a series PV module output voltage must include, cell series resistances, $R_S$ (see Figure 4.3 and Equations 4.12 and 4.15), which yields to:

$$V_{\mathrm{mod}} = N_S (V_d - R_S I) \qquad (4.44)$$

**Example 4.14:** A PV module consists of 36 identical PV cells, wired in series. With 1-Sun insolation (1 kW/m²), the PV cell short-circuit current $I_{SC} = 3.50$ A, and at 25°C, the reverse saturation current, $I_0$ is equal to $10 \times 10^{-10}$ A. Assuming that PV cell parallel resistance, $R_P = 10$ Ω and series resistance $R_S = 0.006$ Ω, compute the voltage, current, and power delivered to the load, when the $V_d$ is 0.45 V

**Solution:** By using Equation (4.13) the output cell current is:

$$I = I_{SC} - I_0 (e^{38.9 V_d} - 1) - \frac{V_d}{R_P} = 3.50 - 10^{-9} (\exp(38.9 \cdot 0.45) - 1) - \frac{0.45}{10} = 3.42 \text{ A}$$

The module voltage (Equation 4.44) is then:

$$V_{\mathrm{mod}} = N_S (V_d - R_S I) = 36 \times (0.45 - 0.006 \times 3.42) = 15.5 \text{ V}$$

Then the delivered power is:

$$P_{out} = V_{\mathrm{mod}} \times I = 15.5 \times 3.42 = 53.0 \text{ W}$$

## 4.3.1 PV Module Efficiency and Characteristics

The electrical output of a PV module depends on solar irradiance, solar cell temperature and efficiency of solar cell, as well as the load resistance. For a given PV cell size, the current increases with increasing solar irradiance, being marginally affected (quite small increase) due to temperature rise. However, a higher solar cell temperature decreases the PV cell output voltage, which in turn is decreasing the power output. Load resistance is decided by the operating point of module; the preferred operating is peak power point. Solar cell efficiency is governed mainly by the manufacturing process and the solar cell material, and it varies from about 9%–20%. Therefore, for better performance, the PV module in an array must operate at the peak power point; the array must be installed in an open place (no shading); and the PV module must be kept cool. In order to estimate the PV module efficiency, we need to define the he packing factor is defined as the ratio of total solar cell area to the total module area, expressed as:

$$F_{pckg} = \frac{\text{Total PV Cell Area}}{\text{PV Module Area}} \tag{4.45}$$

It is clear that the packing factor less than unity (pseudo solar cell), having a maximum value of one when the entire PV module area is covered by the PV cells (e.g., rectangular solar cell). PV module electrical efficiency, taking into account also the transmittance coefficient, $\tau_{glass}$ of the module protection layer, can be expressed as a percentage of the PV cell efficiency by the following relationship:

$$\eta_{mod} = \tau_{glass} \cdot F_{pckg} \cdot \eta_{cell} \tag{4.46}$$

The above relationship is showing that the electrical efficiency of a PV module is less than the electrical efficiency of solar cell due to presence of glass over the solar cell and packing factor. The PV module efficiency is also expressed as:

$$\eta_{mod} = \frac{FF \times I_{SC} \times V_{OC}}{SR \cdot A_{mod}} \tag{4.47}$$

where $A_{mod}$ is the PV module area SR incident solar intensity on the PV module, $FF$, $I_{sc}$ and $V_{oc}$ are the fill factor, the short-circuit current, and the open-circuit voltage, respectively, of the PV module. The maximum value of fill factor ($FF$) of a Si-based PV module is about 0.88. The temperature dependency of the PV module electrical efficiency can be expressed, as for the PV cell by:

$$\eta_{e\,mod} = \eta_{STC-mod}\left[1 - \beta_{ref}\left(T - 298\right)\right] \tag{4.48}$$

where $\eta_{STC-mod}$ is the electrical efficiency of the PV module under standard test conditions (STC), the reference temperature coefficient, as mentioned in previous chapter sections, depends on the PV cell material, having typical values from 0.002 to 0.005.

> **Example 4.15:** Assuming that the STC efficiency of a PV module is 14%, what is the efficiency at an ambient temperature of 38°C?
>
> **Solution:** From Equation (4.45) the actual PV module efficiency is:

$$\eta_{e\,mod} = 0.14\left[1 - 0.05 \times 13\right] = 0.1309 \text{ or } 13.09\%$$

## 4.3.2 PV Module Technology and Types

Single-crystal solar cell module consists of series-connected crystalline solar cells sandwiched between a top glass cover (with high light transmitivity, usually made of low-iron glass), encapsulated in transparent and insulating ethylene vinyl acetate, and a back cover (Tedlar/Mylar/glass). A series, parallel, or series-parallel configurations of these modules forms the crystalline PV panels or PV arrays. Crystalline PV modules are divided into two categories depending on the material of the back cover of the module. If the back cover of the module is made of opaque Tedlar, it is known as a "opaque" PV module, while if the back cover is made of glass, it is known as a "semitransparent" PV module. The amount of light transmitted from the PV module depends on its packing factor, which is affecting the electrical characteristics and the PV module efficiency.

Thin-film PV modules are made of thin-film solar cells, manufactured at lower temperature compared with crystalline solar cells, less energy intensive, and having lower production cost of than that of crystalline solar cells and modules. Furthermore, the thin films can be easily deposited onto different substrates such as glass, metal, or even at plastic, leading to greater interest in manufacturing of thin-film PV modules. Another advantage of thin-film PV modules is that they can bend, therefore, they can be used at different structures of the building facade and other glazing areas. The major drawback of this technology is low energy-conversion efficiency and degradation on exposure to adverse weather conditions. The major challenge is the improvement of the conversion efficiency of commercially thin-film PV modules. However, the thin-film solar cell technology is gaining interest, and large-capacity manufacturing plants are already in operation or in development in many countries.

Multi-junction solar cells utilize a wider range of solar spectra for electricity generation. In multi-junction solar cells, different solar cells placed in a tandem arrangement have different band gaps. Therefore, a multi-junction solar cell utilizes a different range of spectra for electricity generation, which reduces absorption losses and improves efficiency. The PV modules manufactured from multi-junction PV cells are very lightweight panels and are used particularly in space applications. The tandem arrangement can be made by mechanical stacking, by monolithic technique, or by both. The most commonly used dual-junction solar cells include the GaAs cells with efficiency of up to 30% and the multi-junction cell made of GaInP/GaInAs/Ge (gallium indium phosphide/gallium indium arsenide/germanium substrate), with an efficiency (with concentrating sunlight) of about 41%.

These types of PV modules employ the latest and emerging technology of solar cells, namely, organic solar cells (OSC), dye-sensitized solar cells (DSSC), quantum-well solar cells (QWSC), etc. The major research issues and development activities, regarding these technologies are the reduction of the production cost and improving the energy conversion efficiency. PV modules utilizing new and emerging solar cell technology are categorized on the basis of light-absorbing capacity and electricity-generation mechanism. Dye-sensitized solar cells contain porous nano-particles of titanium dioxide, which enhance the light-gathering capacity of the solar cell and hence its electrical efficiency. DSSC has a low production cost because the processing mechanism is simpler and the material cost is also low. The quantum-well solar cell consists of a low-energy band-gap material sandwiched between relatively large-energy band-gap materials such as GaAs.

### 4.3.3 PV System Operation

Solar panels or modules are rarely connected directly to a load, but rather are used to charge energy storage components such as batteries or ultra-capacitors. In most cases, the battery charging voltage determines the solar panel operating voltage. The battery charging voltage is usually not the most efficient operating voltage for the solar cell and therefore the most power is not being extracted from the solar cell. There also exists a possibility of overcharging when the solar panel is connected directly to a battery and overcharging can damage the battery. To avoid these potential problems, a charge controller is inserted between the solar panel and load(s), or the storage units, such as batteries or ultra-capacitors. The most commonly used type of charge controllers include: basic charge controllers and Maximum Power Point Tracker (MPPT) charge controllers. The simplest form of charge controllers is the basic charge controller. These are usually designed to protect overcharging or undercharging of batteries which can cause damage to the battery. Continually supplying a charging current to a fully charged battery increases the battery voltage causing it to overheat and damage. The basic charge controller simply monitors the state of charge of the battery to prevent overcharge. This form of charge controller regulates the voltage supply to the battery and cuts off supply once the battery reaches it maximum charge state. Overcharging some batteries can lead to explosions or leaking. On the other hand, undercharging a battery for sustained periods tend to reduce the life cycle of the battery. The charge controller monitors the state of charge of the battery to prevent it from falling below the minimum charge state. Once a battery reaches its minimum charge state, the charge controller disconnects the battery from any load to prevent the battery from losing any more charge. The basic charge controllers are usually operated by a simple switch mechanism to connect and disconnect the battery from the solar panel or load to prevent overcharging and undercharging.

MPPT charge controllers optimize the power output of the solar cell while also charging the battery to its optimal state. The MPPT constantly tracks the varying maximum operating point and adjusts the solar panel operating voltage in order to constantly extract the most available power. The MPPT charge controller maximizes solar cell efficiency while also controlling the charging state of the battery. The MPPT charge controller is basically a DC-DC converter that accepts a DC input voltage and outputs a DC voltage higher, lower or the same as the input voltage. This capability of the converter makes it ideal for converting the solar panel maximum power point voltage to the load operating voltage. Most MPPT charge controllers are based on either the buck converter (step-down), boost convert (step-up) or buck-boost converter setup. A tracker consists of two basic components, a switch-mode power converter and a control section with tracking capability. The switch-mode power converter is the core of the entire supply. The main component of the MPPT is the DC-DC power (buck, boost or buck-boost) converter that steps down or up the solar panel output voltage to the desired load voltage. To ensure that the solar module operates at the maximum operating point, the input impedance of the DC-DC power converter must be adapted to force the solar module to work at its maximum power point (MPP). Depending on the load requirement, other types of DC-DC converters may be employed in the MPPT design. The control section is designed to determine if the input is actually at the MPP by reading voltage/current back from the switching power converter or from the array terminal and adjust the switch-mode section. Depending on the application, different feedback control parameters are needed to perform maximum power tracking. Most commonly voltage and power feedback controls are employed to control the system and to find the MPP of the array. The solar array terminal voltage is used as the control variable for the system. The system keeps the array operating close to its maximum power point by regulating the array's voltage and matches the voltage of the array to a desired voltage. However, this has the drawback that it cannot be widely applied to battery energy storage systems. The solar array terminal voltage is used as the control variable for the system. The system keeps the array operating close to its maximum power point by regulating the array's voltage and matches the voltage of the array to a desired voltage. However, this has the drawback that it cannot be widely applied to battery energy storage systems.

As discussed in this section, the MPP is obtained by introducing a DC-DC power converter in between the load and the solar PV module. The duty cycle of the converter is changed till the peak power point is obtained.

The power extracted from a PV array Pa is determined by the terminal voltage Va and output current $I_{array}$ of the array. The terminal voltage $V_{array}$ depends on the control of the DC-DC power converter while the output current $I_{array}$ depends on temperature, irradiation level, and the PV array terminal voltage. During a day, solar irradiation and temperature fluctuates over time, causing the MPP of the PV array changes continuously. Consequently, the PV system operating point must be adjusted constantly to maximize the energy produced. There are many different approaches to maximizing the power from a PV system. These range from using simple voltage relationships, to more intelligent and adaptive based algorithms. Different MPPT control algorithms are helping to track the peak power point of the solar PV module automatically. The most methods, used to control the MPPT are:

1. Incremental conductance maximum power point method;
2. Constant voltage maximum power point method;
3. Constant current maximum power point method;
4. Perturb and observe maximum power point method;
5. Fuzzy logic-based maximum power point method
6. Neural network-based maximum power point method; and
7. Neuro-fuzzy-based maximum power point method.

*MPPT* systems are used mainly in systems where source of power is nonlinear such as the solar PV modules or the wind generator systems. *MPPT* systems are generally used in solar PV applications such as battery chargers and grid connected standalone PV systems. However, the primary challenges for maximum power point tracking of a solar PV array include: (1) how to get to a MPP quickly, (2) how to stabilize at a MPP, and (3) how to smoothly transition from one *MPP* to another for sharply changing weather conditions. In general, a fast and reliable *MPPT* is critical for power generation from a solar PV system. In conventional *MPPT* methods for solar PV generators, the *MPPT* control is achieved by varying the terminal voltage applied to the PV generator. Typical *MPPT* techniques are the fixed-step *MPPT*: Short-Circuit Current (*SCC*), Open-Circuit Voltage (*OCV*), Perturb and Observe (P&O) methods, and Incremental Conductance (*IC*) methods. The *SCC* method is based on the observation that $I_{MPP}$ is about linearly proportional to $I_{SC}$ of a PV array, i.e.,

$$I_{MPP}(Sys) = k_{SC} \cdot I_{SC}(Sys) \tag{4.49}$$

where $k_{SC}$ is an empirical constant. According to Equation (4.48), the *SCC* controller generates a control signal to the dc/dc converter based on the error signal between the actual current of the PV array $I_{array}$ and the $I_{MPP}$ calculated from Equation (4.48). This method requires measurements of *ISC*. Therefore, a static switch in parallel with the PV array is needed in order to create the short-circuit condition for each solar irradiation level change, which could cause a large oscillation of PV array output power. Another disadvantage is that the computation of $I_{MPP}$ is very sensitive to $k_{SC}$, and the relation between $I_{MPP}$ and $I_{SC}$ is not 100% linear. Thus, a small deviation of $I_{MPP}$, calculated by the Equation (4.48), can easily reduce the output power of the PV array greatly. However, the method is simple and easy to be implemented, having quite a few applications in low- or very low-power solar harvesters. Similar, the OCV method is based on the observation that $V_{MPP}$ is about linearly proportional to $V_{OC}$ of a PV array, i.e.,

$$V_{MPP}(Sys) = k_{OC} \cdot V_{OC}(Sys) \qquad (4.50)$$

where $k_{OC}$ is a constant. Based on Equation (4.49), a close-loop control scheme is developed to bring the PV array voltage to $V_{MPP}$ calculated from this equation. Similar to the *SCC* technique, the *OCV* method requires measurements of $V_{OC}$. Thus, a static switch in series with the PV array is needed in order to create the open-circuit state for each weather condition change, which can also cause a large oscillation of PV array output power. In addition, since $V_{OC}$ varies with temperature and other factors and the relation of $V_{MPP}$ and $V_{OC}$ is affected by shading, actual $V_{MPP}$ for a practical PV application is quite difficult to get. The *OCV* method has similar applications with *SCC* technique. The *IC* method is based on the principle that ideally the following equation holds at the *MPP*:

$$\frac{\Delta I_{\text{array}}}{\Delta V_{\text{array}}} + \frac{I_{\text{array}}}{V_{\text{array}}} = 0 \qquad (4.51)$$

When the operating point in the P-V plane is to the right (left) of the MPP point, then the expression $\Delta I_{\text{array}} / \Delta V_{\text{array}} + I_{\text{array}} / V_{\text{array}}$ is negative, respectively positive. Thus, the direction in which the MPP operating point must be perturbed can be determined by comparing the instant conductance to the incremental conductance, the second and first terms in Equation (4.50). Using the *IC* method, it is theoretically possible to know when the *MPP* is reached and when the perturbation should be stopped. The P&O method is the most commonly used *MPPT* technique for PV arrays. It operates by periodically perturbing the array terminal voltage or current and comparing the PV output power with that of the previous perturbation cycle. In general, if an increased perturbation of PV array operating voltage causes an increase of output power, the control system moves the operating point in the same direction; otherwise the perturbation is changed to the opposite direction. The process continues until the MPP is reached. There are many different P&O methods available in the literature. In the classic P&O technique, the perturbations of the array operating point have a fixed magnitude. In the optimized P&O technique, an average of several samples of the PV array power is used to adjust the perturbation magnitude. Full discussion of the *MPPT* methods is beyond the scope of this chapter, interested readers are directed to the end-of-chapter references or elsewhere in the literature.

As we can discussed before, when directly coupled, a PV system and load usually leads to a certain mismatch between the actual and the optimum operation voltage (*MPP* voltage) of the solar generator and therefore causes energy losses. The *MPP* of the solar cell (respectively, the solar module or the solar generator) is positioned near the bend in the curve shown in Figure 4.8. The corresponding values of $V_{mpp}$ and $I_{mpp}$ can be estimated, as rule of thumb, based on empirical and test data from the open circuit voltage and the short circuit current, as:

$$V_{mpp} \approx (0.75 \cdots 0.90) V_{OC}$$

$$I_{mpp} \approx (0.85 \cdots 0.95) I_{SC} \qquad (4.52)$$

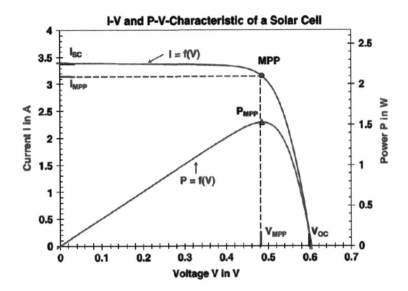

**FIGURE 4.8**
PV cell and I-V and P-V diagrams, showing the location of the *MPP* and maximum power.

Because cell voltage and current depend on temperature (as we discussed in previous chapter sections) the supplied power also changes with temperature. The power of crystalline silicon solar cells drops by about 0.4%–0.5%/K. and the power of amorphous silicon solar modules drops by about 0.2%–0.25%/K. Remember that the rated power of a solar cell or a solar module is measured under internationally specified test conditions (*STC* = Standard Test Conditions) with the following parameters: 25°C, 1000 W/m² solar radiation and calm wind conditions.

## 4.4 PV Systems and Components

PV systems are different from conventional energy systems, because the input in a PV system depends on the insolation, being not set by the load demands. PV outputs can vary as a result of external factors such as local atmosphere conditions, moving clouds, or insulation. In terms of size, PV systems can be very small (less than 5 W), small (5 W–1 kW), lower size (1 kW to a few 10 s of kW), and intermediate size (10 s of kW–100 kW), or a large-scale system (1 MW or larger), which is usually connected to a utility grid for commercial or utility power generation. There are two major types of PV systems: (1) grid-connected (grid-interactive) and (2) stand-alone (see Figure 4.9, for details). There are two possible versions of a stand-alone system, depending on the load power requirements, lighting, communication, charger and resistive loads. In the case of stand-alone PV system with DC loads, a battery storage unit may be is included, while in the case of stand-alone PV system with AC loads, battery storage units are usually required. Battery storage is essential to the success of stand-alone PV systems' design. During the daylight, PV arrays charge the batteries so that they may supply energy at night and on cloudy days.

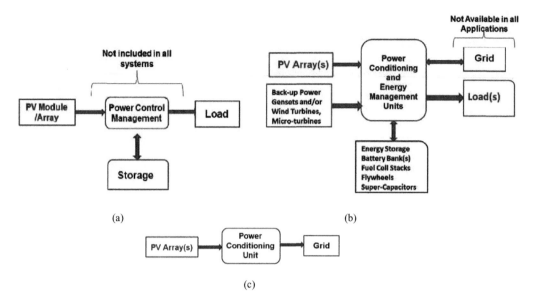

**FIGURE 4.9**
The most common PV systems: (a) PV system connected to a load, (b) hybrid power system, and (c) PV system connected to electric grid.

Photovoltaic systems are classified into two major classes: grid-connected PV systems and stand-alone PV systems. Figure 4.9 illustrates three most common PV system configurations. The simplest system is that of Figure 4.9a in which the PV module or array is directly connected to a DC load, often through a power processing unit and with storage device(s). This configuration is most commonly used with a fan, a water pump, or a lighting system. However, it is likely that even if the load is a water pump, a DC-DC power converter between the PV array and the pump motor is used. The configuration of Figure 4.9b includes a charge controller and storage unit (batteries or other storage devices), and back-up and/or other renewable energy conversion systems, such as wind turbine(s), so the system can produce enough energy that can be used day or night by the loads, or even during the period without insolation. The charge controller serves a dual function. If the load does not use all the energy produced by the PV array, the charge controller prevents the batteries from overcharge. The first two configurations are stand-alone system types, while the one in Figure 4.9b is the so-called hybrid power system. The system shown in Figure 4.9c is a grid-connected (or utility-interactive) system. The inverter of a utility interactive system must meet more stringent operational requirements than the inverter of a standalone PV system. The inverter output voltage and current must be of utility-grade quality, meaning that it must have minimal harmonic content and range fluctuations. Furthermore, the inverter must sense the utility, and if utility voltage is lost, the inverter must shutdown the system until voltage is restored to within normal limits.

Grid-connected photovoltaic systems are composed of PV arrays connected to the grid through a power conditioning unit and are designed to operate in parallel with the electric utility grid as shown in Figure 4.9c. Grid-connected PV energy conversion systems can be grouped into four different types of configurations: centralized configuration for large-scale PV plants (three-phase); string configuration for small and medium-scale PV systems (single phase and three phase); multi-string configuration for small- to large-scale systems

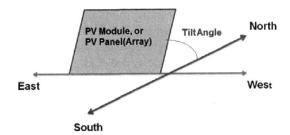

**FIGURE 4.10**
PV array tilt. (Module orientation for maximum solar energy collection year round in the Northern Hemisphere is to tilt the modules south at local (site) latitude angle; the converse holds true for Southern Hemisphere sites.)

(single-phase and three-phase); and AC-module configuration for small-scale systems (commonly single phase). Simplified diagrams of each of these configurations are given in Figure 4.10. The centralized topology (Figure 4.10c) has, as its main characteristic, the use of a single three-phase voltage source inverter to connect the whole PV plant to the grid, or a portion of it, in case the PV plants exceed the power rating of the existing central inverters. The PV system is formed by the series connection of modules (string) to reach the desired DC-link voltage and by several strings in parallel to reach the power rating of the inverter. The advantages of this configuration are its simple structure, single transformer and single control system (single set of sensors, control platform and grid monitoring unit). This comes at the expense of reduced power generation because of a single *MPPT* algorithm for the whole PV system. Also, diode conduction losses are introduced by the string series blocking diodes. Currently, the central configuration is the most widely used topology for large-scale PV plants.

The main advantage of PV systems is their flexibility to be implemented in remote locations where grid connection is either impossible or very expensive to execute. Such systems are called stand-alone PV systems and are described in the following section. A solar PV energy conversion system requires power converters for maximum power extraction and grid integration. At present, many different converter structures have been developed and used in a solar PV system. For all the different converter structures, the energy extraction characteristics and maximum power capture capability for all the converter schemes under even solar irradiation are very similar. However, under shading conditions, the energy extraction depends strongly on what converter structure is used in a PV system. Besides of *MPPT*, designed to control PV arrays to perform efficiently, a number of other components are required to control, convert, distribute and store the energy produced by the array. Such components may vary depending on the functional and operational requirements of the system. They may require battery banks and controller, DC-DC converters, DC-AC inverters, in addition to other components such as overcurrent, surge protection and other power processing equipment.

PV systems are made up of several components, which may include PV array(s), wires, fuses, protection devices, controllers, batteries and other storage devices, Sun trackers, DC-DC power converters and inverters. Included components are determined by the application type. However, the PV systems are modular by nature, thus, systems can be readily expanded, restructured and the components easily repaired or replaced if needed. PV systems are cost effective today for many remote power applications, as well as for small stand-alone power applications even in the proximity to the existing electric grid. These systems must use good electrical design practices, such as the National Electrical

Code (NEC) or its equivalent. Energy flowing through a power system runs through a variety of devices and wires between PV system's components. In a PV system, balance of system (BOS) refers to all of the system components except the PV modules/panels, and they account for half of the system cost and most of the system maintenance. The BOS components may include fuses and disconnect switches to protect the systems, support structures, enclosures, connectors to link different hardware components, switch-gear, ground fault detectors, charge controllers, general controllers, batteries, inverters and meters to monitor the system performance and status. The selection of performant BOS components is as important as the PV module selection. Low-quality BOS is often responsible for many avoidable maintenance problems for PV systems in remote areas, often leading to premature failure and disuse of the whole system. The PV industry goal is to provide PV systems with operational life spans of 25 or more years. Despite this, inexperienced system designers and installers, improperly selecting connectors, cable, protective devices, etc., can significantly reduce the system lifespan. Power electronic converters, a critical component in a PV system, have the following typical architectures: (1) central DC-AC and DC-DC converter structure, (2) central DC-AV inverter and string DC-DC converter structure, (3) central DC-AC converter and DC-DC optimizer structure, (4) detached micro-inverter structure and (5) central and string inverter structures.

Charge controllers manage the electricity flow among the PV array, storage units, loads and eventual grid connection. The appropriate charge control algorithms and charging currents are required for the batteries, storage units, loads and grid connection specifications. High quality charge controllers allow for adjustable regulation voltages, multiple stage charge control, temperature compensation, and equalization charges at specified intervals for flooded batteries, loads and grid connection. The main purpose of a charge controller is to protect system components from damage due to excessive overcharging, discharging or over-currents. Most controllers function by sensing battery voltage and then take action based on voltage levels. Other controllers have temperature compensation circuits to account for the effect of temperature on battery voltage and state-of-charge. Controllers pose more problems than any other component in a stand-alone PV system, because they are complex devices, depending on the battery state-of-charge. The battery's state of charge depends on many factors and is difficult to measure. The controller in a PV system must be sized to handle the maximum current produced. It is recommended to multiply the array short-circuit current by a factor of 1.25 or greater to allow for short-term insolation enhancement by clouds. The maximum current value and system voltage is the minimum needed to specify a controller. The battery charge controller, an electronic device that prevents overcharging and excessive discharging, both of which can dramatically shorten the battery's life is essential for the long life of the battery. In large PV systems, equilibration of the battery charging (so that all battery cells charge equally) should also be incorporated. In hybrid PV systems, which combine photovoltaics and a diesel or wind generator, the control unit must connect and disconnect the different generators according to a plan. Also, loads can be categorized, so that in case of low battery charge and low PV output, some loads can be disconnected while some essential ones are maintained active. Among the most important requirements of charge controllers are: low internal consumption, high efficiency value (96%–98%), load disconnection if deep discharge occurs (current-dependent, discharge cut-off voltage, if possible), regular charging at a higher voltage to promote gassing, temperature compensation of the charging cut-off voltage, breakdown voltage of the semiconductor components at least twice the open circuit voltage of the solar generator, integrated overvoltage protection and good ambient temperature range.

Inverters are a key component to most PV systems installed in grid-connected or distributed generation applications, converting DC voltage to AC. Aside from the modules themselves, inverters are often the most expensive PV system component, and quite often are the critical factor in terms of overall system reliability and operation. Utility-interactive PV systems installed in residences and commercial facilities are a low-power, but important, source of electric generation. This new concept in power generation, a change from large-scale central generation to small-scale dispersed generation is part of the smart grid paradigm. The system, a PV array generates DC power, converted to AC by inverters and delivered to the grid, is very simple, while yet elegant. The converted AC signals can have square, modified-sine, quasi-sine and pure sine wave forms. The simplest inverter converts DC voltage to square waves. Square waves are acceptable for many AC loads, however their harmonic content is very high, and there are situations where square waves are not satisfactory. Other more suitable inverter output waveforms include the quasi-sine wave and the utility-grade sine wave. Both are most commonly created by the use of the multilevel H-bridges controlled by microprocessors. The pure sine wave is high cost, high efficiency, and has the best power quality. Modified sine wave is mid-range cost, quality, and efficiency. Square wave is low cost and low efficiency, and it has poor power quality, being useful for limited applications. Square wave signals can also be harmful to some electronic appliances due to the high-voltage harmonic distortion. Inverters generate electromagnetic interference that may affect electronic systems. One method of attenuating this electromagnetic noise in some cases is by grounding the inverter enclosure, also a code requirement for safety reasons (NEC, 2008). There are three basic configurations for inverters: stand-alone, grid-tied, and UPS type. The stand-alone inverter must act as a voltage source that delivers a prescribed amplitude and frequency RMS sine wave without any external synchronization. The grid-tied inverter is essentially a current source delivering a sinusoidal current waveform, synchronized by the grid voltage. Synchronization is typically sufficiently close to maintain a power factor in excess of 0.9. The UPS inverter combines the features of the stand-alone and the grid-tied inverter, so that if grid power is lost, the unit acts as a stand-alone inverter, supplying power to emergency loads.

The system-mounting structure is also important, in particular, in concentrating systems. In fact, this is often the second most important cost element in concentrating PV systems, after the modules. In contrast, the power conditioning cost is comparable to many other small costs associated with the plant construction. Aside from the PV array(s) or panel(s), the charge controller, and the inverter, as we mentioned before, a number of other components are needed in a code-compliant PV system. For example, if a PV array consists of multiple series-parallel connections, then it is necessary to incorporate fuses or circuit breakers in series with each series string of modules, defined as a source circuit. This fusing is generally accomplished by using a source circuit combiner box as the housing for the fuses or circuit breakers, while the PV output circuit becomes the input to the charge controller, if a charge controller is used. If multiple parallel source circuits are used, it may be necessary to use more than one charge controller, depending upon the rating of the charge controller. When more than one charge controller is used, source circuits should be combined into separate output circuits for each charge controller input. In a utility-interactive circuit with no battery backup, a charge controller is not necessary. The PV output circuit connects directly to the inverter through either a DC disconnect or a DC ground fault detection and interruption (GFDI) device. A GFDI device is required by the National Electrical Code (NEC) whenever a PV array is installed on a residential rooftop. The purpose of the device is to detect current flow on the grounding conductor. The grounding conductor is used to ground all metal parts of the system and for a properly installed and operating system, no current will flow on the grounding conductor. The NEC also requires properly rated disconnects at the inputs and outputs of all power conditioning equipment. An additional disconnect will be needed at the output of a charge controller as well as between any battery bank and inverter input or DC load center.

### 4.4.1 PV Array Orientation

Maximum energy is obtained when the Sun's rays strike the receiving surface perpendicularly. In the case of PV arrays, perpendicularity between the Sun's rays and the modules can be achieved only if the modules' mounting structure can follow the movements of the Sun (i.e., track the Sun). Mounting structures that automatically adjust for azimuth and elevation do exist. These types of structures are called Sun trackers. Usually, the angle of elevation of the array is fixed. In some cases, azimuth-adjusting trackers are used. Depending on the latitude of the site, azimuth-adjusting trackers can increase the annual average insolation received up to 20% in temperate climates. For the case in which a tracker is not used, the array is mounted on a fixed structure as is shown in Figure 4.10. This system arrangement is simple to operate and easy to install. Because the angle of elevation of the Sun changes during the year, the fixed-tilt angle of the array is chosen so that maximum energy generation is guaranteed. In the Northern Hemisphere, the Sun tracks primarily across the southern sky, for this reason, a fixed PV array is inclined (from the horizontal) to face South. The angle of inclination of the array is selected to satisfy the energy demand for the critical design month. If producing the maximum energy over the year course is the desired goal, the value of the tilt angle of the array is equal to the site latitude. Wintertime production can be maximized by tilting the array about 10° more than latitude, while summertime production is maximized by tilting the PV array about 10° less than latitude. Some practitioners recommend that for sites located at higher latitudes than 30°, the tilt angle can be set at the latitude angle plus 15°. This helps to even out the daily electrical output over the year by optimizing the tilt angle for the winter months. For sites at latitudes between 15°S and 15°N (for example, Malaysia or Kenya), a tilt angle of 15° should be used.

### 4.4.2 Factors Affecting PV Output

PV systems produce power in proportion to the intensity of sunlight striking the solar array surface. Thus, there are some factors that affect the overall output of the PV system and are discussed here. Output power of a PV system decreases as the module temperature increases. For crystalline modules, a representative temperature reduction factor is about 89% in the middle of spring or in a fall day, under full-light conditions. Dirt and dust can accumulate on the solar module surface, blocking some of the sunlight and reducing the output. A typical annual dust reduction factor to use is 93%. Sand and dust can cause erosion of the PV surface, which affects the system's running performance by decreasing the output power to more than 10%.Because the power from the PV array is converted back to AC as shown earlier, some power is being lost in the conversion process, in addition to losses in the wiring. Common inverters used have peak efficiencies of about 88%–90%.

The deposition of dust on PV modules is a serious concern because it affects the performance of the PV systems. Dust deposition depends on the weather conditions and the local environment of the particular place. It also depends on the physical and chemical composition of the dust. The surface of the module also plays an important role for dust deposition because dust accumulation is higher for rough surfaces. The optical performance of a PV module is negatively affected due to dust deposition because dust blocks the solar radiation from coming into direct contact with the solar cells of the module. Under outdoor conditions, the surface finish, tilt angle, humidity and wind speed also affect dust accumulation. Tilted PV systems usually accumulate less dust compared with strictly horizontal systems.

PV module and cell performances are also degraded over the time, the so-called "aging effect" of PV modules. The time period during which a PV module yields the rated power is closely connected to the life time of the PV module, which significantly depends on the materials and technology used for fabrication of the PV module. The aging effect is due to the weather conditions, fluctuation in ambient temperature, rain, dust deposition, etc. Discoloration and large strain formation affect the performance and life time of the PV module. It is suggested to optimize the effect of the environment on the operation of modules. By using UV-filtering glass super-stratus, the PV modules can be made more resistant to degradation.

Another serious problem affecting the PV module, panel or array performances is the PV cells' mismatch. Mismatch losses are caused by the interconnection of solar cells or modules which do not have identical properties or which experience different conditions from one another. Mismatch losses are a serious problem in PV modules and arrays under some conditions because the output of the entire PV module under worst case conditions is determined by the solar cell with the lowest output. For example, when one solar cell is shaded while the remainder PV cells in the module are not, the power being generated by the un-shaded (good ones) PV cells can be dissipated by the lower performance cell rather than powering the load. This in turn can lead to highly localized power dissipation and the resultant local heating may cause irreversible damage to the PV module. The impact and power loss due to mismatch depend on: (a) the operating point of the PV module, (b) the circuit configuration, and (c) the parameter (or parameters) which are different from the remainder of the solar cells that are in good operating conditions. The output of a PV module can be reduced dramatically when even a small portion of it is shaded. Unless special efforts are made to compensate for shade problems, even a single shaded cell in a long string of cells can easily cut output power by more than half.

Partial shading and module mismatch, because of differences in fabrication or differences caused by usage and aging, have been identified ones of the main causes of the PV system reduced energy yields. They are also responsible of the creation of hot spots in the PV cells of a module, which are reducing not only energy production but also the module lifespan. In a string of series-connected PV modules, a shaded module has a very low or even zero photo-current, and thus the current generated by the other modules passes through the module shunt resistance, resulting in a negative voltage that are reducing module voltage and increasing the overall module temperature. Shading is one of the major factors affecting the power output of a PV module, panel or array. The PV module power output is reduced dramatically when even a small portion of it is shaded. Unless special efforts are made to compensate for shading problems, even a single shaded PV cell in a long string of cells can easily cut output power by more than half. External diodes added by the PV module manufacturer or by the system designer, can help preserve the PV module performances. The main purpose for such diodes is to mitigate the shading impacts on I–V curves. Such diodes are usually added in parallel with PV modules or blocks of cells within a module. In order to understand this important shading phenomenon, consider the diagrams of Figure 4.11 in which an *n*-cell PV module with current *I* and output voltage *V* shows one cell separated from the others (shown as the top cell, though it can be any cell in the string). In the case of the shaded PV cell (Figure 4.11a), there is no photo-generated current, meaning the top cell, instead of adding to the output voltage, actually reduces it.

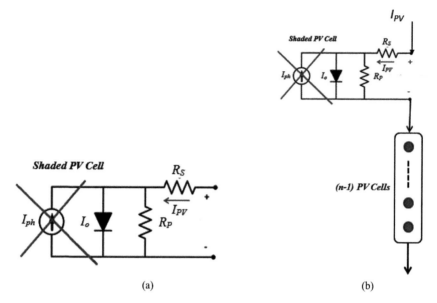

**FIGURE 4.11**
Diagrams of (a) shaded PV cell and (b) *n*-cell PV module with the top cell is in the shade.

The cells having full Sun will generate the same voltage and they have the same current. This means that the voltage, $V_{SH}$, of the entire PV module with one cell shaded drops from the initial $V_{module}$ to:

$$V_{SH} = \left(\frac{n-1}{n}\right) V_{module} - (R_P + R_S)I \qquad (4.53)$$

The drop in voltage $\Delta V$ at any given current $I$, caused by the shaded cell, is given by:

$$\Delta V = V_{module} - V_{SH} = V_{module} - \frac{n-1}{n} V_{module} + (R_P + R_S)I = \frac{V_{module}}{n} + (R_P + R_S)I \qquad (4.54)$$

Since parallel cell resistance is much larger than the series cell resistance the above expression reduces to:

$$\Delta V \simeq \frac{V_{module}}{n} + R_P \cdot I \qquad (4.55)$$

At any given module current the *I*–*V* curve for the module with one shaded cell drops by $\Delta V$, having a large impact on the PV module output voltage, as illustrated in the following example.

**Example 4.16:** The 28-cell PV module described in Example 4.14 has a parallel resistance per cell of $R_p = 5.4 \, \Omega$, and a negligible series resistance. In the full Sun the module current is 1.65 A, and its voltage is 12.0 V. If one cell is shaded and assuming that the current stays the same, estimate the new module output voltage and power, the voltage drop across the shaded cell, and the power dissipated in the shaded cell.

**Solution:** From Equation (4.55) the drop in the module voltage is:

$$\Delta V \simeq \frac{12.0}{28} + 5.4 \times 1.65 = 9.34 \text{ V}$$

The new module output voltage is 12–9.34 = 2.66 V. Power delivered by the module with one cell shaded is then:

$$P_{\text{module}} = V \times I = 2.66 \times 1.65 = 4.39 \text{ W}$$

Notice that, in full sun with all cells operating, the module generates 19.8 W. The voltage drop across the shaded cell and the power dissipated by this cell are:

$$V_{sh-cell} = 5.4 \times 1.65 = 8.91 \text{ V}$$

$$P_{SH-cell} = R_p I^2 = 5.4 \times 1.65^2 = 17.7 \text{ W}$$

All of that power dissipated in the shaded cell is converted into heat, which can cause a local hot spot that may permanently damage the plastic laminates enclosing the cell and eventually the module.

The destructive effects of hot-spot heating may be circumvented through the use of a bypass diode, which is connected in parallel, but with opposite polarity to a solar cell. Under normal operation, each solar cell will be forward biased and therefore the bypass diode will be reverse biased and will effectively be an open circuit. However, if a solar cell is reverse biased due to the a mismatch in short-circuit current between several series connected cells, then the bypass diode conducts, thereby allowing the current from the good PV cells to flow in the external circuit rather than forward biasing each good cell. When a solar cell is in the Sun, there is a voltage rise across the cell so the bypass diode is cut off and no current flows through it. In this way the maximum reverse bias across the poor cell is reduced to about a single diode drop, thus limiting the current and preventing hot-spot heating. In practice, however, one bypass diode per solar cell is generally too expensive and not easy to install. Instead, *bypass diodes* are usually placed across groups of solar cells. The maximum power dissipation in the shaded cell is approximately equal to the total power of all cells in the group, so the maximum group size per diode, without causing damage, is about 15–18 PV cells per bypass diode, for silicon PV cells and modules. For a 36 cell module, tow such bypass diodes are used to ensure the module will not be vulnerable to "hot-spot" damage. In small modules the cells are in placed in series so parallel mismatch is not an issue. Modules are usually paralleled in large PV arrays so the mismatch usually applies at a module level rather than at a cell level. Bypass diodes help current go around a shaded or malfunctioning module within a string. This not only improves the string performance, but also prevents hot spots from developing in individual shaded cells. When strings of modules are wired in parallel, a similar problem may arise when one of the strings is not performing well. Instead of supplying current to the array, a malfunctioning or shaded string can withdraw current from the rest of the array. For this reason, in addition to the use of by-pass diodes to prevent mismatch losses, an additional diode, called *a blocking diode*, may be used to minimize mismatch losses. With parallel connected modules, each string to be connected in parallel should have its own blocking diode. This not only reduces the required current carrying capability of the blocking diode, but also prevents current flowing from one parallel string into a lower-current string and therefore helps to minimize mismatch losses arising in parallel connected arrays. By placing blocking diodes (also called *isolation diodes*) at the top of each string, the reverse current drawn by a shaded string can be prevented.

## 4.5 PV System Design

The general principle of the PV system dimensioning is to always respect the balance between the energy generated by PV system and the consumed energy by the load(s). The balance is achieved for a defined period, usually a day, a week or a month. Presence of the storage units (battery banks, fuel cells or flywheels) allows compensation of the shortfall between the generated and consumed energy, which may be due to weather conditions or overloading on behalf of the consumer. Sizing a PV system requires the following basic steps:

1. Calculation of available solar radiation on the surface of PV module;
2. Setting the voltage level (in many cases this is determined by the user) and thus the type of photovoltaic system (DC, AC, combined DC and AC, grid-connected, stand-alone, with or without a back-up generator);
3. Daily electricity consumption calculation—Ec; this step also includes the determination of the energy demand and optimization of the consumption (this step includes determining the energy demand of the intended consumers as exactly as possible and investigating possibilities for saving energy by reducing the power consumption of the appliances or systems used);
4. Calculating the amount of electrical energy needed to be produced by PV module—*Ep*;
5. Calculation of critical power of PV module—*Pc* and its choice;
6. Calculation of battery capacity—*C* and choice of batteries;
7. Dimensioning the charge controller, cables, wires, and all other system components; and
8. Checking electrical energy consumption and generation balance.

It may be necessary to iterate the previous steps several times in order to optimize the system design and to obtain the most suitable system size. The size of the PV system for a specific application is also determined by considering the available solar insolation, the tilt of the array, and the characteristics of the photovoltaic modules being considered. The available insolation striking a photovoltaic array varies throughout the year and is a function of the tilt of the array. If the load is constant, the designer must consider the time of year with the minimum amount of sunlight (typically December or January). Knowing the insolation available (at tilt) and the power output required, the array can be sized using module specifications supplied by manufacturers. Although it is energy (watt-hours) that is consumed by the loads, PV modules are current-producing devices. Voltages are established by the load or battery bank. Consequently, in systems where the voltage is permitted to vary, the designer should consider current (rather than voltage) in sizing the array. Both "peak" power and current (at peak power) are available from manufacturers' specifications. Using module peak current and daily insolation (in peak sun hours), the average amp-hours of current delivered by a PV module for one day of the "worst" month can be determined. The amp-hours required by the load determined previously are adjusted (upward) because batteries are less than 100% efficient. (Usually 80% battery efficiency is assumed.) Then, knowing the requirements of the load and the capability of a single module at the location during the worst month, the array can be sized. In summary, photovoltaic array sizing is accomplished as follows:

1. Determine available sunlight, and other climatologic conditions.
2. Select PV array, tilt, and candidate modules.
3. Size the PV array (compute daily kWh output).
4. Evaluate the PV system.

As previously discussed, the amount of energy delivered by a PV array or module depends on irradiance and temperature. It is possible to estimate the electrical energy (in kWh/day) expected of an array with known nominal power by using the approximations described here. PV modules, panels or arrays installed on structures anchored to the ground are operating at approximately 55°C during the summer days, while the desert climates may be even hotter yet. This is 30° above standard test conditions (25°C). This means that the real capacity of the array is approximately 15% less than the nominal power rating. The effective capacity, then, is about 85% of the nominal capacity. Expected electrical energy (kWh) is the product of the real capacity of the array (kW) and insolation (peak sun hours) at the angle of elevation of the array. Generated PV energy varies seasonally, as do the levels of insolation. If a Sun tracker is used, annual energy production can increase up to ~20% in temperate climates.

### 4.5.1 Site Selection and Load Estimates

The most important factors taken into account in the selection of a particular site or location to install and build a photovoltaic power plant include: insulation, land area requirements, economic and institutional factors. Insolation is the most important factor used in selecting a site to build a photovoltaic power plant. The energy output of a photovoltaic system is directly proportional to the insolation input. Other climatic and environmental factors such as temperature extremes, precipitation, wind and land topography, will limit and constrain a PV plant but these factors are all secondary when compared with the availability of insolation. Land area requirements are the second most important selection factor, because the photovoltaic systems are requiring large areas for array deployment. Larger areas are required in order to space the arrays for minimalizing the inter-array shadow effects. Simpler arrays such as fixed, south-tilting, east-west arrays require smaller areas compared to the two-degree-of-freedom tracking arrays. Economic consideration in site selection must account for the cost of land required, the cost of site preparation and the cost of access to the site. While all these costs are important, the land and site preparation costs are generally a minor part of the overall cost of a photovoltaic power plant. The access cost, however, has an additional significance since remote access will contribute to the operation and maintenance cost over the years. Institutional criteria involve considerations of land use requirements and social activities. Population density is a prime determinant in choosing sites.

The first task in designing a photovoltaic system is to estimate the load demand and requirements. This is achieved by listing the power demand of all loads, expected number of hours of use per day, and operating voltage. From the load current, time (usually combinedin ampere-hours), and the given operating voltage for each load, the power demand can be calculated. The recommended voltage for a stand-alone photovoltaic system is determined by considering information obtained by grouping the loads by type and operating voltage and calculating the power demand for each group. The operating voltage selected for a stand-alone photovoltaic system is usually the voltage required by the largest loads. When AC loads dominate, the DC system voltage should be selected for compatibility with the inverter input voltage. If DC loads have the largest power demand, the voltage accompanying the largest load is selected. The unit that we measure for the consumption of electricity of a typical electrical appliance is kilo-watt hours (kWh) or watt hours (Wh) for small appliances. To calculate the daily requirement in Wh per day, we as the designer must first list down all the appliances that are expected to be used in the system. For each appliance, first find its power and decide the amount of time in hours that it will be used each day. The power or wattage of an appliance can be found somewhere on the outside of the appliance, electrical tag attached to the cable or in the instruction book (sometimes only its current and voltage are specified). The calculation of daily requirement for each appliance is then as:

$$\text{Appliance Daily Energy Requirements} = \text{Appliance Power (W)} \times \text{Expected}$$
$$\text{Appliance Daily Use (h)} \tag{4.56}$$

Total daily energy demand is then 20.07 kWh. If assume that the peak power of a PV module of 40 W, then the average daily energy generated by one module is about 150 Wh. The total number of PV modules is then:

$$\#\,\text{PV Module} = \frac{20,070}{150} = 133.8 \text{ or } 134 \text{ PV Modules}$$

This is a very large number of solar modules for just one home. Clearly it is not economical viable to use solar electricity for running some of the appliances that are found in a home connected to the mains or another electric generator system and/or other renewable energy sources, such as a wind turbine. When sizing for a solar system, the daily requirement can be significantly reduced the values in the previous example. Reductions are made by carefully decided which appliances need to be run on solar electricity and for how long they really need to be used each day.

> **Example 4.17:** A farmer decides to install a stand-alone PV system in his house. Assuming a full insulation of about 4.5 hours average daily, size the PV power system needed.
>
> **Solution:** The first step that he has to do is to determine the total daily requirement of appliances of his house. The appliances in his house and the daily usage are given in the following table.

| Appliance | Power per Unit (W) | Daily Use (h) | Energy Demand (Wh) |
|---|---|---|---|
| 12 Fluorescent Lamps | 20 | 5 | 1200 |
| 4 Light Bulbs | 100 | 4 | 1600 |
| 10 W DVD Player | 10 | 2 | 20 |
| 2 Radio Sets (20 W) | 20 | 5 | 200 |
| 2 TV Sets (80 W each) | 80 | 5 | 800 |
| 2 Fans | 40 | 5 | 400 |
| Refrigerator | 150 | 24 | 3600 |
| Clothe Iron | 1000 | 0.5 | 500 |
| Electric Cooker | 3000 | 1 | 3000 |
| AC Unit | 1250 | 6 | 7500 |
| Washing Machine | 3000 | 0.25 | 750 |
| Dishwasher | 2000 | 0.25 | 500 |

## 4.6 Summary

Photovoltaic systems constitute a relatively new form of producing electric energy that is environmentally clean and very modular. In stand-alone installations, it must use storage or another type of generator to provide electricity when the Sun is not shining. In grid-connected installations storage is not necessary: in the absence of sunlight, electricity is provided from other (conventional) sources. The power supply of satellites has been the first professional application of photovoltaics (Vanguard I, 1958). PV power generation can be a unique option for many applications of high social value such as providing electricity to people who lack it in remote areas. Often, international donor agencies are providing the funding, as many of the users are very poor. Photovoltaic systems are an excellently suitable solution for low power electricity supply in rural and remote areas in developing countries. Photovoltaic power generation is also very suitable for proving electricity to remote communication equipment or for remote monitoring or applications. However, the system electronics and storage batteries have to be further developed in order to address the market in an optimal way. Electricity produced from PV systems has a far smaller impact on the environment than traditional methods of electrical generation or even than other alternative energy sources. During their operation, PV cells need no fuel, give off no atmospheric or water pollutants and require no cooling water. Unlike fossil fuel (coal, oil and natural gas) fired power plants, PV systems do not contribute to global warming or acid rain. The use of PV systems is not constrained by material or land shortages and the Sun is a virtually endless energy source. The cost of PV systems has decreased more than twenty times since the early 1970s, and research continues on several different technologies in an effort to reduce costs to levels acceptable for wide scale use. Current PV cells are reliable and already cost effective in certain applications such as remote power, with stand-alone PV plants built in regions not reached by the utility networks. Photovoltaic devices and systems are based on the photovoltaic effects, transforming directly the solar energy into electricity. Different technologies utilizing applications of PV cells constitute the field of photovoltaics. The solar radiation incident on the solar cell separates the charge carriers in the absorbing material. The electric fields present at the junctions or inhomogeneities in material provide the required electromotive force for the flow of electric current in the external circuit and hence the power generation. Photovoltaic devices are driven by the flux of solar radiation and acts like a current source. The PV module technology and the type of installation affect significantly the performance of the PV systems. A comprehensive review of PV cell and module characteristics and performances, as well as major photovoltaic technologies comprising of photovoltaic, reliability of PV system, environmental aspects and PV applications are presented and discussed in this chapter. The operating characteristic of the solar cells play a vital role in locating maximum power point. Crystalline Si technology, both monocrystalline and multi-crystalline is today clearly dominant, with over 3 quarters of the market, and will remain dominant for next five years, with the trend is toward the multi-crystalline option. Thin-film technology is one of the candidates to take over from Si technology in the long-term. There are many technological options regarding thin-film materials and methods of deposition but their primary claim to the throne currently occupied by Si is that they can be ultimately produced at much lower cost. Concentration of sunlight is another candidate for mass penetration of photovoltaics, although it will not be easily accepted for the grid-connected houses, one of the most promising applications today. Other options, such as organic photovoltaics, quantum dots and dye-sensitized solar cells, are still in the research phase with very limited commercial applications.

## 4.7 Questions and Problems

1. The area of New Mexico is 121,666 square miles. The average annual insolation (hours of equivalent full sunlight on a horizontal surface) is 2200 h. If one-half of the area of New Mexico is covered with solar panels of 10% efficiency, how much electricity can be generated per year? What percentage of U.S. energy needs can be satisfied? (The total energy consumption of the United States in 2007 was 100 EJ.)

2. Find the area of your state, province or country and the total average annual insulation and estimate the total electricity that can be generated if 10% of the area is covered with PV arrays having an overall efficiency of 12%.

3. Explain the photoelectric effect.

4. A PV solar module is made from a 30-cm × 20-cm piece of silicon. Reverse saturation current $I_0$ is $3.7 \times 10^{-11}$ A/cm$^2$. For $V = 0, 0.30$, and $0.6$ V, calculate the forward current for the solar cell at room temperature (25°C).

5. Determine the value of dark current in the limiting case of the voltage, $V \to 0$.

6. Given the ideal ($\beta = 1$) solar cell diode equation, for an open circuit voltage $V_{oc} = 0.6$ V and a short circuit current $I_{Sc} = 10$ A, determine the photocurrent and dark saturation current parameter values for the model, and estimate to two significant figures the maximum power (in W) and corresponding voltage $V_{mp}$.

7. A solar cell of 100 square centimeter area has a reversed-bias dark current of $2 \times 10^{-9}$ A, and a short-circuit current at one sun (1 kW/m$^2$) of 3.5 A. At room temperature, what is the open-circuit voltage? What is the optimum load resistance ($V_{mp}/I_{mp}$)? What is the PV cell fill factor? What is the maximum power output?

8. Calculate the power output of a PV cell under standard test conditions, $SR(t) = 1000 \text{ W/m}^2$ and $T_c = 25°C$, when cell efficiency is 18% and the fill factor is equal to 0.835, and the aperture area is equal to $0.405 \cdot 10^{-3} \text{ m}^2$.

9. Estimate the cell temperature and power delivered by a 150-W PV module with the following conditions: 0.5%/°C power loss, NOCT equal to 47°C, ambient temperature of 27°C, and insolation of 1-sun.

10. An ideal PV cell with a reverse saturation current of 2 nA has a light-induced current of 0.95 A and is operating at 30°C. Compute the cell output voltage and output power when the PV load draws 0.45 A.

11. Which material used for manufacturing PV cell has the highest reported solar cell efficiency?

12. What is the maximum efficiency of commercial PV cells, reported in the literature?

13. Calculate the dark current of a PV cell, assuming that is operating at 300 K and the saturation current is $10^{-10}$ A.

14. Why do we need to collect PV cells in series, in parallel or in series-parallel configurations?

15. What are a solar module and a PV array?

16. A PV cell with saturation current of 1.15 nA is operating at 36°C. The photocurrent at 36°C is 1.08 A. Compute the maximum power output of the cell, assuming an ideality factor 1.2.

17. Given an ideal PV cell equation ($n = 1$), and for an open-circuit voltage 0.6 V and a short-circuit current of 9.30 A, determine:

    a. The photocurrent and the dark saturation current values, and

    b. The maximum power (in W and with two significant figures) and the corresponding $V_{mp}$.

18. A PV module is composed of 24 PV cells in series. If the light-induced current of each cell at 27°C is 1 A and the saturation current is 1.25 nA, find the module output power when its voltage is 10 V.

19. An ideal PV cell has a reveres saturation current of 1.15 nA, and photocurrent of 1.1 A and is operating at 36°C and is connected to 3.5 Ω load. Compute the power delivered to the load.

20. In above problem assume that the PV cell is non-ideal, having a series resistance of 0.01 Ω. Compute the output power in this case.

21. Compute the diode saturation current and open-circuit voltage for a 1 dm² Si with $I_{SC} = 3.0$ A, operating at 300 K, with energy bad gap 1.2 eV, and ideality factor, $n = 1$.

22. A 1 dm² PV cell operates at 300 K, the open circuit voltage is 0.6 V, the short-circuit current is 3.5 A under 1000 W/m². Assuming an ideal PV cell, what is the its efficiency at maximum power point? What would be the corresponding efficiency, if the cell has a 0.1 Ω series resistance, and a 5 Ω parallel (shunt) resistance?

23. An ideal PV cell (assuming the series resistance equal to zero, and shunt resistance infinite) with $V_{OC} = 0.6$ V and $I_{sc} = -5$ A is mounted outdoor. Weather effects on the electrodes causes the series resistance to increase from 0 on the installation

day to certain value following a linear relationship, $R_s(t) = 3.5 \cdot 10^{-5}$ t, where t is the number of days after the installation. Plot graphs of the peak cell power and corresponding $V_{mp}$ for $0 \leq t \leq 1000$ days.

24. For a PV module operating in a hot desert climate at 65°C, what would be the approximate system percentage temperature derated from STC?

25. Calculate the cell temperature and power delivered by a 90 W PV module with the following conditions. Assume a power loss of 0.45% per Celsius degree.

    a. NOCT = 50°C, ambient temperature of 25°C, insolation of 1000 W/m².

    b. NOCT = 45°C, ambient temperature of 0°C, insolation of 600 W/m².

26. A photovoltaic module has the following parameters: $I_{sc}$ = 2.5 A, $V_{oc}$ = 21 V, $V_{mp}$ = 16.5 V, $I_{mp}$ = 2.35 A (given at standard conditions). The length of the module is $L$ = 90.6 cm and the width of the module is $W$ = 41.2 cm.

    a. Calculate the fill factor, *FF*.

    b. The module is connected to a variable resistive load at an irradiance of $S$ = 800 W/m². The voltage across the resistance was $V$ = 17.5 V and the current flowing in the resistance was $I$ = 1.4 A.

    c. Calculate the module efficiency, $\eta$, where A is the cross-sectional area of the module.

    d. The irradiance increased from $S$ = 800 W/m² to $S$ = 1000 W/m². How much would you increase or decrease the load resistance to extract maximum power from the photovoltaic module?

27. Plot the I-V characteristics of an ideal PV cell that has a saturation current of $10^{-12}$ A and a short circuit current of 1.0 mA for 0–0.7 V in steps of 0.05 V. Assume that the cell is operating at 300 K. Estimate the cell maximum output power, current and voltage and the fill factor.

28. Describe the temperature effect on crystalline PV modules and how it affects power output.

29. A PV module consists of 72 cells connected in series. At 25°C the photocurrent of each cell is 1.15-A and the reverse saturation current is 12.5 nA. Find the module power if the average voltage is 32 V.

30. Given a solar cell with photocurrent density of 0.036 A/cm², and saturation current density $1.5 \cdot 10^{-10}$ A/cm², calculate maximum voltage, current, and fill factor if the ambient temperature is 30°C. Assume and ideality factor of 1.2 and the cell area of 0.25 cm². Plot current and power vs voltage, from 0 to $V_{OC}$ to verify your calculation.

31. Calculate the dark current for a solar PV cell for reverse and forward bias mode, assuming that the cell is operating at 300 K and $I_0$ is 0.1 nA.

32. Calculate the fill factor of a solar module when: open circuit voltage and short circuit current are $V_{OC}$ = 11 V and $I_{SC}$ = 3 A, and maximum voltage and current are $V_{mp}$ = 10 V and $I_{mp}$ = 2 A.

33. Compute the cell temperature and power delivered by a 800 W PV panel with the following conditions:

    a. NOTC = 60°C, ambient temperature of 27°C, and insulation 1 kW/m²; and

    b. NOTC = 40°C, ambient temperature 25°C, and insulation 500 W/m². Assume 0.5%/°C power loss.

34. Determine the number of PV modules needed to provide power for a log-cabin or similar facility in your state, by selecting the optimum type of the needed appliances and electronics and their daily usage. What is the optimum tilt for PV system?

35. A PV module with 36 cells has an idealized, rectangular I-V curve with $I_{SC} = 3.5$ A and $V_{OC} = 18$ V. If a single PV cell has a parallel resistance of 5 Ω and negligible series resistance, draw the I-V curve if one cell is completely shaded. What is the drop in the module voltage?

36. A small field research facility consisting of the small electrical installation made up of: a 40 W data logger, a 80 W automatic weather station, a 40 W LCD television, two compact fluorescent lamps of 20 W each, a 200 W radio receiver and transmitter (RX set), and 300 W cooking equipment. The TV, which is on 2.0 h/day; two lamps working 3 h/day; the rad, RX set is on 2 h/day, data logger 0.5 h/day, the cooking equipment 6 h/day and the weather station is operating 6 h/day. Estimate the energy demand and the size of PV systems needed to power the facility.

## References and Further Readings

1. G. Boyle, *Renewable Energy—Power for a Sustainable Future*, Oxford University Press, Milton Keynes, England, 2012.
2. V. Quaschning, *Understanding Renewable Energy Systems*, Earthscan, London, UK, 2006.
3. R. A. Ristinen and J. J. Kraushaar, *Energy and Environment*, Wiley, Hoboken, NJ, 2006.
4. J. Andrews and N. Jelley, *Energy Science, Principles, Technology and Impacts*, Oxford University Press, Oxford, UK, 2007.
5. E. L. McFarland, J. L. Hunt and J. L. Campbell, *Energy, Physics and the Environment* (3rd ed.), Cengage Learning, Mason, OH, 2007.
6. B. K. Hodge, *Alternative Energy Systems and Applications*, John Wiley & Sons, Hoboken, NJ, 2010.
7. L. E. Chaar, L. A. Iamont, and N. E. Zein, Review of photovoltaic technologies, *Renewable and Sustainable Energy Reviews*, Vol. 15, pp. 2165–2175, 2011.
8. B. Everett and G. Boyle, *Energy Systems and Sustainability: Power for a Sustainable Future* (2nd ed.), Oxford University Press, Oxford, UK, 2012.
9. R. A. Messenger and J. Venter, *Photovoltaic Systems Engineering* (3rd ed.), CRC Press, Boca Raton, FL, 2010.
10. M. R. Patel, *Wind and Solar Power Systems*, CRC Press, Boca Raton, FL, 1999.
11. R. A. Dunlap, *Sustainable Energy*, Cengage Learning, Stamford, CT, 2015.
12. V. Nelson and K. Starcher, *Introduction to Renewable Energy (Energy and the Environment)*, CRC Press, Boca Raton, FL, 2015.

13. B. Parida, S. Iniyan, and R. Goic, A review of solar photovoltaic technologies, *Renewable and Sustainable Energy Reviews*, Vol. 15, pp. 1625–1636, 2011.
14. H. Andrei, V. Dogaru-Ulieru, G. Chicco, C. Cepisca, and F. Spertino, Photovoltaic applications. *Journal of Materials Processing Technology*, Vol. 181, pp. 267–273, 2007.
15. E. E. Michaelides, *Alternative Energy Sources*, Springer, Berlin, Germany, 2012.
16. I. Bostan, A. Gheorghe, V. Dulgheru, I. Sobor, V. Bostan, and A. Sochirean, *Resilient Energy Systems—Renewables: Wind, Solar, Hydro*, Springer, Dordrecht, the Netherlands, 2013.
17. S. C. W. Krauter, *Solar Electric Power Generation—Photovoltaic Energy Systems*, Springer, New York, 2006.
18. R. Foster, M. Ghassemi, and A. Cota (eds.), *Solar Energy—Renewable Energy and the Environment*, CRC Press, Boca Raton, FL, 2010.
19. R. Belu, Tracking Systems: Maximum Power and Sun, in *Encyclopedia of Energy Engineering & Technology* (Online) (eds. Dr. Sohail Anwar, R. Belu et al.), CRC Press/Taylor & Francis, doi:10.1081/E-EEE-120048434/31 pages.
20. C. J. Chen, *Physics of Solar Energy*, John Wiley & Sons, Hoboken, NJ, 2011.
21. A. Luque and S. Hegedus, *Handbook of Photovoltaic Science and Engineering*, John Wiley & Sons, New York, 2011.
22. H. Haberlin, *Photovoltaics System Design and Practice*, John Wiley & Sons, Hoboken, NJ, 2012.
23. D. R. Myers, *Solar Radiation—Practical Modeling for Renewable Energy Applications*, CRC Press, Boca Raton, FL, 2013.
24. N. Enteria and A. Akbarzadeh, *Solar Energy Sciences and Engineering Applications*, CRC Press, Boca Raton, FL, 2014.
25. D. Y. Goswami, *Principles of Solar Engineering* (3rd ed.), CRC Press, Boca Raton, FL, 2015.
26. M. Martín (Ed.), *Alternative Energy Sources and Technologies—Process Design and Operation*, Springer, Switzerland, 2016.
27. G. N. Tiwari, A. Tiwari and Shyam, *Handbook of Solar Energy—Theory, Analysis and Applications*, Springer, Singapore, 2016.
28. S. R. Wenham, M. A. Green, M. E. Watt and R. Corkish, *Applied Photovoltaics*, Earthscan, London, UK, 2007.
29. G. M. Masters, *Renewable and Efficient Electric Power Systems* (2nd ed.), Wiley, Hoboken, NJ, 2014.
30. NFPA 70™, *2008 National Electrical Code (NEC)*, NFPA, Quincy, MA, 2008.

# 5

## Wind Energy Resources

### 5.1 Introduction, Historical Notes

Wind has been utilized as a source of power for thousands of years for sailing ships, grinding grain, pumping water or powering factory machinery. Humans have been harnessing the wind ever since farmers in ancient Persia figured out how to use wind power to pump water. These windmills used to pump water or to grind wheat had vertical axis and used the drag component of wind power (one of the reasons for their low efficiency). Moreover, to work properly, the part rotating in opposite direction compared to the wind has to be protected by a wall, so they can be used only in places with a dominant wind direction. The first windmills built in Europe used instead a horizontal axis rotor, substituting the drag force with the lift force, making their inventors also the unaware discoverers of aerodynamics. During the following centuries many modifications and improvements were applied especially to cope in areas where the wind direction varied a great deal. The best examples are the Dutch windmills, used to drain the water in the lands taken from the sea with the dams. These windmills could be oriented in any wind direction in order to increase their efficiency. The wind turbines used in the USA between the nineteenth century and up until 1930s were mainly used for water pumping. They had a high number of steel-made blades and represented a huge economic potential because of their large quantity.

Wind energy conversion systems are turning the wind kinetic energy into mechanical and eventually into electrical energy that can be used for a variety of tasks. Whether the task is creating electricity or pumping water, wind offers an inexpensive, clean and quite reliable form of mechanical power. First attempts to generate electricity were made at the end of nineteenth century and almost all those models were horizontal rotor type. The world's first wind turbine used to generate electricity was built by a Danish engineer, Poul la Cour, in 1891. It is also interesting to note that La Cour used the electricity generated by his wind turbines to electrolyze water, producing hydrogen for gas lights in the local schoolhouse. In that regard we could say that he was 100 years ahead of his time since the vision that many have for the twenty-first century includes photovoltaic and wind energy conversion systems making hydrogen to generate electricity in fuel cells. In the first half of last century, a French engineer, Darrieus designed one of the most famous and common types of the vertical axis wind turbine, the so-called Darrieus type. Sadly, with the use of coal and oil in the last century, the importance of the wind energy decreased. However, during the last three decades there has been an increased interest due to several reasons, such as: depletion of fossil fuel reserves, environmental concerns, technological advances in turbine design, electric machines, power electronics and control, deregulated energy market or distributed generation research. There are two major types of wind turbines have been built in order to harness this renewable energy source: those with horizontal axis of rotation (HAWT) and those with vertical axis (VAWT). The first type is the most common today, but growing market asks for turbines with different proprieties to fit different requests and to take advantage of the wind characteristics of particular sites.

Worldwide development of wind energy expanded rapidly starting in the early 1990s. The average annual growth rate from 1994 to 2015 of the world installed capacity of wind power has been over 35%, making the wind industry one of the fastest growing. Unlike the last surge in wind power development during 1970s which was due mainly to the oil embargo of the OPEC countries, the current wave of wind energy development is driven by many forces that make it favorable. These include its tremendous environmental, social and economic benefits as well as its technological maturity, the deregulation of electricity markets, public support and government incentives. Even among other applications of renewable energy technologies, power generation through wind has an edge because of its technological maturity, good infrastructure and relative cost competitiveness. Wind energy is expected to play an increasingly important role in the future national energy scene. Differential heating of the Earth's surface by the Sun causes the movement of large air masses on the surface of the Earth, i.e., the wind. Wind energy conversion systems (WECS) convert the kinetic energy of the wind into other forms of energy. At good windy sites, wind energy price is already competitive with that of conventional fossil fuel generation technologies. With this improved technology and superior economics, experts predict wind power can capture 5% of the world energy market by the year 2020. In order to do that, wind turbines must be more efficient, more robust and less expensive than current turbines.

Electricity generation from wind can be economically achieved only where a significant wind resource exists. Either wind power has been acclaimed as one of the most potential and techno-economically viable renewable energy sources of power generation the technical know-how is not yet fully adequate to develop reliable wind energy conversion systems for all wind regimes. For securing maximum power output, wind energy resource assessment and analysis at any prospective site is critical. The available wind energy varies as the wind speed cube, so an understanding of the wind characteristics is essential to all wind energy exploitation aspects, from the suitable site identification and predictions of the wind energy project economic viability through to the design of wind turbines, and understanding their effect on power grid. The most striking characteristic is the wind variability, both geographically and temporally. Furthermore, this variability persists over very wide spatio-temporal scale ranges, sites with small differences in average wind speeds can have substantial available wind energy differences. Therefore, accurate and thorough monitoring of wind resource at potential sites is critical in the wind turbine siting. Accurate measurements of site wind speed frequency spectrum are very important in wind energy potential analysis and assessment. For assessment of the wind power potential, most investigators have used simple wind speed distributions that are parameterized solely by the arithmetic mean of the wind speed. However, an assessment of wind turbine power output is accurate, only if the wind speeds and directions are measured at the turbine hub height.

Knowledge of the local wind capacity remains vital to the industry, yet commercially viable renewable-related geospatial products that meet the wind industry needs are often suspect. There are three stage involved with wind power project planning and operations during which accurate wind characterization plays a critical role: (1) prospecting: uses historical data, retrospective forecasts, and statistical methods to identify potential sites for wind power projects; (2) site assessment: determines the optimum placement of a wind power project; and, (3) operations: uses wind forecasting and prediction to determine available power output for hour-ahead and day-ahead time frames. The most critical is the first one: identifying and characterizing the wind resources. An outline of the state of the art in understanding the wind resource assessment, and discussing the strengths and weaknesses of existing methods is included. Appropriate statistical methods to compute the wind speed probability density function (PDF) are described and examined. In addition, although there has been an increasing awareness of wind energy as a viable energy source, there has not been a concomitant increase in the awareness of the impacts that any spatial and temporal trends in the wind resource may have on long-term production, use, and implementation. Despite environmental benefits and technological maturation, penetration of wind-generated power represents a challenge for reliability and stability of the power grids due to the highly variability and intermittent nature of winds.

Wind energy resources rely on the incident wind speed and direction, both of which vary in time and space due to large-scale and/or small-scale circulation changes, surface energy fluxes, and topography. Since the wind power density is proportional to the cube of the wind speed, any small errors in forecasted wind speeds can result in significant differences between forecasted and actual wind energy output. Consequently, accurate assessment and forecasting of spatial and temporal characteristics of the winds and turbulence remains the most significant challenge in wind energy. The wind energy viability is governed by factors as: the potential for large scale energy production, the predictability of the power to be supplied to the grid, and the expected return on investment. The various wind energy uncertainties impact the reliable determination of these viability factors. Currently the worldwide nameplate capacity of wind-powered generators is approximately 2.5% of the electricity consumption, with a steady annual growth. For wind to play a more prominent role in the future energy market, improvements in the technology are needed, while some advancement can be realized through appropriate quantification of the uncertainties and their explicit consideration in the optimal design. A robust and optimal wind farm planning include: (a) site optimal selection based on the quality of the local wind resources, (b) maximization of the annual energy production and/or minimization of the cost of generated energy and (c) maximization of the reliability of the predicted energy output. The most important activity in a site selection is to determine the wind resource potential, consisting in the estimated local wind probability density function. Another important activity is to determine the turbulence levels and the resulting wind loads at the concerned site, or in selecting the most suitable wind turbines for that site and in optimum life cycle cost prediction. Other site selection criteria include, but not limited to: (1) local topography, (2) distance to electric grid, (3) vegetation, (4) land acquisition issues and (5) site accessibility for turbine transport and maintenance. A planning strategy that simultaneously accounts for the key engineering design factors and addresses the uncertainties in a wind energy project can offer a powerful impetus to the wind energy development. However, the resource itself is highly uncertain, wind conditions, including wind speed and direction, turbulence intensity and air density, show strong temporal variations; in addition, the wind conditions varies significantly from year to year. The resulting ill-predictability of the annual distribution of wind conditions introduces significant errors in the estimated resource potential or the predicted wind farm performances. This chapter presents a review of the wind energy technology, wind resource assessment and analysis, wind statistics, as well as some of the recent developments in wind energy conversion systems, and the wind energy social and environmental benefits and impacts. A reach and updated references are included at the end of this chapter for interested readers, students, engineers and researchers.

---

## 5.2 Wind Energy Resources

Wind energy is a special form of kinetic energy in air as it flows. Wind energy can be either converted into electrical energy by power converting machines or directly used for pumping water, sailing ships, or grinding gain. Wind energy is not a constant source of energy. It varies continuously and gives energy in sudden bursts. About 50% of the entire energy is given out in just 15% of the operating time. Wind strengths vary and thus cannot guarantee continuous power. It is best used in the context of a system that has significant reserve capacity such as hydropower, compressed air storage, or reserve load, such as a microgrid, desalination plant, to mitigate the economic effects of resource variability. Accurate knowledge of the wind regime and characteristics is critical not only to wind energy, but also in other applied science fields. Knowledge of the local wind capacity remains vital to the industry, yet commercially viable renewable-related geospatial products that meet the wind industry needs are often suspect. There are three stage involved with wind power project planning and operations during which accurate wind characterization plays a critical role: (1) prospecting: uses historical data, retrospective forecasts, and statistical methods to identify the best wind energy sites for any project; (2) site assessment: determines the optimum placement of a wind project; and (3) operations: uses wind forecasting and prediction to determine available power output for hour-ahead and day-ahead time frames. The most critical is the first one: identifying and characterizing the wind resources, the wind speed probability density functions are critical in wind resource analysis. In addition, although there has been an increasing awareness of wind energy as a viable energy source, there has not been a concomitant increase in the awareness of the impacts that any spatial and temporal trends in the wind resource may have on long-term production, use, and implementation. Despite environmental benefits and technological maturation, penetration of wind-generated power represents a challenge for reliability and stability of the power grids due to the highly variability and intermittent nature of winds. Moreover, for the purposes of wind energy use and wind turbine design, the wind vector is considered to be composed of a steady wind plus fluctuations about the steady wind. Whereas for designing wind turbines, the steady wind and the wind fluctuations have to be considered, the power and energy obtained from wind can be based only on the steady wind speed. The power available in the wind varies with the cube of the wind speed and depends also on the air density. From kinetic energy of the wind by taking it time derivative the available power in the wind can be expressed as:

$$P_{wind} = 0.5 \rho A v^3 \tag{5.1}$$

Here $\rho$ is the air density (kg/m³), $A$ is the cross-sectional area (m²), and $v$ is the wind speed (m/s). Examination of Equation (5.1) reveals that in order to obtain a higher wind power, it requires a higher wind speed, a longer length of blades for gaining a larger swept area, and a higher air density. Because the wind power output is proportional to the cubic power of the mean wind speed, a small variation in wind speed can result in a large change in wind power.

**Example 5.1:** For an average wind speed 10 mph, a small wind turbine produces 100 W/m². What is the power density for a speed of 40 mph?

**Solution:** From Equation (5.2) the power density is proportional to cube of the wind speed so:

$$\text{Speed Ratio} = \frac{40}{10} = 4$$

$$P_{40} = 4^3 P_{10} = 64 \times 100 = 6.4 \text{ kW}$$

A common unit of measurement is the wind power density, or the power per unit of area normal to the wind direction from the wind is blowing:

$$p_w = \frac{P_{wind}}{A} = 0.5\rho v^3 \tag{5.2}$$

Here $p_w$ is wind power density (W/m²). The ultimate goal of wind energy projects is to extract energy from the wind and not just producing power, an important parameter in site selection is the mean wind power density, expressed if the wind frequency distribution $f_{PDF}(v)$ is known as:

$$\bar{p}_w = 0.5\rho v^3 f_{PDF}(v) \tag{5.3}$$

Weibull or Rayleigh probability distribution, are the most used in wind energy assessment and analysis. Instead of integration the mathematical Weibull function, the mean value of the third power of the wind speeds in appropriate time intervals can also be used. Wind resource maps often estimate the potential of the wind resources in terms of wind power classes referring to the annual wind power density. In a later chapter section the most common PDFs used in wind energy are discussed in some detail.

**Example 5.2:** The diameter of a large offshore wind turbine is 120 m, assuming the air density is compute the available wind power for wind speed of 5, 10, and 15 m/s. What the available wind power density for wind speed of 8.5 m/s?

**Solution:** By using Equation (5.1) the available power densities are:

$$P_5 = 0.5 \cdot 1.2 \left( \pi \frac{120^2}{4} \right) (5)^3 = 847.8 \text{ kW}$$

$$P_{10} = 0.5 \cdot 1.2 \left( \pi \frac{120^2}{4} \right) (10)^3 = 6.7824 \text{ MW}$$

$$P_5 = 0.5 \cdot 1.2 \left( \pi \frac{120^2}{4} \right) (5)^3 = 22.8906 \text{ MW}$$

The power density for a wind speed of 8.5 m/s is then calculated as:

$$p_w = 0.5 \cdot 1.2 (7.5)^3 = 253.12 \text{ W/m}^2$$

Wind results from the movement of air due to atmospheric pressure gradients. Wind flows from regions of higher pressure to regions of lower pressure. The larger the atmospheric pressure gradient, the higher the wind speed and thus, the greater the wind power that can be captured from the wind by means of wind energy-converting machinery. The wind regime is determined due to a number of factors, the most important factors being uneven solar heating, the Coriolis force, due to the Earth's rotation, and local geographical conditions. For example, the surface roughness is a result of both natural geography and man-made structures. Frictional drag and obstructions near the Earth's surface generally retard with wind speed and induce a phenomenon known as wind shear. The rate at which wind speed increases with height varies on the basis of local conditions of the topography, terrain and climate, with the greatest rates of increases observed over the roughest terrain. A reliable approximation is that wind speed increases about 10% with each doubling of height. In addition, some special geographic structures can strongly enhance the wind intensity. Air masses move because of the different thermal conditions, and the motion of air masses can be a global phenomenon (i.e., the jet stream), or a regional or local phenomenon. The regional phenomenon is determined by orography (e.g., the surface structure of the area) and also by global phenomena. Wind turbines utilize the wind energy close to the ground. The wind conditions in this area, known as boundary layer, are influenced by the energy transferred from the undisturbed high-energy stream of the geostrophic wind to the layers below and by regional conditions. Owing to the roughness of the ground, the local wind stream near the ground is turbulent, and wind speed varies continuously as a function of time and height.

Wind energy resources rely on the incident wind speed and direction, both of which vary in time and space due to changes in large-scale and small-scale circulations, surface energy fluxes and topography. Since the wind power density is proportional to the cube of the wind speed, any small errors in forecasted wind speeds can result in significant differences between forecasted and actual wind energy output. Consequently, accurate assessment and forecasting of spatial and temporal characteristics of the winds and turbulence remains the most significant challenge in wind energy production. The wind energy production viability is governed by such factors as: the potential for large scale energy production, the predictability of the power to be supplied to the grid, and the expected return on investment. The various wind energy uncertainties impact the reliable determination of these viability factors. Currently the worldwide capacity of wind-powered generators is approximately 2.5% of the electricity consumption, with a steady annual growth. For wind to play a more prominent role in the future energy market, improvements in the generation technology are needed, while some advancement can be realized in part through appropriate uncertainties quantification and explicit consideration of their influences in the optimal design. A robust and optimal wind farm planning include: (a) site optimal selection based on the quality of the local wind resources, (b) maximization of the annual energy production and/or minimization of the cost of generated energy and (c) maximization of the reliability of the predicted energy output. The most important activity in a site selection is to determine the wind resource potential, consisting in the estimated local wind probability density function. Another important activity is to determine the levels of turbulence and the resulting wind loads at the concerned site, promoting better decision making, in selecting the most suitable wind turbines for that site and in optimum life cycle cost prediction; higher wind loads generally result in higher costs. Among others site selection criteria include: (1) local topography, (2) distance to electric grid, (3) vegetation, (4) land acquisition issues and (5) site accessibility for turbine transport and maintenance. For a better estimate of the wind power potential for any extended time period, you would need to know the frequency distribution of the wind speeds; the amount of time for each wind speed value, or a wind speed histogram; and the number of observations within each wind speed range. It is also important to have information about wind regime characteristics, air density and turbulence intensity.

Wind power density is a comprehensive index in evaluating the wind resource at a particular site. It is the available wind power in airflow through a perpendicular cross-sectional unit area in a unit time period. The classes of wind power density at two standard wind measurement heights are listed in Table 5.1. Some of wind resource assessments utilize 50 m towers with sensors installed at intermediate levels (10 m, 20 m, etc.). For large-scale wind plants, class rating of 4 or higher are preferred. The use of wind power classes to describe the magnitude of the wind resource was first defined in conjunction with the preparation of the 1987 U.S. Department of Energy Wind Energy Resource Atlas. The atlas is currently available through the American Wind Energy Association and is an excellent source of regional wind resource estimates for the United States and its territories. The atlas wind resource magnitude is expressed in terms of the seven wind power classes, as well as the wind velocity. The wind power classes range from class 1, for winds containing the least energy to class 7, for winds containing the greatest energy (see Table 5.1). Mean wind speed estimates hare are based on Rayleigh wind speed probability distribution of equivalent mean wind power density, for standard sea-level conditions, and to maintain the same power density, speeds are increased by 3%/1000 m (5%/5000 ft) elevation.

Wind resource assessment is the most important step in planning a wind project because it is the basis for determining initial feasibility and cash flow projections, being vital for financing. Assessment and project progress through several stages: (1) initial assessment; (2) detailed site characterization; (3) long-term data validation; and (4) detailed cash flow projection and financing. Prediction of wind energy resources is crucial in the development of a commercial (large-scale) wind energy installation. The single most important characteristic to any wind development is the wind velocity. The performance and wind farm power output is very sensitive to uncertainties and errors in wind velocity estimates, so the wind resource assessment must be extremely accurate in order to procure funding and accurately estimate the project economics. Commercial wind resource assessment performed by wind developers is using both numerical and meteorological data. Wind speed and direction measurements are collected by permanent or semi-permanent meteorological towers

**TABLE 5.1**

Classes of Wind Power Density at 10 and 50 m

| Wind Power Class | 10 m—Wind Power Density (W/m²) | 50 m—Wind Power Density (W/m²) |
|---|---|---|
| 1 | <100 | <200 |
| 2 | 100–150 | 200–300 |
| 3 | 150–200 | 300–400 |
| 4 | 200–250 | 400–500 |
| 5 | 250–300 | 500–600 |
| 6 | 300–350 | 600–700 |
| 7 | >400 | >800 |

designed to measure wind velocity using a variety of wind sensors (sonic, Lidar, cup and vane, sonic, etc.), at different hub-heights. An important aspect is to gain an understanding of the wind profile both spatially across the location of interest and in elevation above terrain level. The main factors impacting the wind flows are: orography, surface roughness and the atmospheric stability. The latter represents the thermal effects on the wind flow, being rarely taken into account for the wind power assessments, since the wind statistics are averaged over a long period, and the atmosphere is generally considered as neutral. However this assumption presents some limitations: (1) sites where the average wind speed is low (<6 m/s), the thermal effects are starting to be significant; (2) on offshore site, where the atmospheric stability is predominant over the orography and the roughness; and (3) for short-period simulations (hours or days), for short-term prediction, supervision of operation, and power curves measurement with site calibration, as defined in the IEC 61400-12 standard.

## 5.3 Factors Affecting Wind Energy Estimates and Computations

Since the effects of wind shear, turbulence intensity and atmospheric stability on wind turbine production are not fully understood, wind resource assessment studies can have large uncertainties. The estimation of the magnitude of the uncertainty source is often related to empirical considerations rather than analytical calculations. Some studies suggest probability models for the natural wind resource variability that include air density, mean wind velocity and associated Weibull parameters, surface roughness exponent, and error for prediction of long-term wind velocity. Depending on atmospheric conditions, waking by upstream turbines and terrain, roughness interactions, wind turbines often operate far from the ideal conditions and field-deployed power curves can be very different from certified ones. Better predictions of power output or loads require more representative wind measurements and power computations over the rotor-swept area for individual wind turbines. One of the main challenges in harvesting wind energy is that wind is generally intermittent and variable in speed and direction. Depending on the flow properties and scales of motion, the flow can become turbulent with chaotic and stochastic properties. There are three main aspects that can reduce the intermittency problem: spatial distribution of wind facilities, accurate forecasting methods, and storage systems. A single wind generator is subject to large wind variations, if the wind facilities are spatially distributed, the total output at any time becomes more uniform and reliable.

### 5.3.1 Air Density, Temperature, Turbulence, and Atmospheric Stability Effects

Wind speed is usually increasing with height, higher elevation sites potentially can offer greater wind resources than comparable lower ones, being advantageous to site wind turbines at higher elevations, and taking advantage of higher wind speeds. However, the decrease of air density with height can make an impact on the output power, wind power density being proportional to air density, so a given wind speed therefore produces less power from a particular turbine at higher elevations, because the air density is less. Output power and the power curve depend on the air density. Air density is a function of atmospheric pressure, temperature, humidity, elevation and acceleration due to gravity. For example, the air density values encountered at measurement sites in western Nevada were mostly between 0.936 and 1.025 kg/m$^3$ with a multi-annual mean value of 0.982 kg/m$^3$, significantly lower than the mean standard air density of 1.225 kg/m³. Power curves for various values of the air density effect must be accounted for to improve the power output estimate accuracy. Air density is usually computed from temperature and pressure data, as expressed by:

$$\rho = \rho_0 \left( \frac{T}{T_0} \right)^{-(g/cR+1)} \quad \text{or} \quad \rho = \rho_0 \left( 1 + \frac{c \cdot z}{T_0} \right)^{-(g/cR+1)} \tag{5.4}$$

where $T$ is the local air temperature (°K), $T_0$ is the air temperature at the ground (°K), z is the elevation in $m$, $c = dT/dz$ is the atmosphere thermal gradient (~9.80°C/km), R is the gas constant (287 J/kg-K for air). Alternative relationships to estimate the air density dependence on the elevation are:

$$\rho = 1.229 \frac{P - VP}{760} \frac{273}{T} \ \text{kg/m}^3 \tag{5.5a}$$

$$\rho = \frac{353.049}{T} \exp\left( -\frac{0.034 \cdot z}{T} \right) \tag{5.5b}$$

Here, the atmospheric pressure, $P$ is expressed in mm Hg, VP is the vapor pressure in mm Hg, and $T$ is the local absolute temperature in Kelvin degrees. This relationship yields to a value of 1.225 kg/m³ for dry atmosphere in standard atmospheric conditions. The vapor pressure represents a small correction, around 1%, and can be neglected. High temperatures and low pressures reduce the density of air, which reduces the wind power per rotor area. A major factor for change in density is the change in pressure with elevation. A 1000 m increase in elevation reduces the pressure by 10%, reducing the wind power by 10%, too. If only elevation is known, air density can be estimated by using:

$$\rho = 1.225 - 1.194 \times 10^{-4} z \tag{5.6}$$

**Example 5.3:** A future wind farm site is located at 500 m in a semi-desert area. Compute the wind power density when the air temperature is 35°C and the wind speed is 12.5 m/s.

**Solution:** Using Equation (5.5b) the air density is:

$$\rho = \frac{353.049}{273.15 + 35} \exp\left( -\frac{0.034 \cdot 500}{273.15 + 35} \right) = 1.084 \ \text{kg/m}^3$$

The wind power density, by using Equation (5.2) is:

$$p_w = 0.5 \rho v^3 = 0.5 \times 1.084 \times 12.5^3 = 1058.8 \ \text{W/m}^2$$

Depending on the turbine's control method, either the power or velocity is normalized for power density calculations, such as the velocity normalized, here with the reference air density $\rho_0$:

$$v_{norm} = v \left( \frac{\bar{\rho}}{\rho_0} \right)^{1/3} \tag{5.7}$$

**Example 5.4:** For a pressure of 750 mm Hg, and local air temperature equal to 21.5°C estimate the air density and the normalized value of a wind speed of 10 m/s. Assume VP = 0.

**Solution:** The air density, by using Equation (10.4) is 1.124 kg/m³, and by using Equation (10.4) the normalized value of the wind speed (10 m/s) is equal to: 9.18 m/s.

At today's usual hub-heights at 80 m or so, turbine rotors encounter large vertical gradients of wind speed and boundary layer turbulence. Rotors are susceptible to fatigue damage that results from turbulence. Wind turbulence represents the fluctuation in wind speed in short time scales, especially for the horizontal velocity component. The wind speed $v(t)$ at any instant time $t$ can be in two components: the mean wind speed $V_{mn}$, and the instantaneous speed fluctuation $v'(t)$, i.e.:

$$v(t) = V_{mn} + v'(t)$$

Wind turbulence has a strong impact on the power output fluctuation of wind turbine. Heavy turbulence may generate large dynamic fatigue loads acting on the turbine and thus reduce the expected turbine lifetime or result in turbine failure. In selection of wind farm sites, the knowledge of wind turbulence intensity is crucial for the stability of wind power production, turbine control and proper wind turbine design. For example, understanding of the impact of turbulence on the blades can help in designing long-term operational and maintenance schedules for wind turbines, or can lead to the design of advanced control schemes to mitigate such loads. Quantification of the turbulence effects on wind turbine is done by computing an equivalent fatigue load parameter, as a function of wind fluctuation amplitudes within an averaging period, blade material properties, number of counting averaging bins and a total number of samples. Turbulent fluctuations are the main source of the blade fatigue. The turbulence intensity *(TI)* is a measure of the overall turbulence level, defined as:

$$TI = \frac{\sigma_v}{v} \tag{5.8}$$

where $\sigma_v$ is the wind speed standard deviation (m/s), usually at the nacelle height over a specified averaging period (e.g., 10 min). There also are differences in the output power standard deviations. In the wind speed range 4–15 m/s the standard deviation of certain turbulence intensity classes (4%–8% and 10%–15%) differ up to about 50% with the standard deviation for all turbulence intensities. TI is affected by atmospheric stability, so the theoretical wind turbine power curves (Belu, 2014; Belu and Koracin, 2015). A turbulence intensity correction factor can be expressed as:

$$v_{corr} = v_{norm}\left(1 + 3\left(TI\right)^2\right)^{1/3} \tag{5.9}$$

**Example 5.5:** If the standard deviations for the following wind speeds 6.5, 10, and 13.5 m/s are 0.90, 1.05 and 1.15 m/s respectively. What is the turbulence intensity corrected wind speeds?

**Solution:** From Equation (10.7) the turbulence intensity (TI) levels for these data are:

$$TI_{6.5} = \frac{0.90}{6.5} = 0.1385$$

$$TI_{10.0} = \frac{1.05}{10.0} = 0.1050$$

$$TI_{13.5} = \frac{1.15}{13.5} = 0.0852$$

By using modified Equation (10.6), the corrected wind speeds are:

$$v_{corr} = 6.5\left(1+3(0.1385)^2\right)^{1/3} \qquad = 6.624 \text{ m/s}$$

$$v_{corr} = 10.0\left(1+3(0.105)^2\right)^{1/3} \qquad = 10.11 \text{ m/s}$$

$$v_{corr} = 13.5\left(1+3(0.0852)^2\right)^{1/3} \qquad = 13.60 \text{ m/s}$$

Notice that these are corrected speed and the wind turbine power output is depended of cube of the wind speed.

### 5.3.2 Wind Shear, Wind Profile, Wind Gust, and Other Meteorological Effects

Vertical wind shear is important as wind turbines become larger and larger. It is therefore questionable whether the hub height wind speed is still representative. Various methods exist in the literature concerning the extrapolation of wind speed to the hub height of a wind turbine. There are several theoretical expressions used for determining the wind speed profile. The Monin–Obukhov method is usually used to determine the wind speed $v$ at height $z$ by:

$$v(z) = \frac{u_*}{K}\left[\ln\frac{z}{z_0} - \Psi\left(\frac{z}{L}\right)\right] \qquad (5.10)$$

The function $\Psi(z/L)$ is determined by the solar radiation at the site under survey. This equation is valid for short periods of time, and not for monthly or annual average readings. This equation has proven satisfactory for detailed surveys at critical sites; however, such a method is difficult to use for general engineering studies. Thus the surveys must resort to simpler expressions and secure satisfactory results even when they are not theoretically accurate. Vertical wind shear is an important consideration as wind turbines are becoming larger. Obstacles can cause the displacement of the boundary layer from the ground. The roughness length ($z_0$) describes the height at which the wind is zero by definition, meaning that surfaces with a large roughness length have a large effect on the wind. It ranges from 0.0002 for open sea, 0.005–0.03 for open land, 0.03–0.1 for agricultural land, and 0.5–2 m for very rough terrain or urban areas (see Table 5.2 for details). However, the increase of wind speed with height should be considered for the installation of large wind turbines. Thus the surveys must rely to

**TABLE 5.2**

Roughness Classes

| Roughness Class | Description | Roughness Length [$z_0$ (m)] |
|---|---|---|
| 0 | Water surface | 0.0002 |
| 1 | Open areas | 0.03 |
| 2 | Farm land with few windbreaks more than 1 km apart | 0.1 |
| 3 | Urban districts and farm land with many windbreaks | 0.4 |
| 4 | Dense urban or forest | 1.6 |

simpler expressions and secure satisfactory results even when they are not theoretically accurate. For $h_0 = 10$ m and $z_0 = 0.01$ m, the parameter $\alpha = 1/7$, which is consistent with the value of 0.147 used in the wind turbine design standards (IEC standard, 61400-3, 2005) to represent the change of wind speeds in the lowest levels of the atmosphere. Wind speed is usually recorded at the standard meteorological height of 10 m, while wind turbines usually have hub heights near 80 m.

In cases which lack elevated measurements hub-height wind velocity is estimated by applying a vertical extrapolation coefficient to surface measurements. However, the vertical extrapolation coefficient may contain errors and uncertainties due to terrain complexity, atmospheric stability and turbulence. Various methods exist for the wind speed extrapolation to the wind turbine hub height. The theoretical background of the wind extrapolation methods is based on the Monin-Obukov similarity theory. However, the wind speed $v(z)$ at a height $z$ can be calculated directly from the wind speed $v(z_{ref})$ at height $z_{ref}$ (usually the standard measurement level) by using the logarithmic law (the Hellmann exponential law) expressed by:

$$\frac{v(z)}{v_0} = \left(\frac{z}{z_{ref}}\right)^{\alpha} \tag{5.11}$$

where, $v(z)$ is the wind speed at height $z$, $v_0$ is the speed at $z_{ref}$ (usually 10 m height, the standard meteorological wind measurement level), and $\alpha$ is the friction coefficient, power low index or Hellman index. This coefficient is a function of the surface roughness at a specific site and the thermal stability of the Prandtl layer. It is frequently assumed to be $1/7$ for open land. However, this parameter can vary diurnally and seasonally as well as spatially. It was found that a single power law is insufficient to adequately project the power available from the wind at a given site, especially during night time and also in presence of the low-level jets. There are also significant discrepancies of values for $\alpha$, especially for arid and dry regions, with ranging from 0.09 to 0.120, quite smaller comparing to the standard 0.147 value (Belu, 2014; Belu and Koracin, 2015). Moreover, $\alpha$ can vary for one place to other, during the day and year. Another formula, known as the logarithmic wind profile law and widely used across Europe, is the following:

$$\frac{v}{v_0} = \frac{\ln\left(\dfrac{z}{z_0}\right)}{\ln\left(\dfrac{z_{ref}}{z_0}\right)} \tag{5.12}$$

The other formula for estimating wind speed with height, slightly different of the one above to is:

$$\frac{v}{v_0} = \frac{\ln\left(1+\dfrac{z}{z_0}\right)}{\ln\left(1+\dfrac{z_{ref}}{z_0}\right)} \tag{5.13}$$

where $z_0$ is called the roughness coefficient length and is expressed in meters; it depends basically on the land type, spacing and height of the roughness factor (water, grass, etc.) and it ranges from 0.0002 up to 1.6 or more, see Table 5.2 for the most common values.

> **Example 5.6:** A meteorological tower (mast) is located close to the edge of a hill. If the wind speed is 8.5 m/s at standard 10 m observation level, what is the speed at 80 m? Use Equations (5.12) and (5.13) and assume $z_0 = 1.2$.
>
> **Solution:** Wind speeds at 80 m height computed with the two equations are:

$$v = 8.5 \frac{\ln(80/1.2)}{\ln(10/1.2)} = 16.8 \text{ m/s}$$

$$v = 8.5 \frac{\ln(1+80/1.2)}{\ln(1+10/1.2)} = 14.2 \text{ m/s}$$

And compared with 11.5 m/s using the 1/7 power law.

In addition to the land roughness, these values depend on several factors, varying during the day and at night, and even during the year. For instance, the reading or monitoring stations can be within farming land; it follows that the height/length of the crops will change. However, once the speeds have been calculated at other heights, the relevant equations can be used for calculating the useful energy potential via different methods. If the type of ground cover is known, the wind speed at other heights can be estimated. For practical evaluation, roughness is divided into classes so that there are classes for landscape forms. The classes range from open sea through flat open country to urban community areas, as shown in Table 5.2. Equations (5.11) through (5.13) describe basically very similar profiles. There is an interconnection between the Hellmann height exponents $\alpha$ and the ground roughness $z_0$. This is not uniform relationship, being dependent upon the height of the reference measurement, and an approximation is given by:

$$\alpha = \frac{1}{\ln\left(\dfrac{15.25}{z_0}\right)} \tag{5.14}$$

The ideal wind turbine site is a location with smooth wind, constant speed, and prevailing wind direction and without or minimum turbulence, which is only possible in locations with roughness class 0 or 1, water surfaces and open terrains with smooth surfaces, where wind speeds are constant and there is no turbulence because there are no obstacles. Due to very low roughness, the wind speed at 30 m above water surface is not significantly different from its speed at 50 m height. Moreover, due to the low turbulence, the lifetime of the turbine is extended. However, there are great variations in wind speed and direction, depending on the weather and local surface conditions, with the greatest variations happening during the daytime. Because of low turbulence, lots of wind turbines are situated offshore, in sites about 5–10 km in the sea. Wind shade (turbulence due to the obstacles in lands like forests, hills, etc.) does not affect wind turbines that are far in the sea. Therefore, based on their siting, there are two main types of wind turbines: onshore (site is on the land) and offshore (site is in the sea).

**Example 5.7:** The wind velocity measured at 10 m height at a weather station is 7.5 m/s. Find out the velocity at 50 m height at a wind turbine site having similar wind profile. The roughness heights at the weather station and the wind turbine site are 0.03 and 0.1 m, respectively.

**Solution:** Using Equation (5.12), the wind speed at turbine height is

$$v = 7.5\frac{\ln\left(\dfrac{50}{0.01}\right)}{\ln\left(\dfrac{10}{0.03}\right)} = 11.0 \text{ m/s}$$

Aside from ground level to hub height shear, the wind shear over the rotor disc area can also be significant in the case of large wind turbines spanning over lager height. The standard procedure for power curve measurements is given by the IEC standard (IEC Standard, 6-1400-12-1, 2005) where the wind speed at hub height is considered to be representative of the wind over the whole turbine rotor area. This assumption can lead to considerable wind power estimate inaccuracies, since inflow is often non-uniform and unsteady over the rotor-swept area. In most studies about the effect of wind shear on power performance, the wind speed shear is described by the shear exponent, obtained from the assumption of a power law profile. By integrating the wind profile over the rotor span, the corrected wind speed at the turbine nacelle can be obtained:

$$U_{avrg} = \frac{1}{2R}\int_{H-\frac{D}{2}}^{H+\frac{D}{2}} v(z)dz = v(H)\cdot\frac{1}{\alpha+1}\cdot\left(\left(\frac{3}{2}\right)^{\alpha+1} - \left(\frac{1}{2}\right)^{\alpha+1}\right) \tag{5.15}$$

where $H$ is the nacelle height and $D = 2R$ is the rotor diameter. From (5.7), it is obvious that the hub height wind speed $z(H)$ is $\alpha$ corrected based on the profile it is experiencing. It is observed that both corrections have more or less the same effect. For wind speeds in the range 5–20 m/s (the useful wind turbine speed regime) the corrected power differs in general less than 5% from the uncorrected power. However, in all cases the corrected power is larger than the uncorrected power.

**Example 5.8:** Estimate the wind power density available for a wind turbine, if the wind speed of 14.5 m/s at the hub height using Equation (5.14) to corrected the average wind aped across the rotor area and uncorrected wind speed. Assume the power index is 0.147 and the air density 1.2 kg/m³.

**Solution:** Corrected wind speed is:

$$U_{avrg} = 14.5\cdot\frac{1}{0.147+1}\cdot\left(\left(\frac{3}{2}\right)^{0.147+1} - \left(\frac{1}{2}\right)^{0.147+1}\right) = 14.42 \text{ m/s}$$

Using Equation (5.1), the wind power densities are:

$$p_{uncor} = 0.5\times1.2\times(14.5)^3 = 1829.2 \text{ W/m}^2$$

$$p_{corr} = 0.5\times1.2\times(14.5)^3 = 1799.1 \text{ W/m}^2$$

An additional wind property that can make the impact on wind turbine operations is wind gustiness. Proper design and operation of a wind turbine for a specific wind climate requires knowledge of wind extremes and gustiness, often defined by a wind gust factor. This is especially true in areas where wind climate is determined by inherently strong gusty winds, such as downslope windstorms (Belu and Koracin, 2015; Emeis, 2013). In sites with high ambient turbulent intensity and gusty winds, turbines are subject to extreme structural loading and fatigue. The gust factor ($G$) is defined as:

$$G = \frac{u_g}{U} - 1 \tag{5.16}$$

where $u_g$ is the gust speed and $U$ is the mean daily wind speed. One expects higher gusts to be associated with higher mean speeds; however, one may also expect that the normalized gust speed $u_g/U$ and, consequently, the gust factor, $G$, decreases with the increasing mean speed. The following equation relates the gust factor to the mean daily wind speed:

$$G = AU^n \tag{5.17}$$

where the parameters $A$ and $n$ are obtained by using a least-square fit of the logarithm of $G$ vs. the logarithm of the mean daily wind speed. While gusts generally decrease as wind speed increases, in extreme cases the wind gusts can easily reach over twice the strongest wind speeds ($v > 20$ m/s) and damage a wind turbine. However, wind gusts over 25 m/s, the upper wind speed limit of a large wind turbine, are quite unlikely in many areas. Belu and Koracin (2009) used four and half years (2003–2009) of composite data sets and found that winds over 25 ms$^{-1}$ occurred only about 2% of time. Gusts associated with stronger winds may cause considerable losses by reducing the energy production of the wind turbine which would otherwise operate at nominal output power. Another effect of wind gusts is the additional stress on the wind turbine structure, which may reduce its lifespan.

The low-level jet is a mesoscale phenomenon associated with the nighttime very stable boundary layer that can have a width of hundreds of kilometers and a length of a thousand kilometers. They have been observed worldwide. During nighttime and over land, ground surface cools at a faster rate than the adjacent air and stable stratification forms near the surface and propagates upward. Downward mixing of the winds is reduced and winds aloft become decoupled from the surface and accelerate. The maximum wind speeds are usually 10–20 m/s or more at elevations mainly at 100–300 m and occasionally as high as 900 m above ground. Consequently, it is not possible to accurately estimate winds aloft at hub and blade heights from routine surface measurements. Additionally, a strong wind shear and associated turbulence develop at the bottom and top of the jet layer.

## 5.4 Instrumentation and Measurement Techniques, Data Validation

Wind resource assessment measurement campaigns have traditionally been conducted with mast based instrumentation consisting on cup anemometers and wind vanes. 92% of the wind analysts with experience in measurement campaigns use this type of instruments as a baseline. Additionally, 90% of them are also used to other meteorological instruments like temperature, pressure and humidity sensors. Wind monitors (propeller anemometers that measure both the wind speed and direction) are only used by 18%

of them, while sonic anemometers and remote sensing instruments (Lidar, SODAR and satellites) are used by 35% and 50% of them respectively, of which 75% are consultants or researchers. Wind energy developers and manufacturers stick to the standard mast configuration for long-term measurement campaigns. Sonic anemometers or remote sensing instruments are still far from being standard instruments in wind assessment although they are being used more and more by consultants and researchers for detailed measurement campaigns or power performance testing. Given the limited time span of the measurement campaigns it is necessary to extrapolate to a period of at least 5 years in order to predict the long-term average energy yield. To this end, Measure-Correlate-Predict (MCP) methods are most used ones. Surprisingly, in spite of its primary importance for wind energy developers, only 70% of them declared in the questionnaire that they used MCP methods. This percentage is raised to 80% in the case of consultants. Numerical models are used in wind resource assessment to spatially extrapolate the wind measurements horizontally to obtain wind maps, and vertically to reach hub-height wind fields. Additionally, numerical weather prediction models are being used to build virtual historical time series that can be used, via statistical downscaling, to extend limited periods of measurements to longer time spans in a similar way as it has been traditionally done with MCP methods.

### 5.4.1 Ecological Indicators for Wind Energy

Aeolian features may be used as indicators for strength of the wind prevailing in an area. Aeolian features are the formations on land surface due to continuous strong wind. Sand dunes are one example for the aeolian formations; the wind is carrying the sand particles, being deposited back when the wind speed is lower. The size of the particle thus carried and deposited along with the distance can give us an indication on the average strength of the wind in that region. Other types of aeolian features are playa-lake, sediment plumes and wind scours. Another way to identify a windy site is to observe the biological indicators. Trees and bushes get deformed due to strong winds. The intensity and nature of this deformation depends on the wind strength. This method is specifically suitable to judge the wind in valleys, coasts and mountain terrains. Vegetation can indicate regions of high wind speed where there are no measurements available. Deformation or flagging of trees is the most common indicator. In some cases the flagging of trees is a more reliable indicator of the wind resource than the data available. The Griggs and Putnam Index for flagging of coniferous trees, $G$ is related to the annual mean wind speed by this empirical relationship:

$$\bar{v} = 0.95G + 2.6 \tag{5.18}$$

The use of trees as an indicator of wind speed is subject to practical limitations with the greatest concern is the tree's exposure to the wind. The deformation must be viewed perpendicularly to the prevailing wind direction, so the full flagging and throwing effects of are taken into account. Hence, trees selected as wind indicators are well exposed to the prevailing winds. An index for broad leave trees is the deformation ratio, $D$, representing the amount of crown asymmetry and trunk deflection of trees caused by the wind often used is:

$$D = \frac{A}{B} = \frac{C}{45°} \tag{5.19}$$

This relationship is used for both coniferous and hemispherical crowned trees. For coniferous trees, $A$ is the angle formed by the crown edge and the trunk on the leeward side, $B$

is the angle formed by the crown edge and the trunk, and $C$ is the average angle of trunk deflection. In the case of hemispherical crowned trees, $A$ is the distance between the trunk and crown perimeter on the leeward side, $B$ is the distance between the trunk and crown perimeter on the windward side, and $C$ is the angle between the crown perimeter and trunk on the leeward side. The ratio $A/B$ is assumed to be between 1 and 5, so the minimum $D$ value is 1, corresponding to no crown asymmetry, or trunk deflection, $C$. Since the maximum deflection is 90° for a tree growing along the ground, then the maximum deformation ratio, $D$ is 7. Regression analysis of the deformation ratio vs the mean annual wind speed estimated for Douglas fir or Ponderosa pine trees gives the following empirical relationship:

$$\bar{v} = 0.95D + 2.3 \tag{5.20}$$

Photographs are often used to determine the deformation ratio in lieu of direct examination, which are expensive and time consuming. The deformation ratio and Griggs–Putnam Index give similar ranges of wind speeds.

## 5.4.2 Wind Measurement Techniques

Nowadays a series of measurement techniques is available for on-site wind resource measurement ranging from point measurements performed at different heights using cup anemometers or ultrasonic sensors to profiling techniques like sonic detection and ranging (SODAR) or light detection and ranging (Lidar) systems. Until now, the overwhelming majority of measurement campaigns for commercial wind developments or even for wind energy research projects rely on cup anemometry and occasionally on ultrasonic sensors, where the latter is often preferred in research applications. Remote-sensing techniques like SODAR or Lidar are increasingly explored as complementary techniques, providing high quality wind vertical profiles at higher sampling rates. In large wind projects the profiling instrument can be conveniently relocated within the project area for wind resource measurements at different site points, following the initial calibration period where the profiler is operating in conjunction with a conventional tower-based measurement system. Remote sensed wind speed measurements are needed to supplement and extend tall met mast measurements, on- and off-shore, and to evaluate various wind flow models for a number of purposes, including: wind resource assessments, wind energy development projects, power curve measurements, wind resource uncertainty evaluation, etc. The common denominator in most of these issues is high accuracy, and with a demand for reproducible certainty to more than 99% of what can be achieved with a corresponding calibrated cup anemometer.

A significant source for uncertainty with remote sensing instrumentation relative to a cup anemometer, and for SODARS in particular, is the remote instruments relative big measurement volumes. A SODAR measuring the wind speed from say 100 meters height probes a total sampling volume of more than 1000 $m^3$ whereas a cup anemometer essentially is essentially a point measurement device. In addition the SODAR measured wind components are displaced in space and time, making the interpretation of measured turbulence by a SODAR impaired. In addition the huge sampling volumes will be putting restrictions on measurements in non-uniform flow regimes such as found near forest edges on offshore platforms, and over hilly or complex terrain. SODAR remote sensing is also in demand for direct turbine control integration, wind power optimization and turbine mounted gust warning systems, but here the demand on accuracy and reliability is correspondingly high.

For about 150 years, the primary sensor used for wind speed measurements has been the vertical axis cup anemometer. The sensor must be positioned in the wind stream and kept free of ice and debris and the internal friction must remain constant. Because a structure is required to hold the sensor in place, a disturbance is introduced into the free-stream wind field and invariably some wind data must be "corrected" or discarded. Cup anemometers are known to be influenced by turbulence, air temperature, air density, and flow inclination. A significant vertical component to the wind vector can cause over-speeding of cup anemometers. More recently, ultrasonic anemometers have been used to more precisely measure wind fields. These sensors detect Doppler shifts in an ultrasonic wave transmitted between near-field (~20 cm) transducers. Ultrasonic systems also are subject to structure flow disturbance effects, however, as the device must be mounted on a "lightning cage" structure to ground out lightning, and the sensed volume must be bracketed by transducer elements. All anemometers have operational characteristics that are subject to external conditions that may influence the wind speed measurement and introduce error. Despite these shortcomings, the error of class I 15 anemometers is extremely low, less than 0.1 m/s at wind speeds below 16 m/s (IEC61400-12-1, ISO 16622). A cup anemometer is a simple device relying on the different aerodynamic drag of the convex and concave surfaces of a suitably designed cup. Typically two to four cups are mounted symmetrically on a vertical axis and allowed to rotate freely. If placed in a constant speed air stream, the anemometer will eventually spin at a frequency proportional to the wind speed. If coupled to a small electric generator, the corresponding electrical signal can be conveniently registered by data conditioning and logging. Since only the signal frequency information is of interest, a signal forming device is required for processing in data loggers, as the amplitude of the signal inconveniently varies proportionally to the frequency. A common choice is to convert the analog signal into a train of fixed amplitude pulses which can be counted by the digital device. The velocity ratio, related to the drag coefficients in the x and y direction is defined as:

$$\lambda = \frac{\omega R}{v} \tag{5.21}$$

And the velocity ratio can be related to the drag coefficient ratio, $\mu$, through:

$$\lambda = \frac{\mu+1}{\mu-1} - \sqrt{\left(\frac{\mu+1}{\mu-1}\right)^2 - 1} \tag{5.22}$$

In Equations (5.16) and (5.17), $R$ is the anemometer radius, $\omega$ is the anemometer angular velocity, and $v$ is the wind speed. For typical values of the drag coefficients for the concave and convex surfaces of 1.4 and 0.4, respectively, the steady-state speed ratio is calculated to be 0.303, i.e., the cups will rotate at about a third of the wind speed. The equations above assumed the wind speed to be uniform horizontally. If a horizontal wind shear is considered, then corrections have to be applied to the apparent wind speeds at the concave and convex surface, respectively. Sonic anemometers measure the wind speed in 2- or 3-D based on a comparison of the flight times of two anti-parallel sonic pulses. In modern sonic anemometers each measurement path consists of a pair of transducers, both capable

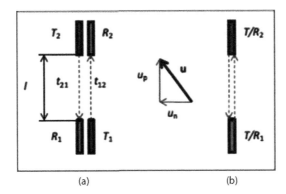

**FIGURE 5.1**
Geometry of the ultrasonic transceivers (one measurement path): (a) separate transmitters and receivers for each direction and (b) integrated transceiving unit.

of transmitting and receiving. For a uniform and stationary wind field the wind velocity component in the direction of the measurement direction, $v_M$ can then be calculated from:

$$v_M = \frac{l}{2}\left(\frac{1}{t_d} - \frac{1}{t_r}\right)$$ (5.23)

where $l$ is the distance between transceivers and $t_d$ and $t_r$ are the direct and reverse travel times (see Figure 5.1 for details). Two or three orthogonal measurement paths can be combined in one instrument, in order to measure the full wind velocity vector. Sonic anemometers have a series of advantages over cup anemometers, such as the absence of moving parts, faster response to fluctuations and avoiding over-speeding. Furthermore, the response is linear over a large frequency range, and the measurement is relatively independent of the flow properties, such as spatial and time variations, temperature, density, etc. Sonic anemometers are absolute instruments that do not require individual calibration. Some drawbacks do exist, such as the influence of the finite measurement path, path separation, and transducer shadows. For estimating wind farm power production, the relevant wind field includes the entire rotor disk plane. Since these data are rarely obtainable, the usual alternative is to erect a tall offshore tower to mount sensors and measure the wind speeds up to hub height. The cost to erect an onshore tall (90–100 m) meteorological mast is on the order of ~$250,000, but for offshore, it can cost in range of millions of dollar, depending on water depth, tower facilities and other factors.

Cup and/or sonic anemometers are installed at specific levels, while temperature, pressure, humidity and solar radiation are measured at standard meteorological level. The use of conventional anemometers on a fixed met tower is thus problematic for the reasons, such as: (1) limited spatial coverage and data that can be collected cost effectively; (2) increased cost and time required for resource assessment and project development; (3) it creates a barrier to competition among developers, limiting the pool to well capitalized firms that can risk the cost of a met tower; and (4) high potential environmental impacts, view-shed impacts, and human use conflicts.

Both mechanical and sonic anemometers measure wind velocity at a specific location in space, requiring the use of several instruments for the assessment of vertical profiles, remote sensing techniques are capable of providing an almost instantaneous photograph of the complete vertical wind velocity profile up to a certain height. One such technique, becoming increasingly popular in the wind energy industry, is SODAR. SODAR is a remote sensing methodology for measurements of the wind speed and direction aloft at various heights in the atmosphere. SODAR's are ground-based instruments that transmit a sequence of short bursts of sound waves at audible frequencies (2000–4000 Hz) upward in three different inclined directions into the atmosphere. The SODAR measurement technology was well established and in operational use for decades by now, starting in the 1980s ties where they served environmental protection issues and has been extensively applied to atmospheric research for environmental protection air pollution prediction measures well before the present burst in wind energy research and application. SODAR was originally developed for atmospheric research and air traffic safety but is increasingly deployed where knowledge of the wind velocity profile up to greater heights than provided by anemometer towers (generally limited to 60 or 80 m) is required, or the assessment of different locations within a vast wind project development is desirable.

Commercially available systems tailored for wind resource assessment are capable of providing information on the atmospheric boundary layer up to heights of 200 or 300 m, thereby covering the full range of heights swept by typical rotor blades. This type of information is particularly useful when dynamic structural analysis of the turbine rotor is performed, since both wind shear over the rotor diameter and the evolution of turbulence with height significantly impact on the prediction of rotor stress, and their precise knowledge avoids the use of sometimes oversimplifying assumptions. As the sound waves from a SODAR propagate forward a small fraction of the transmitted sound energy is scattered and reflected in all directions from temperature differences and turbulence in the atmosphere. A very small fraction of this scattered energy reaches back into the SODAR's detector, which in principle is a directional-sensitive microphone. The height at which the wind speed is measurement is usually determined by the time delay in the backscatter from the transmitted pulse. Under standard atmospheric conditions with sound propagation speed of about 340 m/s backscatter from a SODAR measurement at 170 meters height above the ground will reach back into the detector after 1 s delay time. The wind speed component in the transmitted beam direction is subsequently determined from the Doppler shift observed as frequency difference between the transmitted frequency and the frequency of the received backscattered sound wave. By combining the measured wind speed components obtained in this way from three differently inclined sound path directions, e.g., from one vertical and two inclined sound paths, the three-dimensional wind vector including wind speed and direction and tilt can be measured by SODAR from preset heights from the ground and up to the limit determined by the SODARS lowest acceptable Carrier-to-Noise (C/N) ratios. The above description is for a mono-static system, where transmitter and receivers are co-located on the ground. But alternative configurations, e.g., in the form of so-called bi-static SODAR configurations exists as well, where the transmitter and receivers are separated e.g., 100–200 meters on the ground. Bi-static configurations have significant S/N-ratio advantages over mono-static configurations for wind energy applications. Received backscatter in a bi-static configuration is not limited to direct (180°) backscatter from temperature (density) fluctuations only, but enables also backscatter contributions from the atmospheric turbulence.

And the higher the wind speed the more turbulence. This becomes in particular relevant during strong wind situations, where the background noise level increases with the wind speed. A particular configuration considered for wind energy applications is therefore the bi-static Continuous Wave (CW) SODAR configuration. Alternatively to the range gating in a pulsed system, the range to the wind speed measurement in a CW system can be determined by well-defined overlapping transmission and receiving antenna functions.

A recent adaptation of a common remote sensing technology can help provide the type of wind data required to measure the energy field on the relevant time and spatial scales.

Although other remote sensing methods exist for measuring wind speeds (e.g., SODAR, radar), using a coherent laser provides the most accurate and versatile way to provide remote measurements. Lidar has been in use for decades to accurately measure distances and generate digital elevation models for topography and mapping. In the last ten years or so it has been adapted to measure wind speed and direction, and in the last few years it has begun to appear in the offshore wind industry. Wind lidar instruments are capable of providing diverse benefits, including better resource assessment and better turbine control systems. Lidar measures reflected light just as radar measures reflected radio waves. Doppler Lidar principle (the dominant technology for wind speed measurement) is the measurement of the Doppler shift of the reflected radiation from a coherent laser. The laser beam, at frequency $\omega_o$, hits natural aerosols carried on the wind and is reflected and scattered. Some of the light is reflected back at a frequency altered by the Doppler shift ($\Delta f$), and the Doppler-shifted frequency of the reflected light is detected by a sensor (see Figure 5.1 for details). Wave interference between the two signals creates a "beat" frequency proportionally to the wind speed vector component along the laser. By probing the laser along three or more radial vectors, the wind direction can be resolved, providing an accurate estimate of the average wind speed and direction at the focal distance sampled. Although the detectors may be focused at a set distance, they actually sense the backscatter from a probed volume defined by depth of the focal field. This results in a narrow Gaussian distribution of Doppler shift which must be interpreted with algorithms. Much like SODAR, modern Lidar devices designed for wind energy purposes relay on the detection and frequency analysis of backscattered waves. As in SODAR, the wind velocity component along with the observation direction is obtained through a Fast Fourier Transformation of a Doppler-shifted signal. However, instead of sound waves laser beams are used and backscattering is caused by interaction with particles (aerosols) and molecules in the atmosphere instead of density fluctuations as in the case of SODAR. Scattering mechanisms include molecular processes such as Rayleigh (elastic) and Raman (inelastic) scattering, with Rayleigh scattering being by far the dominant process. Interaction with particles is also known as Mie scattering. For eye safety, commercial units use infrared laser light at a wavelength of 1.55 µm. In order to obtain the three vector components of the wind velocity, the laser beam has to be inclined at an angle $\theta$ with respect to the normal, much like in SODAR units. As opposed to SODAR, however, where beam steering relies on phase shifts between the units contained in the antenna, requiring a relatively complex electronic control scheme, laser beams can be easily rotated by means of a rotating wedge, in order to acquire redundant information by performing a full 360° scan around the vertical axis.

The line-of-sight velocity component obtained from the Doppler shift of the laser frequency then becomes a function of the azimuth angle $\varphi$. Although each Lidar system is slightly different, they are primarily characterized by their laser emission waveform, which is either continuous wave (CW) or pulsed laser (PL). This distinction is the most germane to understanding the subtleties of Lidar wind measurement. A good technical description and comparison of the strengths and weaknesses inherent in PL and CW Lidar can be found in. The summary below draws information from these three sources. Continuous Wave (CW): The CW laser emits a continuous (non-pulsed) beam and optically focuses the receiver at the target distance, resulting in a distribution of return signal gain around the focal distance, as shown in Figure 5.2.

Due to its optical focus, the CW probe length increases with the square of the range. This larger sample volume boosts the signal while the longer distance attenuates it, resulting in a fairly constant carrier to noise ratio (CNR) over the target range of the unit. Although there is bigger potential for range (height) error, with the longer probe

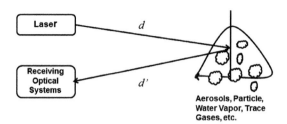

**FIGURE 5.2**
Doppler Shift Lidar, Beat Detection Schematic.

volume at greater heights, because the wind profile is generally more vertical at greater heights, this may not be significant. Beyond several hundred meters, however, the probed volume becomes too large to render a meaningful point estimate, and clouds and other factors come into play. In pulsed laser (PL) technology discrete laser pulses are sending, while setting the receiver timing gates to capture the pulse reflection from the target range. Thus the probe length is proportional to pulse duration (fixed) and the reflected signal gain exhibits a Gaussian distribution. PL can probe several ranges near-simultaneously through the use of multiple range gates, being so able to capture turbulence structures. On the downside, the PL CNR decreases with distance since the probed volume does not increase to offset the signal fade. This can only be overcome by increasing laser power. Also, PL must use a minimum pulse duration related to the Nyquist frequency, thus a minimum pulse length and a minimum probe length, usually the distance between targets. PL is therefore ill-suited to ranges below the minimum probe length. There is no simple, single trade-off between the two technologies, but in general, CW is better at short ranges (<30 m), while PL is at long ranges (>200 m). The transitional region lies between 100–150 m, where various factors could drive selection either way.

### 5.4.3 Data Validation and Editing

After the field data about the wind speed are collected and transferred to the computing environment, the next steps are to validate and process data, and generate reports. Data validation is defined as the inspection of all the collected data for completeness and reasonableness, and the elimination of erroneous values. Data validation transforms raw data into validated data. The validated data are then processed to produce the summary reports required for analysis. This step is also crucial in maintaining high rates of data completeness during the course of the monitoring program. Therefore data must be validated as soon as possible, after they are transferred. The sooner the site operator is notified of a potential measurement problem, the lower the risk of data loss. Data can be validated either manually or by using computer-based techniques. The latter is preferred to take advantage of the power and speed of computers, although some manual review will always be required. Validation software may be purchased from some data-logger vendors, created in-house using popular spreadsheet programs: e.g., Microsoft Excel, or adapted from other utility environmental monitoring projects. An advantage of using spreadsheet programs is that they can also be used to process data and generate reports. Data validation implies visual data inspection, missing data interpolation, outliers and questionable data rejection, saving data in an appropriate file format for further processing.

## 5.5 Wind Resource Statistics

In a meteorological sense, winds are movements of air masses in the atmosphere due to the temperature differences generated by uneven solar heating from the Sun. The equatorial region is more irradiated than the polar ones. Consequently, the warmer and lighter air of the equatorial region rises to the outer layers of the atmosphere and moves toward the poles, being replaced at the lower layers by a return flow of cooler air coming from the polar Earth regions. This air circulation is also affected by the Coriolis forces associated with the rotation of the Earth. In fact, these forces deflect the upper flow toward the east and the lower flow toward the west. The effects of differential heating dwindle for latitudes greater than 30°N and 30°S, where westerly winds predominate, due to the rotation of the Earth. These large-scale (synoptic) air flows that are taking place all over the atmosphere constitute the geostrophic winds. The lower layer of the atmosphere is known as surface layer and extends to a height of 100 m. In this layer, winds are delayed by frictional forces and obstacles altering not only their speed but also their direction. This is the origin of turbulent flows which cause wind speed variations over a wide range of amplitudes and frequencies. Additionally, the presence of seas and large lakes causes air masses circulation similar in nature to the geostrophic winds. All these air movements are called local winds. The wind in a given site near the surface of the Earth results from the combination of the geostrophic and local winds. Therefore, it depends on the geographic location, the climate, the height above ground level, the roughness of the terrain and the obstacles in the surroundings. These are the winds for the wind energy interests of which the wind turbines interact with.

The wind is characterized by its speed and direction which is affected by several factors, including: geographic location, climate characteristics, height above ground, vegetation and surface topography. Wind turbines interact with the wind capturing part of its kinetic energy and converting it into usable energy. This energy conversion is the result of several phenomena that are explored briefly in this chapter. The study of geographic distribution of wind speeds, characteristic parameters of the wind, topography and local wind flow and measurement of the wind speed, are all essential in wind resource assessment and analysis for the successful wind turbine applications. Wind availability, the influence of the turbine height installation above ground, the wind gusting effect and the WECS micro-sitting are the main influences of the energy output and are the basis for the wind energy assessment. There are two aspects of wind resource assessment: (1) determination of the general wind power potential, and (2) determination of wind power potential and predicted energy production for wind farms. The general wind power potential was determined from the wind speed data available, and then wind maps were developed. In general, the measured wind speed data that were available were at heights of 6 to 20 m, more often at 10 m height. However, some anemometers were on top of buildings or airport control towers, or at meteorological masts, which influences the accuracy of the data. Wind classes were developed for 10 m height because that was the standard for world meteorological data, and then the wind power potential at 50 m was double that at 10 m due to the assumption that the wind shear exponent was 1/7 for all locations.

Global pressure and wind patterns, upper air wind data, and boundary layer meteorology were also used to obtain a consistent estimate of the wind energy resource. The knowledge of the quasi-steady mean wind speeds that can be expected at a potential site is crucial to determine the economic viability of a wind energy project. These data are also essential in selecting the WECS in order to maximize efficiency and durability. The probability distribution of mean wind speed is predicted from measurements collected during several years. All these data are usually arranged in a histogram. Where actual wind speed frequency distribution was available, that was used. If such data was not available, wind speed probability distributions, such as Weibull or Rayleigh distributions, are used to estimate the wind power potential. The highly dependent nature of power production on wind speed necessitates accurate predictions in any proposed location. However, the wind is not highly predictable and a simple average wind speed for a location does not suffice in providing enough information for determining the annual energy production of a turbine. It is important to know both the average speed and the distribution of wind speeds for a location for an accurate energy production calculation.

Wind resource assessment and analysis represent the methods and techniques of measuring and quantifying the quality of wind power potential at a particular location or site. This is a critical preliminary step in developing new wind power locations, and is necessary because wind resources can vary significantly even over relatively small geographic scales. Wind velocities have usually strong inter-annual and seasonal variabilities, meaning that wind resource assessments should ideally be conducted over a multiple year span to provide more realistic potential power production estimates. The available wind energy in an area varies appreciably, even inter-annually. The uncertainties in wind resource potential are significant and therefore thorough wind resource assessment and analysis are necessary for determining long-term wind conditions and regimes at the site of interest to the wind developers. Determining and predicting long-term wind conditions serves two important objectives: (i) analyzing the wind energy conversion site quality, and (ii) designing and setting optimum wind plant layout, including selecting appropriate wind turbine types for that site. A comprehensive wind resource assessment and analysis usually entails the following tasks: (1) identification of a suitable wind energy site using wind maps and power system maps; (2) characterization of the on-site wind resource by measuring the winds (and other meteorological parameters) for at least one year continuously at or the closest hub height, with temporary meteorological masts (possibly completed or supplemented with remote sensing instruments); (3) process and extrapolate the date using appropriate techniques at hub heights and for longer time periods, by correlating with the available long-term wind records from neighbor weather stations, observation sites or numerical model output; (4) setting of the turbine locations in the wind farm, relative to the wind resource estimation and site topography; (5) estimate of the potential wind energy production over a year and over the projected project lifetime, for the entire site; (6) perform the evaluation of the losses due to various causes (e.g., equipment scheduled and unscheduled maintenance, collection array losses, etc.); and (7) conduct the most accurate possible evaluation of the uncertainty associated with every step above. Physical uncertainties in wind energy may be broadly classified into long-term and short-term uncertainties. Long-term uncertainties are mainly introduced by: variation of wind conditions, wind turbine design, and other environmental, operational and financial factors. Short-term uncertainties are mainly introduced by boundary layer turbulence and other flow variations that occur over small time scales (order of minutes to hours). Accurate evaluations and reductions of the uncertainties are of particular importance to secure financing and ensuring the wind plant investor's confidence. The uncertainties are driving the probability distribution of the expected wind energy production. These values depend on many factors, including the project size, the topography complexity and the availability of historical wind data.

### 5.5.1 Wind Statistics and Probability Distribution Functions

In the following subsection we considered the most common wind speed distributions used by the wind energy community. The wind probability distribution (PDF) functions have been investigated, employed and explained by many researchers and engineers involved in the wind energy. In many wind power studies the features and characteristics of such distributions are used for assessment and analysis of wind resources, wind power plant operation, grid integration as well as for design purposes. Both analytical and numerical simulation methods can be carried out, although they are generally used with features of wind power and not output power in mind. However, things can be planned from a different point of view, as similar distribution functions can be described for power, if wind distribution functions are taken into account, together with WT features, on the basis of data provided by the manufacturers. Usually the time series of wind speeds and directions are rather large, differences among parameter estimation methods will not be nearly as important as differences among distributions. There are several estimators of PDF parameters, such as the Method of Moment (MOM), Maximum Likelihood Estimators (MLE), Least-Square (LS) and Percentile Estimators Methods.

These estimators are unbiased, so there is no reason to give preference to any of them. The choice of specific estimators is usually based on the existing wind speed observations, computing availability and user preference. The rule of thumb is to select a number of estimators of the PDF parameters, while the parameters are usually computed by taking the averages of the estimates found by these methods. We preferentially use MLE, MOM and LS estimators for the large samples, and the averages are the PDF parameters. However, when using MOM, we calculate the sample mean $\bar{v}$, standard deviation ($S$), and skewness ($Sk$) as:

$$\bar{v} = \frac{1}{N}\sum_{i=1}^{N} v_i \tag{5.24}$$

$$S = \sqrt{\frac{1}{N-1}\sum_{i=1}^{N}\left(v_i - \bar{v}\right)^2} \tag{5.25}$$

and

$$Sk = \frac{1}{N} \cdot \frac{\sum_{i=1}^{N}\left(v_i - \bar{v}\right)^3}{S^3} \tag{5.26}$$

where $N$ is the number of observations in $v$, our random variable, the wind speed.

**Example 5.9:** In a location, the hourly wind speeds (m/s) for a 24-hour interval are: 1, 0, 2, 1.5, 4, 5, 8, 3.8, 10, 12, 9.5, 8.5, 7, 4, 5, 5.5, 6, 3, 1, 0.5, 2.7, 3, 4.5, 6.5, 3. Calculate the mean wind speed, standard deviation and skewness for this data set.

**Solution:** By using Equations (5.24) through (5.26) the required wind speed statistics are:

$$\bar{v} = \frac{1}{N}\sum_{i=1}^{N}v_i = \frac{116.5}{24} = 4.85 \text{ m/s}$$

$$s = \sqrt{\frac{1}{N-1}\sum_{i=1}^{N}(v_i - \bar{v})^2} = 3.10 \text{ m/s}$$

$$Sk = \frac{1}{N}\cdot\frac{\sum_{i=1}^{N}(v_i - \bar{v})^3}{s^3} = 0.522$$

Wind speeds are normally measured in integer values, each integer value is usually observed several times during a specific observation period, such as one year of observations. The numbers of specific wind speed observations $v_i$ is defined as $m_i$. The mean wind speed is computed in this case as:

$$\bar{v} = \frac{1}{N}\sum_{i=1}^{p}m_i\cdot v_i \tag{5.27}$$

Here $p$ is the number of different values of wind speed observed and $N$ is still the total number of observations. Then the standard deviation (and the variance) is given by:

$$s = \sqrt{\frac{1}{N-1}\left[\sum_{i=1}^{p}m_i v_i^2 - \frac{1}{N}\left(\sum_{i=1}^{p}m_i v_i\right)^2\right]} \tag{5.28}$$

**Example 5.10:** A wind monitoring system is located in the trade-winds on the northeast coast of a Caribbean island was measured 1 m/s for 36 times, 2 m/s for 28 times, 3 m/s for 32 times, 4 m/s for 72 times, 5 m/s for 48 times, 6 m/s 24 times, 7 m/s 54 times, and 8 m/s 44 times, 9 m/s for 20 times and 10 m/s for 12 times during a given period. Find the mean and standard deviation.

**Solution:** The total number of observations in this case is $N = 370$, and $p = 10$. The mean wind speed and standard deviation are computed by using Equations (5.27) and (5.28), respectively:

$$\bar{v} = \frac{1}{N}\sum_{i=1}^{p}m_i\cdot v_i = 5.11 \text{ m/s,}$$

$$s = 2.486 \text{ m/s}$$

For a better understanding on wind variability and characteristics, the measurement data are often presented in the form of frequency distribution. This gives us the information on the number of hours for which the velocity is within a specific range. For developing the frequency distribution, the wind speed sample is divided into equal intervals (e.g., 0–1,

1–2, 2–3, etc.) and the number of times the wind record is within this interval is counted. An example for the monthly frequency distribution of wind speed is given in Table 5.2. If the velocity is presented in the form of frequency distribution, the mean (average) and standard deviation are given by:

$$\bar{V} = \frac{\sum_{i=1}^{N} f_i \cdot v_i}{\sum_{i=1}^{N} f_i} = \frac{1}{N}\sum_{i=1}^{N} f_i \cdot v_1 \tag{5.29}$$

and

$$S = \left[\frac{1}{N-1}\sum_{i=1}^{N} f_i\left(v_i - \bar{V}\right)^2\right]^{1/2} \tag{5.30}$$

Here $f_i$ is the frequency and $v_i$ is the mid value of the corresponding interval and the $N$ is the number of total measurements in the sample. It is noticed the basis that the distribution function of the wind speed in a certain location depends just on the mean wind speed, a distribution function of the wind power can be obtained for a given WT by using its power curve. Once the wind power distribution function is obtained, the mean power available is deduced. So as not to depend on the type of WT, this will be shown per unit of surface (mean power density). This process is performed in four different ways: (1) obtaining of the wind power; (2) Betz' law considerations; (3) consideration of realistic values, remembering that Betz' law is an upper efficiency limit; and (4) consideration of WT parameters such as *cut-in* and *cut-out wind speed, rated speed,* and *rated power.* The goal of any wind energy assessment and analysis is to obtain expressions that allow us to give response to questions about the mean value of the statistical distribution of the maximum power obtainable from the wind, regardless of the WT chosen, and also taking into account its features, when the only input value is the mean wind speed.

As we mentioned, the wind velocity changes rapidly with time. In tune with these changes, the power and energy available from the wind also vary. The variations may be short time fluctuation, day-night variation, weekly, monthly, seasonal, yearly or multi-decadal variations. Statistical descriptions, tables or histograms are useful tools to characterize and asses the wind regime. However, it is more convenient for a number of theoretical reasons to model the wind speeds and directions by continuous mathematical functions rather than for example table of discrete values. Various probability functions were fitted with the field data to identify suitable statistical distributions for representing wind regimes. Once a probability distribution function is found the wind regime and wind power characteristics can be computed. For example, wind power density (WPD), a useful way to evaluate the available wind resource at a potential site can be calculated. WPD (measured W/m²) indicates how much energy is available at the site, and is a nonlinear function of the probability density function of wind speed. If we characterize the actual wind data by a probability distribution function $f(v)$, then the average power in the wind is:

$$P_{WPD} = 0.5 \cdot \rho \cdot \int_0^{V_{max}} v^3 \cdot f_{PD}(v)dv \tag{5.31}$$

where $v$ represents the wind speed; $V_{max}$ is the maximum possible wind speed at that location; and $f_{PD}(v)$ is the pdf of the wind speed. Existing wind probability distribution functions are classified into: (i) univariate and unimodal distributions of wind speed (such as Weibull, Rayleigh, and Gamma distributions), and (ii) bivariate and unimodal distributions of wind speed and wind direction. For univariate wind speed distribution, the most widely used is the 2-parameter Weibull distribution. Other distributions used to characterize wind speed include the 1-parameter Rayleigh distribution, 3-parameter generalized Gamma distribution, 2-parameter Lognormal distribution, 3-parameter Beta distribution, 2-parameter inverse Gaussian distribution, singly truncated normal Weibull mixture distribution, and the maximum entropy probability density functions. Wind temporal characteristics are originating from the fact wind velocity (wind direction and wind speed magnitude) varies significantly over relatively short period of time. The mean wind speed, $\bar{V}$ and the mean wind power speed, $V_P$ are used to characterize wind velocity and available power variations, and are defined as:

$$\bar{V} = \frac{1}{\Delta t} \int_{t_1}^{t_2} v(t)dt \tag{5.32}$$

And

$$V_P = \frac{1}{\Delta t} \left[ \int_{t_1}^{t_2} v^3(t)dt \right]^{1/3} \tag{5.33}$$

Here, $\Delta t = t_2 - t_1$, and the mean wind power velocity can be used to estimate the available wind power density in this interval. Typically the time interval is taken as one year and the two speeds are the annually averaged values. From empirical observations and measurements the tow variables are related as:

$$V_P \approx 1.25 \cdot \bar{V} \tag{5.34}$$

And the available average power density can be estimated, from Equation (5.2), as:

$$P_{WPD} \approx 0.5\rho \cdot V_P^3 \tag{5.35}$$

Often wind speed or velocity may vary from zero to values significantly higher than the mean wind speed; however the very high values are occurring over short time periods and not very often, so their contributions to the total generated power are quite low, even during that short period power values are significant. Notice also that the integrals of Equations (5.32) and (5.33) are computed from a large number of wind velocity measurements, being usually approximated with sums of the integrants, leading to expression similar to ones of Equations (5.24) or (5.29).

### 5.5.2 Weibull Probability Distribution

The Weibull probability density distribution function is commonly used statistical distribution to model wind speed distribution. The Weibull curve is a probability density function and indicates both the frequency and magnitude of a given wind speed over a period of time. To employ the probability distribution function to define, characterize, and fit the field data has had a long history. It has been established that the Weibull distribution is often used to characterize wind speed regimes, and it is commonly employed to estimate and to assess wind energy potential. Although efforts have been made over the years to fit the field wind data to other distributions such as exponential distribution, Pearson type VI distribution, logistic distribution, etc., the Weibull distribution is well accepted and the most used for wind data analysis, and is given by:

$$f_{WB} = k \frac{v^{k-1}}{c^k} \exp\left(-\left(\frac{v}{c}\right)^k\right) \tag{5.36}$$

The Weibull distribution is a function of two parameters: the shape parameter, $k$, and the scale factor, $c$, defining the shape or steepness of the curve and the mean value of the distribution. These coefficients are adjusted to match the wind data at a particular site. For modeling wind, typical $k$ values range from 1 to 3 and can vary drastically from site to site. The scale parameter corresponds to the site average wind speed. The main inaccuracy of the Weibull distribution is that it always has a zero probability of zero wind speed, which is not the case, since there are frequently times in which no wind is blowing. However, the fault is virtually without consequence because most turbines will not operate in speeds below 3 m/s and the distribution is more accurate, compared to measured data, within the zone most used by turbines: 8 to 14 m/s. The higher the $k$ value, the sharper the increasing part of the curve is. The higher $c$ values correspond to a shorter and fatter distribution, with a higher mean value. Ideally the mean value would correlate with the rated wind speed of the turbine: producing rated power for the greatest period of time annually. The cumulative probability function for Weibull distribution is given by:

$$F(v) = 1 - \exp\left[-\left(\frac{v}{c}\right)^k\right] \tag{5.37}$$

The availability of high quality wind speed distributions is crucial to accurate forecasts of annual energy production for a wind turbine. Statistical distributions suffice for early estimations, while the actual wind speed measurements are necessary for accurate predictions. The probability of the wind speed $v$ being equal to or greater than $v_a$ is:

$$P(v \geq V) = \int_V^\infty f_{WB}(v) \cdot dv \tag{5.38}$$

The probability of the wind speed being within an interval $(V - \Delta v, V + \Delta v)$ centered on the wind speed $V$ is given by:

$$P(V - \Delta v \leq v \leq V + \Delta v) = \int_{V - \Delta v}^{V + \Delta v} f_{WB}(v) \cdot dv \simeq f_{WB}(V) \tag{5.39}$$

The factors $k$ and $c$ are determined for each measurement site. The $f_{WB}(v)$ is the probability of observing the particular wind speed, $v$. There are several estimators of Weibull parameters, such as the Moment, Maximum Likelihood, Least-Square, and Percentile Estimators methods. These estimators are unbiased, although some of them, such as the Method of

Moments, may have large variances, so there is no reason to prefer any of them. The choice of specific estimators is usually based on the existing wind speed observations, computing availability and user preference. The rule-of thumb is to select a number of estimators of the Weibull parameters, such as the standard least-square, maximum likelihood or variants of the maximum likelihood methods (MLE), while the shape and scale parameters are usually computed by taking the averages of the estimates found by these methods. If sufficient wind speed observations are available, one of the most used estimation method is the Method of Moments (MOM) or its variants. It is based on the numerical iteration of the following two equations while the mean ($\bar{v}$) and standard deviation ($s$) of the wind speeds is available from the following observations:

$$\bar{v} = c \cdot \Gamma\left(1 + \frac{1}{k}\right) \tag{5.40}$$

$$s = c\left[\Gamma\left(1 + \frac{2}{k}\right) - \Gamma^2\left(1 + \frac{1}{k}\right)\right]^{1/2} = (\bar{v})^2\left[\frac{\Gamma(1 + 2/k)}{\Gamma^2(1 + 1/k)} - 1\right] \tag{5.41}$$

where $\bar{v}$ is the wind speed data set (sample) mean, as defined in Equation (5.24), $s$ is the wind speed data set (sample) standard deviation (Equation (5.25)), and $\Gamma()$ is the Gamma function expressed as:

$$\Gamma(x) = \int_0^\infty t^{x-1}\exp(-t)dt \tag{5.42}$$

Usually, the wind data collected at a site is used to directly calculate the mean (average) wind speed. If we want to find $c$ and $k$ from the observation data, there are several methods available, as discussed above. However, a good estimate for $c$ can be obtained quickly from Equation (5.34) by considering the ratio $c/\bar{v}$ as a function of $k$. For values of $k$ above 1.5 and less than 3 or even 4, however, the ratio $c/\bar{v}$ is essentially a constant, with a value of about 1.12. This means that the scale parameter is directly proportional to the mean wind speed for this range of $k$, as given bellow. Most good wind regimes will have the shape parameter $k$ in this range, so this estimate of $c$ in terms of $u$ will have wide application.

$$c = 1.12 \cdot \bar{v} \tag{5.43}$$

**Example 5.11:** The Weibull parameters at a given site are $c = 6.5$ m/s and $k = 1.89$. Estimate the number of hours per year that the wind speed will be between 5.0 and 9 m/s, number of hours per year that the wind speed is greater than or equal to 12.5 m/s (the so-called rated wind speed of a wind turbine), and the average wind speed at that site.

**Solution:** From Equation (5.35), the probability that the wind is between 5.0 and 9.0 m/s is just $f_{WB}(7)$, ($v = 7$ m/s is the middle of this interval) which can be evaluated from Equation (5.36) as:

$$f_{WB}(7) = 1.89\frac{7^{1.89-1}}{6.5^{1.89}}\exp\left(-\left(\frac{7}{6.5}\right)^{1.89}\right) = 0.098311$$

This means that the wind speed is in this interval 9.831% of the time, so the number of hours per year (≈8760 hours) with wind speeds in this interval is:

$$0.098311 \cdot 8760 = 861.2 \, \text{h/yr}$$

From Equation (5.34), the probability that the wind speed is greater than or equal to 12.5 m/s is

$$P(v \geq 12.5) = \exp\left[-\left(\frac{12.5}{6.5}\right)^{1.89}\right] = 0.032015$$

This is giving: $0.032015 \cdot 8760 = 280.45$ hours/year

From Equation (5.39) an estimate at the mean (average) wind speed is:

$$\bar{v} = \frac{c}{1.12} = \frac{6.5}{1.12} = 5.80 \, \text{m/s}$$

A special case of the moment method is the so-called empirical method, proposed by Justus 1976. Weibull shape parameter, $k$ is estimated in this method by following relationship:

$$k = \left(\frac{s}{\bar{v}}\right)^{-1.086} \tag{5.44}$$

Then the scale parameter, $c$ is computed by using the following relationship:

$$c = \frac{\bar{v}}{\Gamma\left(1 + \dfrac{1}{k}\right)} \tag{5.45}$$

Both, the moment and empirical method require a reasonable wind speed observations data set to be available.

**Example 5.12:** For the wind speed data of Table 5.3 compute the average wind speed, standard deviation, the Weibull scale and shape parameters and the wind power density.

**Solution:** From Equations (5.29) and (5.30) the average wind speed and standard deviation are: 9.10 and 4.32 m/s. The Weibull shape and scale parameters are then computed by using:

$$k = \left(\frac{4.32}{9.10}\right)^{-1.086} = 2.246$$

and

$$c = \frac{9.10}{\Gamma\left(1 + \dfrac{1}{2.246}\right)} = 10.28 \, \text{m/s}$$

**Note:** If we calculate the scale parameter from Equation (5.43), $c = 1.12 \cdot 9.10 = 10.19$ m/s, almost the same as the previous scale parameter value.

**TABLE 5.3**

Frequency Distribution of Wind Speed

| No | Speed (m/s) | Hours per Month | Frequency |
|----|-------------|-----------------|-----------|
| 1  | 0–1   | 12 | 0.0167 |
| 2  | 1–2   | 10 | 0.0139 |
| 3  | 2–3   | 25 | 0.0347 |
| 4  | 3–4   | 36 | 0.0500 |
| 5  | 4–5   | 42 | 0.0583 |
| 6  | 5–6   | 50 | 0.0694 |
| 7  | 6–7   | 58 | 0.0806 |
| 8  | 7–8   | 65 | 0.0903 |
| 9  | 8–9   | 71 | 0.0986 |
| 10 | 9–10  | 77 | 0.1069 |
| 11 | 10–11 | 81 | 0.1125 |
| 12 | 11–12 | 52 | 0.0722 |
| 13 | 12–13 | 33 | 0.0458 |
| 14 | 13–14 | 24 | 0.0333 |
| 15 | 14–15 | 18 | 0.0250 |
| 16 | 15–16 | 14 | 0.0194 |
| 17 | 16–17 | 11 | 0.0153 |
| 18 | 17–18 | 8  | 0.0111 |
| 19 | 18–19 | 7  | 0.0097 |
| 20 | 19–20 | 6  | 0.0083 |
| 21 | 20–21 | 5  | 0.0069 |
| 22 | 21–22 | 4  | 0.0056 |
| 23 | 22–23 | 4  | 0.0056 |
| 24 | 23–24 | 3  | 0.0042 |
| 25 | 24–25 | 2  | 0.0028 |
| 26 | 25–26 | 1  | 0.0014 |
| 27 | 26–27 | 1  | 0.0014 |
| 28 | 27–28 | 0  | 0.0000 |

Another Weibull parameter estimator, based on the concept of least squares, is the graphical method. In which a straight line is fitted to the wind speed data using lease squares minimization, where the time-series data must be sorted into bins. Taking a double logarithmic transformation, the equation of cumulative distribution function can be rewritten as:

$$\ln\left\{-\ln\left[1 - F(v)\right]\right\} = k\ln(v) - k\ln(c) \tag{5.46}$$

Plotting $\ln(v)$ against $\ln\{-\ln[1-F(v)]\}$, the slope of the straight line fitted best to data pairs is the shape parameter; the scale parameter is then obtained by the intercept with y-ordinate. The application of the graphical method requires that the wind speed data be in cumulative frequency distribution format. Time-series data must therefore first be sorted into bins. The line of best fit can be drawn by hand or determined using a least-squares regression. This method is referred in literature as the "graphical method" even though the least-squares regression can be performed without producing a graph of the data. In essence it a is a variant of the moment method, consisting of the calculation from the time series of the observations of statistical estimators, such as means of the wind speeds and of the square wind speeds. Then the Weibull parameters are calculated as follows:

$$\bar{v} = \frac{c}{k}\Gamma\left(\frac{1}{k}\right) \tag{5.47}$$

and

$$\overline{v^2} = \frac{2c^2}{k}\Gamma\left(\frac{2}{k}\right) \tag{5.48}$$

After computing the mean values from the observations, Equations (5.47) and (5.48) are solved and the Weibull parameters are estimated. The Weibull distribution can also be fitted to time-series wind data using the maximum likelihood method. The shape factor $k$ and the scale factor $c$ are estimated using the following two equations:

$$k = \left(\frac{\displaystyle\sum_{i=1}^{N} v_i^k \ln(v_i)}{\displaystyle\sum_{i=1}^{N} v_i^k} - \frac{\displaystyle\sum_{i=1}^{N} \ln(v_i)}{N}\right)^{-1} \tag{5.49}$$

$$c = \left(\frac{1}{N}\sum_{i=1}^{N} v_i^k P(v_i)\right)^{1/k} \tag{5.50}$$

where $v_i$ is the wind speed in the bin $i$ and $N$ is the number of nonzero wind speed data points (the actual wind speed observations), $P(v_i)$ is the frequency with which the wind speed falls within bin $i$. Equation (5.46) is solved numerical, usually through iterative processes, with $k = 2$ as initial guess. After which Equation (5.43) can be solved explicitly. Care must be taken to apply Equation (5.46) only to the nonzero wind speed observations. When wind speed data are available in frequency distribution format, a variation of the maximum likelihood method can be applied. One of the parameter estimator not very often used is the energy pattern factor method. In this approach, the energy pattern factor for a given wind speed data is defined as:

$$E_{pf} = \frac{\overline{v^3}}{\bar{v}^3} \tag{5.51}$$

here $\overline{v^3}$ is the mean of the cubes of the wind speed. Notice that the factors in Equation (5.42) are related to the wind energy estimates. Weibull shape parameter can be estimated with the following equation:

$$k = 1 + \frac{3.69}{E_{pf}} \tag{5.52}$$

Then, the scale parameter is estimated using Equation (5.45). The above method is included because it is often necessary to estimate the Weibull parameters in the absence of suitable information about the distribution of wind speeds. For example, only annual or monthly averages may be available. In such a situation, the value of $k$ must be estimated. The value of $k$ is usually between 1.5 and 3, depending on the variability of the wind. Smaller $k$ values correspond to more variable (more gusty) winds. Interested readers can find more about the estimation of the Weibull parameters in the references included at the end of this chapter, or elsewhere in the literature. To analyze the efficiency of the aforementioned methods, the following tests are used: RMSE (root mean square error), $\chi^2$ (chi-square), $R^2$ (analysis of variance or efficiency of the method) and the Kolmogorov–Smirnov test. These tests are used to examine whether a probability density function is suitable to describe the wind speed data or not. The RMSE test is defined by:

$$RMSE = \left[ \frac{1}{N} \sum_{i=1}^{N} (y_i - x_i)^2 \right]^{1/2} \tag{5.53}$$

where $y_i$ are the actual values at time stage $i$, $x_i$ the values computed from correlation expression for the same stage, and $N$ is the number of data. The nest two tests are defined by:

$$\chi^2 = \frac{\sum_{i=1}^{N} (y_i - x_i)^2}{N - n} \tag{5.54}$$

and

$$R^2 = \frac{\sum_{i=1}^{N} (y_i - z_i)^2 - \sum_{i=1}^{N} (y_i - x_i)^2}{\sum_{i=1}^{N} (y_i - z_i)^2} \tag{5.55}$$

where $N$ is the number of observations, $y_i$ is the frequency of observations, $x_i$ is the frequency of Weibull, $z_i$ is the mean wind speed, and $n$ is the number of constants used. The Kolmogorov–Smirnov test is defined as the max-error between two cumulative distribution functions:

$$Q = max \left| F_T(v) - F_O(v) \right| \tag{5.56}$$

where $F_T(v)$ and $F_O(v)$ are the cumulative distributions functions for wind speed not exceeding $v$ computed by using estimated Weibull parameters and by observed (or randomly generated) time-series, respectively. The critical value for the Kolmogorov–Smirnov test at 95% confident level is given by:

$$Q_{95} = \frac{1.36}{\sqrt{N}} \tag{5.57}$$

If $Q$ value exceeds the critical value then one can say that there is significant difference between the theoretical and the time-series data under the given confident level. Figure 5.3 is showing the wind speed histograms and the fitted Weibull probability distributions

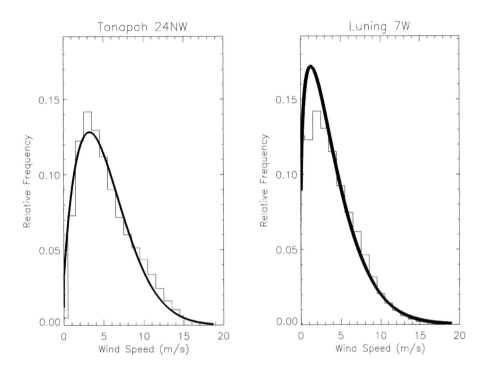

**FIGURE 5.3**
Wind speed frequency histograms and fitted Weibull probability distributions, composite 2003–2008 data sets at two towers, Tonopah experiment, western Nevada (Belu and Koracin, 2009 and 2013).

for 2003–2008 composite wind speed data set at two locations (meteorological towers), Tonopah experiment, western Nevada (Belu and Koracin, 2009 and 2013). The data were collected at 50-m meteorological towers for over four years. Several methods for determine the Weibull parameters are used, and the shape and scale parameters are taken as averages. Figure 5.3 is showing a very good agreement between fitted PDF and actual data. The energy that a wind energy conversion system is generating depends on both its power curve (representing the nonlinear relationship between the wind speed and turbine power output) and the site wind speed distribution diagram, which is essential a graph or histogram showing the number of hours for which the wind blows at different wind speeds during a given period of time. If derived from long-term (multi-annual) wind speed data sets the histograms, the shape of the probability distribution functions characterize the wind regime at that site. The histogram takes into account the seasonal variation and year-by-year variations for the years covered by the statistics.

If the probability distribution function at a site is determine the available wind power density is determined from Equation (5.27). It can be shown that when probability distribution is the Weibull density function, the average power density is:

$$P_{WPD} = 0.5\rho \cdot \bar{v}^3 \, \frac{\Gamma(1+3/k)}{2\left[\Gamma(1+1/k)\right]^2} \tag{5.58a}$$

or an alternative relationship:

$$P_{WPD} = 0.5\rho \cdot c^3 \Gamma(1+3/k) \tag{5.58b}$$

Three other very important wind speed parameters used for wind energy assessment or in turbine design can also be estimated, once the Weibull parameters are computed: the most probable wind speed, the wind speed carrying the maximum energy and the energy pattern factor. The most probable wind speed ($v_{MP}$), denoting the most frequent wind speed for Weibull distribution, is given by:

$$v_{MP} = c\left(\frac{k-1}{k}\right)^{1/k} \tag{5.59}$$

The wind speed carrying the maximum energy ($v_{MaxE}$) is expressed by:

$$V_{MaxE} = c\left(\frac{k+2}{k}\right)^{1/k} \tag{5.60}$$

The energy pattern factor, *EPF*, used in turbine aerodynamic design being defined as the total amount of power available in the wind divided by the power calculated from cubing of the average wind speed is given by:

$$EPF = \frac{\overline{v^3}}{\overline{v}^3} = \frac{\Gamma(1+3/k)}{\Gamma(1+1/k)} \tag{5.61}$$

**Example 5.13:** For the site characterize of the Weibull probability distribution function of Example 5.12 (average wind speed is 9.10 m/s, $k = 2.24$ and $c = 10.28$ m/s) calculate, the available wind power density, the most probable wind speed, the wind speed carrying the maximum energy and the most frequent speed. Assume standard air density of 1.2 kg/m³.

**Solution:** From Equation (5.56), the average available wind power density is:

$$P_{WPD} = 0.5 \times 1.2 \times 9.10^3\, \frac{\Gamma(1+3/2.28)}{2\left[\Gamma(1+1/2.28)\right]^2} = 344\ \text{W}/\text{m}^2$$

From Equations (5.59) and (5.60), the most probable wind speed, the wind speed carrying the maximum energy and the most frequent speed are:

$$v_{MP} = 10.28\left(\frac{2.24-1}{2.24}\right)^{1/2.24} = 7.91\ \text{m}/\text{s}$$

$$v_{MaxE} = 10.28\left(\frac{2.24+2}{2.24}\right)^{1/2.24} = 13.65\ \text{m}/\text{s}$$

### 5.5.3 Other Probability Distribution Function Used in Wind Energy

Another used probability distribution is the Rayleigh distribution, which is a special case of the Weibull distribution where $k = 2$. The Rayleigh distribution is simpler because it depends only on the mean wind speed, and is given by:

$$f_{RL}(v) = \frac{\pi}{2}\frac{v}{c^2}\exp\left[-\frac{\pi}{4}\left(\frac{v}{c}\right)^2\right]$$

$$\tag{5.62}$$

These two probability distribution functions are the most commonly used for wind energy analysis and assessment. The simpler of the two is the Rayleigh distribution which has a single parameter $c$. The Weibull distribution shown before has two parameters $k$ and $c$. The Rayleigh distribution is actually a special case of the Weibull distribution with $k = 2$. Setting $k = 2$ in the Weibull distribution gives the Rayleigh distribution. For both distributions, $V_{min} = 0$ and $V_{max} = \infty$. Setting $k = 2$ in this result gives the cumulative Rayleigh distribution.

$$F(v) = 1 - e^{-(v/c)^2} \tag{5.63}$$

For the Rayleigh distribution the single parameter, $c$, relates the following three properties:

$$c = V_{mp}\sqrt{2} = \frac{2\mu}{\sqrt{\pi}} = \sigma\sqrt{\frac{4}{8-\pi}} \tag{5.64}$$

Here $V_{mp}$ and $\sigma$ are the mean wind speed and the standard deviation. The Rayleigh distribution can be written using $V_{mp}$ or the mean velocity, $\mu$. The usual determination of the mean and standard deviation from experimental data for the normal distribution are well known. The minimum-least-squares-error (MLE) estimate of the mean of the normal distribution is the arithmetic mean (the sum of all values divided by the number of values). The formula for the MLE estimate of the variance is also familiar. The parameter $c$ in the Rayleigh distribution can be evaluated from a set of $N$ data points on wind velocity, $V_i$. When experimental data are used to determine parameters in probability distributions, the computed result is called an estimate of the true parameter. Here we use the symbol to indicate that the equation below gives us only an estimate of the true distribution parameter, $c$.

$$\hat{c} = \sqrt{\frac{1}{2N}\sum_{i=1}^{N}V_i^2} \tag{5.65}$$

The cumulative Rayleigh distribution is often re-written in terms of a given wind speed and the average wind speed, $\mu$, in a more convenient form for wind energy analysis, as:

$$F(v) = 1 - \exp\left[\left(-\frac{\pi}{4}\right)\left(\frac{v}{\mu}\right)^2\right] \tag{5.66}$$

The cumulative distribution function is used to compute the probability the wind speed will be at or below a given value $U$, knowing the average wind speed:

$$P(\text{wind speed} \le U) = 1 - \exp\left[\left(-\frac{\pi}{4}\right)\left(\frac{U}{\mu}\right)^2\right] \tag{5.67}$$

**Example 5.14:** Using Rayleigh distribution and $\mu = 5.57$ m/s calculate the probability the wind speed is between 4 and 10 m/s. The probability the wind is in the range is difference between the probability of wind at the maximum value and the probability of the minimum value.

**Solution:** From Equation (5.44) the required probabilities are estimated as:

$$P\left(\text{Wind Speed } \leq 4 \text{ m/s}\right) = 1 - \exp\left[\left(\frac{\pi}{4}\right)\left(\frac{4}{5.57}\right)^2\right] = 0.3333 \text{ or } 33.33\%$$

$$P\left(\text{Wind Speed } \leq 10 \text{ m/s}\right) = 1 - \exp\left[\left(\frac{\pi}{4}\right)\left(\frac{10}{5.57}\right)^2\right] = 0.9205 \text{ or } 92.05\%$$

Therefore the probability that the wind speed is in the range 4 to 10 m/s is: 92.05−33.33 = 58.72%.

According to Rayleigh probability distribution the average available wind power density is computed as:

$$P_{WPD} = \rho \int_0^\infty v^3 \cdot f_{RL}\left(v\right) dv = \rho \int_0^\infty \frac{\pi}{2} \frac{v^4}{c^2} \exp\left[-\frac{\pi}{4}\left(\frac{v}{c}\right)^2\right] dv = \frac{6}{\pi} \rho \cdot V_{mp}^3 \tag{5.68}$$

Notice since $6/\pi \approx 1.91$, implying that in Rayleigh distribution the average available wind power density is the one in Equation (5.35). It is also noted that this PDF has a maximum for $V/V_m = (8/\pi)^{1/2} \approx 1.60$. From Equations (5.33) and (5.62) the most probable wind power speed, for Rayleigh PDF is:

$$V_P = \frac{2}{\pi} V_{mn} \simeq 0.637 V_m \tag{5.69}$$

**Example 5.15:** Determine the wind power speed and the most probable available wind power for a Rayleigh PDF having the mean wind speed 12.5 m/s.

**Solution:** From Equations (5.68) and (5.69) the required values are:

$$V_P = 0.637 \cdot 12.5 = 7.96 \text{ m/s}$$

$$P_{WPD} = 1.91 \cdot (7.96)^3 = 964.2 \text{ W/m}^2$$

The 3-parameter Weibull (W3) is a generalization of the 2-paramter Weibull distribution, where the location parameter s establishes a lower bound (which was assumed to be zero for the 2-parameter Weibull distribution). It was found that for some areas the W3 fits wind speed data better than the 2-parameter Weibull model. The W3 probability distribution and cumulative distribution functions are expressed as:

$$f(v,k,c,\tau) = \frac{kv^{k-1}}{c^k} \exp\left[-\left(\frac{v-\tau}{c}\right)^k\right] \tag{5.70}$$

and

$$F(v,k,c,\tau) = 1 - \exp\left[-\left(\frac{v-\tau}{c}\right)^k\right] \tag{5.71}$$

Equations (5.70) and (5.71) are valid for $v \geq \tau$. It is recommended to use MLEs for the parameter estimation:

$$\frac{\sum_{i=1}^{N}(v_i - \hat{\tau})^{\hat{k}} \ln(v - \hat{\tau})}{\sum_{i=1}^{N}(v_i - \hat{\tau})^{\hat{k}}} - \frac{1}{\hat{k}} - \frac{1}{N}\sum_{i=1}^{N}\ln(v_i - \hat{\tau}) = 0 \tag{5.72}$$

and

$$\hat{c} = \left(\frac{1}{N}\sum_{i=1}^{N}(v_i - \hat{\tau})^{\hat{k}}\right) \tag{5.73}$$

$$\hat{\tau} + \frac{\hat{c}}{N^{1/\hat{k}}}\Gamma\left(1 + \frac{1}{\hat{k}}\right) = U_{min} \tag{5.74}$$

where $N$ is the number of observations in the sample $v$, and $U_{min}$ indicates the minimum values in the $v$ time series. Parameters of the W3 distributions are then found iteratively solving the Equations (5.72) through (5.73).

### 5.5.3.1 Gamma Probability Distribution

Gamma PDF can be expressed with the following function:

$$g(v; x; \beta) = \frac{v^{x-1}}{\beta^2\Gamma(x)}\exp\left[-\frac{v}{\beta}\right] \quad \text{for } v, x, \beta > 0 \tag{5.75}$$

where $x$ and $\beta$ are the shape and scale parameter, respectively, $\Gamma(x)$ is the Gamma function, as expressed in Equation (5.42). The parameters of the Gamma distribution can be estimated using graphical, moment or maximum likelihood methods, similar to one presented above in the Weibull case. For more information of fitting measurement data sets to Gamma distributions the reader is advised to follow. However, the Gamma PDF is usually employed in a mixture of distributions in connection with Weibull PDF.

### 5.5.3.2 Lognormal Probability Distribution

Another PDF used in wind energy assessment, especially in offshore applications even not quite often is the lognormal distribution. The 2-parameter lognormal PDF is given by:

$$f(v, \mu, \sigma) = \frac{1}{\sigma v\sqrt{2\pi}}\exp\left[\frac{-(\ln(v) - \mu)^2}{2\sigma^2}\right]$$

$$\tag{5.76}$$

And its cumulative distribution function is expressed as:

$$F(v,\mu,\sigma) = \frac{1}{2} + \frac{1}{2} erf\left[\frac{\ln(v) - \mu}{\sigma\sqrt{2}}\right] \tag{5.77}$$

where $erf()$ is the error function from the Normal distribution, and the parameters $\mu$ and $\sigma$ are the mean and standard deviation of the natural logarithm of $v$. MOM method is usually employed to estimate the PDF parameters because it showed considerable better fits to the data and because it enables to handle samples with zeros. The estimators are given by:

$$\hat{\mu} = \ln\left(\frac{\bar{v}}{\sqrt{1 + \frac{s^2}{\bar{v}^2}}}\right)$$

$$\hat{\sigma} = \sqrt{\ln\left(\frac{s^2}{\bar{v}^2}\right)}$$

### 5.5.3.3 *Truncated Normal Probability Distribution*

The truncated normal distribution is the probability distribution function of a normally distributed random variable whose values are either bounded below, above or both. Since the wind speed is only positive, the most common is the single truncated normal distribution, suitable for nonnegative case:

$$n(v,\mu,\sigma) = \frac{1}{I(\mu,\sigma)\sigma\sqrt{2\pi}}\exp\left[-\frac{(v-\mu)^2}{2\sigma^2}\right] \quad \text{for} \quad v > 0 \tag{5.78}$$

where $\mu$ and $\sigma$ the date mean and standard deviation, and $I(\mu,\sigma)$ is the normalized factor that leads the integration of the truncated normal distribution to one, which can the cumulative distribution evaluated in its domain de definition. The normalized factor is given by:

$$I(\mu,\sigma) = \frac{1}{\sigma\sqrt{2\pi}}\int_0^\infty \exp\left[-\frac{(v-\mu)^2}{2\sigma^2}\right]dv \tag{5.79}$$

The distribution function parameters can be determine using graphical, moment or maximum likelihood methods or a combination of them. In recent years, in order to improve the accuracy of wind statistics, mixtures of PDFs were employed. Distribution function mixed with Gamma, Weibull, or Normal distribution functions can be used to describe the wind statistics. For example, Gamma and Weibull mixture applied to wind energy assessment is given by:

$$h(v;w;x,\beta,k,c) = wg(v,x,\beta) + (1-w)f(v,k,c) \tag{5.80}$$

where $0 \leq w \leq 1$ is the weight parameter indicating the mixed proportion of each distribution included in the mixture. Again, the five parameters, in the Equation (5.80) can be estimated using graphical, moment, maximum likelihood methods or combination them, as discussed in the Weibull probability distribution case.

Notice that the two parameters Weibull probability function at many sites is characterized by $k = 2$, which is the *Rayleigh wind speed distribution*. The actual measurement data taken at many sites compare well with the Rayleigh probability distribution function. The Rayleigh distribution is then a simple and accurate enough representation of the wind speed with just one parameter, the scale parameter $c$. Summarizing the characteristics of the Weibull probability distribution function, for $k = 1$ makes it an exponential distribution, where $\lambda = 1/c$, $k = 2$ makes it the Rayleigh distribution, while values of $k > 3$ makes it approach a normal bell-shaped (normal) probability distribution function. It is also worth to remember that most of wind energy sites have the scale parameter ranging from 10 to 20 mph (or about 5 to 10 m/sec) and the shape parameter ranging from 1.5 to 2.5 (quite unlikely 3.0 or higher). Notice also, that regardless of the shape and the scale parameters, use of the mode or the mean speed in the power density equation would introduce a significant error in the annual energy or power density estimate, sometimes off by several folds, making the estimates completely useless. Only the RMS speed in the power equation always gives the correct average power over a period.

## 5.5.4 Wind Direction Analysis and Statistics

To ensure the most effective use of a wind turbine it should be exposed to the most energetic wind. Though the wind may blow more frequently from one direction more wind energy may come from a different direction if those winds are stronger, being noticed for many areas that strong winds usually come from a particular direction. These wind direction changes are due to the general atmospheric circulation of atmosphere, on an annual basis (seasonal) to the mesoscale horizons (4–5 days). The seasonal changes of prevailing wind direction could be as little as 30° in trade wind regions to as high as 180° in temperate regions. For example, in the United States plains, the predominant directions of the winds are from the south to southwest in the spring and summer and from the north in the winter. There can also be changes in wind direction on a diurnal basis. Notice that, wind shear changes in wind direction with height are generally nonexistent or quite small, except for very short time periods as weather fronts move through. In areas where weather patterns tend to be dominated by the passage of fronts over a time scale of a day or two, they determine the coarse structure of the wind. Within these daily patterns, variations may occur over periods of tens of minutes and these are very significant in terms of windmill performance. Changes over such time scale can influence the amount of time a conventional generating plant may be needed to respond rapidly to changes in wind power output. Daily, seasonal and annual variations also occur and can have a significant impact on wind energy economics. On a daily basis, the pattern of wind speeds, and thus the energy available from the winds, is not entirely random. Above land, atmospheric heat loss is greatest in early afternoon and this is particularly pronounced in the summer months. This is reflected in both higher mean wind speeds and greater turbulence at this time. At sea, the position is much less clear; there are little diurnal variations in water surface temperature and an inverse relationship between onshore and offshore atmospheric stability is to be expected. However, this will be significantly affected by land and sea breeze effects, distance offshore, wave height, seasonal variations and wind direction. These diurnal changes in wind are important because there exists well-defined variations of demand for electricity during the day and this affects the value of the wind energy. The fact, for example, that power demand peaks in late afternoon in most of the locations, means that the statistically significant extra wind energy produced at this time by wind power plants sited on land may slightly increase its value to the electricity supply system. The wind direction is measured by a wind vane, which is counterbalanced by a weight fixed on the other end of a rod. However, in the case of propeller anemometers, the vane is a part of the propeller's axis. The vane requires a minimum force to initiate movement. The threshold wind speed for this force, usually, is of the order of 1 m/s. The motion of the vane is normally damped to prevent rapid changes of directions. Wind vanes generally produce signals either by contact closures or by potentiometers. The accuracy obtained from potentiometers is higher than that obtained from contact closures, but the latter are less expensive. The wind direction is determined by averaging the direction over 2- to 10-minute periods. When the wind direction sensor(s) is out of service, at designated stations, the direction may be estimated by observing the wind cone or tee, movement of twigs, leaves, smoke and so on, or by facing into the wind in an unsheltered area. The wind direction may be considered variable if, during the 2- or 10-minute evaluation periods, the wind speed is 6 knots or less. Also, the wind direction should be considered variable if, during these periods, it varies by 60 degrees or more when the average wind speed is greater than 6 knots. To show the information about the distributions of wind speeds, and the frequency of the varying wind directions, the so-called wind roses as one shown in Figure 5.4 are used. The wind roses and the histograms of wind direction are constructed on the basis of meteorological observations of wind speeds and wind directions at specific site or area. We have to keep in mind that it is very important to find out which directions have the best winds for electricity production. The wind direction distributions are critical important for the evaluation of the possibilities of utilizing wind power, for wind power system operation. Determining the wind speed according to wind direction is important to conduct the wind energy researches and also displays the impact of geographical features on the wind regime. As we pointed out usually strong winds usually come

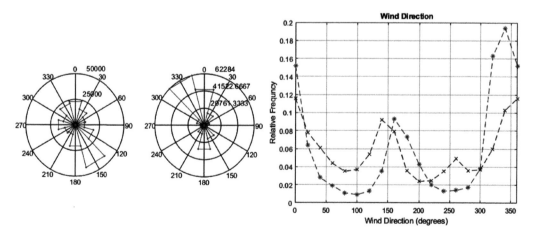

**FIGURE 5.4**
An example of wind roses and wind direction histogram.

from a particular direction; to show the information about the distributions of wind speeds, and the frequency of the varying wind directions, we can draw a so-called wind rose on the basis of meteorological observations of wind speeds and wind directions. The distributions of wind speed and direction are conventionally given by wind roses. Often the wind rose presents a summary of annual wind data. The circular space can e.g., be divided into 12 sectors, one for each 30 degrees of the horizon representing major directions from which wind might come. A wind rose can also be drawn for 8 or 16 sectors, but 12 sectors tend to be the standard set for example by the European Wind Atlas. The radius of the 12 outermost, wide wedges gives the relative frequency of each of the 12 wind directions (i.e., how many percent of the time is the wind blowing from that direction). The number of segments may be more but it makes the diagram difficult to read and interpret. The concentric circles having percent values represent the probability of wind coming from any particular direction. A wind rose chart, which is generated from your wind observations or from the data used in the wind resource assessment, is a helpful tool to determine wind direction and its distribution. Traditionally, wind direction changes are illustrated by a graph, which indicates percent of winds from that direction, or the wind rose diagram, as the ones shown in Figure 5.4.

The wind rose diagrams and wind direction frequency histograms provide useful information on the prevailing wind direction and availability in different wind speed bin. Notice that a wind vane points toward the source of the wind, not where the wind is blowing. Wind direction is reported as the direction from which the wind blows, not the direction toward which the wind moves. A north wind blows from the north toward the south, meaning that a site has 'North wind,' signifying that wind is coming from the North and not going to the North. A wind rose can have several different useful data such as wind frequency, mean wind speed, and mean cube of the wind speed, which are important to calculate potential wind power. The wind direction varies from station to station due to differential local features (topography, altitude, orientation, distance from the shore, vegetation, etc.). It is important to remember that usually there are also changes in the wind directions on diurnal, seasonal or annual basis. Further to note is that wind patterns may vary from year to year, and the energy content may vary (typically by some 10%) from year to year, so it is best to have, as we pointed out in other chapter sections observations from several years to calculate a credible average. Wind roses from neighboring areas are quite often fairly similar, so in practice it may sometimes be safe to interpolate (take an average) of the wind roses from surrounding observations. The wind rose, once again, only tells us the relative distribution of wind directions, not the actual level of the mean wind speed.

The wind direction can also be analyzed using continuous variable probability models to represent distributions of directional wind speeds, such as von Mises circular statistics. The model usually composed of a finite mixture of the von Mises distributions. The presentation of the wind circular statistics and probability distribution is beyond the scope of this chapter however interested readers are directed to use the end-of-chapter references. While obtaining the average wind speed at a given location is useful, it does not accurately portray the power of the wind resource. The statistical distribution of wind speeds vary depending on climate conditions, landscape, and surface roughness. Strong gale force winds are rare while consistent low speed velocities are most common. The Weibull distribution is a statistical tool which is a measure of the variation, skewness and frequency of observed winds speeds. Further use of the Weibull distribution is detailed in the wind speed validation process to provide a tool for calculating the power density of the wind resource. The wind rose diagrams and wind direction frequency histograms provide useful information on the prevailing wind direction and availability of directional wind speed in different wind speed bins. The wind roses were constructed using the composite data sets of measurements of wind velocities. The wind direction is usually analyzed using a continuous variable probability model to represent distributions of directional wind speeds. The model is composed of a finite mixture of the von Mises distributions (vM–PDF). The parameters of the models are estimated using the least square method. The range of integration to compute the mean angle and standard deviation of the wind direction is adjusted to minimum variance requirements. The proposed probability model $mvM(\theta)$ is comprised of a sum of $N$ von Mises probability density functions, $vM_j(\theta)$, as:

$$mvM(\theta) = \sum_{j=1}^{N} w_j vM_j(\theta) \qquad (5.81)$$

where $w_j$ are nonnegative weighting factors that sum to one:

$$0 \leq w_j \leq 1 (j = 1, \ldots, N) \text{ and } \sum_{j=1}^{N} w_j = 1$$

A random variable function has a von Mises distribution $vM$–PDF if its probability is defined by the equation:

$$vM_j(\theta; k_j, \mu_j) = \frac{1}{2\pi I_0(k_j)} \exp\left[ k_j \cos\left(\theta - \mu_j\right) \right],$$

$$0 \leq \theta \leq 2\pi \qquad (5.82)$$

where $k_j \geq 0$ and $0 \leq \mu_j \leq 2\pi$ are the concentration and mean direction parameters, respectively. In this paper, the angle corresponding to the northerly direction is taken as 0°. Note that in meteorology, the angle is measured clockwise from north. Here, $I_0(k_j)$ is a modified Bessel function of the first kind and order zero and is given by:

$$I_0\left(k_j\right) = \frac{1}{2\sqrt{\pi}} \int_0^{2\pi} \exp\left[ k_j \cos\theta \right] d\theta \approx \sum_{p=0}^{\infty} \frac{1}{(p!)^2} \left( \frac{k_j}{2} \right)^{2p} \qquad (5.83)$$

The distribution law $mvM(\theta)$, given by Equation (5.81), can be numerically integrated between two given values of $\theta$ to obtain the probability that the wind direction is found within a particular angle sector. Various methods are employed to compute the 3N parameters on which the mixture of von Mises distribution depends, the least squares (LS) method being the most common. Figure 5.5 is showing the fitted von Mise distributions to the wind direction time series collected at two locations in western Nevada, for the period from 2003 to 2008.

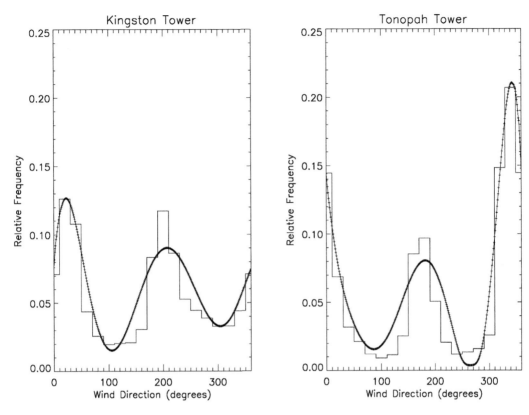

**FIGURE 5.5**
Wind direction (von Mise) probability distributions fitted to wind direction histograms. (Data collected in Western Nevada, 2003–2008.)

## 5.6 Wind Energy Potential Assessment

The extent to which further wind plant projects can be carried out in the future depends upon a number of different factors, including the energy policy, the support schemes available to renewable energy sources (in most cases, electricity producers from wind still want some incentives to get enough profit from their undertakings) and the attitude of local authorities as far as plant permitting procedures are concerned. But is goes without saying that the presence of suitable windiness, in areas where all the other conditions for wind farm sites also exist, is the essential pre-requisite that has to be ascertained to estimate a country's wind energy potential. Wind resource assessment, being the most important step in planning a wind project because it is the basis for determining initial feasibility and cash flow projections, and is ultimately vital for acquiring the investment financing. Wind energy site assessment gauges the potential for a site to produce energy from wind turbines. Assessment and wind energy project progress through four assessment stages: (1) initial assessment, (2) detailed site characterization, (3) long-term validation of data and (4) detailed cash flow projection and acquiring financing. Prediction of wind energy resources is crucial in the development of a commercial (large-scale) wind energy installation. The single most important characteristic to any wind development is the wind speed. The performance and power output of a wind farm is very sensitive to uncertainties and errors in estimating the wind speed. When any wind energy development is under consideration, a site assessment is undertaken. Specifically, wind energy site assessment is the process of evaluating the wind resource at a potential wind turbine or wind farm location, then estimating the energy production of the proposed project. On the other hand, the evaluation of the wind energy potential (onshore and offshore) of a region or a country is a complex analysis that has to take into account many different factors and information to be properly combined. In particular, three are the main items: evaluation of wind resources (i.e., annual average wind speed and specific energy production), evaluation of the characteristics of the territory that could determine the possibility or not to install wind farms (i.e., morphology, sea depth and land cover) and knowledge of environmental constraints and regulatory and permitting aspects.

Wind resource at a site directly affects the amount of energy that a wind turbine can extract, and therefore the success of the venture. The quality of the wind resource is primarily quantified by the mean wind speed at the site, although the turbulence intensity, probability distribution of the wind speed and prevailing wind direction are also important factors. Once the wind resource has been assessed at a site, the expected annual energy production, AEP, of a selected wind turbine is calculated. However, the evaluation of a wind resource and the subsequent estimation of the annual energy production is a process, with a high degree of uncertainty. Uncertainty arises at all phases in the process, from measuring the wind speed, selecting and fitting the most appropriate wind probability distribution function to the uncertainty in the wind turbine power curve, (the graph of the wind turbine power versus wind speed, representing the electrical power generated by the turbine at each wind speed). A proper assessment of uncertainty is critical for judging the feasibility and risk of a potential wind energy development. This calculation combines the expected wind resource with the wind turbine(s) power curve and the expected energy losses in order to estimate how much energy the wind turbine(s) will actually produce at the site. The AEP will ultimately help determine the profitability of the wind energy project. The accuracy and precision of the wind resource assessment and AEP calculation must also be determined when evaluating a potential site. Wind resource assessment is an uncertain process, and several factors ranging from wind speed measurement errors to the inherent physical wind variations contribute to this uncertainty. Overall, these various individual sources of error must all be accounted for to provide an estimate of the total uncertainty of the wind resource. Accurate analysis of wind data is crucial in estimating the amount of potential electricity to be produced by wind at proposed areas or locations. The first criteria of interest when surveying a potential area for electricity production with wind turbines are the wind speed and direction. The overall annual mean wind speed is crucial as much as the daily, monthly, seasonal and annual variations. It is important to know the overall amount of energy that can be produced over a certain period and the periodic variations due to the variability of the wind. Moreover, for proper planning the possible cut-off times when the turbine will not be functioning due to either very strong winds, normally above 25 m/s, or when there is hardly any wind, normally below of about 3 m/s, need to be known.

Variations of wind speed in time can be divided into the categories of inter-annual, annual, diurnal, and short-term. Inter-annual variations in wind speed occur over timescales greater than 1 year. They can have a large effect on long-term wind turbine production. Meteorological and atmospheric science researches are showing that to characterize accurately long-term values of weather or climate over 25 years of observations are needed, and at least 5 years of data to have a reliable estimate of average annual wind speed at a given location or site. However, this is not meaning that data spanning shorter time periods are useless. Shorter wind velocity time series can be correlated with longer time series from neighboring weather stations to infer the wind regime at the location of interests. It is worth to point out that annual variations refer to significant variations in seasonal or monthly averaged wind speeds, being common over most of the world. Large diurnal variations in wind occur quite often in tropical and temperate latitudes. This type of wind variation is mainly due to differential heating of the Earth's surface during the daily radiation cycle, being especially true in temperate latitudes over relatively flat areas. A typical diurnal variation is an increase in wind speed during the day, with the wind speeds lowest during the hours from midnight to sunrise. Usually the largest diurnal changes generally occur in spring and summer, and the smallest ones during the winter. However, the diurnal variation may vary with location and altitude above sea level, being strongly influenced by the local topography and atmospheric circulation. Although gross diurnal cycle features can be established with a single year of data, a good characterization of detailed features, such as the amplitude of the diurnal oscillation and the time of day that the maximum winds occur, requires multiple years of data. The main factors impacting the wind flows, as discussed in the previous chapter sections are the orography, the roughness and the atmospheric stability. The latter represents the thermal effects on the wind flow. It is rarely taken into account for the wind power assessments, since the wind statistics are averaged over the year, and the atmosphere is thus generally considered as neutral. However this assumption presents some limitations: on the sites where the average wind speed is low (<6 m/s), the thermal effects are starting to be significant; on offshore site, where the atmospheric stability is predominant over the orography and the roughness; and for simulation on short periods, by hour or by day (short-term prediction), supervision of operation and power curves measurement with site calibration, as defined in the norm IEC 61400-12.

The results of the wind speed variation and the prevailing wind directions which characterized the location under investigation are further analyzed with respect to the corresponding mean wind power density. The amount of energy available in the wind (the wind energy resource) at a site is the average amount of power available in the wind over a specified period of time, at least one year period, usually longer time series. When a series of measurement is conducted in the site, the following method can be used, since it uses directly the collected data. By expressing the energy pattern factor $EPF_W$, as the ratio of the available wind power density to the wind power corresponding to the cube of the mean wind speed, and following Equation (5.61), as:

$$EPF_W = \frac{1}{NV_{mean}^3} \sum_{k=1}^{N} v_k^3 \tag{5.84}$$

$N$ is the number of measurements, $v_k$ the wind speed and $V_{mean}$ the mean wind speed, then the mean wind power density in the site will be given by:

$$\bar{P}_{WD} = 0.5 \cdot PF_{WE} \cdot \rho V_{mean}^3 \tag{5.85}$$

If a wind speed probability distribution at a site or location, together with the local air density, the available wind power (as given in Equation (5.30)), or often the available average wind power density, for a specific site is estimated as:

$$P_{AWE} = 0.5 \int \rho \cdot v^3 f_{WB}(v) dv \ \left(W/m^2\right) \tag{5.86}$$

With a discrete wind speed distribution and the local air density, the $P_{AWE}$, the average wind power availability, for a specific site or location is then estimated as:

$$P_{AWE} = 0.5 \sum_{k=1}^{N} \rho f(U_k) \cdot U_k^3 \Delta U_k \ \left(W/m^2\right) \tag{5.87}$$

where $N$ is the number of wind speeds included in the distribution $f(U_k)\Delta U_k$ is the probability of the wind speed occurring in the wind speed range $\Delta U_k$, centered on wind speed $U_k$. Once the wind power density is estimated, the wind energy density for a period of time is calculated as:

$$E_{AWE} = P_{AWE} \times T \tag{5.88}$$

where $T$ is the time period. For the annual wind energy density estimation the value of 8640 h is used. However, the wind energy potential ($E_{WER}$) or potential gross annual wind energy production for a specific site and a specific wind turbine, very important in the analysis and assessment of the wind development vial ability can be calculated with a wind speed distribution and the turbine power curve, properly adjusted for the local air density. For a discrete wind speed distribution, the available annual potential wind energy is estimated then as:

$$E_{WER} = 8760 \sum_{k=1}^{N} f(U_k)\Delta U_k \cdot PC(U_k) \ (Wh) \tag{5.89}$$

Here $PC(U_k)$ is the electrical power produced by the turbine at wind speed $U_k$, the center of the range (bin) $\Delta U_k$. In the case that wind regime is characterized by the Weibull probability distribution function, with shape and scale factors, $k$ and $c$, the available wind power density and subsequently the annual energy can be estimated by:

$$\bar{P}_{WD} = 0.5 \cdot \rho \cdot c^3 \left(1 + \frac{3}{k}\right) \ (W/m^2) \tag{5.90}$$

The annual wind energy is then computed by using Equation (5.88). If the wind energy resources are characterized by a Rayleigh probability distribution (Equation 5.65), with the parameter $v_m$, the available wind power density is given by:

$$\bar{P}_{WD} = \frac{3}{\pi} \rho \cdot v_m^3 \ (W/m^2) \tag{5.91}$$

Where $\rho$ is the air density (the standard air density value is: $\rho = 1.225$ kg/m³ dry air at 1 atm and 15°C). Again annual available wind energy is estimated by using Equation (5.88).

**Example 5.16:** For a site characterized by a Weibull probability distribution of the wind speed, having a $k = 1.95$ and $c = 6.45$ m/s calculate the available power density and estimate the monthly wind energy yield. Assume the standard air density for this location.

**Solution:** From Equation (5.90) the available average power density at this site is:

$$\bar{P}_{WD} = 0.5 \cdot 1.225 \cdot 6.45^3 \left(1 + \frac{3}{1...95}\right) = 417.2 \ W/m^2$$

Assuming that a 30-day month, $T = 720$ h, the monthly average energy density is:

$$E_{MWE} = 720 \times 417.2 = 300.392 \text{ kWh per Square Meter of Rotor}$$

### 5.6.1 Long-Term Assessment and MCP (Measure-Correlate-Predict) Methods

Measure-correlate-predict (MCP) algorithms are used to predict the wind resource at target sites for wind power development. MCP methods model the relationship between wind data (speed and direction) measured at the target site, usually over a period of one year, and concurrent data at a nearby reference site. The model is then used with reference site long-term data to predict wind speed and direction distributions at the target site. In order to be most useful for wind power development, the prediction uncertainties need to be understood and estimate. Since typical wind farm assessments last anywhere from one to three years, with important decisions to be taken often only after several months, there is an obvious need for a prediction of the performance of a planned wind farm during its expected life time (20 years or more). Such an assessment is an important part of the wind power financing process. While the measurement campaign may correspond to an untypically high or low period, correlations with nearby reference stations should help detect such trends and provide a corrected long-term estimate of the wind speed at the development site and its inter-annual variations. Moreover, since the power output of wind turbine depends on the wind speed in a non-linear way, the distribution of the wind speed values needs to be predicted correctly. These methods proceed by "measuring" the winds at a target site, "correlating" them with winds from a nearby reference site, and then by applying these correlations to historical data from the reference site, to "predict" the long-term target site wind resource. Various MCP algorithms have been studied using wind data from a number of potential wind farm sites. The general methodology of the MCP process proceeds as follows: (1) collect wind data at the predictor site for an extended period; (2) identify a reference site, for which high-quality, long-term records exist, in the vicinity of the predictor site, and which has a similar exposure, the so-called "reference" site; (3) obtain wind data from the reference site for the same time period as for the predictor site, the so-called "concurrent period"; (4) establish a relationship between the data from the reference and predictor sites for the concurrent period; (5) obtain wind data from the reference site for a historic period of over 10 years duration or the longest possible one, the so-called "historic" period; (6) apply the relationship determined in step 4 to the historic data from the reference site to "predict" what the winds would have been at the predictor site over that period. Note that this is a prediction of the winds that would have been observed had measurements been made at the predictor site for the same period as the historic data, rather than a prediction of winds that will be observed in future. The key factor to any MCP technique is the algorithm used in step 4. Most MCP techniques use regression analysis to establish a relationship between wind speed and direction at the reference site and the wind speed at the potential wind farm site. The general approach is to look for a relationship between the wind speed variables $v_{site}$ and $v_{met}$ of the site under development and a suitable reference station, respectively:

$$v_{site} = f(v_{met}) \tag{5.92}$$

Often, it may be suitable to consider several reference stations with concurrent data sets for a given development site; Equation (5.49) has then to be generalized to:

$$v_{site} = g\left(v_{met}^1, v_{met}^2, \ldots, v_{met}^N\right) \tag{5.93}$$

Currently different MCP methods currently implemented in the WindPRO, WAsP software packages or other wind software applications. Wind speed time series can be analyzed irrespective of wind direction.

## 5.6.2 Wind Atlas in Context of Wind Energy

In many parts of the world the amount of wind speed data was limited to daily or even monthly averages. The most comprehensive, long-term source of information on wind speeds, pressure, and temperature is data collected at National Weather Stations. Other sources in the United States on record at the National Climatic Center, Asheville, North Carolina, are from Federal Aviation Administration stations, U.S. air bases, Coast Guard, etc. Such large information was summarized in the form of wind energy maps, the Wind Atlas. The wind energy resources of larger regions or countries are often mapped in terms of the wind speed, wind direction, the wind power density in $W/m^2$, or the wind energy potential in $kWh/m^2$ per year. Often the wind energy resources are mapped in all three forms, including numerical data. The data is usually represented by contour curves, being the most useful and easily to interpret or understood mapping technique. Along the contour line, the plotted parameter remains constant. For example, the map plots the contour lines connecting sites having the same annual wind speed or available wind power density. The wind resource maps of many countries and larger regions have been prepared in such contour format. Estimates of wind resource are expressed in wind power classes ranging from wind power class 1 to 7 (see Table 5.1). Areas designated class 4 or higher are the most suitable for wind energy, being also the most compatible with existing wind energy conversion technologies. Power class 3 areas may become suitable with future wind energy technology, while class-2 and class-1 areas are marginal or unsuitable for wind energy development and exploitation. The two most known wind energy maps (Wind Atlas) are the European Union and U.S. model. American (US) model includes wind climatology in mountainous and complex terrain areas, requiring however large computing power, for this reason is not commercial available directly. To create the Wind Atlas at any site ere meteorological or weather stations are located requires the following initial information: wind regime primary data for periods over 10 years, site area topographical, climatologic and vegetation description, and the digital map of the region.

In the context of wind energy industry, the concept of wind atlas is much broader. It contains not only maps, graphs or images, but also numerical data in tabular or list formats on the wind speed, direction, wind power density ($W/m^2$) and other weather and climatic parameters. Wind Atlas is developed to data on wind energy resources in a given area (by using weather station observations, numerical model outputs, etc.), providing adequate information needed to estimated wind energy potential in surrounding regions, with purpose to identify suitable locations and/or areas most suitable for wind energy projects. In a Wind Atlas the wind climatology (wind speed and direction, variability and patterns) and wind energy regime are presented in a graphical format, including histograms of wind distributions, wind roses. The wind data and characteristics in tabular format are also included in a Wind Atlas. The detailed wind maps of many U.S. states are being prepared and many of them are available on the Internet. Notice that U.S. wind resources are large enough to generate over 3.4 trillion kWh of electricity each year. US Department of Energy (DOE) estimates that almost 90% of the usable wind resource in the U.S. lies in the wind belt spanning the 11 Great Plains states, spanning from Montana, North Dakota, and Minnesota in the North to Oklahoma and Texas in the South. However, as more data have been collected in the last decades, specifically for wind power potential for countries, states, provinces and/or regions, digital wind maps are available with better resolution than the older wind maps, additional wind energy information, and the values are more accurate, as data above 10 m have become available. However, the data collected by private wind energy developers are not usually available to the public, so data at 20–50 m heights are still being collected to provide regional wind data bases. The Wind Energy Resource Atlas covers the United States and its territories, while wind power potential by year and season were also estimated for each state and region. The wind power classes (Table 5.1) were estimated for a grid of 20 min longitude by 15 min latitude (27 by 25 km, 16 by 15 miles). New computer tools and technical analyses, which use satellite, weather balloon and meteorological tower data, are being used to create better and more accurate wind maps for assessing the wind power potential and helping the wind energy development process. The European Union (EU) has also made efforts on implementing wind resource assessment beginning with the publication of the European Wind Atlas in 1989 and continuously after. It provides an overall view of the wind resources, information for determining the wind resource and the local siting of wind turbines, descriptions and statistics for more than 200 weather stations in the EU countries, includes methods for calculating the influence on the wind due to landscape features such as coastlines, forests, hills, and buildings, and also includes information explaining the meteorological background and analysis, and the physical and statistical bases for the models. The wind power classes in the EU Wind Atlas are somewhat different from those of the U.S. wind power classes, being also classified for terrain types. Wind power maps and contour lines of wind speed are available for a number of countries and regions around the world, as wind energy has become part of many national energy policies.

## 5.7 Summary

This chapter provides the fundamental concepts of wind energy resources, assessment and analysis methods, wind statistics, and wind measurement techniques. Previous studies of the behavior of the wind were done by meteorologists who were mainly interested in weather and, for research, turbulence and momentum transfer. Over last four decades, numerous studies have been funded on the wind characteristics as they pertain to wind energy potential and the effects on wind turbines. Development of techniques for accurate assessment of wind energy potential at a site is gaining increased importance. This is because of the fact that the planning, developing and establishment of wind energy conversion systems and wind farms depend upon factors like variation of wind speed and wind direction distributions, topography and weather characteristics, turbine-speed type and performances, namely, wind farm layout and hub height. Once the details of the wind resource for a potential site are known, an efficient design of a wind-energy system demands an optimum matching of the wind turbines to the potential wind site, in order to obtain the higher power generation and performances. The field of wind resource assessment is evolving rapidly, responding to the increasingly stringent requirements of large-scale wind farm projects often involving investments of several hundred million dollars. Traditional cup anemometry is being complemented with ultrasonic sensors providing information on all three components of the wind velocity vector and enabling a better assessment of turbulence. Many factors influence accurate assessment and prediction of wind energy production. A primary issue is adequate understanding of the effects of wind variability, atmospheric stability and turbulence on energy production. Non-negligible error is incurred when the effects of shear, TI, and atmospheric stability on the wind turbine power performance are ignored, as in the IEC standard, 61400-12-1 (2005). The standard procedures are valid only for ideal neutral conditions and a small wind turbine. Besides the dominant cubic dependence of the wind speed on the wind power density, there are smaller but still important corrections to the air density that are important to harvesting wind energy at high-elevation sites. In this chapter a comprehensive and detailed discussions of the wind regime characteristics, factors affecting wind regime and wind energy potential, such as air density, turbulence intensity, wind shear, extrapolation methods are presented. A large part of the chapter focused on the wind speed and direction statistics, the presentation and characterization of the most common probability distribution functions for wind speed and direction, instrumentation used in the wind observations, wind energy potential estimates, analysis and assessment. Last chapter section is dedicated to the wind energy assessment issues, including a short description of the measure-correlate-predict (MCP) methods and approaches. In the last section are also included briefly discussion on the energy estimate and wind energy mapping and atlases.

---

## Questions and Problems

1. What is the wind power class for your hometown or city?
2. If there is a wind map for your state or country, what is the wind speed or wind power potential near your location?
3. What is the information typically provided by a wind rose?
4. What are the major factors affecting the wind regime of an area?
5. What is the air density difference between sea level and a height of 1000 and 2500 m?
6. What is the air density difference between sea level and a height of 2500 m?
7. Describe how turbulence may affect the wind turbine operation.
8. Do the same for wind shear and wind gust.
9. Which are the factors affecting air density?
10. A wind turbine is rated at 300 kW in a 10 m/s wind speed in air at standard atmospheric conditions. If assume that the power output is directly proportional to air density, what is the power output of the turbine in a 10 m/s wind speed at elevation 1500 m above sea level at a temperature of 20°C?
11. Calculate the power, in kW, across the following areas for wind speeds of 5, 10, and 20 m/s. Use diameters of 10, 30, 50, and 100 m for the area, assuming standard air density is 1.225 kg/m$^3$.
12. Solar power potential is about 1 kW/m$^2$, for mid-latitudes. What wind speed gives the same power potential?
13. If there any examples of vegetation indicators of wind in your region or in any areas that you may know, what wind speed do they indicate?
14. From a standard 10 min period, the mean wind speed is 7.3 m/s and the standard deviation is 1.65 m/s. What is the turbulence intensity?

15. Calculate the factor for the increase in wind speed if the original wind speed was taken a height of 50 m. New heights are at 80 and 100 m. Use the power law with an exponent of 0.20.

16. Calculate the factor for the increase in wind speed if the original wind speed was taken at a height of 10 m. New heights are at 30, 60, and 90 m. Use the power law with an exponent equal to: (a) 0.1 and (b) 0.20.

17. Repeat the calculations in problems 10 and 11, by using Equation (5.12) and $z_0 = 1.05$.

18. An investor is planning to install a wind farm in a hilly area near order with Canada. The elevation is 720 m, the average winter temperature is −5°C and the average summer temperature is 21°C. Compute the average wind power density assuming an average wind speed of 13.5 m/s in winter and 7.5 m/s in summer.

19. If there are any examples of vegetation indicators of wind characteristics in your region or in any area that you are known for such indicators? What wind speed do they indicate, based on the chapter information and any additional information available from local sources?

20. With a remote sensing system for measuring wind speed, such as Lidar or SODAR you do not need a tower. What are the reasons for not employing such system?

21. At a wind farm, very high winds with gusts of over 60 m/s were recorded. An average value for 15 min was 40 m/s with a standard deviation of 8 m/s. What was the turbulence intensity?

22. List the most common probability distribution functions used in wind energy.

23. List the instruments, devices and techniques used in wind velocity measurements.

24. Calculate the wind speed distribution using the Rayleigh distribution for an average wind speed of 6.3 m/s. Use 2 m/s bin widths. Calculate the wind energy density, the monthly energy density availability, the most frequent wind velocity, and the wind velocity corresponding to the maximum energy, for Rayleigh wind speed distribution of previous problem, assuming and air density 1.225 kg/m³.

25. Compute the wind power density for a site located at an elevation of about 500 m when the air temperature is 30°C and the wind speed is equal to 10, 12.5 and 15 m/s.

26. An automatic wind observation system located at future wind farm site, measures 3 m/s 24 times, 4.5 m/s 72 times, 6 m/s 18 times, 7.2 m/s 21 time, 9 m/s 85 times, 10 m/s 48 times, and 11 m/s 9 times during a given period. Find the mean, standard deviation and skewness.

27. Calculate the wind speed distribution for a Weibull distribution for $c = 7.2$ m/s and $k = 1.8$. Use 1 m/s bin widths. How many hours per day the wind speed is between 5 m/s and 12 m/s, the cut-in speed and the rated speed of a medium-size wind turbine.

28. From the wind speed probability distribution of the previous problem, calculate the available wind power density and the important speed to characterize the wind power potential at that site.

29. Calculate the wind speed distribution for a Weibull distribution for $c = 7.5$ m/s and $k = 2.7$.

30. Calculate the factor for the increase in wind speed if the original wind speed was taken at the standard level of 10 m. New heights are at 60, 80, 100, and 120 m. Use the power law with an exponent of 0.20.

31. A 60-m meteorological mast (tower) observes an average wind speed of 6.5 m/s (wind speed is sampled by averaging over 10-min interval) and a corresponding standard deviation of 1.15 m/s. What is the turbulence intensity? If the wind speed, extrapolated at 90 m hub height is 7.4 m/s what will be the turbulence intensity at hub height?

32. Calculate the wind speed distribution for a Weibull distribution for $c = 7.20$ m/s and $k = 2.5$. For this distribution compute the mean wind speed, the most probable wind speed, and the wind speed carrying most of the energy. Compute also the available wind power density.

33. In many areas of Mid-west there are wide temperature differences between summer (100°F) and winter (−20°F). What is the difference in air density? What is the change in the available average wind power? Assuming the same elevation, the same average wind speed, and pressure is the same.

34. Calculate the increase in wind speed if the original wind speed was taken at a height of 10 m. New heights are at 30, 60, and 80 m. Use the power law with exponents 0.147 and 0.21. Compare with the Equation (10.10) for $z_0 = 1.15$.

35. If the Weibull shape and scale parameters are 2.1 and 6.3 m/s, what is the average wind velocity?

36. Calculate and plot the wind speed distribution for a Weibull distribution for $c = 7.8$ m/s and $k = 1.85$. Use 1 m/s bin widths.

37. The Weibull parameters at a given site are $c = 5.5$ m/s and $k = 1.7$. Estimate the number of hours per year that the wind speed will be between 6.0 and 9.0 m/s. Estimate the number of hours per year that the wind speed is greater than or equal to 12.5 m/s.

38. Calculate the wind energy density, the monthly energy density availability, the most frequent wind velocity, and the wind velocity corresponding to the maximum energy, for Weibull wind speed distribution of previous problem, assuming and air density 1.225 kg/m³.

39. A wind turbine is planned to be installed at a site having a multiannual average wind speed of 8.4 m/s. Plot throughout the year, by using the appropriate probability distribution function for wind speed based on the available information. What is the probability of having a 12 m/s wind speed?

40. A wind turbine is about to be installed at a site in which the average wind speed is 6.75 m/s. Assuming that the wind is modeled by a Rayleigh probability distribution, plot the probable wind speed distribution throughout the year. What is the probability having a wind speed of 12 m/s? What is the most probable wind speed and the most probable average available wind power density?

41. For Weibull probability distribution shape and scale parameters of Problems 36 and 37, estimate the available power density and the monthly and the annual energy.

42. Repeat problem 41, but for the Rayleigh probability distribution of Problem 40.

## References and Further Readings

1. N. Jenkins, J. Walker, *Wind Energy Technology*, Wiley, Chichester, UK, 1997.
2. E. L. Petersen, N. G. Mortensen, L. Landberg, et al., Wind power meteorology. Part I: Climate and turbulence, *Wind Energy*, Vol. 1(1), pp. 25–45, 1998.
3. E. L. Petersen, N. G. Mortensen, L. Landberg, et al., Wind power meteorology, Part II: Siting and models, *Wind Energy*, Vol. 1(2), pp. 55–72, 1998.
4. N. Jenkins, T. Burton, D. Sharpe, E. Bossanyi, *Wind Energy Handbook*, John Wiley & Sons, Chichester, UK, 2001.
5. T. Burton, D. Sharpe, N. Jenkins, and E. Bossanyi, *Wind Energy Handbook*, John Wiley & Sons, Chichester, UK, 2001.
6. J. Manwell, J. McGowan, and A. Rogers, *Wind Energy Explained: Theory, Design and Application*, John Wiley & Sons, New York, 2002.
7. B. K. Hodge, *Alternative Energy Systems and Applications*, Wiley, Chichester, UK, 2010.
8. L. Landberg, L. Myllerup, O. Rathmann, E. L. Petersen, B. H. Jørgensen, J. Badger, and N. G. Mortensen, Review wind resource estimation: An overview, *Wind Energy*, Vol. 6, pp. 261–271, 2003.
9. M. R. Patel, *Wind and Solar Power System*, 2nd Edition, 2005, 143–181, CRC Press, Boca Raton, FL.
10. L. Freris, and D. G. Infield, *Wind Energy in Power Systems*, John Wiley & Sons, Chichester, UK, 2008.
11. A. Schaffarczyk, (Ed.), *Understanding Wind Power*, Wiley, Chichester, UK, 2014.
12. M. Martín, (Ed.), *Alternative Energy Sources and Technologies*, Springer, Heidelberg, Germany, 2016.
13. V. Lyatkher, *Wind Power*, Wiley/Scrivener Publishing, Boston, MA, 2014.
14. M. C. Brower, (Ed.), *Wind Resource Assessment, A Practical Guide*, Wiley, Hoboken, NJ, 2014.
15. R. Belu, Wind energy conversion and analysis, In *Encyclopedia of Energy Engineering and Technology* (Ed. Sohail Anwar), Vol. 3, 27 p, Taylor & Francis Group, Boca Raton, FL, 2014.
16. J. Andrews, and N. Jelley, *Energy Science, Principles, Technology and Impacts*, Oxford University Press, Oxford, 2007.
17. E. L. McFarland, J. L. Hunt, and J. L. Campbell, *Energy, Physics and the Environment (3 ed.)*, Cengage Learning, Boston, MA, 2007.
18. R. Belu, and D. Koracin, Wind energy analysis and assessment, *Advances in Energy Research*, Vol. 20(1), pp. 1–55, 2015.
19. S. Emeis, *Wind Energy Meteorology*, Springer, Cham, Switzerland, 2013.
20. L. Landberg, L. Myllerup, O. Rathmann, E. L. Petersen, B. H. Jørgensen, J. Badger, and N. G. Mortensen, Review wind resource estimation: An overview, *Wind Energy*, Vol. 6, pp. 261–271, 2003.
21. O. Probst, and D. Cárdenas, State of the art and trends in wind resource assessment, *Energies* Vol. 3, pp. 1087–1141, 2010.
22. H. Basumatary, E. Sreevalsan, K. K. Sai, Weibull parameter estimates: A comparison of different methods, *Wind Engineering*, Vol. 29, pp. 309–315, 2005.
23. T. P. Chang, Estimation of wind energy potential using different probability density functions, *Applied Energy*, Vol. 88, pp. 1848–1856, 2011.
24. T. P. Chang, Performance comparison of six numerical methods in estimating Weibull parameters for wind energy application, *Applied Energy*, Vol. 88, pp. 272–282, 2011.
25. V. Nelson and K. Starcher, *Introduction to Renewable Energy (Energy and the Environment)*, CRC Press, Boca Raton, FL, 2015.
26. R. A. Dunlap, *Sustainable Energy*, Cengage Learning, Boston, MA, 2015.
27. R. Bansal (Ed.), *Handbook of Distributed Generation: Electric Power Technologies, Economics and Environmental Impacts*, Springer, Cham, Switzerland, 2017.
28. R. Belu, *Industrial Power Systems with Distributed and Embedded Generation*, The IET Press, London, UK, 2018.
29. R. Belu and D. Koracin, Wind characteristics and wind energy potential in Western Nevada, *Renewable Energy*, Vol. 34(10), pp. 2246–2251, 2009.

# 6

## *Wind Energy Conversion Systems*

### 6.1 Introduction, Wind Turbine Configurations

Since ancient times, people have utilized wind energy to propel boats along the Nile River as early as 5000 BC or help Persians pump water and grind grain between 500 and 900 BC, by using windmills. Over the centuries, the use of windmills spread from Persia to the surrounding areas in the Middle East, where windmills were used extensively in food production or to pump water. Eventually, in the tenth century, wind power technology spread north to European countries such as the Netherlands, which adapted windmills to help drain lakes and marshes in the Rhine River Delta. The windmills built in Europe were of horizontal axis rotor type; they were made more efficient by using the lift force, in place of drag force as the Middle-eastern ones. Wind power being the first industrial type energy source in our history is used for navigation and generate mechanical power. Through history, the use of wind power has waxed and waned, being more evident in the last century and a half, when several remarkable technological advances have been made over this period of time. In the USA during the nineteenth century and until the 1930s of twentieth century, wind turbines were mainly used for irrigation and pumping water. They had a high number of steel-made blades, being a huge economic potential because over 8 million were built and operated during the nineteenth country. The world's first wind turbine used to generate electricity was built by a Danish engineer, Poul la Cour, in 1891. It is interesting to note that La Cour used the electricity generated by his wind turbines to electrolyze water, producing hydrogen for gas lighting of the local schoolhouse. He was over 100 years ahead of his time since the twenty-first-century vision for renewable energy includes photovoltaic and wind energy conversion systems making hydrogen by electrolysis to generate electricity in fuel cells. Wind power systems convert wind kinetic energy into mechanical energy, and eventually electrical power is used for a variety of tasks. Whether the task is, the wind offers inexpensive, clean and reliable form of energy. First attempt to generate electricity was made at the end of nineteenth century and almost all those models had a horizontal axis. However, in the same time period (1931) Georges Jean Marie Darrieus designed one of the most famous and common types of the vertical axis wind turbine, which still bears his name. However, due to the use of cheap and abundant coal and oil in the last two centuries, and due to the technological advances in thermal engines, the importance of the wind energy decreased significantly. Over the last four decades there are significant increases into the research and applications of wind energy technologies.

The earliest windmills in antiquity rotated about a vertical axis and they were driven by drag wind component. Modern vertical axis turbines use vertical symmetrical airfoils and the driving force is produced by lift developed by the blade in the moving air stream. However, the only vertical axis turbine which has been manufactured commercially at any volume is the Darrieus machine. The conventional Darrieus turbine has curved blades connected at the top and at the bottom and rotates like an "egg whisk," as illustrated in Figure 6.1. Vertical axis wind turbines have the advantages that no tower is needed, they operate independently of the wind direction (a yawning mechanism is not needed) and heavy gearboxes and generators can be installed at ground level. But they have many disadvantages: they are not self-starting, the torque fluctuates with each revolution as the blades move into and away from the wind, and speed regulation in high winds can be difficult. Vertical axis turbines were developed and commercially produced in the 1970s until the end of the 1980s. But since the end of the 1980s the research and production of vertical axis wind turbines has practically stopped worldwide. Another common type of a VAWT is the Savonius type rotor used mainly for water pumping. By the utilization of wind energy, the wind energy converters are designed to generate electrical energy. The generated electrical energy is usually fed into the electrical grid or sometime directly to local loads. The kinetic energy of flowing air masses is converted into rotational mechanical energy with the help of uplift or air resistance effects at the wind rotors. This energy is transmitted through a drive shaft and, in most cases, additionally through a gear box, to feed an electric generator. The generator transforms the mechanical energy into electrical energy. Depending on the generator system, the energy is then either directly or through a power electronic interface fed into the electrical grid. The recent development led to the realization of a great variety of the wind turbine types and models, both with vertical and horizontal

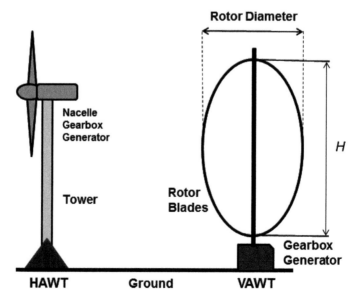

**FIGURE 6.1**
Horizontal axis and vertical axis wind turbines configurations.

axis, with rated power from the few kW to the 5 MW and more for the latest constructions. However, in the electricity generation market the HAWT type has currently a large predominance.

As we discussed briefly in previous chapter, wind energy offers many advantages, which explains why it is the fastest-growing energy source in the world. Research efforts are aimed to address the challenges and technical issues to increase the use of wind energy for electricity generation and other applications. Among the major wind energy advantages are: wind energy is free and available as domestic energy source, does not pollute the environment like fossil fuel power plants, has one of the lowest-priced renewable energy technologies available today, and wind turbines can be install on farms or ranches, thus benefiting the economy in rural areas, where often most of the best wind sites are found. Wind energy main disadvantages include: wind power has to compete with conventional power generation sources on a cost basis, wind energy requires higher initial investments, wind is intermittent and variable, often it does blow when electricity is needed, wind energy cannot be stored, very often good wind sites are often located in remote locations, far from cities where the electricity is needed, requiring significant power transmission investments, wind energy system developments may compete with other uses for the land and those alternative uses may be more highly valued than electricity generation, and there are, even minor some environmental concerns of the wind energy use, such as noise produced by the rotor blades and aesthetic (visual) impacts. However, most of these problems and wind energy issues have been resolved or greatly reduced through technological development s or by properly siting wind plants and wind farms.

### 6.1.1 Wind Turbine Configurations and Classifications

Wind turbines are mechanical devices designed to convert part of the kinetic energy of the wind into useful mechanical energy. Several designs have been devised and proposed throughout the times. Most of them comprise a rotor that turns round propelled by lift or drag forces, which result from its interaction with the wind. Depending on the position of the rotor axis, wind turbines are classified into vertical axis and horizontal axis ones. Wind generators need to be mounted in places where the wind power is significant, wind speed being higher up than near the ground. The optimal site needs to be far by tall structures to keep off turbulences, which are the link between changing speed of wind when it hits the actual obstacles. Horizontal axis wind turbines harvest the wind energy against the actual direction associated with wind. This type of wind power converter consists of only a few aerodynamically optimized rotor blades, which for the purpose of regulation can usually be turned about their long axis (so-called pitch control, discussed later in this chapter). Another cheaper way of regulation consists in designing the blades in such a way that the air streaming along the blades will go into turbulence at a certain wind speed (so-called stall control). These converters can deliver power ranging from about 10 kW to a few MWs. The energy production, regardless of the wind turbine types depends on the interaction between the rotor and the wind, so major aspects of wind turbine performance like power output and loads are determined by the aerodynamic forces generated by the wind. The wind turbine rotor interacts with wind stream, resulting in what is the turbine aerodynamics, depending on the rotor blade profile, rotor and blade other characteristics, number of blades, control methods, or tower type. The most successful vertical axis wind turbine is the Darrieus rotor. The most attractive feature of this type of turbine is that the generator and transmission devices are located at ground level. Additionally, they are able to capture the wind from any direction without the need to yaw. However, these advantages are counteracted by lower efficiency since the rotor intercepts winds having less kinetic energy. Despite having the generator and transmission at ground level, their maintenance is not simpler, requiring often the rotor removal. In addition, these types of rotors are supported by guy-ropes taking up large land areas. By these reasons, the use of vertical axis wind turbines has declined during the last decades.

Wind turbine operation and characteristics can only be understood with a deep comprehension of the aerodynamics of steady state operation. Accordingly, this chapter focuses primarily on steady state aerodynamics. The turbine aerodynamics describes the forces developed on a wind turbine by airflow. The two major approaches to derive aerodynamic models for wind turbines are the actuator disc theory and the blade element theory (discussed in some detail in the later chapter sections). The former explains in a simple manner the energy extraction process. Also, it provides a theoretical upper-bound to the energy conversion efficiency. The latter studies the forces produced by the airflow on a blade element. This theory is more suitable to explain some of their aerodynamic phenomena and to study the aerodynamic loads. VAWTs capture the wind speed regardless of its direction, so there is thus no need for rotor orientation equipment, as in the case of horizontal axis wind turbines. However, winds are quite low near the ground, showing also stronger turbulence, because of the strong impacts of the ground topography and characteristics. There are two types of vertical turbines: the Savonius rotor relying on the differential drag principle and the Darrieus rotor, relying on cyclic incidence (or lift) variation. The generated power is lower in comparison to a centralized capture at a greater height. A summary of the advantages and disadvantages of VAWTs and HAWTs follows.

**Advantages of vertical wind turbines**

- Wind turbines are easier to maintain because most of their moving parts are located near the ground. This is due to the vertical wind turbine's shape. The airfoils or rotor blades are connected by arms to a shaft that sits on a bearing and drives a generator below, usually by first connecting to a gearbox.

- As the rotor blades are vertical, yaw devices are not needed, reducing the need for bearings and cost.

- Vertical wind turbines have a higher airfoil pitch angle, giving improved aerodynamics while decreasing drag at low and high pressures.

- Hilltops, ridgelines and passes can have higher and more powerful winds near the ground than up high because of the speed up effect of winds moving up a slope or funneling into a pass combining with the winds moving directly into the site. In these places, VAWTs placed close to the ground can produce more power than HAWTs placed higher up.

- Low height is useful where laws do not permit structures to be placed high.

- Smaller VAWTs are easier to transport and install.

- There is no need for free standing towers, being less expensive and stronger in high winds.

- They usually have a lower Tip-Speed ratio so less likely to break in high winds.

## Disadvantages of vertical wind turbines

- There may be a height limitation to how tall a vertical wind turbine can be built and how much sweep area it can have.
- Most VAWTS need to be installed on a relatively flat piece of land and some sites could be too steep for them but still be usable for HAWTs.
- Most VAWTs produce energy at only 50% of the efficiency of HAWTs in large part because of the additional drag that they have as their blades rotate into the wind.

## Advantages of horizontal wind turbines

- The blades are to the side of the turbine's center of gravity, helping stability.
- There is the ability to wing warp, which gives the turbine blades the best angle of attack. Allowing the angle of attack to be remotely adjusted gives greater control, so the turbine collects the maximum amount of wind energy for the time of day and season.
- There is the ability to pitch the rotor blades in a storm, to minimize damage.
- The tall tower allows access to stronger wind in sites with wind shear. In some wind shear sites, every ten meters up, the wind speed can increase by 20% and the power output by 34%.
- The tall tower allows placement on uneven land or in offshore locations.
- It can be sited in forests above the tree line.
- Most HAWTs are self-starting systems, being cheaper because of higher production volume, larger sizes and, in general higher capacity factors and efficiencies.

## Disadvantages of horizontal wind turbines

- HAWTs have difficulty operating in near ground, turbulent winds because their yaw and blade bearing need smoother and more laminar wind flows.
- The tall towers and long blades are difficult to transport and to install on the sea and on land. Transportation can cost 20% of equipment costs, while expensive cranes and skilled operators are needed for installation.
- The FAA has raised concerns about large HAWTs effects on radar in proximity to air force bases.
- Height is a safety hazard for low-altitude aircraft, while offshore towers can be a navigation problem.
- Downwind wind turbines suffer from fatigue and structural failure caused by turbulence.

Large-scale electricity generation is dominated by the rugged three-bladed HAWT types, mainly manufactured by Danish and German wind turbine companies. These turbines have high efficiency, and low torque ripples and oscillations which contribute to good reliability. Ones of the world's largest turbines deliver up to 6 MW in operation, an overall height of 186 m and a diameter of 126 m. Most of these wind turbines are omnidirectional systems, meaning they direct their rotor according to the wind, either passively (rudders) or actively (gears). The conventional wind turbine's rotors are positioned facing the wind and integrate equipment and controllers, which enable positioning to face the wind. Based on driving aerodynamic force, wind turbines are classified as lift and drag machines. Turbines that work predominantly by the lift force are called the lift machines and that by the drag force are called drag turbines. It is always advantageous to utilize the lift force to run the turbine, having higher efficiencies. Wind turbines can be divided into a number of broad categories in view of their rated capacities: micro, small, medium, large and ultra-large wind turbines. Though a restricted definition of micro wind turbines is not available, it is accepted that a turbine with the rated power less than several kilowatts can be categorized. The classification depending on the amount of the generated power is as follows: small wind turbines (less than 25 kW), medium (range power from 25 to 100 kW), large turbines (range power from 100 to 1000 kW), and very large wind turbine (above 1 MW).

Depending on the number of blades, horizontal axis wind turbines are further classified as single-bladed, two-bladed, three-bladed and multi-bladed types. Single-bladed turbines are cheaper due to savings on blade materials. The drag losses are also lower for these types of wind turbines. However, to balance the blade, a counterweight has to be placed opposite to the hub. Single-bladed designs are not very popular due to problems in balancing and visual acceptability. Two-bladed rotors also have these drawbacks, but to a lesser extent. Most of the present commercial turbines used for electricity generation have three blades. Three-bladed wind turbines are more stable as the aerodynamic loading will be relatively uniform. Machines with more blades (6, 8, 12, 18 or even more) are also available, but rear used in electricity generation. The Savonius rotor, one of the VAWT type, often used in low power applications is designed to exploit differential dragging of the two-rotor sides. The efforts exerted by wind on each of the faces of a hollow body have varying intensities. Figure 6.2 shows the assembly of two half cylinders on an axis, subjected to the incoming wind. Convex side of one of the half cylinder and the concave side of the other are facing the wind at a time as shown in Figure 6.2. The basic driving force of Savonius rotor is drag. The drag coefficient of a concave surface is more than the convex surface. The air flow is modified, and the force resultants will create a resultant force $F_1$ on the concave face which is higher than that of the convex face $F_2$. This "differential drag" principle creates a torque, leading to the rotation of the whole.

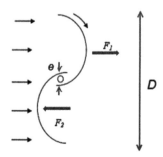

**FIGURE 6.2**
Operating principle of a Savonius rotor.

Being drag machines, Savonius rotors have relatively lower power coefficient. However, some experimental rotors have shown power coefficient up to 35 percent. These rotors have high solidity and thus high starting torque, and are characterized by a low rotational speed and a high torque. The wind speed that enables the machine startup is quite low: about 2–3 m/s. The power coefficient reaches at maximum the value 0.3 for lower tip speed ratios, example the one close to 1. The first models are described as two half-cylinders, whose axes are out of line from one another (distance e in Figure 6.2). They are thus quite easy to carry out. Very good performances are obtained by taking the ratio, e/D = 0.33, as can be found in the literature. These turbines have evolved toward devices using several levels of half-cylinders or a twist form. They enable a continuous circulation of the flows inside the turbine and have a starting torque that is independent from the position of the blades in relation to the wind direction.

The other common VAWT type, the Darrieus rotors are based on the cycle incidence variation principle. A profile positioned in an air flow according to various angles is subjected to forces of variable intensities and directions. The resultant of these forces then generates a torque leading to the device rotation.

## 6.1.2 Wind Turbine Terminology

*Solidity represents* the ratio between the total actual blades' area, to the swept area of a rotor. Hence, multi-bladed rotors are also called high-solidity rotors. These rotors can start easily as more rotor area interacts with the wind initially. Some low-solidity designs may require external starting. These rotors can start easily as more rotor area interacts with the wind initially. Some low-solidity designs may require external starting. Most lift devices use airfoils for blades similar to propellers or airplane wings; however, other concepts have been used. Using lift, the blades can move faster than the wind and are more efficient in terms of aerodynamics and amount of material needed for the blades. The tip speed ratio is the speed of the tip of the blade divided by the wind speed. At the point of maximum efficiency for a rotor, the tip speed ratio is around 7 for a lift device and 0.3 for a drag device. For a lift device the ratio of amount of power per material area is around 75, again emphasizing why wind turbines using lift are used to produce electricity. The optimum tip speed ratio also depends on the solidity of the rotor. Solidity is the ratio of blade area to rotor swept area. Some wind turbine applications, on the other hand, like water pumping require high starting torque. For such systems, the torque required for starting goes up to 3–4 times the running torque. Starting torque increases with the solidity. Hence to develop high starting torque, water pumping windmills are made with multi-bladed rotors. Based on the direction of receiving the wind, HAWT can be also classified as upwind and down wind turbines as shown in Figure 6.2a. For the upwind turbines, as the wind stream passes the rotor first, they do not have the problem of tower shadow. However, yaw (part of wind turbine control, discussed later in this chapter) mechanism is essential for such designs to keep the rotor always facing the wind. For the downwind rotors, as the rotors are placed at the lee side of the tower, there may be uneven loading on the blades as it passes through the shadow of the tower. On the other hand, downwind machines are more flexible and may not require a yaw mechanism.

The blade profiles and the settings have strong effect on the WECS efficiency because the power developed by a rotor at a certain wind speed depends significantly on the relative velocity between the rotor (blade) tip and the wind speeds. Two cases can be analyzed: (a) *slow rotor, fast wind velocity* and (b) *fast rotor-slow wind*. Consider a situation in which the rotor is rotating at a very low speed and the wind is approaching the rotor with a very high velocity. Under this condition, as the blades are moving slow, a portion of the air stream approaching the rotor may pass through it without interacting with the blades and thus without energy transfer. Similarly if the rotor is rotating fast and the wind velocity is low, the wind stream is deflected from the turbine and the energy is lost due to turbulence and vortex shedding. In both of cases, the interaction between the turbine rotor and the wind stream is not efficient and thus results in poor power coefficient. The power coefficient is ratio of the actual power that is generated by the wind turbine to the available power in the wind. This means that the ratio of the rotor speed to the wind speed, the tip speed ratio (TSR) is an important factor in wind turbine design. If the wind turbine rotor spins too slowly, most of the wind passes straight through the gap between the blades, therefore giving it no or very little power. But if the rotor spins too fast, the blades blur and act like a solid wall to the wind, and the rotor blades create turbulence as they spin through the air, further reducing the captured power. If the next blade arrives too quickly, it hits turbulent air and reduces the collected power. Wind turbines must be designed with optimal TSRs to get the maximum amount of the wind power. TSR is expressed as:

$$TSR = \lambda = \frac{\omega R}{v} = \frac{\omega D}{2v} \tag{6.1}$$

where $\omega$ is the angular velocity of the rotor, and $v$ is the wind speed, $D$ and $R$ are the outer turbine rotor diameter and radius, respectively. Note that often is used, the rotational speed in rot/min (RPM), $n$, which is related to the frequency, $f$ (in Hz), through:

$$n = \frac{60f}{2\pi}$$

**Example 6.1:** A 1.5 MW wind turbine has the rated wind speed 12 m/s and a rotor diameter of 72 meters and a rotational speed ranging from 13.6 to 21.8 RPM. Estimate the wind turbine TSR range.

**Solution:** The rotor angular speed is:

$$\omega = 2\pi f = \frac{2\pi n}{60} = \frac{2\pi(n_2 - n_1)}{60} = \frac{2\times3.14\times(21.8 - 13.8)}{60} = 2.28 \cdots 1.44 \text{ rad/s}$$

The range of its rotor's tip speed can be estimated as:

$$\lambda = \frac{\omega R}{v} = \frac{(2.28 \cdots 1.44)\times 36}{12} = 6.84 \cdots 4.33$$

The total energy conversion system consists of the wind turbine and load, while the rotor represents the shaped airfoil and blades designed to convert to wind kinetic energy into mechanical energy available at rotor shaft for further use. For example, the horizontal wind turbine components include: blade and rotor, converting the energy in the wind to rotational shaft energy, a drive-train, usually including a gearbox and a generator, a tower that supports the rotor and drive-train, and additional equipment, controls, electrical cables, ground support equipment and interconnection equipment. The *gearbox* is transferring the shaft energy to the load and adapting its characteristics to the ones required by the load. The *protection and control system* consists of the equipment and devices ensuring the correct and

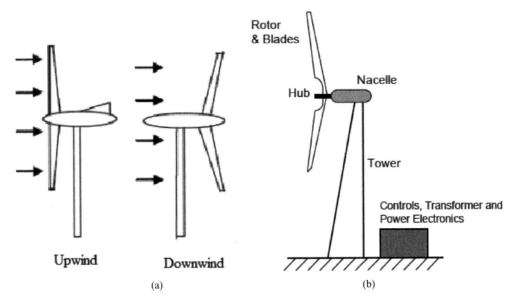

**FIGURE 6.3**
(a) Upwind and downwind turbines (classification of the wind turbines based on the direction of the receiving of the wind) and (b) horizontal wind turbine configuration and components.

safety operation of the all wind turbine components and devices. A typical wind turbine consists of the rotor (blades and hub), speed increaser (gearbox), conversion system, controls and tower (Figure 6.3). The output of the rotor, the rotational kinetic energy, can be converted to electrical, mechanical or thermal energy. *Rotor blade configuration* may include a non-uniform platform (blade width and length), twist along the blade and variable (blades can be rotated) or fixed pitch. The *chord* is the line from the nose to the tail of the airfoil. The drag and lift forces and the overall performance of a wind turbine depend on the construction, shape and orientation of the blades. An important parameter of a blade is the *pitch angle*, the angle between the chord line of the blade and the plane of rotation. The *chord line* is the line connecting the leading and trailing edges of an airfoil, while the rotation plane is the plane in which the blade tips lie as they rotate. In fact, the blade tips trace out a circle which lies on the plane of rotation. Full power output is normally obtained when the wind direction is perpendicular to the plane of rotation. The pitch angle is a static angle, depending only on the blade orientation. Another important blade parameter is the *angle of attack*, the angle between the blade chord line and the relative wind or the effective air flow direction, a dynamic angle, depending on both the speed of the blade and the wind speed. A blade with twist has a variation in angle of attack from hub to tip because of the variation of $\omega R$ with distance from the hub. The lift and drag have optimum values for a single angle of attack so a blade without twist is less efficient than a blade with the proper twist to maintain a nearly constant angle of attack from hub to tip. However, a straight blade is easier and cheaper to build and the cost reduction may more than offset the loss in performance. When the blade is twisted, the pitch angle changes from the hub to the blade tip. In this situation, the pitch angle measured at three fourths of the distance out from the hub is selected as the reference.

The *electrical generators*, converting the shaft mechanical power into electricity are alternating current (AC) generators or direct current (DC) machines. For units connected to the utility grid, 50 or 60 Hz, the generators can be synchronous or induction type connected directly to the grid, or a variable-frequency alternators or DC generators connected indirectly to the grid through a power converter. Most of DC generators and permanent magnet alternators used on small wind turbines do not have a gearbox. The *nacelle* is the covering or enclosure of the drivetrain, control and protection equipment, generator, etc. The *tower* is connecting the turbine rotor to the foundation, supporting the rotor, the nacelle and the equipment located in the nacelle. For HAWT the tower also contains ways to access the nacelle equipment or the rotor. The tower, the structure on which a wind generator is installed is not just a support framework. It in addition raises the particular wind turbine so that it is blades safely and securely clears the floor and to attain the stronger winds at increased elevations. The tower height is different for offshore and onshore turbines. The higher towers are more appropriate for wind energy harvesting, since winds contain less turbulence in higher altitudes. However, the stability issues limit the tower height. Onshore wind systems have higher towers than offshore wind turbines, the land having higher roughness than the water surface. On the water surface, there are almost no obstacles hence the lower towers may be sufficient to capture the wind. In onshore applications, there may be some objects around the tower that may block the wind speed. In areas with high roughness, high turbine towers are required to avoid the effect of wind blocking objects such as buildings, hills, trees, and so on. However, the maximum tower level is optional in many cases, except in which zoning restrictions apply. Deciding exactly what height tower should be mounted is based on the cost regarding taller towers versus the amount of the increase in power production caused. Towers could be constructed of simple tubing, wood poles or lattices connected with tubes, supports, and angle bracket. Large wind generators may be mounted in lattice towers, tube towers or guyed systems. Installers can suggest the better type connected with tower on wind turbine, must be strong and ample to offer the wind turbine and to sustain shake, wind packing and the complete weather components for its lifetime. Tower expenses are changing widely, being a function associated with design and height, while fewer wind turbines that are marketed complete with tower. More often, however, podiums are marketed separately. The *foundation* supports the entire system, making sure that is well fixed onto the ground or the roof, in the case of building integrated turbines. It consists of concrete assembly around the tower to maintain its integrity. Blade configuration may include a non-uniform platform (blade width and length), twist along the blade, and variable (blades can be rotated) or fixed pitch. The pitch is the chord angle at the tip of the blade to the plane of rotation. Most large wind turbines, which are pitch controlled, have blade control, often electric motors are used to rotate, change the blade pitch. All blades must have the same pitch for all operational conditions. Podiums for small wind methods are generally "guyed" styles, anchored to the floor on three or four sides of the tower. These podiums cost lower than free-standing systems, but require more acreage to ground tackle the man wires. A few of these guyed towers are built by tilting them upwards. This function can possibly be quickly achieved using just a winch. This shortens not only installation, but maintenance as well.

## 6.2 Wind Turbine Aerodynamics

### 6.2.1 Power of the Wind

The turbine aerodynamics describes the forces developed on a wind turbine by the airflow. The two major approaches to derive aerodynamic models for wind turbines are the actuator disc theory and the blade element theory. The former explains in a simple manner the energy extraction process. Also, it provides a theoretical upper-bound to the energy conversion efficiency. The latter studies the forces produced by the airflow on a blade element. This theory is more suitable to explain some aerodynamic phenomena such as stall, as well as to study the aerodynamic loads. This model is based on the momentum theory. The turbine is regarded as an actuator disc, which is a generic device that extracts energy from the wind. Consider the actuator disc is immersed in an airflow, which can be regarded as incompressible. Since the actuator disc extracts part of the kinetic energy of the wind, the upstream wind speed $V$ is necessarily greater than the downstream speed $V_{-\infty}$. Consequently, for the stream tube just enclosing the disc, the upstream cross-sectional area $A_\infty$ is smaller than the disc area $A_D$, which in turn is smaller than the downstream cross-sectional area $A_{-\infty}$. This is because, by definition, the mass flow rate must be the same everywhere within the tube. From the expression for kinetic energy in flowing air follows the power contained in the wind through an area $A$ with wind velocity $v_1$ is expressed as:

$$P_W = 0.5\rho A v_1^3 \tag{6.2}$$

Here $\rho$ is the specific air mass density, depending on air pressure and moisture, for practical calculation is assumed as: $\rho \approx 1.225$ kg/m³. The air streams in the axial direction through the cross-sectional area, A of the wind turbine rotor. The useful mechanical power is expressed as by means of power coefficient as:

$$P = 0.5 C_P \rho A v_1^3 \tag{6.3}$$

In the case of homogeneous of wind flow, the wind velocity, $v_1$ before the wind turbine plane is reduced due to the power extraction to a speed $v_3$, as shown in Figure 6.6. Simplified theory claims that the wind velocity in the rotor plane has an average value:

$$v_2 = \frac{v_1 + v_3}{2} \tag{6.4}$$

On this basis, Betz (1929) has shown by a simple calculation that maximum useful power is obtained for $v_3/v_1 = 1/3$, where the power coefficient is:

$$C_P = \frac{16}{27} \approx 0.59 \tag{6.5}$$

And the wind speed corresponding to maximum power, as determined in the next chapter section is:

$$v_d = \frac{2}{3} v_1 \tag{6.6}$$

However, the actual wind turbines display power coefficients in the range 0.35–0.50, due to various losses (profile loss, tip loss and loss due to radiation wake). In order to determine the mechanical power available to the turbine load (electrical generator, pump, etc.), the power expression (Equation 6.3) has to be multiplied with the drivetrain efficiency, which is taking into account losses in bearings, couplings and gearboxes. A common expression for the expression for the maximum power that could be extracted from a wind stream is given by:

$$P_{WT(max)} = \frac{16}{27} \frac{\rho}{2} \frac{\pi D^2}{4} v^3 \text{ (W)} \tag{6.7}$$

Here, $D = 2R$ is the swept area (basically wind turbine rotor) diameter. The most important implication from the Betz Limit is that there must be a wind speed change from the wind turbine upstream to the downstream in order to extract energy from the wind stream. If no change in the wind speed occurs, energy cannot be efficiently extracted from the wind. Realistically, no wind machine can totally bring the air to a total rest, and for a rotating machine, there is always some air flowing around it. Thus a wind turbine can only extract a fraction of the kinetic energy of the wind. The wind speed on the rotors at which energy extraction is maximal has a magnitude lying between the upstream and downstream wind velocities. The Betz Limit reminds us of the Carnot Cycle Efficiency in Thermodynamics suggesting that a heat engine cannot extract all the energy from a given heat reservoir and must reject part of its heat input back to the environment. Another important concept relating to the power of wind turbines is the optimal TSR. If a rotor rotates too slowly, it allows too much wind to pass through undisturbed, and thus does not extract as much as energy as it could, within the limits of the Betz Limit. On the other hand, if the rotor rotates too quickly, it appears to the wind as a flat (solid) disc, which creates a large amount of drag. TSR depends on the blade airfoil profile used, the number of blades, and the type of wind turbine. In general, power in a mechanical system is the product of the angular velocity and torque ($P = T{\cdot}\omega$), the torque coefficient can be related to the power coefficient through:

$$C_T(\lambda) = \frac{C_P(\lambda)}{\lambda} \tag{6.8}$$

The torque at rotor shaft then can be expressed from Equation (6.3) by the expression:

$$T = C_T \frac{\rho}{2} A v^2 \tag{6.9}$$

**Example 6.2:** Consider a wind turbine with 30 m diameter. The rotor is rotating with 36 RPM at 10 m/s wind velocity, and its power coefficient is 0.35. Assume the standard air density is 1.225 kg/m³, calculate: (a) the tip-speed ratio, (b) torque coefficient, and (c) torque available on the rotor shaft.

**Solution:** The rotor area, angular velocity and TSR are:

$$A = \frac{\pi 30^2}{4} = 706.858 \text{ m}^2$$

$$\omega = \frac{2\pi \times 36}{60} = 3.77 \text{ rad/s}$$

$$\text{TSR} = \lambda = \frac{3.77 \times 15}{10} = 5.65$$

Then, from Equation (6.8) the turbine torque coefficient is:

$$C_T = \frac{C_P}{\lambda} = \frac{0.35}{5.65} = 0.062$$

For this torque coefficient the torque available on the rotor shaft (from Equation 6.9) can be calculated:

$$T = 0.062 \times \frac{1.225}{2} \times 706.858 \times 10^2 = 2679.7 \text{ Nm}$$

Rotor power curves are typically plotted against the non-dimensional tip speed ratio (the ratio of the speed of the blade tip and the free-stream wind speed, as defined above). These curves are unique for a given rotor with fixed pitch and can be performance-tailored by modifying the airfoil section characteristics, pitch angle of the blades, and the blade twist over the span of the rotor. It is worth to notice that high-speed rotors with low solidity (as defined above, the solidity is the ratio of actual blade airfoil area, length times average chord, to rotor disk-swept area) have higher conversion efficiency. Note that the torques varies with the square of velocity, while the turbine power depends on the cube of the wind speed. Figure 6.4 is showing typical power coefficients for different WT rotors. There are notable differences in the region of lower TSR, indicating the current preference for three-blade rotors. Fast running wind turbines display larger values of power coefficients, with poor starting torques. Since one or two blade rotors are also problematic with respect to torque variations and acoustic noise. The three-blade wind turbines are currently predominant in modern wind industry. Rotors are normally designed to values of TSR between 5 and 8, in the case of three-blade wind turbines, with 7 being the most widely used value. In addition to the factors mentioned above, other concerns dictate the TSR to which a wind turbine is designed. The wind turbine aerodynamic behavior analysis can be performed without any specific turbine design, by considering only the energy extraction process. A simple model, as the actuator disc model, can be used to calculate the power output of an ideal turbine rotor and the wind thrust on the rotor. Additionally more advanced methods including momentum theory, blade element theory and finally blade element momentum (BEM) theory are briefly discussed.

BEM theory is employed to determine the optimum blade shape and to predict the rotor performance parameters, for the ideal steady operating conditions. BEM theory combines two methods to analyze the aerodynamic performance of a wind turbine. These are momentum

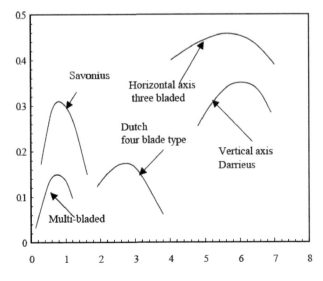

**FIGURE 6.4**
Typical power coefficients for different wind turbines vs. tip speed ratio.

theory and blade-element theory which are used to outline the governing equations for the aerodynamic design and power prediction of a WT rotor. Momentum theory analyses the momentum balance on a rotating annular stream tube passing through a turbine, while blade-element theory examines the forces generated by the airfoil lift and drag at various sections along the blade. Combining these theories gives a series of equations that can be solved iteratively, and current design codes for wind turbine rotors are based on the BEM theory, an elegant simplicity, with modest calculation requirements, and reasonably successful in predicting performances and loads. However, it is not able to predict all flow conditions with sufficient accuracy. Due to its simplicity, BEM is easily extended by engineering rules to cover these deficiencies. Models are still being developed, since old problems are still unsolved, and new aerodynamic problems occur with the increasing size of wind turbines. Theoretical background in energy extraction generalities and more specifically rotor aerodynamics of horizontal axis wind turbines (HAWTs) is developed in this chapter. Some prior knowledge of fluid dynamics in general and as applied to the analysis of wind turbine systems is assumed, in particular basic expressions for energy in a fluid flow, Bernoulli's equation, definitions of lift and drag, some appreciation of stall as an aerodynamic phenomenon and BEM theory in its conventional form as applied to HAWTs. Nevertheless some of this basic knowledge is also reviewed, more or less from first principles. The aim is to express particular insights that will assist the further discussion of issues in optimization of rotor design and also aid evaluation of various types of innovative systems, for example those that exploit flow concentration.

### 6.2.2 Actuator Disk Theory

The function of a wind turbine rotor is to extract the kinetic energy of the incoming flow field by reducing the velocity abaft the rotor, while a thrust in the direction of the incoming flow field is produced with a magnitude directly related to the change in kinetic energy. The actuator disk is a physical "trick" to represent the pressure drop as the rotor plan. Figure 6.5 presents the different plane used in the BEM theory in order to obtain theoretical loss of velocity and thus, the theoretical maximum power that can be harvested from the wind, known as the Betz limit. The actuator disk concept, introduce a gap of pressure through the disk, whereas the velocity is continuous. With the rotational movement and the frictional drag of the blades, the flow field is furthermore imparted by a torque which contributes to the change in kinetic energy. Thus, the flow field and forces related to operating wind turbine rotors are governed by the balance between the thrust and torque on the rotor and the kinetic energy of the incoming flow field. The behavior of a wind turbine rotor in a flow field may conveniently be analyzed by introducing the actuator disc principle. The basic idea of the actuator disc principle is to replace the real rotor with a permeable disc of equivalent area where the forces from the blades are distributed on the circular disc. The actuator disk distributed forces change the local velocities through the disc and the entire flow field around the rotor disc. The balance between the applied forces and the changed flow field is governed by the mass conservation law and the balance of momenta, which for a real rotor is given by the axial and tangential momentum equations. Figure 6.5 displays an actuator disc where the expanding streamlines are due to the reaction from the thrust. The classical Rankine-Froude theory considers the balance of axial-momentum far up- and downstream the rotor for a uniformly loaded actuator disc without rotation, where the thrust $T$ and kinetic power $P_{kin}$ in terms of the free stream $V_0$ and far wake velocity $u_1$ reduces to:

$$T = \frac{dm}{dt}(V_0 - u_1) \tag{6.10}$$

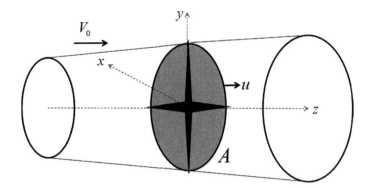

**FIGURE 6.5**
Flow field around an actuator disc.

and

$$P = 0.5 \frac{dm}{dt} \left( V_0^2 - u_1^2 \right) \tag{6.11}$$

Here the mass flow through the disc is given by:

$$\frac{dm}{dt} = \rho A_1 u_1 \tag{6.12}$$

where $A_1$ is the far wake area given by the limiting streamline through the edge of the disc and $\rho$ is the air density. From the mass conservation through the disc (configuration), shown in Figure 6.5, we have:

$$uA = u_1 A_1 \tag{6.13}$$

By combining the mass conservation, Equation (6.12) with Equations (6.10) and (6.11) yields that the kinetic power extracted from the flow field by the thrust is given by:

$$P_{kin} = 0.5 \left( V_0 + u_1 \right) T = uT, \text{ and } u = 0.5 \left( V_0 + u_1 \right) \tag{6.14}$$

Equation (6.14) is showing that the velocity at the disc $u$ is the arithmetic mean of the free stream $V_0$ and the slipstream velocity $u_1$. The importance of this result is seen with the evaluation of the aerodynamic blade forces. For convenience Equation (6.14) is usually presented in non-dimensional form by introducing the *axial interference factor*:

$$a = 1 - \frac{u}{V_0} \tag{6.15}$$

This analysis is performed by using the free stream dynamic pressure and rotor area. To go into details in the correlation of the above parameters with the power coefficient, the force balance on the rotor blades needs to be analyzed. The thrust, experienced by the rotor is:

$$F = 0.5 \cdot \rho \cdot A \cdot v^2 \tag{6.16}$$

While the torque is expressed by:

$$T = \frac{1}{2}\rho \cdot A \cdot V^2 \times R \tag{6.17}$$

where $R$ is the radius of the torque, and Equation (6.16) is the maximum theoretical torque. In practice the rotor shaft can develop only a fraction of this maximum torque, expressed by the torque coefficient. The non-dimensional power and coefficients $C_P$ and $C_T$ are established as:

$$C_P = \frac{0.5\rho Au\left(V_0^2 - u_1^2\right)}{0.5\rho AV_0^3} = \frac{Rotor\ Power}{Power\ in\ the\ Wind} = 4a(1-a)^2 \tag{6.18}$$

$$C_T = \frac{\rho uA\left(V_0 - u_1\right)}{0.5\rho AV_0^2} = \frac{Trust\ Force}{Dynamic\ Force} = 4a(1-a) \tag{6.19}$$

While the power and trust coefficients are related through the TSR, as:

$$C_T = \frac{C_P}{\lambda}$$

**Example 6.3:** Consider a wind turbine with 10 m diameter. The rotor is rotating with 120 RPM at 10 m/s wind velocity. $C_P = 0.37$. Calculate (a) the tip-speed ratio; (b) torque coefficient; and (c) the available torque. Assume $\rho = 1.24$ kg/m³.

**Solution:** Area of the wind turbine rotor is:

$$A = \frac{\pi \times 10^2}{4} = 78.75\ \text{m}^2$$

The rotor angular velocity and the TSR are:

$$\omega = \frac{2 \times \pi \times 120}{60} = 12.56\ \text{rad/s}$$

$$\lambda = \frac{5 \times 12.56}{10} = 6.28$$

Then the torque coefficient and the torque are:

$$C_T = \frac{C_P}{\lambda} = \frac{0.35}{6.28} = 0.057$$

$$T = 0.5 \times 0.057 \times 1.24 \times 78.75 \times 10^2 \times 5 = 1391.5\ \text{N} \cdot \text{m}$$

The maximum of $C_P$ is obtained by taking its derivative of $C_P$ vs. $a$, and setting it to zero, yielding to:

$$a_{max} = \frac{1}{3}$$

Substituting it in Equation (6.17) yields to the maximum power coefficient, the Betz limit:

$$C_{P,max} = \frac{16}{27} = 0.5926 \qquad (6.20)$$

The maximum possible power extraction from the wind stream by a wind turbine is then:

$$P_{max} = \frac{16}{27} \cdot \left( 0.5 \rho A v^3 \right) \qquad (6.21)$$

Notice that, the axial inference factor (inflow factor) $a$ represents the proportion by which the air slowed in its approach from the free air-stream to the wind turbine disk, and in fact it is the ratio by which the cross-sectional air-stream area contracts when approaching the actuator disk (i.e., as the same proportion as the air-stream speed-up). The wind speed magnitude or inflow factor reduction, achieved by a wind turbine is dependent on the design and operation of the rotor blades. Blades that are designed and operated at the *"optimal attack angle"* to the incoming air-stream, $a = 1/3$ are extracting maximum wind power, while the ones that are not optimally oriented (for example, blades that are intentionally stalled by adjusting their pitch away from their *optimal attach angle*) have lower $a$, and in consequence they extract less wind power. They also are experiencing less trust.

> **Example 6.4:** Calculate the maximum power (the Betz limit of the extracted power) for a horizontal axis wind turbine whose blades sweep a circular area with a diameter of 80 m, and for an upstream wind speed of $v = 12$ m/s.
>
> **Solution:** From Equation (6.18), for this wind turbine and considering air density to be 1.225 kg/m³ the maximum power extracted for the wind is:
>
> $$P = \frac{16}{27} \cdot 0.5 \cdot 1.225 \cdot \pi \left( 40 \right)^2 \left( 12 \right)^3 = 3.151 \text{ MW}$$
>
> While the maximum output power density of this turbine is then:
>
> $$P_d = \frac{P}{A} = \frac{3.151 \times 10^6}{\pi \left( 40 \right)^2} = 627.2 \text{ W/m}^2$$

At this point of the wind turbine power estimates, it is worthwhile to briefly discuss the Betz limit interpretation. It quite often (rather not accurately) that this is the maximum possible wind turbine efficiency. However, the system efficiency is defined as the ratio of the output power to the input system power. A closer look to the Equation (6.21) reveals that this expression is not quite the representative of the turbine output power, even it resembles the wind available power, Equation (6.1), and however this is not the representative power flowing through any cross-sectional area of a flow tube. For example, the input power of wind stream is rather expressed as:

$$P_{in} = 0.5 \rho \cdot \pi R^2 \left( \frac{v_2}{v_1} \right) v_1^3 = \frac{2}{3} \left( 0.5 \rho \cdot \pi R^2 v_1^3 \right) \qquad (6.22)$$

This expression yields to a maximum efficiency, $\eta = (16/27)/(2/3) = 8/9$ or 88.9%, meaning that when a wind turbine is operating at Betz limit captures about 89% of the available flow tube power. In fact, an alternative interpretation of the 59% factor is that it is the ratio of power extracted by the turbine over *the power of the unperturbed wind stream* that exist before erecting the wind turbine at the same location. Of course, actual wind turbines can never reach the Betz limit. However, the Betz limit is caused not by any deficiency in design, but because the stream-tube expand downstream of the rotor, so the flow tube area, where the air is at full, free-steam velocity is smaller than the disk area. In practice there are three effects leading to a decrease in maximum power coefficient: the wake rotation behind the rotor, finite number of blades and the associate tip losses, and the non-zero aerodynamic drag. Note than the overall wind turbine efficiency is a function of both the rotor power coefficient and the mechanical and electrical efficiency of the wind energy conversion systems. The Betz limit is theoretically the rotor maximum power coefficient. The wind turbine shaft power output is not normally used directly the turbine usually is coupled to a load through a transmission or gear box. The load may be a pump, compressor, grinder, electrical generator and so on. To illustrate this, the load considered here is an electrical generator, as shown in Figure 6.6. The power in the wind passes through the turbine, converted into mechanical power, at the turbine angular velocity, is then supplied to the transmission system, the drivetrain and later converted by the generator into electricity. The actual turbine efficiency, including the mechanical (drivetrain) efficiency is lower 60%–80% of the maximum value, the Betz limit. With notation of Figure 6.6 the following relations relates the power in the wind with electrical converted power, as shown below.

$$\eta_{rotor-drivetrain} = \eta_{mech} \cdot C_P \tag{6.23}$$

And the actual power output (electric power for the system in Figure 6.6) of a wind turbine is then:

$$P_{out} = 0.5\rho A v^3 \left( \eta_{mech} \cdot C_P \right) \tag{6.24}$$

When the wind turbine is used to generate electricity, the overall wind energy conversion efficiency, includes also the electric generator efficiency:

$$\eta_{overall} = \eta_{mech} \cdot \eta_{el} \cdot C_P \tag{6.25}$$

And the overall output power of wind turbine generator system is then:

$$P_{electric} = 0.5\rho A v^3 \left( \eta_{mech} \cdot \eta_{el} \cdot C_P \right) \tag{6.26}$$

**Example 6.5:** Estimate the actual available power density and the power output for the wind turbine in the Example 6.3. Assume that wind turbine site is characterized by a Weibull wind speed probability distribution, having a shape and scale coefficients, 1.85

**FIGURE 6.6**
Wind turbine-generator system.

and 11.5 m/s, a turbulence intensity of 10%. The turbine mechanical efficiency is 90%, and the electric generator efficiency is 95%. The air density is 1.225 kg/m³.

**Solution:** By using Equation (6.21) the available power density is equal to: 937 W/m², and the Betz limit is then, 552.8 W/m². The actual wind turbine output power, given by Equation (6.22) is then:

$$P_{out} = \eta_{mech} \cdot \eta_{el} \cdot \pi (40)^2 \times 552.8 = 2.37462 \text{ MW}$$

However, the previous approach to estimate the output wind turbine power did not account for the rotating turbine blades, which induce angular momentum in the rotor wake region reducing the overall power output. The above expressions are not the actual maximum wind turbine power density. By included the effects of the turbulence intensity, wind shear, etc. the theoretical maximum power density $P_d$ (W/m²) that can be extracted by a wind turbine is limited not only by the Betz equation, but also by the impacts of such wind characteristics on the extracted power. For Weibull distribution case of the wind speeds, the wind power density can be expressed by a relationship as (Belu, 2014):

$$P_d = \frac{116}{227} \rho \left(1 + 3\sigma_i^2\right) c^3 \Gamma \left(1 + \frac{3}{k}\right) \text{ W/m}^2 \qquad (6.27)$$

where $\rho$ is the air density and $\sigma_i$ is the turbulence intensity (usually $\sigma_i = 0.2$ at 10 m in open areas). Once the power density per unit of area $P_d$ is computed, the available annual energy $E$ per unit of area of the turbine can be calculated [Belu, 2014].

> **Example 6.6:** A 25 m diameter, wind turbine rotor generates 600 kW at rated wind speed 15 m/s and 135 kW at 9.6 m/s, assuming the standard atmospheric conditions. Compute the overall efficiency at each wind speed.
>
> **Solution:** The rotor area is:
>
> $$A = \pi R^2 = 3.14 (25)^2 = 1962.5 \text{ m}^2$$
>
> The overall efficiencies at 9 m/s and rated wind speed, 15 m/s, respectively are:
>
> $$\eta_{9.6\,m/s} = \frac{135,000}{0.5 \times 1.225 \times 1962.5 \times (9.6)^3} = 0.127 \text{ or } 12.7\%$$
>
> and
>
> $$\eta_{15\,m/s} = \frac{600,000}{0.5 \times 1.225 \times 1962.5 \times (15)^3} = 0.148 \text{ or } 14.8\%$$

## 6.2.3 Vertical Axis Wind Turbines

Vertical axis wind turbines, the main rotor shaft is arranged vertically, have the main advantages that such turbines do not need to be pointed into the wind to be effective. This is an advantage on sites where the wind direction is highly variable, a VAWT can utilize winds from varying directions. Moreover, the generator and gearbox can be placed near the ground, being not supported by the turbine tower, having easier maintenance access. However, their major drawbacks are that some VAWT designs produce pulsating torque and have lower efficiency than HAWTs. Savonius rotors are an example of extremely simple vertical axis wind energy devices that are operating entirely because of the thrust force of the wind. The basic equipment is a two-halve vertically cylinder. The two parts are attached to the two opposite sides of a vertical shaft. The wind blowing into the assembly meets two different surfaces—convex and concave, and different forces are exerted on them, generating torque to the rotor. Providing a certain overlap between drums increases the torque because wind blowing on the concave side turns around and pushes the inner surface of the other drum, which partly cancels the wind thrust on the convex side. An overlap of one-third of the drum diameter gives best results. In a Darrieus turbine rotor, two or more flexible blades are attached to a vertical shaft. The blades bow outward taking a parabola shape and are of a symmetrical airfoil section. In stationary states no torque is produced, so the Darrieus wind turbines are not self-starting, needing external means to start. At each rotor blade position the lift force has a positive component in the direction of rotation, giving rise to net positive torque. The torque is different in different directions. It varies from zero to maximum in about a quarter of a revolution. The torque makes two complete excursions from zero to maximum and back in each revolution-both in positive sense. The shaft torque pulsations can be minimized by using a three-blade system. However, the two-blade rotor design has lower erection cost. The torque here is function of speed of rotation and the wind speed. The torque increases with rotational speed (zero at zero rotational speed) and with wind speed up to a certain value and then falls off at very high wind speeds. Therefore, such wind turbines must have inbuilt protection from stormy weather, the rotor tends to stall at high winds.

The momentum by Betz limit indicates the physical based ideal limit value for the extraction of the mechanical energy from free-stream airflow, with considering the energy convertor design. However, extracted power under real conditions cannot be independent from the energy convertor characteristics and design. For real conditions, the actual extracted power depends on which aerodynamic forces are used for power production. All bodies exposed to airflow experience aerodynamic forces, having a drag component in the direction of the flow, and a lift component perpendicular on the airflow direction. The actual (real) power coefficients obtained vary greatly in dependence on whether drag or lift is used. The power captured, $P_{ST}$ by an aerodynamic drag device, such as Savonius rotors can be calculated from the aerodynamic drag, area and its velocity. The maximum value of the power coefficient is about 4/27 times drag coefficient. The maximum value of the aerodynamic drag coefficient of a concave surface is 1.3 or lower, giving maximum power coefficient for a drag device 0.18–0.20. The most common used to estimate the maximum extracted power relation of a Savonius rotor (as one in Figure 6.2) is:

$$P_{ST} = 0.18 \cdot \rho \cdot D \cdot h \cdot v^3 \text{ (W)} \tag{6.28}$$

Here, $D$ is the rotor blade diameter and $h$ is the rotor blade height. The torque at the shaft is then computed from, the turbine power and rotor angular speed, as:

$$\tau_{ST} = \frac{P_{ST}}{\omega} \text{ (Nm)} \tag{6.29}$$

**Example 6.7:** A Savonius rotor has the rotor blade diameter of 0.6 m and height of 1.0 m. The wind speeds are equal to 4 m/s, 8 m/s, 12 m/s, 16 m/s, and 20 m/s. Assuming the air density 1.20 kg/m³, TSR equal with 1.2, computed the maximum power generated by this Savonius rotor and the torque at the rotor shaft.

**Solution:** By using Equations (6.28) and (6.29), and the equation for angular speed of the maximum generated power and torque at the rotor shat for the given wind speeds are listed in table. The rotor blade radius is r = 0.6/2 = 0.3 m.

| Wind speed (m/s) | 4 | 8 | 12 | 16 | 20 |
|---|---|---|---|---|---|
| Angular rotor speed (rad/s) | 13.33 | 26.67 | 40 | 53.33 | 66.67 |
| Generated Power (W) | 23.33 | 55.3 | 186.62 | 442.37 | 864 |
| Torque at rotor shaft (Nm) | 1.75 | 2.07 | 4.67 | 8.29 | 12.96 |

From aerodynamic point of view, vertical axis wind rotors have several aspects distinguishing them from the horizontal axis rotors. The vertical axis rotor blades are rotating on a surface that is at a right angle to the wind direction, so the aerodynamic attack angle varies during the rotation, leading to higher torque pulsations. Moreover, with other effects not discussed here, the maximum power coefficient of the Darrieus wind turbine is about 0.24, somewhat lower than the HAWT ones. The swept area of a Darrieus wind turbine is a little bit more complicated to compute as in the case of Savonius rotors or HAWTs. One approximation to the swept area is about two-thirds the area of a rectangle with width equal to the maximum rotor width, D and height equal to the vertical extent of the blades, $H$ (as shown in Figure 6.1), expressed as:

$$A_{Dr} = \frac{2}{3} D \cdot H \tag{6.30}$$

The maximum power generated by the Darrieus vertical axis rotors is expressed by:

$$P_{Dr} = 0.21 \cdot \rho \cdot A_{Dr} \cdot v^3 \tag{6.31}$$

The aerodynamic characteristics of VAWT exhibits other difference compared to the HAWT ones, the optimum power coefficient is achieved at relatively low TSRs, due to the higher drag components of the rotor blades, i.e., poor lift/drag ratio shifts the optimum power coefficients to lower tip-speed-ratios, 5 or lower (about half of the ones of the HAWTs). Vertical axis turbines rotate at lower speeds and their power must be generated at higher torques which requires higher rotor weight and higher manufacturing and installation cost. However, VAWT are not often used for electricity generation. For example, Savonius turbines are in wide use in rural areas for water pumping, attic ventilation, and water agitation to prevent freeze-up during the winter. The best configuration for the half-barrels, with notation of Figure 6.2 is given by the relationships:

$$D_{blade} = 2 \cdot R = \frac{D - e}{2} - 0.5 \cdot S \quad \text{and} \quad S = 0.1 \times D_{blade} \tag{6.32}$$

The area of the Savonius rotor exposed to the wind is expressed as:

$$A = 2R \cdot H \tag{6.33}$$

Here $R$ is the projection radius of the area exposed to the wind (the effective half rotor cylinder or the blade radius), and $H$ is the rotor height. The Savonius rotor torque is due to the pressure difference between the blade concave and convex surfaces and by the recirculation of wind coming from behind the convex surface. Its efficiency reaches 31%, but it presents disadvantages with respect to the weight per unit of power, because its constructional area is totally occupied by material. A Savonius rotor needs 30 times more material than is needed by a conventional horizontal axis wind turbine of the same power.

## 6.2.4 Blade Element Model

The momentum theory is tried to be explained on account of HAWT rotor design, but it does not consider the effects of rotor geometry characteristics like chord and twist distributions of the blade airfoil. For this reason BEM theory needs to be added to the design method. The blade element theory is useful to derive expressions of developed torque, captured power and axial thrust force experienced by the turbine. This theory is based on the analysis of the aerodynamic forces applied to a radial blade element of infinitesimal length. To carry out the analysis, the stream tube just containing the turbine swept area is divided into concentric annular stream tubes of infinitesimal radial length, each of which can be treated independently. In BEM, the swept area of the rotor is considered as a set of annular areas, swept by each blade element. The blade is divided span-wise into a set of elements which are assumed to be independent of each other so that balance of rate of change of fluid momentum with blade element forces can be separately established for each annular area. In order to apply blade element analysis, it is assumed that the blade is divided into N sections. This analysis is based on some assumptions including no aerodynamic interactions between different blade elements and the forces on the blade elements are solely determined by the lift and drag coefficient. Figure 6.7 illustrates a blade element and a transversal cut of the blade element viewed from beyond the tip of the blade, with the aerodynamic forces acting on the blade element are also depicted. The blade element moves in the airflow at a relative speed $V_{rel}$, which as a first approach can be imagined as the composition of the upstream wind speed $V$ and the tangential blade element speed $\Omega_r$. It is worth to mention that some corrections are often introduced to account for the local flow variations caused by the wind turbine, either how $V_{rel}$ is actually composed is immaterial for the moment

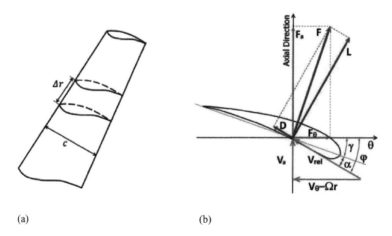

(a)      (b)

**FIGURE 6.7**
Blade element (a) and (b) element cross section and the forces acting on the element.

element. The airflow establishes a differential pressure around the blade element, resulting in a force perpendicular to the local air movement direction, the lift force $F_L$ and a drag force $F_D$ in the flow direction. The drag force is caused by fluid friction and is never zero. Lift and drag values can be found by integrating the pressure values along the surface of the body (along the perimeter of a section parallel to the flow). Streamlined bodies (airfoils) are used to maximize the ratio between lift and drag force. They achieve very high lift force with very small drag force at certain (small) angles of attack. The lift and drag forces per unit length are generally expressed in terms of the lift and drag coefficients $C_L$ and $C_D$:

$$F_L = 0.5\rho C_L V_{rel}^2 C_L(\alpha) \tag{6.34}$$

and

$$F_D = 0.5\rho C_D V_{rel}^2 C_D(\alpha) \tag{6.35}$$

where $C_L$ is the lift force coefficient, and $C_D$ is the drag force coefficient, and is caused by air friction at the surface of the profile. Both lift and drag coefficients are functions of the incidence angle $\alpha$ defined as the angle that the flow makes with the chord. The relationship between the two forces is given by the form ratio, $E_G$ of the lift and drag coefficients, expressed as:

$$E_G = \frac{C_L}{C_D} \tag{6.36}$$

The above relationship is given by the aerodynamic blade quality, determined by the blade design and its surface quality. The lift and drag forces can be resolved into axial and tangential components. The former is called axial thrust force, and is supported by the rotor, tower and foundations. On the other hand, the tangential force develops a rotational torque that produces useful work. This torque per unit length is given by:

$$\tau_R = 0.5\rho C V_{rel}^2 \left( C_L(\alpha)\sin(\phi) - C_D(\alpha)\cos(\phi) \right) \tag{6.37}$$

Both lift and drag forces contribute to the axial thrust force. Further, the lift force develops useful torque whereas the drag opposes it. So, a high form ratio $C_L/C_D$ is desirable to achieve high conversion efficiency. In fact, the higher the form ratio $C_L/C_D$, the higher will be the useful work. During stall, an abrupt drop of this ratio takes place. This is the basis for one of the most common methods to limit the captured power at winds exceeding the rated wind speed of the turbine. To compute the contribution of each blade element to the global thrust force and rotational torque, the magnitude and direction of the relative air movement are needed. The thrust force acting on the entire rotor and the total useful torque developed by the turbine are obtained by integrating Equations (6.21) and (6.22) along the blade's length. Commonly, thrust force, torque and power are expressed in terms of non-dimensional thrust, $C_T$, torque, $C_\tau$ and power, $C_P$ coefficients as:

$$F_T = 0.5\rho\pi R^2 C_T(\lambda,\beta) V^2 \tag{6.38}$$

$$\tau_R = 0.5\rho\pi R^3 C_\tau(\lambda,\beta) V^2 \tag{6.39}$$

and

$$P_R = 0.5\rho\pi R^2 C_P(\lambda, \beta) V^3 \qquad (6.40)$$

$R$ is the rotor radius, $V$ is the mean wind speed, $\lambda$ is the TSR, and $\beta$ is the pitch angle. These coefficients are written in terms of the pitch angle and TSR, $\lambda$. The TSR parameter is extremely important in turbine control, and together with $\beta$ in the case of variable-pitch rotors, determines the operating condition of a wind turbine. The torque and power coefficients are of special interest for design and control purposes.

## 6.2.5 Factors Affecting Power Curve

Betz limit states that an ideal wind turbine can extract 59% from the available wind power, without making any reference to what turbine operating system or the rotor configuration must have in order to achieve this limit. In this section we are quantitatively analyze, factors such number of rotor blades, rotor diameter or wind turbine tower shadow effects on its power curve. However, due to the intermittent nature of wind, wind turbines do not make power all the time. Thus, a capacity factor of a wind turbine is used to provide a measure of the wind turbine's actual power output in a given period (e.g., a year) divided by its power output if the turbine has operated the entire time. A reasonable capacity factor would be 0.25–0.30 and a very good capacity factor would be around 0.40. In fact, wind turbine capacity factor is very sensitive to the average wind speed, as well as to other factors. *Availability* is the time that the wind turbine is in operational mode, and it does not depend on whether the wind is blowing. Availability is related to reliability of the wind turbine, which is affected by both the quality of the turbine and operation and maintenance. The most important implication from the Betz Equation is that there must be a wind speed change from the upstream to the downstream in order to extract energy from the wind, in fact by breaking it using a wind turbine. If no change in the wind speed occurs, energy cannot be efficiently extracted from the wind. Realistically, no wind machine can totally bring the air to a total rest, and for a rotating machine, there will always be some air flowing around it. Thus a wind machine can only extract a fraction of the kinetic energy of the wind. The wind speed on the rotors at which energy extraction is maximal has a magnitude lying between the upstream and downstream wind velocities. The Betz limit is reminding us of the Carnot cycle efficiency, suggesting that a heat engine cannot extract all the energy from a given heat reservoir and must reject part of its heat input back to the environment. In practice, a number of factors or effects are leading to the reduction of the actual efficiency, among these are: finite number of blades and associated tip losses, the rotation of the wake behind the turbine rotor and the non-zero aerodynamic drag. Notice that the overall turbine efficiency depends on the rotor efficiency, and the efficiencies of the mechanical and electrical wind turbine components. In summary, the efficiency of a wind turbine would depend on several factors such as the ambient air velocity, which is variable, the rotational design speed of the wind turbine, $\omega$, the type of wind turbine and the size of the wind turbine, or blade diameter $D$. However, a large number of empirical and filed observations with several types of wind turbines have shown that a single parameter, the tip-speed ratio, can define quite very well the efficiency of a wind turbine (as shown in Figure 6.4).

The wind turbine power curve is a diagram that shows the specific turbine output power at various wind speeds, being very important in wind turbine design. Power curves are made by a series of measurements for one turbine with different wind speeds. In order to obtain accurate power curves, the wind should be non-turbulent, and the measurements are conducted at standard meteorological conditions. The "cut-in" and "cut-out" wind speeds can be found from the power curve. The "cut-in" shows the minimum wind speed needed to start the turbine (which depends on turbine design) and to generate output power. Usually it is 3 m/s for smaller turbines and about 5 m/s for large wind turbines. The "cut-out" wind speed represents the speed point where the turbine should stop rotating due to the potential damage that can be done if the wind speed increases more than that. The power curve can be divided into a few regions. These regions are presented in Figure 6.7 and are separated by cut-in, nominal and cut-out speeds. Once the rotor of the turbine starts spinning, it can be assumed. Wind turbines must be designed to operate at their optimal wind tip speed ratio in order to extract maximum power possible from the wind stream. When a rotor blade passes through the airstream it leaves a turbulent wake in its path. If the next blade in the rotating rotor arrives at the wake when the air is still turbulent, it is not be able to extract power from the wind efficiently, and is also subjected to high vibration stresses. If the rotor rotated slower, the air hitting each rotor blade would no longer be turbulent. This is another reason for the tip speed ratio to be selected so that the rotor blades do not pass through turbulent air. Wind power conversion is analogous to other methods of energy conversion such as hydro-electric generators and heat engines. In a heat engine, the heat energy cannot be extracted from a totally insulated reservoir. Only when it is allowed to flow from the high temperature reservoir, to a low temperature one where it is rejected to the environment; can a fraction of this energy be extracted by a heat engine. Totally blocking a wind stream does not allow any energy extraction. Only by allowing the wind stream to flow from a high speed to a low speed region can energy be extracted by a wind turbine. In thermodynamics, the ideal heat cycle efficiency is expressed by the Carnot cycle efficiency. In a wind stream, the ideal aerodynamic cycle efficiency is expressed by the Betz limit. The inefficiencies and losses encountered in the operation of wind turbines include the blade number losses, turbine wake losses, end losses, the airfoil profile losses and tower shadow losses.

Horizontal axis wind turbines always have a tower supporting the rotor, transmission and generator. Regardless the type or shape, the towers are obstacles that affect significantly the airflow around them. Although some down-wind designs were developed in the past to simplify the yaw mechanism, most of the modern wind turbines have the rotor upstream of the tower, the streamlines deviate just in front of the tower of an up-wind turbine. Towers are obstacles that affect appreciable the airflow, and the streamlines are deviating just in front of the tower of an up-wind turbine. Although some down-wind designs were developed in the past to simplify the yaw mechanism, modern wind turbines have the rotor upstream of the tower. The airflow takes a lateral speed whereas its axial speed decreases. This effect is called tower shadow. Draft and lift are primarily affected by down-stream wind speed. In general, tower shadow is the most important deterministic load. It is worthy to mention that the 'shadow' caused by other obstacles, such as plants, buildings or other wind turbines, has similar effects and can be treated likewise, though the quantification may result much more involved. Tower may have important effects on the wind turbine power output. For example, the tower shadow effect is caused by the periodical passing by the wind turbine over the wind turbine tower, which is giving a mechanical torque drop, transferred to the generator shaft resulting in voltage drop. To account for this effect a periodic torque pulse, equal to the number of blades times the rotor frequency is added to the aerodynamic model output torque. The torque magnitude depends on the type of wind turbine. The tower shadow effect is more significant tot the in the down-wind turbines, being smaller in the case of upwind turbines. This torque pulse magnitude is about 10% of the rated wind turbine torque.

### 6.2.6  Operating Regions, Energy Production, and Capacity Factor

The power curve for a wind turbine indicates the net electrical energy output from a wind turbine as a function of the wind velocity at hub height. The power curve is determined either by theoretical calculations or by field tests, carried out according to international guidelines and recommendations such as those produced by the International Energy Agency or International Electrotechnic Commission [IEC 61400]. The operating region of a variable-speed variable-pitch wind turbine can be illustrated by their power curve, which gives the estimated power output as function of wind speed as shown in Figure 6.8. Three distinct wind speed points can be noticed in this power curve:

- Cut-in wind speed: The lowest wind speed at which wind turbine starts to generate power.
- Rated wind speed: Wind speed at which the wind turbine generates the rated power, which is usually the maximum power wind turbine can produce.
- Cut-out wind speed: Wind speed at which the turbine ceases power generation and is shut down (with automatic brakes and/or blade pitching) to protect the turbine from mechanical damage.
- Rated power is the nominal maximum continuous power output of the wind turbine.

Once the wind regime is determined and analyzed, we can estimate the average power generated by a wind turbine. The average power can be calculated by:

$$P_{avrg} = \int_{0}^{\infty} P_e(u) \cdot f(u) \, du \tag{6.41}$$

Here, $f(u)$ is the probability density function of the wind speeds, such as Weibull distribution function. Figure 6.8 shows a typical wind turbine power curve and it can be

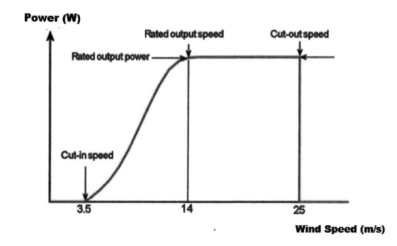

**FIGURE 6.8**
Typical power curve of a wind turbine.

separated into four regions. Wind turbine starts to produce energy in second region, when the wind speed is higher than the cut-in speed. In third region turbine works its rated power for all models and types of wind turbine, so energy output is similar for all models. In forth region, wind turbines stop to prevent damages because of high wind speeds. Selected wind turbines second part power curve models are given in following equations. There are several relationships used to estimate the electrical power output of the wind turbine for the region of power curve between the cut-in and rated wind speed. They include polynomials, exponential, splines and other curve fittings, such as:

$$P_{o,1} = P_R \left( \frac{v - v_{in}}{v_R - v_{in}} \right), \qquad v_{in} \le v_R \tag{6.42a}$$

$$P_{o,2} = P_R \left( \frac{v^2 - v_{in}^2}{v_R^2 - v_{in}^2} \right), \qquad v_{in} \le v_R \tag{6.42b}$$

$$P_{o,3} = P_R \left( \frac{v^3 - v_{in}^3}{v_R^3 - v_{in}^3} \right), \qquad v_{in} \le v_R \tag{6.42c}$$

$$P_{o,4} = P_R \left( a_{k-1} v^{k-1} + a_{k-2} v^{k-2} + \cdots + a_0 \right), \qquad v_{in} \le v_R \tag{6.42d}$$

where, $a_{k-1}$ ($k$ = 3, 4, 5 the polynomial index) are the regression constants, $P_R$ is wind-turbine-rated power, $v_{in}$ and $v_R$, are cut-in and rated wind speeds, respectively. The electrical power output of the wind turbine over all the possible ranges (the whole power curve) is then computed as:

$$P_{out} = \begin{cases} 0, & \text{if } v < v_{in} \\ P_{o,i}, \ i = 1,2,3,4 & \text{if } v_{in} \le v \le v_R \\ P_R & \text{if } v_R \le v \le v_{ct} \\ 0, & \text{if } v > v_{ct} \end{cases} \tag{6.43}$$

Here $P_{o,i}$ is power output of the wind turbine for the region of the power curve between the cut-in and rated speed, computed with one of the (6.42) relationships or the similar ones, and $v_{ct}$ is the cut-out wind speed. *Capacity factor (CF)* is one element in measuring the productivity of a wind turbine or any other power production facility and to compare different locations with each other, comparing the plant's actual production over a given period of time, e.g., one year, with the amount of power the plant would have produced if it had run at full capacity for the same amount of time. The CF of wind energy conversion system is the ratio of average delivered power to its theoretical maximum power. It can be computed for a single turbine, a wind farm consisting of dozens of turbines or an entire country consisting of hundreds of farms. Although geographical location determines in great part the capacity factor of a wind farm, it is also a matter of turbine design. Capacity factors for wind turbine power curve model can be obtained as:

$$CF = \frac{1}{P_R} \int_{v_{in}}^{v_R} P_{o,i} \cdot f(v)dv + \int_{v_R}^{v_{ct}} f(v)dv \tag{6.44}$$

The average output wind turbine power, by using capacity factor ($CF$) can be expressed as:

$$P_{out,avrge} = P_R \times (CF) \tag{6.45}$$

The yearly energy production for a wind turbine is given by:

$$W_{annual} = (CF) \cdot P_R (8760) \tag{6.46}$$

**Example 6.8:** What is the capacity factor of a 2 MW wind turbine that are generating, due to the wind conditions only 4200 MWh?

**Solution:** In one year (8,760 h), a 2 MW wind turbine can theoretically generate energy:

$$E_{year} = 8760 \times 2 \times 10^6 = 17,520 \text{ MWh}$$

The capacity factor, from Equation (6.43) is:

$$CF = \frac{4200}{17,520} = 0.239 \text{ or } 23.9\%$$

Annual energy production is the most important factor for wind turbines and critical parameter in any wind development project. However, the annual energy production is combined with cost and revenue analysis (economics 0 to determine feasibility for installation of wind turbines and wind farms. Approximate annual energy can be estimated by the following methods, involving the wind turbine-generator size (the rated power), rotor area and wind map (or if known the wind velocity probability distributions), and the manufacturer's curve of energy versus wind speed. The annual energy production is then estimated from Equation (6.46), if the rated power is used, while in the case of wind map is:

$$W_{annual} = CF \times A_{Rotor} \times WM \times 8760 \tag{6.47}$$

Here $WM$ represents the estimated wind power density per area (W/m²) from the wind maps of the region or country. The effects of the wind regime and the rated power for the rated wind speed distribution can be estimated by changing the capacity factor. The capacity factor is estimated from energy production over a selected time period, usually one year. In general, how we mentioned the capacity factors are quoted on an annual basis, although in some cases the capacity factors are calculated for shorter time intervals or periods. Capacity factors can also be calculated for wind farms, and they should be close to the same values as capacity factors calculated for individual wind turbines. However, if the wind farm is composed of different wind turbines, this must be taken into account and proper adjustments must be made. Notice that capacity factor is like an average efficiency.

**Example 6.9:** Estimate the annual energy production of a wind turbine having the rotor diameter 72 m, and a capacity factor 0.265, if the wind turbine is installed in an area with WM equal to 250 W/m².

**Solution:** From Equation (6.47) the wind turbine annual energy production is:

$$W_{annual} = CF \times \left( \frac{1}{4} \pi D^2 \right) \times WM \times 8760 = 0.265 \times 4069.44 \times 250 \times 8760 = 2361.7 \text{ MWh}$$

In some EU countries, for example Germany, another term, *load duration,* is used to indicate the capacity factor, being the product of the total number of hours in one year and the capacity factor. The capacity factor is usually much lower than 1, so the load duration also comes out to be much less than 8760 hours. For example, if the capacity factor is 20%, the load duration would be 8760 times 0.2, which is 1752 h. The significance of load duration is that it expresses that number of hours for which the wind turbine can be considered to be virtually operating at its rated capacity in one year. It is also important to distinguish between plant availability factor and plant capacity factor. The *plant availability* term refers to a fraction of one complete year for which the plant is available for use, irrespective of the availability of wind. For example, if a plant is under maintenance for 300 h in one year out of a total of 8760 h of one year, the availability factor would then be: (8760–300)/8760, i.e, 0.9658 or 96.58%. *Availability* is the time that the wind turbine is in operational mode, and it does not depend on whether the wind is blowing. Availability is related to reliability of the wind turbine, affected by both the quality of the turbine and operation and maintenance. Experimental values of the wind turbine availability in the past were poor for first production models; however, availabilities of 98% are now commonly reported for modern wind turbines, which have a good program of ongoing maintenance. The issues and problems regarding the wind turbine performances discussed above can be summarized and related to the design of a wind energy converter through the following principles: A high airfoil (blade) form ratio, lift to drag coefficient leads a high tip to speed ratio and therefore a large power coefficient $C_p$, modern wind energy converters with a good and proper aerodynamic profile rotate quickly. On the other hand, simple profiles with smaller profile form ratios have a small tip speed ratio. Therefore, the area of the rotor radius that is occupied by blades must be increased in order to increase the power coefficient. Slow rotating wind turbines have poor aerodynamic profiles and a high number of blades. The profile form ratio and the tip speed ratio have a considerably greater influence on the power coefficient than the number of blades. For modern converters with a good aerodynamic profile, the number of blades is not so important for higher power coefficients.

## 6.3 Wind Turbine Components

The main parts of the systems that comprise these wind turbines are the tower, the rotor control subsystem(s), the rotor blades, the gearbox, the electric generator and the power electronics unit(s). In order to increase the wind energy captured, the wind turbines are placed on top of a *tower* at a significant height above the ground. The height of the tower depends on the diameter of the blade and is of the order of magnitude of the blade diameter, D, allowing a clearance of D/2, between the ground and the lower part of the blade. Thus, tower heights are commonly between 30 and 120 m high. The tower is a simple structural element, usually made of reinforced concrete, which is designed to withstand the axial force and resulting moment generated by the wind turbine. It is typically thicker at the lower part and is usually designed as a hollow structure to allow easy access to the top for engine repairs at the turbine hub. Because the wind turbine must rotate to face the instantaneous direction of the wind, the entire electricity producing system is pivoted on strong bearings that allow the rotation of the system around a vertical axis, through the *yaw bearings and yaw's break*. The drag force on a downstream rotating vane or a simple rudder provides the force for this rotation. In order to avoid overshooting in the rotation of the electricity generating system and unnecessary power fluctuations, the yaw brake system slows the rotational motion by providing damping. *Rotor blades* are the most important part of the generating system, where the wind energy is imparted to the engine. They are very long, typically between 30 and 100 m in diameter. The rotor blades are designed aerodynamically with pitch angles that vary with the distance from the hub and they are made of low weight and strong materials. Low density woven composites are now typically used for the turbine blades, which are typically hollow. The blades are connected to the hub, which extends to a horizontal metal shaft that becomes the prime mover of the engine. The shaft is supported by a series of bearings. In the more advanced and better optimized engines, a mechanism is put in place that changes the pitch of the blades to produce maximum power at the instantaneous wind velocity. These mechanisms are made of sensors and actuators, which measure the magnitude of the instantaneous wind velocity, adjust the position of the base of the blades inside the hub and, thus change the pitch of the entire blade. The actuator mechanisms are attached to the blades, rotate with them and are supported by their own pitch-control bearings. In order to minimize the centrifugal stresses, the rotational speed of the blades at operating conditions is fairly low, typically of the order of 100 RPM, so a *gearbox* steps up the rotational speed of the prime mover to reach a range 2000–3000 RPM and transmits the power to a secondary high rpm shaft, which is connected to the generator. A small fraction of the blade power is dissipated in the gear box by friction. For this reason, larger wind power systems may require a cooling system for their gearbox.

The mechanical power from the wind turbine rotor is transferred to the electric generator through the drivetrain or powertrain, through the main shaft, the gearbox and the high-speed shaft. It is not appropriate to just drive the electric generator directly with the power from the main shaft unless the wind turbine is equipped with a multi-pole generator and power conditioning unit. If an ordinary generator with two or four poles (which makes one pair of poles) is used without a gearbox, and the machine was directly connected to a 50 or 60 Hz AC three-phase electric grid, the wind turbine rotor must be turned at about 50 Hz or about 3000 RPM. For a 50 m or a 60 m rotor diameter, this speed of rotation implies a tip speed of the rotor of far more than twice the speed of sound, a condition which is not supported by the wind turbine mechanical strength capabilities. With a gearbox, slowly rotating, high torque power obtained from the wind turbine rotor can be converted into high speed power, which would be required for the electric generator. The gearbox of a wind turbine usually does not "change gears," normally having a single gear ratio between the rotation of the rotor and the generator. In some wind energy conversion systems, the gear ratio ranges from 30 to 200. One typical issue for a wind turbine gearbox is cooling and heating of the lubricant. In cold climates, during winters, the lubricating oil can freeze due to low temperatures, so special heating settings are needed to keep the oil sufficiently warm. The situation in hot climates is reverse, arrangements are needed to prevent the lubricating oil from becoming too hot, when the oil viscosity decreases with the temperature increases and therefore at high temperature the oil does not serves its purpose of lubricating very well. Recent developments in the field of gearboxes for modern wind power converters lead to the hydrodynamic gearbox designs, a combination of a conventional mechanical gearbox and a torque changer. There is a variable speed drive inside and a constant drive outside of the system. The main advantage of this type is continuously variable input but constant output speed adjustment over a wide range.

A wind turbine generator converts the wind mechanical energy to electrical energy. Wind turbines are like inverse fans, so the generator is essentially a motor that is running backward. In fact, with the right input generators can often be made to run like an electric motor. Both permanent magnet generators and conventional electric synchronous and asynchronous generators are used for the conversion of the turbine mechanical power to electricity. The main concern with generators in wind turbines is that they must produce electricity compatible with that in the electrical grid at any given site. In U.S., the grid has a 60 Hz frequency, while in most parts of Europe the grid runs at 50 Hz. The electricity must have the same characteristics (voltage and frequency), be of sufficient quality, and be connected in such a way as to not interrupt the existing current flow. Wind turbines have one of two types of generators: synchronous or the more common asynchronous (induction) generators. Usually, current flowing through solenoids coiled around iron cores provides the magnetic field. Only a few generators use permanent magnets, since the required size makes them prohibitively expensive, bigger in size and over time they demagnetize.

The *nacelle* is sitting on top of the tower and contains the wind turbine electrical components, the gearbox, the brake, the wind speed and director monitoring unit, the yaw mechanism, part of the power condition unit and the electric generator.

*Rotor blades* are one of the major wind turbine components. The diameter of the blades is a crucial and critical element in the generated wind turbine power. The longer are the blades, the higher is the turbine output. However, the design of the blades and the materials incorporated by them are also key design elements, by determining the overall manufacturing, installation and maintenance costs. Blades are often made of fiberglass reinforced with polyester or wood epoxy. Vacuum resin infusion is a new material connected to a technology presented by manufacturers. Typically blades rotate at 10–30 RPM, either at a constant speed (the more traditional solution) or at a variable speed, for the new and more efficient design solutions.

*Gearboxes* and *direct drives* are other important wind turbine components. The mechanical connection between an electrical generator and the turbine rotor can be direct or through a gearbox. Most wind turbines use gearboxes, whose function is to increase the rotational speed required by generators. Some new technologies are exploring direct drives generators to dispense with the expensive gears. In fact, the gearbox allows the matching of the generator speed to that of the turbine. The use of gearbox is dependent on the kind of electrical generator used in WECS. However, disadvantages of using a gearbox are reductions in the efficiency and, in some cases, reliability of the system.

The *brake system* is a critical part in the turbine operation and lifetime. A disk is used to stop the rotor blades in emergencies and to ensure the safety of the turbine in case of very high damaging winds or other exceptional situations.

*Controller* and *power conditioning unit* are other important parts of the wind turbine designed to generate electricity. They consist of a set of electrical components, circuits and devices designed to control the starting, the stopping, and the turbine rotor blade speed in order ensure the safety turbine operation and desired quality of the power generated. Typically, in the constant wind speed model the controller starts up the turbine at wind speeds around 8 to 14 mph and stops the machine at around 55 mph (or in the rage from about 3.5 to 25 m/s), to avoid the damage caused by turbulent high winds. The power electronic (PE) converter has an important role in modern WECS with the variable-speed control method. The constant-speed systems hardly include a PE converter, except for compensation of reactive power. The important challenges for the PE converter and its control strategy in a variable-speed WECS are:

1. Attain maximum power transfer from the wind, as the wind speed varies, by controlling the turbine rotor speed, and
2. Change the resulting variable-frequency and variable-magnitude AC output from the electrical generator into a constant-frequency and constant-magnitude supply which can be fed into an electrical grid.

*Generators* are the devices responsible for the production of 60-cycle alternating current (AC) electricity. The function of an electrical generator is providing a means for energy conversion between the mechanical torque from the wind rotor turbine, as the prime mover, and the local load or the electric grid. Different types of generators are being used with wind turbines. Small wind turbines are equipped with DC generators of up to a few kilowatts in capacity. Modern wind turbine systems use three-phase AC generators. The common types of AC generator that are possible candidates in modern wind turbine systems are as follows:

1. Squirrel-cage rotor induction generator (SCIG),
2. Wound-rotor induction generator (WRIG),
3. Doubly-fed induction generator (DFIG),
4. Synchronous generator (with external field excitation), and
5. Permanent magnet synchronous generator (PMSG).

For assessing the type of generator in WECS, criteria such as operational characteristics, weight of active materials, price, maintenance aspects and the appropriate type of power electronic converter, are used. Historically, the induction generator (IG) has been extensively used in commercial wind turbine units. Asynchronous operation of induction generators is considered an advantage for application in wind turbine systems, because it provides some degree of flexibility when the wind speed is fluctuating. There are two main types of induction machines: squirrel-cage (SC), and wound-rotor (WR). Another category of induction generator is the DFIG; the DFIG may be based on the squirrel-cage or wound-rotor induction generator. The induction generator based on SCIG is a very popular machine because of its lower cost, mechanical simplicity, robust structure, and resistance against disturbance and vibration. The wound-rotor induction generator is suitable for speed control purposes. By changing the rotor resistance, the output of the generator can be controlled and also speed control of the generator is possible. Although the WRIG has the advantage described above, it is more expensive than a squirrel-cage rotor. The DFIG is a kind of induction machine in which both the stator windings and the rotor windings are connected to the source. The rotating winding is connected to the stationary supply circuits via power electronic converter. The advantage of connecting the converter to the rotor is that variable-speed operation of the turbine is possible with a smaller and therefore cheaper converter. The converter power rating is often about 1/3 the generator rating. Another type of generator that has been proposed for wind turbines in several research articles is a synchronous generator. This type of generator has the capability of direct connection (direct drive) to wind turbines, with no gearbox. This advantage is favorable with respect to lifetime and maintenance. Synchronous machines can use either electrically excited or permanent magnet (PM) rotor. The PM and electrically excited synchronous generators differ from the induction generator in that the magnetization is provided by a permanent magnet pole system or a dc supply on the rotor, featuring providing self-excitation property. Self-excitation allows operation at high power factors and high efficiencies for the PM synchronous. It is worth mentioning that induction generators are the most common type of generator use in modern wind turbine systems.

The *yaw mechanism* of wind power generators: In more typical wind turbines, the yaw mechanism is connected to sensors (e.g., anemometers) that monitor wind direction, turning the tower head and lining up the blades with the wind. *Tower* is another critical part of any wind turbine. A tower supports the nacelle and rotor. The electricity produced by the generator comes down cables inside the tower and passes through a transformer into the electricity network. *Base* or *foundation* is the part that is supporting the tower. Large turbines are built on a concrete base foundation. When a wind turbine ceases production, it is a simple task to dig these out or cover them, leaving little trace behind.

## 6.4 Operation and Control of Wind Energy Conversion Systems

Wind is a highly intermittent energy source for causing overall fluctuation in wind power generation. Electricity generated from wind turbines strongly depends on the local weather and geographic conditions that fluctuates a great deal more than with some renewable energy sources such as hydropower. The impacts of wind power to a power grid depend on the level of wind power penetration, grid size and generation mix of electricity in the grid. Undoubtedly, there is no problem for low wind power penetration in a large power grid. However, integrating large utility-scale wind power presents unique challenges, such as how to integrate large wind power into the grid or how to ensure system compatibility. In order to understand, the challenge that the operation and eventually integration of wind power poses are illustrated, by using, the balance power equation between the demand (industries and households that consume power) $P_D$, wind power $P_W$ generated by the wind energy conversion system or wind power plant, electrical losses, $P_L$, and the additional power, $P_G$, that is produced at another location. Power balance requires that there is always a power balance in this system, as expressed by:

$$P_G = P_D + P_L - P_W \tag{6.48}$$

Equation (6.48) is valid for any situation and for any system type, stand-alone or grid integrated, small or large size, it does not matter whether we look at a short time period (e.g., minutes) or a long period (e.g., a year). Equation (6.48) also implies that electricity cannot be stored within an electric power system, and no storage is included here. Hence any change in electricity demand (or wind power) must be simultaneously balanced by other generation sources within the power system.

### 6.4.1 Control Methods

With the evolution of WECS, due to the technology advancements in turbine aerodynamics, generators, control, power electronics during the last decades, many different control methods have been developed and employed. The control methods developed for WECS are usually divided into the following two major categories: (a) constant-speed methods, and (b) variable-speed methods. In constant-speed turbines, there is no control on the turbine shaft speed. Constant speed control is an easy and low-cost method, but variable speed brings the following advantages: maximum power tracking for harnessing the highest possible energy from the wind, lower mechanical stress, less variations in electrical power, and finally the reduced acoustical noise at lower wind speeds.

Theoretically, higher wind speeds enable us to obtain significant powers, but they are not very common. Wind turbines are thus not oversized for these speeds, because this would lead to additional costs. Turbines are generally designed to reach their nominal power at a wind speed of about 15 m/s and are automatically stopped at about 25 m/s or lower cut-off speeds lower power settings. The operation of a wind turbine involves starting the turbine from rest, regulating the power while the system is running, and stopping the turbine if and when the wind speed becomes excessive. Start-up of most turbines usually means running the generator as a motor to overcome initial resistive torque until sufficient power is generated at cut-in speed assuming that a power source is available. The anemometer and wind monitoring equipment are critical elements of wind turbines. They gauge the wind speed and direction and then send the information to the controller that in turn provides necessary data to critical elements of the system. The controller essentially directs the yaw motor to turn the rotor to face toward or away from the wind, depending on wind direction. The gear box, the heaviest element of the system, converts the slow rotation (revolutions per minute or RPM) of the low-speed rotor shaft to higher RPM of the high-speed shaft which is mechanically coupled to a generator that produces the electricity.

To limit the power produced at high wind speeds although the aerodynamic gross wind power keeps on increasing with the cube of wind speeds, there are two categories of power regulation systems: (a) the *stall* (the natural aerodynamic stall system), for which blades have a profile that is designed to naturally reduce the power coefficient $C_P$, when the wind speed increases, i.e., when the tip speed ratio $\lambda$ decreases, this is the stall control principle; and (b) the *pitch* or variable pitch angle system, with mobile blades around their longitudinal axis, enabling to reduce the lift and therefore the $C_P$ coefficient for significant wind speeds. In pitch-control system, an electronic system monitors the generator output power. If the power exceeds the rated power, the pitch angle of the rotor blades is actively adjusted by the machine control system, so the turbine blades is adjusted to shed some wind. The electronic system controls a hydraulic system to slowly rotate the blades about the axes and turn them a few degrees to reduce the wind power. In conclusion, this strategy is to reduce the blade's angle of attack when the wind speeds over the rated wind speed. The blade pitch control has the advantage that the blades have built-in braking which brings the blades to rest. A blade is at variable pitch angle or at variable step, if the orientation of the blade in relation to its initial position can be modified during operation. Blades are equipped with actuators (hydraulic jacks or electric motors), which are responsible for changing their orientation. Given the inertia of a blade and the importance of the forces exerted on it, orientation is slow. The value of the aerodynamic lift, which is generated on each blade, can thus be modified. This is influencing the power coefficient $C_P$ of the wind turbine. A zero-pitch angle corresponds to a blade facing the wind, while a negative value of $C_P$ means that the turbine is in brake operating mode (excessive rotational speed).

To extract the maximum power (operation at speeds lower than the speed, beyond which attempt to reduce the lift), and because of the highly variable nature of wind, the adjustment of the tips speed ratio requires to optimize the lift and to adapt the turbine speed, in order to maintain the energy conversion efficiency at its maximum. The regulation system of the pitch angle enables active control of the recovered power, which is mainly used for two tasks: power or rotor speed control and aerodynamic braking for stopping the turbine.

Thus, the power can be limited to the nominal power and guiding the blades in the direction of the wind for high speed winds and thus protect the machine. On the other hand, if the angle of attack is too large, the air flow no longer manages to follow the trajectory that has been imposed by the (highly sloping) profile, a smaller deflection from the trajectory and a less significant acceleration of the flow on the upper side of the blade. Thus, the decrease of pressure and lift are less significant. In order to exploit this phenomenon, blades have particular forms in order to progressively *uncouple* from the nominal wind speed. At high speeds, intrusive turbulences appear and naturally make the power drop by causing an efficiency loss. The *stall control* concept is that the power is regulated through stalling the blades after rated speed is achieved and does not require any external control system. It is thus less expensive and more reliable than the variable pitch angle control, since have only a small number of elements in motion, being extensively used in low-power wind turbines. However, it is more difficult to optimize the wind turbine operation, which is equipped with a stall control system, since we have one less control parameter (the pitch angle) in comparison to the variable pitch angle system. Passive stall control is basically used in wind turbines in which the blades are bolted to the hub at a fixed installing angle. In a passive stall-regulated wind turbine, the power regulation relies on the aerodynamic features of blades. In low and moderate wind speeds, the turbine operates near maximum efficiency. At high wind speeds, the turbine is automatically controlled by means of stalled blades to limit the rotational speed and power output, protecting the turbine from excessive wind speeds.

For the *stall-controlled* machines, the turbine blades can reduce the efficiency automatically when the winds exceed the rated speed. In this control method, there are no moving parts, so this way is a kind of passive control. Most of the modern, large wind turbines use this passive, stall-controlled approach.

For large (above 1.0 MW), when the wind speed exceeds the rated wind speed, the turbine machine will not reduce the angle of attack but increases it to induce stall. The *active stall control* technique has been developed for large wind turbines. An active stall wind turbine has stalling blades together with a blade pitch system. Since the blades at high wind speeds are turned toward stall, in the opposite direction as with pitch-control systems, this control method is also referred to as negative pitch control. Compared with passive stall control, active control provides more accurate control on the power output and maintains the rated power at high wind speeds. However, with the addition of the pitch-control mechanism, the active stall control mode increases the turbine cost and decreases operation reliability. In order to maximize the wind turbine power output and minimize the asymmetric loads acting on the rotor blades and the tower, a horizontal axis wind turbine must be oriented with rotor against the wind by using an active yaw control system. Like wind pitch control systems, *yaw control systems* are driven electrically or hydraulically, and the hydraulic yaw systems were used in the earlier time of the wind energy. In modern wind turbines, yaw control is done by electric motors. The yaw control system usually consists of an electrical motor with a speed reducing gearbox, a bull gear which is fixed to the tower, a wind vane to gain the information about wind direction, a yaw deck and a brake to lock the turbine securely in yaw when the required position is reached. For a large wind turbine with high driving loads, the yaw control system may use two or more yaw motors to work together for driving a heavy nacelle. In practice, the yaw error signals obtained from the wind vane are used to calculate the average yaw angle in a short interval. When this average yaw angle exceeds the preset threshold, the yaw motor is activated to align the wind turbine with the wind direction however the actions of yaw control are rather limited and slow.

In order to capture the highest wind power, a turbine must operate with a rotational speed such that the power coefficient is at its maximum (or near its maximum). However, for wind turbines to always harness the maximum possible wind power must operate at an angular speed that is adjustable based on the wind speed. In practice, however, this is very difficult to manage because of the way electric generators operate. For example, a synchronous generator must operate at a constant rotational speed and an induction generator can only slightly deviate from synchronous speed. In addition, continuously changing the rotational speed of a turbine not only requires additional control devices, which is costly, it introduces technical difficulties such as torque and shaft pulsations, extra shocks and vibrations. Consequently, a better solution is based on the fact that the wind turbine aerodynamics states that the rotor performance is controlled mainly by the blade pitch angles. By adding such capabilities to adjust (control) the blade pitch angles in a wind turbine implies additional cost and system complexity. However, the benefits introduced by the variable pitch angle control compensate for and justify the extra turbine and maintenance costs. Today most of the wind turbines are equipped with devices or subsystems able to rotate each rotor blade about its longitudinal axis and modifying the blade pitch angle, being the so-called variable pitch angle turbines. Usually, all the blades in a wind turbine with a variable pitch angle control are rotated simultaneously and by the same magnitude (in future such designs this may change based of newer technologies). For each pitch angle, the performance of the rotor alters, and a different characteristic curve represents its aerodynamic characteristics. The effect of changing the blade pitch angles is to decrease or increase the wind turbine power coefficient, so to modify the power capture capability of the turbine. In a wind turbine with variable pitch angle, all the blades are rotated simultaneously and by the same amount. Another application of changing the blade pitch angle is to limit the power capture by a wind turbine, when such things are required. The *blade pitch control* mechanism rotates the blades about their longitudinal axis with respect to the hub. Such mechanism, usually must operate from inside the wind turbine hub, is done by employing either an electric or a hydraulic actuator.

Notice that pitching the whole blade requires large actuators and bearings, increasing the weight and expense of the wind energy conversion system. Alternative solution to this problem is to use partial span blade pitch control where only the outer one third of the blade span is pitched. There are two main power output control methods from the rotor blades. On the other hand, stall regulation, sometimes described as passive control, since it has the inherent aerodynamic properties of the blade, which determine power output with no moving parts to adjust. The twist and thickness of the rotor blade vary along its length in such a way that turbulence occurs behind the blade whenever the wind speed becomes too strong. By pitching the rotor blades around their longitudinal axis, the relative wind conditions, and subsequently, the aerodynamic forces are affected in a way so that the power output of the rotor remains constant after rated power is reached. Notice that the pitching system in medium and large grid-connected wind turbines is usually based on a hydraulic system, controlled by a computer. The rotor thrust on the tower and foundation is much lower for pitch-controlled turbines than for stall-regulated wind turbines, allowing for the reductions in material and weight. Pitch-controlled turbines achieve a better yield at low-wind sites than stall-controlled turbines, as the rotor blades can constantly be kept at optimum angle even at low wind speeds. Stall-controlled turbines have to be shut down once a certain wind speed is reached, whereas pitch-controlled turbines can gradually change to a spinning mode as the rotor operates in a no-load mode, i.e., it idles at the maximum pitch angle. Another advantage of stall-regulated turbines consists in that, in strong winds, when the stall effect becomes effective, the wind oscillations are converted into power oscillations that are smaller than those of pitch-controlled turbines in a corresponding regulated mode. Pitch control has the potential for producing the highest level of interaction because of the diesel and wind turbine control loops presence. The pitch control system consists of a power measurement transducer, a manual power set point control, an adaptive feedback, and a hydraulic actuator, which varies the pitch of the blades. Turbine blade pitch control has a significant impact on the dynamic behavior of the system. This type of control only exists in horizontal axis machines. Variable pitch turbines operate efficiently over a wider range of wind speeds than fixed pitch machines. However, cost and complexity are higher. In addition to the conventional stall control, there is also a more marginal intermediary system, the so-called *active stall*, which consists of adjusting the stall effect by a very small variation of the pitch angle. This thus makes the mechanical rotation device of the blades lighter and more robust. For small size wind turbines, there are a variety of techniques to spill wind. The common way is the passive yaw control that can cause the axis of the turbine to move more and more off the wind. Another way relies on a wind vane mounted parallel to the plane of the blades. As winds get stronger, the wind pressure on the vane is rotating the machine away from the wind.

## 6.4.2 Electrical Aspects of Wind Energy Conversion Systems

The modern wind turbine is a sophisticated system with an aerodynamically designed rotor, powertrain, control unit, and power electronics interface and an efficient power generation, transmission and regulation components. Size of such wind turbines ranges from a few watts to several megawatts, the current trends for larger systems. The wind turbines may be grouped into wind farms, feeding power to utility, with its own transformers, transmission lines and substations. Standalone wind energy conversion systems (WECS) catering the needs of smaller communities or individual users are also quite common. As wind is an intermittent and fluctuating phenomenon, hybrid systems with energy storage units are also popular, especially in remote areas. For efficient and reliable performance of a WECS, all its components are to be carefully designed, crafted and integrated. Wind energy conversion systems can be divided into two brad types: (1) those which depend on aerodynamic drag and (2) those which depend on aerodynamic lift. The early vertical axis wind wheels utilized the drag principle. Savonius rotors are the most common of the last types of wind wheels. Drag devices, however, have a very low power coefficient, with a $CP_{max}$ of around 0.16. Modern wind turbines are predominately based on the aerodynamic lift. Lift devices use airfoils (blades) that interact with the incoming wind. The force resulting from the airfoils body intercepting the air flow does not consist only of a drag force component in the direction of the flow but also of a force component that is perpendicular to the drag: the lift forces. The lift force is a multiple of the drag force and therefore the relevant driving power of the rotor. By definition, it is perpendicular to the direction of the air flow that is intercepted by the rotor blade, and via the leverage of the rotor, it causes the necessary driving torque. Wind turbines using the aerodynamic lift can be further divided according to the orientation of the spin axis into horizontal axis and vertical axis type turbines, the so-called Darrieus wind turbines. The Darrieus wind turbine offers an advantage over the horizontal axis wind turbine because of the structural simplicity due to the independence with respect to the wind direction. This feature makes the control system unnecessary to direct the rotor.

The horizontal axis or propeller type approach dominates the wind turbine applications. This consists of a tower and a nacelle that is mounted on the top of a tower. The nacelle contains the generator, gearbox and the rotor. Different mechanisms exist to point the nacelle toward the wind direction or to move the nacelle out of the wind in case of strong wind speeds. On small turbines, the rotor and the nacelle are oriented into the wind with a tail vane. On large turbines, the nacelle with rotor is electrically yawed into or out of the wind, in response to a signal from a wind vane. Horizontal axis wind turbines typically use different blade numbers, depending on the purpose of the wind turbine. Two or three bladed turbines are usually used for electricity power generation. Turbines with twenty or more blades are used for mechanical water pumping. However, electricity generation is the most important application of wind energy today. The major components of a wind energy conversion system are tower, rotor, high-speed and low-speed shafts, gear box, electric generator, sensors and yaw drive, power regulation and control system, and safety systems. The following parameters are used to specify a wind turbine: *hub height* (the height of the hub above the ground), *swept area* (the area defined by the rotating rotor disk), *solidity* (the ratio of the area of the blades to the swept area), *tip speed ratio* (the ratio of the speed of the blade tip to the wind speed), and *rated power* (maximum continuous power output at the electrical connection point).

Tower supports the rotor and nacelle of a WECS at desired heights. The main tower types used in modern wind turbines are lattice, tubular steel and guyed towers. However, most of the recent WECS installations are using tubular towers. The rotor is the most important component of a wind turbine. The size of the rotor depends on the turbine rated power. Components of a wind turbine rotor are blades, hub, shaft, bearings and other internals. The wind turbine rotor may be mounted on the tower upwind side or down-wind side. Blades have different airfoil sections and shapes and are made from materials ranging from wood to composite materials. The wind turbine rotors may have a different number of blades, from one to several blades in the case of small size turbines. Currently, three-blade wind turbines dominate the market for the horizontal axis wind turbines. Two-bladed wind turbines have the advantage that the tower top weight is lighter and, therefore, the whole supporting structure can be built lighter at lower costs. Three-bladed rotors have the advantage that the moment of inertia is easier to understand and, therefore, often better to handle than the rotor moment of inertia of a two-bladed turbine. The mechanical power generated by the rotor blades is transmitted to the generator by a transmitted system located in the nacelle. This consists of a gearbox, sometimes a clutch, and a braking system to bring the rotor to rest in an emergency when not in operation.

The rotational motion of the wind turbine rotor is transmitted to the electrical generator by means of a drive train (mechanical transmission system). Its structure and components strongly depend on each particular WECS technology. For example, wind turbines employing synchronous generators may use direct drive transmission, in which the rotor and generator are coupled on the same shaft. However, most of the wind energy conversion systems use gearbox for mechanical power transmission. Therefore, the electrical generator is experiencing an increased rotational speed and a reduced electromagnetic torque. The mechanical connection between an electrical generator and the turbine rotor may be direct or through a gearbox, which is an important component of drive (power) train of the wind turbines. In fact, the gearbox allows the matching of the generator speed to that of the turbine. However, the use of gearbox is dependent on the kind of electrical generator used in WECS. The disadvantages of using a gearbox are reductions in the efficiency and, in some cases, reliability of the system, and an increased overall wind energy conversion system cost, as well as the operation and maintenance costs. Speed of a typical rotor may be 30–50 RPM whereas the optimum speed of rotation for a generator may be between 1000 and 1500 RPM. An ideal gearbox should be designed to work smoothly and quietly even under adverse climatic and loading conditions throughout the life span of the wind turbine. Bearings of different points of the gearbox are selected based on the nature of the loads to be transmitted. Gears are designed on the basis of duration and distribution of loads on individual gear teeth.

A modern large electric WECS may generate up to 2.5 MW. The cost, weight and maintenance needs of mechanical gearing between the wind turbine and the electrical generator pose a serious limitation to the further increase in WECS power ratings. Wind power plants are, generally, used to convert wind energy into electrical energy. They consist of a wind turbine, an electrical generator and a control system. Generator is one of the most important components of a wind energy conversion system. The electrical system of the wind turbine includes all components for converting mechanical energy into electrical power. Conceptual changes will incorporate new generators, variable-speed operation, and more flexible wind turbines. A promising development is the direct drive generator, which makes it possible to exclude the gearbox. Furthermore, the direct drive generator could be combined with variable-speed operation, which will increase energy capture and reduce loads, i.e., an improvement in rotor efficiency by allowing the rotor speed to vary with wind speed. Wind turbines with variable-speed operation will be increasingly competitive as the cost of solid-state electronic decreases. Permanent magnet excitation is favored for developing new designs because of higher efficiency, high-power densities, availability of high-energy permanent magnet material at reasonable costs, and the possibility of rather smaller turbine diameter. Other advantages include the absence of brushes, slip rings, excitation windings, and excitation losses. For DC power generation, the lack of excitation control is not a limitation for terminal voltage and power control, since a diode rectifier and a DC–DC converter system, with various control strategies, permit load voltage and/or load power control.

Electrical systems must function within close limits on operating parameters such as voltage, frequency and harmonic content, which are also important in the operation of wind energy conversion systems. In contrast with the generators used in other conventional energy systems, a generator of a wind turbine has to work under fluctuating power levels, due to the intrinsic nature of the wind energy. Different types of generators are used with wind turbines. Smaller wind turbines are in general equipped with DC generators of a capacity ranging from few W to kW, and larger systems use AC electric generators. Synchronous and induction generators are widely used in wind energy systems. Each type of these machines has its own advantages and disadvantages and its own control method, whether mechanical or electrical, to obtain constant magnitude and frequency voltage, in order to be connected to the grid. All grid-connected wind energy conversion systems drive three-phase AC generators. Transformers, an integrated part in the power system, are used in wind energy conversion systems to interlink the turbine generator to the utility grid. Several techniques and methods for the grid-related problems and integration of wind energy systems have been intensively studied and analyzed over the years. Various factors affecting wind power, siting of the electric generator and selection of equipment and grid connection methods and problems associated with grid inter-connections had been discussed in the literature.

The function of an electrical generator is providing a means for energy conversion between the wind turbine mechanical torque (the prime mover), and the local load or the electric grid. Different generator types are used with wind turbines. Small wind turbines are usually equipped with DC generators of up to a few kilowatts or tens of kilowatts in capacity. Modern wind turbine systems use three-phase AC generators. For assessing the type of generator in WECS, the most suitable for a specific application, criteria and such as operational characteristics, weight of active materials, price, maintenance aspects and the appropriate type of power electronic converter, are used. Most wind turbine manufacturers use six-pole induction (asynchronous) generators, while others may use directly driven synchronous generators, often of a permanent magnet type. In the power industry, in general, induction generators are not very common for power production. The power generation industry uses almost exclusively large synchronous generators, as these generators have the advantage of a variable reactive power generation, i.e., voltage control. Synchronous generators with 500 kW to 2 MW are significantly more expensive than induction generators with a similar size. In addition, direct grid-connected synchronous generators have the disadvantage that the rotational speed is fixed by the grid frequency and the number of pairs of generator poles, hence, the power output fluctuations, e.g., due to the wind gusts or to the wind fluctuations, lead to a high torque on the drivetrain as well as high power output fluctuations, if other means, e.g., softer towers, are not used to reduce the impact of gusts. Therefore, directly grid-connected synchronous generators are usually not used for grid-connected wind turbines. They are applied in stand-alone systems sometimes, where the synchronous generator can be used for reactive power control in the isolated network. An option for the utilization of synchronous generators for wind turbines is the decoupling of the electric connection between the generator and the grid through an intermediate circuit. This intermediate circuit is connected to a three-phase inverter that feeds the grid with its given voltage and frequency. Induction generators have a slightly softer connection to the network frequency than synchronous generators, due to a changing slip speed. This softer connection slightly reduces the torque between rotor and generator during gusts. However, this almost fixed-speed operation still leads to the problem that overall efficiency during low wind speeds is very low.

The controller is provided with display unit for the display of instantaneous position of generation, different temperature of generator, rotor rpm and generator rpm, different grid parameters active and reactive consumption, etc. It also has data storage capacity and memory to keep records of different faults. A brief review of these control systems has been presented here. However, interested readers are strongly advised to read the reach literature to learn more about on the wind turbine control methods and techniques included at the end of this chapter or elsewhere in the literature. One of the main goals of wind turbine control is to increase power production and reduce loads with a minimum number of control inputs required for turbine measurement. Wind turbines reach the highest efficiency at the designed wind speed, which is usually between 10 m/s and 16 m/s. At this wind speed, the power output reaches the rated capacity. Above this wind speed the power output of the rotor must be limited to keep the power output close to the rated capacity and thereby reduce the driving forces on the individual rotor blade as well as the load on the whole wind turbine structure. If the wind speed reaches the cut-out wind speed (usually 25 m/s), the wind turbine shuts off and the entire rotor is turned out of the wind to protect the overall turbine structure. Because of this procedure possible energy that could have been harvested will be lost. The wind turbine operation involves starting the turbine from rest, regulating the power while the system is running, and stopping the turbine if and when the wind speed becomes excessive. Start-up of most turbines usually means running the generator as a motor to overcome initial resistive torque until sufficient power is generated at cut-in speed assuming that a power source is available. The angle of the rotor blades is actively adjusted by the machine control system. This, known as blade pitch control, has the advantage that the blades have built-in braking which brings the blades to rest. Pitching the whole blade requires large actuators and bearings, increasing the weight and expense of the system. One solution is to use partial span blade pitch control where only the outer one third of the blade span is pitched.

There are also different configurations for connecting wind energy conversion systems to the grid, depending on the wind turbine generator size, type, generator, or transformer if any is used, control and the auxiliary electrical components and equipment. The connection of the wind turbine to the grid depends on the type of electrical generator and power electronic converter used. Based on the application of PE converters in the WECS, the wind turbine configurations can be divided into three topologies: (a) directly connected to the grid without any PE converter, connected via full-scale the PE converter, and (b) connected via partially rated PE converter. In the following, the generator and power electronic converter configurations most commonly used in wind turbine systems are discussed. As a simple, robust and relatively low-cost system, a squirrel-cage induction generator (SCIG), as an asynchronous machine, is connected directly to the grid. For an induction generator, using a gearbox is necessary in order to interface the generator speed and turbine speed. The capacitor bank (for reactive power compensation) and soft starter (for smooth grid connection) are also required. The speed and power are limited aerodynamically by stall or pitch control. The variation of slip is in the range of 1%–2%, but there are some wind turbines based on SCIG in industry with increased rotor resistance and, therefore, increased slip (2%–3%). This scheme is used to allow a little bit of speeding up during wind gusts in order to reduce the mechanical stresses. However, this configuration based on an almost fixed speed is not proper for a wind turbine in a higher power range and also for locations with widely varying wind velocity.

The power electronic (PE) converter has an important role in modern WECS with the variable-speed control method and in the WECS grid integration. The constant-speed systems often are not including a PE converter, except for compensation of reactive power. The important challenges for the PE converter and its control strategy in a variable-speed WECS are: attain maximum power transfer from the wind, as the wind speed varies, by controlling the turbine rotor speed, and change the resulting variable-frequency and variable-magnitude AC output from the electrical generator into a constant-frequency and constant-magnitude supply which can be fed into an electrical grid. As a result of rapid developments in power electronics, semiconductor devices are gaining higher current and voltage ratings, less power losses, higher reliability, as well as lower prices per kVA. Therefore, PE converters are becoming more attractive in improving the performance of wind turbine generation systems. It is worth mentioning that the power passing through the PE converter (that determines the capacity the PE converter) is dependent on the configuration of WECS. In some applications, the whole power captured by a generator passes through the PE converter, while in other categories only a fraction of this power passes through the PE converter. The connection of the wind turbine to the grid depends on the type of electrical generator and power electronic converter used. Based on the application of PE converters in the WECS, the wind turbine configurations can be divided into three topologies: directly connected to the grid without any PE converter, connected via full-scale the PE converter, and connected via partially rated PE converter.

## 6.5 Setting and Design of Wind Power Plants

Wind turbines require unique installation specifications to operate with optimum efficiency. These strict installation requirements must not be characterized as major disadvantages. For example, wind turbine installers and designers carefully consider turbine sites and wind farm conditions. The optimum parameters for rotor blades must be selected, spacing between turbine units, height of the turbine and tower structures, and other conditions that will yield reliable, safe, and efficient operation with minimum maintenance and operation costs. The large-scale use of small wind systems depends primarily on economics. For wide-spread use, life-cycle costs will need to be comparable to costs from the utility. In some states there are credits and/or subsidies for purchase of small wind systems. Presently there is net energy billing for systems 50 kW and smaller in Texas, however this has not increased the use of small wind systems. The primary infrastructure requirement for wind power is electricity transmission from the windy areas to the load centers. Wind farms can provide rural economic development with the primary benefit being long-term stable income to the landowner. Representative economic values are for a 100 MW wind farm using capacity factors of 30% in wind class 3 and 35% in wind class 4. A 100 MW wind farm would require 6000 acres, which can include 10–30 landowners (Exhibit 4–23). The three main considerations for development of wind farms are: (1) sites or areas with good or very good wind resources, (2) access to transmission and (3) a power purchase agreement. Power purchase has been driven by federal (today, the production tax credits) and state regulations (renewable portfolio standards).

The factors that need to be considered when installing a wind energy conversion system are land availability, distance to the power grid (transmission lines), in the case of grid connected systems, site accessibility, terrain and soil characteristics and conditions, and lightning frequency. After the wind resource are assessed at a particular site or location, the next factor that should be considered is the land availability and cost. The area of the land required depends on the wind farm size, as well as the local topography. In order to optimize the wind energy conversion system power output at a given site, information about wind speed and direction statistics and characteristics, vegetation, topography, ground characteristics (for example ground roughness), or convenient access to the wind farm site are critical. For example, if the land area is limited or is at a premium price, one optimization study that must be conducted in an early stage of the wind farm design is to determine the number of turbines, their size, and the spacing for extracting the maximum energy from the farm annually. Large turbines cost less per megawatt of capacity and occupy less land area. On the other hand, fewer large machines can reduce the MWh energy production per year, as downtime of one machine would have larger impact on the energy output. A certain turbine size may stand out to be the optimum for a given wind farm from the investment and energy production cost points of view. Tall towers are beneficial, but the height must be optimized with the local regulations and constrains of the terrain and neighborhood. Nacelle weight and structural dynamics are also important considerations.

### 6.5.1 Wind Farms

Usually, a single wind turbine is installed and used only for providing energy to a particular site, such as an off-grid home in rural areas, a residence with wind turbine connected or not to the grid or in off-shore areas. In a good or excellent windy site, is desirable that several wind turbines to be installed, in what is often called as a wind farm or a wind park. The advantages of a wind farm are reduced site development costs, simplified connections to transmission lines, generated more power, and more centralized access for operation and maintenance. In order take the advantage of a wind park or farm, the wind turbine must be installed properly not affecting the operation of each other. If the wind turbines

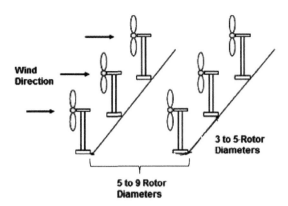

**Wind Direction**

3 to 5-Rotor Diameters

5 to 9 Rotor Diameters

**FIGURE 6.9**
Optimum distances between turbine towers in wind farm.

are located too close, the upwind turbines are interfering with the wind received by those located downwind. However, if the wind turbines are located too far, it means that the site space is not properly utilized, increasing the wind energy project overall cost. When the wind passes the turbine rotor, the energy is extracted by the rotor and the power which is available to the downwind machines is reduced. Experimental and theoretical studies show that the wind turbine performance is degraded when the wind turbines are too close to each other. Figure 6.9 shows the optimum distance between the towers, as resulted from such studies is estimated to be 3–5 rotor diameters between wind turbines within a row and 5–9 diameters between rows.

> **Example 6.10:** Typically the wind turbines in a large wind farm are positioned in a rectangular grid configuration, oriented to face the prevailing wind direction for the site. Assuming that the wind farm consists of 20 units, and that the wind farm grid dimensions are: (1) 3 and 5 diameters apart perpendicular to the prevailing wind and 5 and 10 rotor diameters apart parallel to the prevailing wind. This distance is selected to minimize the interference between turbines, and maximize the energy obtained. We are interested here to evaluate the land usage for this wind farm. If each wind turbine is rated at 1.8 MW and has a diameter of 100 m and a capacity factor of 0.36, what is the wind farm annual energy yield? What is the average annual generated energy per unit of land for minimum grid turbine separation?
>
> **Solution:** Minimum land usage per wind turbine
>
> $$\text{Minimum Land Area per Wind Turbine} = 3\times100\times5\times100 = 15\cdot10^4 \text{ m}^2$$
>
> $$\text{Maximum Land Area per Wind Turbine} = 5\times100\times10\times100 = 50\cdot10^4 \text{ m}^2$$
>
> Wind farm annual energy yield, computed by using Equation (6.41), multiplying the result with number of wind turbines in this wind farm, as:
>
> $$\text{Wind Farm Annual Energy Yield} = 8670\times2\times0.35 = 6069 \text{ MWh}$$
>
> The average annual energy generated per unit of land is then:
>
> $$\text{Average Energy per Land Unit} = \frac{6069\times10^3 \text{ kWh}}{15\times10^4} = 40.46 \text{ kWh/m}^2$$

## 6.6  Economic and Environmental Aspects of Wind Energy Projects

Generating electricity from the wind makes economic as well as environmental senses. Producing and selling electricity from the wind is no different from any other businesses. To be economically viable, the cost of making the electricity has to be less than its selling price. In every country, the price of electricity depends not only on the cost of generation, but also on many different factors that affect the market, such as energy subsidies and taxes. Generally, the cost of generating electricity is made up from: (1) capital cost—building the power plant and connecting it to the grid; (2) operating costs—operating, fuelling and maintaining the plant; and (3) financing—the cost of repaying investors and banks. The decision whether to implement a wind energy project should be based on a feasibility study. The study extent depends on the actual project size. The purpose of the feasibility study is to evaluate a project based on information on all aspects of implementation and operation of the project. The data to be collected to the feasibility study can be divided into: (1) wind resource assessment; (2) electrical system and agreements; (3) land availability; (4) soil conditions; (5) load pattern; (6) implementation expenses and capital costs; (7) operation and maintenance; (8) financing; and (9) organizational data and information. However, a reliable assessment and analysis of the wind regime and characteristics is critical for the project success and requires the reasonable wind data set for the actual area or site.

It is also evident from the previous sections that the WECS performance at a site depends heavily on the efficiency with which the wind turbine interacts with the wind regime. Hence, it is essential that the characteristics of the turbine and the wind regime at which it works should be properly matched. The capacity factor of the system can be a useful indicator for the effective matching of the turbine and wind regime. For turbine with the same rotor size, rated power and conversion efficiency, the capacity factor is influenced by the availability of the turbine to the prevailing wind. In other words, the turbine should be individually designed for a specific wind regime, and the turbine characteristics should be defined according to the site characteristics, which is not practical. However, several wind turbines of different ratings and functional velocities are available in the market. A wind energy project planner can choose a system, best suited for a specific site or location, from these available options. The performance estimation methods discussed above can be used for such an analysis. Environmental issues associated with wind generation are related to birds, bats, noise and visual impacts. In California there was a problem with raptors, especially with truss towers as perches, however after numerous studies this problem has been alleviated. Noise from gearboxes and blades has been reduced to less than ambient noise. It is still noticeable at the tower because the wind turbine noise is not random. Then the other major problem is that some people do not like the visual impact of wind turbines, especially if they are on ridges and mountains. Most farmers and ranchers want wind farms on their land, as it is a long-term source of income. However there are residents who are opposed, the not in my backyard group.

There is no doubt that the purpose of all types of energy generation ultimately depends on the scale of economics. Wind power generation costs have been falling over recent years. It is estimated that wind power in many countries is already competitive with fossil fuel and nuclear power if social/environmental costs are considered. The installation cost of a wind system is the cost of wind turbine, land, tower and its accessories and it accounts for less than any state or federal tax credits. The maintenance cost of the wind system is normally very small and annual maintenance cost is about 2/3% of the total system cost. The cost of financing to purchase the wind system is significant in the overall cost of wind system. Furthermore, the extra cost such as property tax, insurance of wind system and accidents caused from the wind system. One of the main advantages of generating electricity from the wind system is that the wind is free. However, the cost of the wind system just occurs once. On the other hand, the cost of non-renewable energies is more and more expensive, which is required for renewable energies such as wind power. Nowadays, research and development make the wind power generation competitive with other non-renewable fuels such as fossil fuel and nuclear power. Lots of efforts have been done to reduce the cost of wind power by design improvement, better manufacturing technology, finding new sites for wind systems, development of better control strategies (output and power quality control), development of policy and instruments, human resource development, etc.

The most significant environmental advantage of wind energy is its minimal adverse impacts on the environment in contrast with conventional power plants. The wind energy does not pollute the air or water with harmful gases and materials. Being a non-depletable source of energy, extracting power from the wind does not pose the treat of overexploiting the limited natural resources like oil, gas or coal. However, as in the case of any human activities, wind energy generation is also not totally free from environmental consequences. The major environmental problems of the wind energy include: the avian morality due to collision with turbines and related structures, noise emission of the wind turbine during the operation, and visual impacts on the landscape. However, it should be noted that these environmental impacts are not global and thus can be monitored and resolved at local level. However, the modern wind farms today may contain a large number of large-size wind turbines. Therefore, their impacts on the environment cannot be ignored. One of the impacts is that poorly sited wind energy facilities may block bird migration routes and hurt or kill birds. Though blade rotation speeds are rather low for large wind turbines at their normal operation, the tangential speeds at the blade tips could be higher than 70 m/s. At such high speeds, birds flying through the blade sweeping areas may be easily hurt or killed by colliding with blades. However, recent studies have revealed that fossil-fuelled power stations appear to pose a much greater threat to avian wildlife than wind and nuclear power technologies. On the other hand, building wind farms changes the character of local landscape, modern large wind turbines are about or over 100 m tall and thus can be seen at a far distance. In practice, the visual effect for local residents is a significant consideration and is always scrutinized for wind projects. To minimize it, the wind turbines usually use neutral colors such as light grey or off-white. Strategies to minimize visual effects involve the spacing, design, and uniformity of turbines, markings or lighting, roads and service buildings. With the extensive build-up of wind power plants and the population growth all over the world, the influence of wind turbine noise to the nearby residents becomes a problem not to be neglected. Wind turbine noise consists of aerodynamic noise from rotating blades and mechanical vibration noise from gearboxes and generators. For a modern large wind turbine, aerodynamic noise from the blades is considered to be the dominant noise source. There are two components in aerodynamic noise: (1) airfoil self-noise, that is, the noise produced by the blade in an undisturbed inflow and is caused by the interaction in the boundary layer with the blade trailing edge; and (2) inflow turbulence noise which is caused by the interaction of upstream atmospheric turbulence with the blade and depends on the atmospheric conditions. Both airfoil self-noise and inflow turbulence noise mechanisms are dependent on a number of parameters such as wind speed, angle of attack, radiation direction, and airfoil shape.

## 6.7 Summary

Modern wind turbine technology is advancing rapidly and both the system and the generated kWh costs are decreasing rapidly. The value of a clean, a non-fossil fuel-based power generation technology is also a positive factor in wind power development. The fact that wind power is not dispatchable, but determined by the variances in the wind, is seen as a drawback to dominance of the wind energy technology in the electricity industry. Wind turbines must be designed to operate at their optimal tip speed ratio in order to extract maximum possible power as possible from the wind stream. When a rotor blade passes through the air stream it leaves a turbulent wake in its path. If the next blade in the rotating rotor arrives at the wake when the air is still turbulent, it is not able to extract power from the wind efficiently, and is also subjected to high vibration stresses. If the rotor rotated slower, the air hitting each rotor blade would no longer be turbulent. This is another reason for the tip speed ratio to be selected so that the rotor blades do not pass through turbulent air. The performance of a wind energy conversion system depends of its subsystems or components, such as wind turbine aerodynamics, drivetrain (mechanical transfer subsystem), and electrical generator. The kinetic energy extracted from the wind by any wind energy conversion system is strongly influenced by the geometry of the rotor blades. Determining the aerodynamically optimum blade shape, or the best possible approximation to it, is one of the main tasks of the wind turbine designer. The basic of BEM method assumes that the blade can be analyzed as a number of independent elements in span-wise direction. The induced velocity at each element is determined by performing the momentum balance for an annular control volume containing the blade element. Then the aerodynamic forces on each element are calculated using the lift and drag coefficient from the empirical two-dimensional wind tunnel test data at the geometric angle of attack of the blade element relative to the local flow velocity. BEM theory-based methods have aspects by reasonable tool for designer, but they are not suitable for accurate estimation of the wake effects, the complex flow such as three-dimensional flow or dynamic stall because of the assumptions being made. The wind energy technology, the basic components of a wind turbine system, types of wind turbines, wind turbine power scales and installation location are discussed in quite details in this chapter. The equations of output power considering ideal and practical wind turbines are derived. The stall and pitch control mechanisms, among others, and the factors affecting power output and turbine performances, such as wind shear and tower shadow are explained and discussed.

## Questions and Problems

1. What are the pros and cons of horizontal vs. vertical axis wind turbines?
2. Which U.S. states have the greatest availability of wind energy?
3. What are the cut-in, rated and cut-off speed of a wind turbine?
4. Evaluate the advantages and disadvantages of wind power technology.
5. Identify the basic functions and classifications of wind turbines.
6. Identify major horizontal axis wind turbine (HAWT) components and their function.

7. Describe a Savonius rotor, its main advantages and disadvantages.

8. What is a drag-type wind turbine? What is a lift-type wind turbine? Which has higher efficiency?

9. Explain the term power coefficient of a wind turbine? What is the usual value for the power coefficient of a well-designed turbine?

10. List the major wind turbine control methods.

11. What is the maximum theoretical power that can be produced by a horizontal axis wind turbine of 6-m diameter in a wind of 10 m/s? What do you expect the actual power to be?

12. If the wind is blowing at 6.7 m/s, what is the tip speed ratio of this turbine? The rotor diameter is 12 m, and the rotor is spinning at 45 RPM.

13. What is the difference between a horizontal axis wind turbine and a vertical axis wind turbine in terms of how they are installed?

14. Briefly describe the advantages and main disadvantages of vertical axis wind turbines.

15. Briefly describe the advantages and main disadvantages of horizontal axis wind turbines.

16. According to the "optimal" tip speed ratio for a three-blade wind turbine, are the turbine blades moving too fast or too slow?

17. An offshore wind turbine with three 60-meter blades rotates at a leisurely 15 RPM. The wind is whipping along at 18 m/s. What is the tip speed ratio for this turbine?

18. An offshore wind turbine with three 90-meter blades rotates at a leisurely 10 RPM. The wind is whipping along at 15 m/s. What is the tip speed ratio for this turbine? How does this compare to the "optimal" tip speed ratio for this turbine?

19. What are the advantages of a horizontal axis wind turbine over a vertical axis wind turbine?

20. Why is it impossible to reach the Betz limit of wind turbine efficiency? Discuss.

21. At what wind speed you have maximum efficiency?

22. A large turbine is rated at 1.25 MW, located at sea level and at standard atmospheric conditions (0°C and 101.3 kPa). What is its rated power at the same rated wind speed if the temperature is 25°C and the turbine was located at 1350 m above sea level?

23. A wind power plant has 100 turbines, each one rated for 1.5 MW. The capacity factor is 35%. What is the plant's annual energy yield?

24. Explain what is meant by "power coefficient" in a turbine? What is the usual value for the power coefficient of a well-designed turbine?

25. What are the principal components of a horizontal axis wind turbine?

26. What is meant by nacelle, and what it is its role in a wind turbine?

27. What are the most common electric generators used in wind energy? List some of the advantages of each major generator type.

28. Find the power in the wind for a circular area of a 150 ft diameter, if the air density is 0.082 lb/ft$^3$ and wind speed is 33 ft/s. Convert and express everything in SI units.

29. A turbine operates at a constant speed of 18 rpm. If at a certain time the power generated by the rotor is 1000 kW, how much is the torque on the rotor shaft?

30. An electric utility decides to add 50 MW of wind generation to its system. If the individual wind energy conversion systems are rated at 2 MW in a 13.5 m/s wind at standard conditions and have efficiencies $C_P = 0.32$, $\eta_{mech} = 0.94$, and $\eta_{el} = 0.96$, what is the required swept area A of each rotor? What is the electric power generated by each unit? How many wind turbines are needed?

31. Determine the maximum power density a horizontal wind turbine may produce at the following wind speeds: 1, 5, 10, 15, and 20 m/s.

32. For a simple and rudimentary Savonius wind turbine, having the radius rotor, 0.75 m, and a rotor height 0.96 m, calculate the effective rotor radius of the area exposed to the wind and the output power for wind speeds ranging from 2 to 20 m/s.

33. Repeat the previous problem assuming there are two Savonius turbine configurations one consisting of the two similar rotors in top of each other, and the other of three similar rotors. Plot the output power vs. wind speeds.

34. What is the TSR and angular speed of a 90 m diameter wind turbine that rotates at 70 rpm when the wind speed is 13.5 knots?

35. For a Darrieus wind turbine, having geometry of 36 m diameter and 45 m height, estimate the annual energy output for two different U.S. states. Select one state with very good wind energy potential and the other one with medium wind energy potential. Use the U.S. wind atlas or other source of wind energy potential information.

36. A Darrieus wind turbine produces a low speed shaft power of 1800 kW at the rated speed of rotation 22.5 RPM in a 13.5 m/s wind speed. Determine rated rotor torque from the values given for measured power and rated rotor speed.

37. A wind turbine rotor reaches maximum power coefficient, $C_P$ at a tip speed ratio of 7.5. Calculate the rotor RPM for three different wind turbines, having diameters of 30 m, 50 m, and 90 m, respectively at wind speeds of 8 m/s, 15 m/s, and 24 m/s.

38. What is the annual energy yield of a wind turbine with $D = 90$ m, $C_P = 0.35$, at rated wind speed 12 m/s and with 1800 full-load hours?

39. Assume that a wind turbine rated at 300 kW is adequately modeled by Equations (6.39) through (6.41), a rated wind speed of 8 m/s, a cut-in speed of 4.0 m/s, and a furling speed of 21 m/s. Determine the capacity factor and the monthly energy production in kWh (assuming 30 days in a month) for the sites, characterized by Weibull distribution with shape and scale parameters: (a) $c = 6.85$ m/s and $k = 1.93$; and (b) $c = 8.5$ m/s and $k = 2.4$.

40. For a Darrieus wind turbine, having a 36 m diameter and a 45 m height, estimate annual energy output for two different regions from European wind map (class 4 and 5). Assume that is its capacity factor is equal to: 0.23.

41. A wind power plant has 30 wind turbines, each one rated for 1.5 MW. The capacity factor is 35%. What is the plant's annual energy yield? What is the average power per unit of land, assuming that the wind power plant consist of 25 wind turbines, each having diameter of 90 m and are placed in a rectangular grind to the prevailing wind with a spacing of 4 diameters by 9 diameters (in depth)?

42. For a conventional HAWT, radius of 40 m, estimate annual energy output for a good wind region (use class 4, 5, or 6) from U.S. wind power map. Assume the turbine capacity factor is equal to 0.285.

## References and Further Readings

1. G. Boyle, *Renewable Energy—Power for a Sustainable Future*, Oxford University Press, Oxford, UK, 2012.
2. H. K. Hodge, *Alternative Energy Systems and Applications*, Wiley, Hoboken, NJ, 2010.
3. L. L. Freris, Wind Energy Conversion Systems, Prentice Hall, Hemel Hempstead, UK, 1990.
4. K. H. Bergey, 6, *Lanchester-Betz Limit*, American Institute of Aeronautics and Astronautics, Vol. 3, pp. 382–384, 1980.
5. E. E. Michaelides, *Alternative Energy Sources*, Springer, Berlin, Germany, 2012.
6. I. Bostan, A. Gheorghe, V. Dulgheru, I. Sobor, V. Bostan, and A. Sochirean, *Resilient Energy Systems—Renewables: Wind, Solar, Hydro*, Springer, Berlin, Germany, 2013.
7. R. Belu, *Wind Turbine Dynamic Modeling and Control*, (doi:10.1081/E-EEE-120048431/29 pages), in Encyclopedia of Energy Engineering & Technology (Online) (Ed: Dr. Sohail Anwar, R. Belu et al.), CRC Press/Taylor & Francis Group, 2014.
8. R. Belu, *Wind Energy Conversion and Analysis*, (doi:10.1081/E-EEE-120048430/27 pages), in Encyclopedia of Energy Engineering & Technology (Online) (Ed: Dr. Sohail Anwar, R. Belu et al.), CRC Press/Taylor & Francis Group, 2014.
9. M. R. Patel, *Wind and Solar Power Systems*, CRC Press, Boca Raton, FL, 1999.
10. P. Gipe, *Wind Energy Comes of Age*, John Wiley & Sons, New York. 1995.
11. G. L. Johnson, *Wind Energy Systems (Electronic Edition)*, Kansas University Reprints, Manhattan, KS, 2006.
12. Paraschivoiu, I., *Wind Turbine Design with Emphasis on Darrieus Concept*, Polytechnic International Press, Montreal, QC, 2002.
13. Burton, T., D. Sharpe, N. Jenkins, and E. Bossanyi, *Wind Energy Handbook*, John Wiley & Sons, Berlin, Germany, 2001.
14. R. Gasch and J. Twele, *Wind Power Plants: Fundamentals, Design, Construction and Operation*, James and James, London, and Solarpraxis, Berlin, Germany, 2002.
15. J. G. McGowan, J. F. Manwell, and A. L. Rogers, *Wind Energy Explained: Theory, Design and Applications*. John Wiley & Sons, Chichester, UK, 2002.
16. T. Ackermann, (Ed.), *Wind Power in Power Systems*, Wiley, Hoboken, NJ, 2005.
17. S. Mathew, *Wind Energy Fundamentals, Resource Analysis and Economics*, Springer, Berlin, Germany, 2005.
18. E. Hau, *Wind Turbines, Fundamentals, Technologies, Application, Economics* (2nd ed.), Springer, Berlin, Germany, 2005.
19. M. Sathyajith, *Wind Energy: Fundamentals, Resource Analysis and Economics*, Springer Verlag, Berlin, Germany, 2006.
20. M. O. L. Hansen, *Aerodynamics of Wind Turbines* (2nd ed.), Earthscan, London, UK, 2008.
21. M. Stiebler, *Wind Energy Systems for Electric Power Generation*, Springer, Berlin, Germany, 2008.
22. V. Nelson, *Wind Energy, Renewable Energy and the Environment*, CRC Press, Boca Raton, FL, 2009.
23. A. R. Jha, *Wind Turbine Technology*, CRC Press, Boca Raton, FL, 2011.
24. P. Jain, *Wind Energy Engineering*, McGraw-Hill, New York, 2011.
25. S. Muyeen, (Editor), *Wind Energy Conversion Systems—Technology and Trends*, Springer, Berlin, Germany, 2012.
26. D. Rekioua, *Wind Power Electric Systems—Modeling, Simulation and Control*, Springer, Berlin, Germany, 2012.
27. M. H. Ali, *Wind Energy Systems: Solutions for Power Quality and Stabilization*, CRC Press, Boca Raton, FL, 2012.
28. H. Wagner and J. Mathur, *Introduction to Wind Energy Systems*, Springer, Berlin, Germany, 2013.
29. V. Lyatkher, Wind Power, Wiley/Scrivener Publishing, Boston, MA, 2014.
30. P. M. Pardalos, S. Rebennack, M. V. F. Pereira, N. A. Iliadis, and V. Pappu, (Editors), *Handbook of Wind Power Systems*, Springer, Berlin, Germany, 2014.

31. A. P. Schaffarczyk, *Introduction to Wind Turbine Aerodynamics*, Springer, Berlin, Germany, 2014.
32. V. Lyatkher, *Wind Power—Turbine Design, Selection, and Optimization*, Wiley—Scrivener Publishing, Hoboken, NJ, 2014.
33. A. P. Schaffarczyk, (Ed.), *Wind Power Technology—Theory, Deployment and Optimization*, John Wiley & Sons, Hoboken, NJ, 2014.
34. R. Bansal (Ed.), *Handbook of Distributed Generation—Electric Power Technologies*, Economics and Environmental Impacts, Springer, Berlin, Germany, 2017.
35. R. Belu, *Industrial Power Systems with Distributed and Embedded Generation*, The IET Press, London, UK, 2018.
36. IEC 61400 Standard– Wind Turbines, International Electrotechnical Commission, IEC 2005, Geneva, Switzerland.

# 7

## Geothermal Energy

## 7.1 Introduction, Earth Internal Structure

Almost all countries are increasingly focused on the energy supply diversification, evaluating all energy alternatives, particularly those that are significant and well-distributed nationally. One such option, often ignored, is the geothermal energy. Geothermal energy comes from Greek words: thermal meaning heat and geo meaning Earth, being the energy contained as heat into the Earth's interior. The Earth gives us the impression that it is dependably constant, because over the human time scale, little seems to change. However, every Earth cubic centimeter is in motion, and has been since its formation. Earth is a dynamic entity, with time scales spanning from seconds for the earthquakes, years that volcanoes appear and grow, over millennia that landscapes slowly are evolving, to over millions of years the continents rearrange on the planet's surface. The energy source to drive these processes is heat, with a constant flux from every square meter of the Earth's surface. Earth average heat flux is 87 mW/m$^2$, or for the total global surface area of $5.1 \times 10^8$ km$^2$, this heat flux is equivalent to about $4.5 \times 10^{13}$ W. For comparison, it is estimated that the total power used by all human activity in one year is approximately $1.6 \times 10^{13}$ W. Clearly, the Earth heat has the potential to significantly contribute to satisfying world energy needs. This heat, the source of geothermal energy, is contained into the rock and fluid inside the Earth's layers. Its origin is linked with the Earth's internal structure and composition, and the associated physical processes. Despite the fact it is present in huge, inexhaustible quantities into the Earth's crust or deeper layers, it is unevenly distributed, seldom concentrated, and most often at depths too great to be exploited industrially. There are almost 6300 km from the Earth' surface to its center and the deeper it is the hotter it gets. The outer layer of the Earth, the Crust, is about 70 km thick and insulates the surface from the hot interior. The interior generated heat moves toward the surface where it dissipates, the Earth's temperature increases with depth, proving that a geothermal gradient of about 30°C/km of depth exists.

During the last century, several countries started to use geothermal energy, as it becomes economically competitive with other energy sources. Moreover, the geothermal energy is in some regions, the only energy source available locally. Geothermal energy is present on Earth from two main sources:

1. Heat, flowing upward and outward across the entire Earth' surface from its interior radioactive decay. However, for most of Earth areas, this energy flux is too small to be useful for any application.
2. The localized heat resulting from the movement of magma into the Earth's crust. In some areas, the localized heat, with higher temperatures and heat fluxes, is found between the surface and about 3500 m (or 10,000 ft) depth. When they meet the requisite conditions, geothermal energy can be used for multiple purposes such as generating power and providing heat and hot water for buildings and industrial processes.

**Example 7.1:** Estimate the available power for two 2000 km² areas, one having the average geothermal flux, and the second one (an active geothermal energy area) has the geothermal flux of 200 mW/m².

**Solution:** The available power of each of the two different geothermal flux areas is:

$$P_{active} = 2 \times 10^9 \times 200 \times 10^{-3} = 400 \text{ MW}_t$$

$$P_{average} = 2 \times 10^9 \times 87 \times 10^{-3} = 164 \text{ MW}_t$$

A good understanding of the characteristics of the geothermal energy sources is critical in order to use the Earth available heat. Some of the geothermal energy applications can be pursued anywhere, others require special circumstances, but full source characterization is needed for an economically and environmentally development. To understand the geothermal energy distribution and characteristics a description of the Earth's physical structure is needed. Earth is compositionally inhomogeneous, consisting of an iron–nickel core, a dense rocky mantle, and a thin and low-density rocky crust. Because of its complex composition, large differences in physical properties exist (liquid or molten vs. solid, or brittle vs. ductile deformation). Extending outward from Earth's center, over its radius of about 6370 km, significant changes occur in both composition and rheological behavior (material physical properties, such as changes from solid to liquid or brittle to ductile). The Earth structure is shown in Figure 7.1, and it consists of three main layers — *the crust*, a relatively thin region of low-density silicates, *the mantle*, a thick region of higher-density iron-rich silicates, and the core, a *central region (core)* of iron-nickel mixture, each subdivided further into sub-layers. Earth's layers are depicted as concentric spheres, in ultra-simplified schematics. However, the interfaces are irregular and the boundaries very fuzzy. The crust has continental regions, made of even lower-density aluminum-rich silicates and oceanic regions, made of much richer and denser iron-rich silicates. The mantle is divided into upper mantle within the iron-rich silicates are gradually compressed from lower-density mineral structures, to higher density and more compact mineral structures, and lower mantle, where the mineral

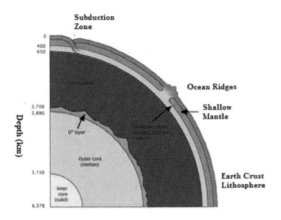

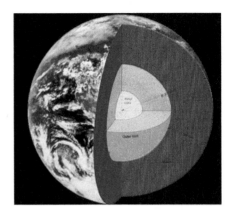

**FIGURE 7.1**
Earth internal structure and layers.

structures have been compacted to their densest forms and increase in density, due to continually increasing weight compressing them. The core consists of a molten (liquid) outer core layer, primarily made of iron but with substantial percentages of sulfur and silicon, lowering its melting temperature, and a solid inner core, consisting of almost certainly of a crystalline mixture of iron and nickel. Core temperature estimates range from about 2700°C to about 6000°C. The mantle is made partly rock and partly magma is about 2900 km thick, from the core to about 100 km, while the core is extending from its center to a depth of about 2900 km. The mantle's volume makes up the largest part of Earth's interior, and its temperature decreases upward from about 5000°C to less than 1500°C. The last layer is the crust, consisting of a thin shell, varying from about 80 km thick under parts of continents to less than a few kilometers thick under the ocean floor.

Drilled wells are giving direct access only to the upper crust, up to 12 km depths, and the information from deeper layers is obtained through indirect and inverse methods. Seismic wave studies have shown that the crust is thinner beneath the oceans than beneath the continents, and that seismic waves travel faster in oceanic crust than in continental crust. In part because of this velocity difference, it is assumed that the two crust types are made up of different kinds of rock. The denser, oceanic crust is made of basalt, whereas the continental crust is often referred to as being largely granite. The mantle lies closer to the Earth's surface beneath the ocean (about 7 km), than beneath the continents (20–65 km), extending from the crust for about 2900 km. Based on the accepted hypothesis about the mantle composition is that it consists of ultrabasic rock (very rich in Fe and Mg) such as peridotite, a heavy igneous rock. The crust and the uppermost mantle form the lithosphere, the outer shell of the Earth that is relatively rigid and brittle. The lithosphere is divided into several large blocks at continental scale, the lithospheric plates of the plate tectonic theory. The lithosphere is about 70 km thick beneath the oceans and up to 125 km thick beneath the continents. Mantle rocks in the asthenosphere are weaker than they are in the overlying lithosphere, and the asthenosphere can deform easily by plastic flow, and convection is taking place within it and within the lower mantle. The lithosphere seems to be in continual movement, likely as a result of the underlying mantle convection and the lithosphere plates probably move easily over the asthenosphere, which may act as a lubricating layer below.

The core extends from 2900 to 6370 km (the Earth's center), about 3470 km in thickness, with temperatures around 4000°C and higher, and the pressure at the Earth's center 3.6 million bar (360,000 MPa). The plate tectonic theory, accepted by most geologists, implies that the Earth rigid outer shell, the lithosphere, is divided into separate plates, the *lithospheric plates* (Figure 7.2). These plates move slowly across the Earth's surface, at a speed of a few centimeters per year, which means that the continents and sea floors are moving, sliding on top of the underlying plastic asthenosphere. These plates pull away from each other, slide past each other or move toward each other. The Earth crust, unlike the more compositionally homogeneous core and mantle, consists of oceanic and continental zones. Oceanic crust consists of denser basalt is relatively thin, reaching a maximum of 7 km to a minimum of a kilometer below mid-ocean ridges. Continental crust is comprised mainly of lower density, lighter colored igneous and metamorphic rocks, such as granite and gneiss. The continental crust igneous and metamorphic rocks are capped in places by a veneer of sedimentary rocks, including sandstone and limestone. Being less dense than oceanic crust, sits higher compared to oceanic crust, explaining why the continents for the most part lie above sea level. Continent and ocean floor plates drift apart and push against each other at about 2.5 cm/yr (1 in./yr), in the plate tectonics process, causing the crust to become faulted, fractured, or thinned, allowing plumes of magma to rise up into the crust. This magma can reach the surface and form volcanoes, but most remain underground where they can underlie regions as large as mountain ranges. To cool the magma can take hundreds of years, as its heat is transferred to surrounding rocks by conduction. In areas where there is underground water, the magma can fill rock fractures and porous rocks. The water becomes heated and can circulate back to the surface, creating hot springs or can be trapped underground, forming deep geothermal reservoirs. The Earth's crust is composed of various types of rock which contain some radioactive isotopes, such as uranium, thorium, and potassium. The heat released by these nuclear reactions is in part responsible for the natural heat, reaching the Earth surface.

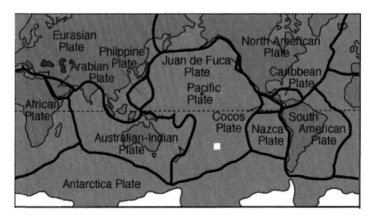

**FIGURE 7.2**
Earth's major tectonics plates.

### 7.1.1 Geothermal Energy Origins

Geothermal energy has provided commercial base-load electricity around the world for more than a century, often ignored or given less attention compared with other energy sources in national energy projections. Its origin is linked with the Earth's internal structure and the physical and chemical processes occurring there. Despite the fact that this heat is present in huge, inexhaustible quantities in the crust, not to mention the deeper Earth layers, is unevenly distributed, seldom concentrated, and often at depths too great to be exploited industrially. There are, however, crust areas accessible by drilling, and where the Earth heat gradient is well above the average. This occurs when, at a few kilometers depth, there are magma bodies cooling, still in fluid state or solidification process. In other areas the heat accumulation is due to particular crust geological processes that lead to very high values of the geothermal gradient. Extraction of these large heat quantities requires a carrier to transfer the energy at accessible surface depths. The heat is transferred from depth to sub-surface regions by conduction and then by convection, with geothermal fluids acting as the carrier. These fluids, usually rain water, penetrated into the crust from the recharge areas, has been heated on contact with the hot rocks, accumulated in aquifers, usually at high pressures and temperatures (about 300°C). These aquifers or reservoirs are the essential parts of most geothermal fields. Geothermal resources can be considered renewable on the human time-scales, not requiring the huge recovery geological times of the fossil fuels. Moreover, the recovery of high-enthalpy reservoirs is accomplished at the same site from which the fluid or heat is extracted. The environmental impacts of geothermal power generation and direct use are usually minor, controllable, or negligible.

The Earth geothermal energy originates from the original formation of the planet and from radioactive decay of materials (roughly in equal proportions). The Earth's heat flow is the heat amount released from the interior through a unit area in a unit of time, measured in $mW/m^2$. It varies from place to place on the surface, and it has varied with time at any particular place during the course of history. The Earth's heat flow originates from the primordial heat, generated during the Earth's formation, or generated by the decay of long-lived radioactive isotopes. Although all radioactive isotopes generate heat as they decay, only isotopes that are relatively abundant and have half-lives comparable to the Earth age (4.5 billion years) have been significant heat producers throughout geological times. Four radioactive isotopes are important heat producers: $^{40}K$, $^{232}Th$, $^{235}U$, and $^{238}U$. The average heat flow through the continental crust is 57 $mW/m^2$, and through the oceanic crust is 99 $mW/m^2$. The Earth's average heat flow is 82 $mW/m^2$, while the total global output is over $4 \times 10^{13}$ W, over four times the present world energy consumption of about $10^{13}$ W. Continental heat flow appears to be derived from radiogenic decay within the upper crust, together with the heat generated in the most recent magmatic episode and the heat coming from the mantle. In the oceanic crust, the concentration of radioactive isotopes is very low, so the heat flow largely derives from heat flowing from the mantle below the lithosphere.

The geothermal gradient, due to the temperature difference between the planet core and its surface, drives a continuous conduction of thermal energy in the form of heat to the surface. This could be a result of the widespread perception that the total geothermal resource is often associated with identified high-grade, hydrothermal systems that are too few and too limited in their distribution to make a long-term, major impact at a national level. In general terms, geothermal energy consists of the thermal energy stored in the Earth's crust. Since the temperature at the base of the crust is about 1100°C, the temperature gradient between the surface, 20°C or so and the crust bottom is 31.1°C/km. This is the so-called normal conductive temperature gradient. Good geothermal prospects occur where the thermal gradient is several times greater than the normal gradient. The rate of natural heat flow per unit area, called the normal heat flux, is roughly $1.2 \times 10^6$ cal/cm²·s, in non-thermal Earth areas. The Earth's conductive heat flow is the product of the geothermal gradient and the thermal conductivity of rocks. The geothermal gradient is measured in shallow holes, while the conductivity of rocks is measured in laboratory on samples, taken from that part of the well where the gradient was measured. Two forms of heat transfer occurring within the Earth are conduction and convection. Conduction involves the transfer of random kinetic energy between molecules without the material transport, being the primary heat transfer mode in solids. Metals are very good conductors of heat, whereas most crust rocks are relatively poor conductors. Convection is the common heat transfer process in fluids and consists of the movement of hot fluid from one place to another. Because motion of material occurs, convection is a vastly more efficient process of heat transfer than conduction. Studies of the Earth thermal behavior imply the determination of how temperature varies with depth, and how such temperature variations may have changed throughout geological time. However, studies of this kind are based entirely on measurements made on, or within, a few km of the Earth's surface. The thermal gradient values lower as 10°C/km are found in ancient continental crust, while very high values about 100°C/km are found in active volcanic areas. Once the gradient has been measured, the upward heat rate through a particular part of the Earth's crust can be determined.

**Example 7.2:** In order for a viable geothermal power plant development, for electricity generation the geothermal fluids must exceed 150°C. If the average surface temperature of 10°C, how deeper the wells must be drilled to reach such temperature at a location having a normal gradient.

**Solution:** If we are assuming a normal average gradient of 30°C per km, and for the 140°C (150°C–10°C) temperature difference between the surface and the well bottom, the well depth is:

$$\Delta H = \frac{140}{31} = 4.516 \text{ km}$$

The main mechanism of geothermal energy transfer is conduction, and the heat flow rate is proportional to the geothermal gradient and to a proportionality constant, the thermal conductivity of rocks defined as the amount of heat conducted per second through an area of 1 m², when the temperature gradient is 1°C/m perpendicular to that area. The thermal conductivity unit is W/(m·K). Thermal gradient is measured in wells with electrical thermometers, a quick and relatively inexpensive method. The rock thermal conductivity is best measured in laboratory, as there are no reliable downhole methods. If the geothermal gradient is expressed in °C/km and conductivity in W/(m°C), then heat flow rate is in W/m². Geothermal energy is distributed between the constituent host rock and the fluid, contained in its fractures and pores at temperatures above ambient levels. This fluid is mostly water with varying amounts of dissolved salts, typically, in their natural in situ state as a liquid phase or sometimes may be a saturated, liquid-vapor mixture or superheated steam vapor. The amounts of hot rock and contained fluids are larger and more widely distributed compared to oil and natural gas contained in sedimentary rock formations.

Geothermal fluids have been used for cooking and bathing since before the beginning of recorded history; but it was not until the early twentieth century that geothermal energy was harnessed for industrial and commercial purposes. In 1904, electricity was first produced using geothermal steam at the vapor-dominated field in Larderello, Italy. Since then, other hydrothermal developments, such as the steam field at The Geysers, California, the hot-water systems at Wairakei, New Zealand, Cerro Prieto, Mexico, Reykjavik, Iceland, in Indonesia and the Philippines, led to an installed world generating capacity of nearly 10,000 MWe and a direct-use, nonelectric capacity of more than 100,000 MWt at the beginning of the twenty-first century. The source and transport mechanisms of geothermal heat are unique to this energy source. The geothermal energy is a renewable resource that can contribute significantly to the world's sustainable and diversified energy mix in the twenty-first century. With continually improving technology for development, geothermal energy is destined to become a major factor in solving the world's increasingly complex energy equation. The heat flowing from the Earth's interior is estimated to be equivalent to 42 million megawatts, significantly greater than the all fossil fuels energy combined. It is estimated that even a small fraction of this heat would supply the world's energy needs for centuries. Geothermal energy is also permanently available, unlike the solar and wind energy, which are dependent upon weather, daily and seasonal fluctuations. Electricity from geothermal energy is more consistently available, once the resource is tapped. Different geothermal systems can be exploited: (a) hydrothermal systems, (b) enhanced geothermal systems (EGS), (c) conductive sedimentary systems, (d) hot water produced from oil and gas fields, (e) geo-pressured systems and (f) magma bodies. Convective hydro-thermal systems for commercial exploitation for electric generation are in place for over a century, however with quite limited distribution. There are two basic convective system types, depending on the thermal energy source: volcanic and non-volcanic. A volcanic convective system drives its thermal energy from a convecting magma body, while a non-convective system drives its thermal energy from meteoric water that has heated through a deep circulation in Earth high heat flow areas, with no magma associated with them. The installed power capacity exploiting such systems totals about 10,000 MW worldwide and 3000 MW in the U.S. only. The reserve base in such systems in the U.S. is estimated to be in the 20,000 MW. It has been suggested that a positive correlation exists between the geothermal resource base potentially available from convective systems of the volcanic type and the number of active volcanoes in the country. Even if a country has no active volcanoes, an exploitable geothermal resource may exist in the form of non-volcanic convective systems. The heat moves from the Earth's interior toward the surface where it dissipates, although this fact is generally not noticed, being aware of its existence because the temperature of rocks increases with depth, proving that a good geothermal gradient exists. There are areas of the Earth's crust, accessible by drilling, where the gradient is well above the average. This occurs when, a few kilometers bellow surface, there are magma bodies undergoing cooling, in a fluid state or in the solidification process, and releasing heat. In other areas, where there is no magmatic activity, the heat accumulation is due to particular crust geological conditions, such that the geothermal gradient reaches anomalously high values.

Experimental power generation installed in Larderello, Italy, on July 4, 1904, and the first commercial geothermal power plant (250 kW) in 1913, and the first large-power installation in 1938 (69 MW). It would be 20 years before the next large geothermal power installation was built in Wairakei, New Zealand, commissioned in 1958, that grew to 193 MW of installed capacity by 1963. In the United States, the installation of the first unit, 11 MW, become the largest geothermal power complex in the world, The Geysers, in Sonoma, California in 1960. Over the next quarter-century, a total of 31 turbine generator sets were installed at The Geysers, with a capacity of 1890 MW. Plant retirements and declining steam supply have since reduced its generation capacity to an annual average of 1020 MW from 1421 MW of installed capacity, still the largest geothermal plant. Since those early efforts, a total of 2564 MW of geothermal power generation capacity is currently installed in the United States, generating approximately 2000 MW each year. In 2005, worldwide annual geothermal power was estimated at 56,875 GW h from 8932 MW of installed capacity. Geothermal energy is also utilized in direct-heat uses for space heating, recreation and bathing and industrial and agricultural uses. Geothermal energy in direct-use application is estimated to have an installed capacity of 12,100 MW thermal, with an annual average energy use of 48,511 GW h/year. This excludes ground-coupled heat pumps (GCHP). In the US, the first district geothermal heating was installed in Boise, Idaho, in 1892, still in operation today. GCHP units are reported to have 15,721 MW of installed capacity and 24,111 GWh, representing 56.5% worldwide direct use, respectively.

The Earth heat is transferred from depth to sub-surface regions by conduction and by convection, with geothermal fluids acting as the carrier. These fluids are essentially rain water that has penetrated into the crust, being heated by the contact with the hot rocks, and accumulated into aquifers, occasionally at high pressures and temperatures (above 300°C). These aquifers are essential parts of most of the geothermal fields. In most cases the reservoir is covered with impermeable rocks that prevent the hot fluids from reaching the surface and keep them under pressure. From such reservoirs superheated steam, steam mixed with water, or hot water only, depending on the hydrogeological situation and the temperature of the rocks present can be obtained for industrial applications. Wells are drilled into the reservoir to extract the hot fluids, and the application depends on the fluid temperature and pressure: for electricity generation, the high-temperature uses or for space heating and industrial processes, the low-temperature uses. Geothermal fields are usually systems with continuous heat and fluid circulation, fluid entering the reservoir from the recharge zones and leaves through discharge areas. During industrial exploitation fluids are recharged to the reservoir by re-injecting through wells the waste fluids from the utilization plants. The reinjection process may compensate, at least part of the extracted fluid and to a certain limit prolong the commercial field lifetime. Geothermal energy is therefore to a large extent a renewable energy source, hot fluid production rates tend however to be much larger than recharge rates.

EGS implies a man-made reservoir created by hydro-fracturing impermeable or very "tight" rock through wells. By injecting normal temperature water in a well in such an artificially fractured reservoir and producing from another well, the water heated up by heat from the rocks it is possible to extract thermal energy. The EGS systems represent conductive systems that have been enhanced their flow and storage capacity by hydro-fracturing or other methods. Theoretically EGS can be developed, anywhere by drilling deep enough to encounter a commercially viable temperature level. However, the EGS technology is still experimental and posing technical challenges, such as: (a) creating a pervasively fractured large rock volume, (b) securing commercially productivity, (c) minimizing the time cooling rate of the produced water, (d) minimizing the loss of the injected water through fractures and (e) minimizing any induced micro-seismicity. Another type of geothermal energy resource, considered for exploitation is the heat contained in the water produced from deep oil and gas wells (co-produced with oil) or from abandoned oil or gas wells. While there are no significant challenges to exploiting such resources, the cost of power from it may not be attractive due to relatively low water temperature and production rates. A type of geothermal energy resource of very restricted distribution worldwide is the *geo-pressured* systems, which are confined to sedimentary reservoirs with pressures much higher than the local hydrostatic pressure. This high pressure allows the exploitation of the kinetic energy of the produced water in addition to its thermal energy. Furthermore, because of its high pressure, such a system may contain attractive amounts of methane gas dissolved in the water; that may be used to the electricity generation. Thus an ideal geo-pressured well can provide thermal, kinetic, and gas-derived energy. No commercial geo-pressured project has been developed to date, due to technical challenges to make this energy source commercial viable. Geothermal energy can be used for direct heat use not only for electricity generation.

## 7.2 Geothermal Energy Resources Characterization and Assessment

Geothermal resources can reasonably be extracted at costs competitive with other energy forms, are confined to the Earth crust areas where heat flow higher than in surrounding areas heats the water contained in permeable rocks, the reservoirs. Usually resources with the highest energy potential are concentrated on the plate boundaries, where significant geothermal activity exists, such as hot springs, stem vents, fumaroles or geysers. Active volcanoes are also a geothermal activity, on a particularly and more spectacular large scale. Geothermal activity in an area is the first significant indicator that sub-surface rocks in the area are warmer than the norm. The local heat source could be a magma body at 600°C–1000°C, intruded within a few kilometers of the surface. Geothermal fields can also form in regions unaffected by recent shallow magmatic intrusions. The anomalous higher heat flow may be due to particular tectonic settings, such as crust thinning, implying the upwelling of the crust-mantle boundary and higher temperatures at shallower depths. However, more than a thermal anomaly is needed to have a productive geothermal field. A reservoir with a large body of permeable rocks and thermal fluid at an accessible depth is also needed. The reservoir, bounded by cooler rocks hydraulically connected to the hot reservoir by fractures and fissures that are providing channels for rain-water to penetrate underground. These cooler rocks crop out at the surface where they represent the geothermal reservoir recharge areas. Thermal waters or steam are mainly rain-water that infiltrates into the recharge areas and proceeds to depth, becoming hotter while penetrating the reservoir hot rocks (Figure 7.3). Convection moves water inside the reservoir, transferring heat from the lowest reservoir levels to its upper levels, resulting in quiet uniform reservoir temperature. Convection, implying matter transfer, is more efficient heat transfer than conduction, the other heat transfer mechanism typical for less permeable rocks. Heat is transferred by conduction from the magma body toward the permeable reservoir rocks, the reservoir, filled with fluids. Hot fluids often escape from the reservoir, reaching the surface, produce the visible geothermal activity.

### 7.2.1 Definition of a Geothermal Resource (Heat, Permeability, Water)

For a new geothermal power plant to be installed or an existing one to continue operating, three elements are required: a good geothermal resource, access to the electrical grid and economic competiveness. The development of new geothermal power plants,

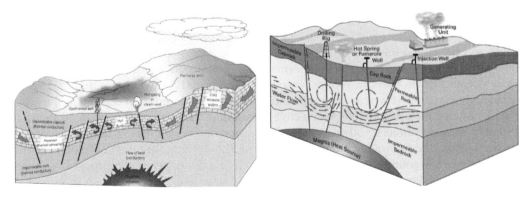

**FIGURE 7.3**
Pictorial diagram of a geothermal steam-field with its elements.

the expansion and maintenance of existing one depends on the three elements: heat, sufficient reservoir permeability, and water. Hydrothermal systems (geothermal reservoirs) are classified as: water-dominated or vapor-dominated, the latter one has higher energy content per unit fluid mass. Water-dominated fields are further divided into hot water fields, producing hot water, and wet steam fields, producing mixtures of water and steam. Geothermal resources are the ones that energy could reasonably be extracted at costs competitive with other energy forms. Geothermal resources are generally confined to crust areas where heat flow higher than in surrounding areas heats the water contained in permeable rocks (reservoirs). The resources with the highest energy potential are usually concentrated on the plate boundaries, where visible geothermal activity frequently exists.

1. Heat is the first element for any commercial power generation. Most commercial geothermal resources produce fluids with a temperature of at least 160°C (320°F), the resource temperature is a good indicator of economic viability. For example, the flow requirements for a 50 MW plant are 1890 kg/s of 150°C geothermal liquid, but only 510 kg/s if the reservoir temperature is 230°C.

2. Permeability describes the ability of the reservoir fluid (water or steam) to flow through the rock formation, allowing deep-seated geothermal heat sources to create a geothermal resource through the convection. The convection of the geothermal fluids through the reservoir heats a large rock volume, thereby storing a large energy over thousands of years. Geothermal reservoir permeability is dynamic, the hotter fluids dissolving minerals and increasing permeability, while cooling fluids depositing minerals and reducing it. Reservoir permeability allows the extraction of the stored heat through fewer wells.

3. Water is the motive fluid for geothermal power production. It may be brought to the surface as steam from one of the few (typically large) geothermal steam fields, such as Larderello (Italy) or The Geysers (California). A water-steam combination may also be produced to the surface from high-temperature liquid-dominated or liquid-and-steam reservoirs. In low- or moderate-temperature resources, pressurized water may be pumped to the surface using downhole pumps.

However, these three conditions occur simultaneously in relatively few places around the world, but there remain many thousands of megawatts of undeveloped worldwide resources. Over the long term, the natural resource base for geothermal power could be supplemented with human intervention to create new systems or to enhance the existing systems. The heat source, the reservoir, the recharge area, and the connecting paths through which cool superficial water penetrates the reservoir and, in most cases, escapes back to the surface, compose the hydrothermal system are the geothermal energy application ingredients. The type of hydrothermal system that can support economic geothermal developments, particularly in electricity generation, is where magmatic intrusions are emplaced high enough in the crust that can induce the groundwater convective circulation. The heat output of hydrothermal systems varies with time. Geothermal resources embrace a wide range of heat sources from the Earth; from the hydrothermal resources currently economically well developed to the non-hydrothermal sources stored more deeper in the Earth and available almost anywhere. Although conventional hydrothermal resources are used effectively for both electric and nonelectric applications in the United States and other countries, they are confined in specific locations and their ultimate potential for supplying electricity is then limited. For many years, because of technological constraints, the geothermal energy exploitation has been limited to very specific regions and those below certain depths considered not economically viable. However, with the EGS introduction, by improving technology and increasing the depth at which resources can be considered economical, the potential for developable geothermal resources expands significantly. Several methods of geothermal system development previously deemed unprofitable may be viable at the present time. It is also imperative to explore the feasibility of using existing oil and gas infrastructure and technology for geothermal energy development. Using existing infrastructure and knowledge are dramatically reducing the up-front cost required for the development of new geothermal operation. It is becoming important to explore the potential of an EGS in this context. This approach, designed to extract and utilize the Earth thermal energy, has enormous capabilities for primary energy recovery using heat-mining technology. EGS methods have been tested at a number of sites around the world and have been improving steadily. Much of the exploration and development technology used for geothermal has been adapted from oil and gas industry. Furthermore, geothermal resource exploration and development requires large amounts of risk capital, and petroleum producing companies are logical sources of the funds and specialized technology necessary to bring this resource to market. Geothermal energy resources from hydrocarbon setting can be divided into two categories: geo-pressured fluids and coproduced fluids. Thus, coproduced hot water from oil and gas production could be developed and provide a first step to more classical EGS exploitation. There appear to be five features that are essential to making a hydrothermal (i.e., hot water) geothermal resource commercially viable: a large heat source; a permeable reservoir; a water supply; an overlying layer of impervious rock; and a reliable recharge mechanism.

**Example 7.3:** Calculate the geothermal power potential of a site that covers 30 km² with a thermal crust of 2 km, where the temperature gradient is 240°C. At this depth the specific heat of the rocks is 2.5 MJ/m³, and the mean surface temperature is measured at 10°C. Assuming that the only 2% of the available thermal energy of the geothermal mass could be used for electricity generation, how much takes to produce $5 \times 10^4$ MWh?

**Solution:** From the estimate of the slab volume, its stored heat is then:

$$Q = VC\Delta T = 2 \times 30 \times 10^9 \times 2.5 \times 10^6 (240 - 10) = 34.5 \times 10^{18} \text{ J}$$

The total to generate power is:

$$W = 0.02 \times Q = 69 \times 10^{16} \text{ J}$$

$$\text{Time} = \frac{69 \times 10^{16} \text{ J}}{5 \times 10^{10} \times 3600 \text{ J}} = 3833.3 \text{ s or } 1.0648 \text{ hr}$$

Further the geothermal systems are classified as: convective, liquid- and vapor-dominated hydrothermal reservoirs, lower temperature aquifers, and conductive, hot rock and magma over a wide temperature range. Lower temperature aquifers contain deeply circulating fluids in porous media or fracture zones, with no localized heat source, being further sub-divided into systems at hydrostatic pressure and systems at pressure higher than geo-pressured systems. Resource utilization technologies are grouped under types for electrical power generation and for direct use of the heat. GHPs are a direct use subset, and EGS, where the fluid pathways are engineered by rock fracturing, are a subset of both types. A geothermal system requires heat, permeability, and water, so EGS technologies make up for reservoir deficiencies, enhancing the existing rock fracture networks, introduce water or another working fluid, or otherwise build on a geothermal reservoir that would be difficult or even impossible to derive energy from by using conventional technologies. However, the most widely exploited geothermal systems for power generation are continental type hydrothermal systems. In magmatic intrusion areas, temperatures above 1000°C often occur at less than 10 km depth. Magma typically involves mineralized fluids, mixed with deeply circulating groundwater, and a hydrothermal convective system is established whereby local surface heat-flow is significantly enhanced. Such shallow systems can last thousands of years, and the gradually cooling magmatic heat sources can be replenished periodically with fresh intrusions from a deeper magma. Finally, geothermal fields with temperatures as low as 10°C are also used for direct heat pumps. Subsurface temperatures increase with depth according to the geothermal gradient, and if hot rocks within drillable depth can be stimulated to improve permeability, using hydraulic fracturing, chemical or thermal methods, they form a potential EGS that can be used for power generation or direct use applications. EGS resources may occur in all geothermal environments, but are likely to be economic in geological settings where the heat flow is high enough to permit exploitation at depths of less than 5 km.

**Example 7.4:** Calculate the geothermal power potential of a site that covers 30 km² with a thermal crust of 2 km, where the temperature gradient is 240°C. At this depth the specific heat of the rocks is 2.5 MJ/m³ and the mean surface temperature is measured at 10°C. Assuming that only 2% of the available thermal energy of the geothermal mass could be used for electricity generation, how much does it take to produce $5 \times 10^4$ MWh?

**Solution:** From the estimate of the slab volume, its stored heat is then:

$$Q = VC\Delta T = 2 \times 30 \times 10^9 \times 2.5 \times 10^6 (240 - 10) = 34.5 \times 10^{18} \text{ J}$$

The total capacity to generate power is:

$$W = 0.02 \times Q = 69 \times 10^{16} \text{ J}$$

$$Time = \frac{69 \times 10^{16} \text{ J}}{5 \times 10^{10} \times 3600 \text{ J}} = 3833.3 \text{ s or } 1.0648 \text{ hr}$$

## 7.2.2 Reservoir Characteristics

Direct-use of geothermal energy is one of the oldest, most versatile, and common forms of geothermal energy uses. Unlike geothermal power generation, in which heat is converted to electricity, direct-use applications use heat energy directly to accomplish a broad range of uses. The temperature range of these applications is from about 10°C to about 150°C. Given the ubiquity of this temperature range in the shallow subsurface, these types of applications of geothermal energy have the potential to be installed almost anywhere that has sufficient fluid available. Approximately 5.4 × 1027 J of thermal energy is available in the Earth continents, of which nearly a quarter is available at depths shallower than 10 km. To be useful directly, the heat must be significantly above ambient surface temperatures and transferrable efficiently. Such conditions have traditionally been satisfied in places where warm or hot springs emerge at the surface or in locations where high thermal gradients allow shallow drilling to access heated waters. Such sites are restricted in their distribution, being concentrated in regions where volcanic activity or where rifting of continents has occurred. For these reasons, a small fraction of the large amount of heat contained within the continents can be economically employed for geothermal direct-use applications. The proportion of the heat that is readily available is not well known because thorough assessment efforts to quantitatively map the distribution of such resources have thus far been limited. As drilling technology improves and fluid circulation to support heat harvesting, and the continental thermal resource that can be accessed significantly expand. As of 2015, about 130 TWh/yr of thermal energy was worldwide used for direct-use purposes, which was derived from an installed capacity of about 55,000 MW. For comparison, global electricity consumption in 2006 was 16,378 TWh/yr. The growth in installed capacity of direct-use applications reflects its rapid growth in international development. In 1985, 11 countries reported using more than 100 MW of direct-use geothermal energy. In 2010, that number had increased to 78. The global distribution of these systems reflects the diversity of applications for which they have been engineered.

A geothermal system that can be developed for beneficial uses requires heat, permeability, and water. When hot water or steam is trapped in cracks and pores under a layer of impermeable rock, it forms a geothermal reservoir. However, the exploration of a geothermal reservoir for potential development includes exploratory drilling and testing for satisfactory conditions to produce useable energy, particularly temperature and flow of the resource. Water is a critical component of geothermal systems. The water used, which comes from the geothermal system, is re-injected back into the reservoir to maintain reservoir pressure and prevent reservoir depletion. On the other hand, rainwater and snowmelt generally continue to feed underground thermal aquifers, naturally replenishing geothermal reservoirs. Reinjection keeps the mineral-rich, saline water found in geothermal systems separate from ground water and fresh water sources to avoid cross-contamination. Injection wells are encased by thick borehole pipe and are surrounded by cement. Once the water is returned to the geothermal reservoir, it is reheated by the Earth's hot rocks and can be used over and over again to produce electricity or to provide heat. The key ingredients for geothermal energy production can be summarized by the following equation:

$$P_{conv} = C_P \cdot \left(T_{R\,svr} - T_{Rjec}\right) \cdot F_{rate} \cdot \eta - P_{Loss} = C_P \cdot \Delta T \cdot F_{rate} \cdot \eta - P_{Loss} \tag{7.1}$$

where $C_p$ is the specific heat of the working fluid, $F_{rate}$ is the flow rate from the production well (in Kg/s), $\Delta T$ is the sensible heat that can be extracted from the fluid produced by the production hole ($T_{reservoir} - T_{rejection}$), $\eta$ is the efficiency with which the heat energy can be used, and $P_{Loss}$ represents the fluid transfer and conversion losses. The goal in any geothermal system project is to optimize as many of these parameters as possible to increase electrical or heat output relative to the capital costs of developing the geothermal energy system. Based on the current experience in power generation from convective hydrothermal resources, the minimum amount of net energy produced by a well in a geothermal power system is around 4 MW. For most of the geothermal systems the working fluid is water with varying degrees of salinity or other dissolved materials. The specific heat is more or less constant for all of the geothermal resources. The $\Delta T$ is often in the range of 50°C–150°C, and the efficiency of power cycles is about 10%. Ignoring parasitic losses, well needs to have a flow rate of minimum of 70 kg/s to be economical viable. This rate is orders of magnitude higher than average flows in the U.S. oil industry, and at the upper end of production rates for wells, at the depths needed to access high temperatures. The flow problem is not as significant in convective hydrothermal resources as these typically produce steam rather than water. Although steam specific heat and density is lower than that of water, higher flow rates are achieved because of steam's low viscosity and density, allowing wells to produce usually without pumping. Exploration targets two components of Equation (7.1), $\Delta T$ and $F_{rate}$. Higher $\Delta T$ differences are targeted by looking for areas with anomalously high thermal gradients. Flow is targeted by looking for areas with high natural permeability or with characteristics that are suitable for geothermal enhancement. However, the exploration for natural resources is inherently high risk. These risks can be significantly reduced by the selection of suitable exploration targets. Targets need to be developed at all scales to enable the selection of appropriate basins, selection of tenements within these fields, the development of prospects within tenements and the locating the best exploration wells within prospects.

A geothermal development project starts with the system location, the steam or hot water production by means of strategically drilled wells. As might be presumed, most (but not all) hydrothermal systems indicate the best location through surface thermal manifestations. If any one of the features listed as needed for a viable hydrothermal resource is lacking, the field is likely not be worth for exploitation. Four types of geothermal systems are classified: *hydrothermal*, *hot dry rock*, *geo-pressured*, and *magmatic*. The commonly systems exploited at present are the hydro-thermal systems. The other three may be exploited industrially in the future as the technology evolved. Hydrothermal systems (geothermal reservoirs) are classified as *water-dominated* and *vapor-dominated*, the latter having higher energy content per unit fluid mass. Water-dominated fields are further divided into *hot water fields*, producing hot water, and the *wet steam field*, producing mixtures of water and steam. Hot water fields are able of producing hot water at the surface at temperatures up to 100°C. They are the geothermal fields with the lowest temperature, and the reservoir contains water in liquid phase. The reservoir may not have a cover of impermeable rock acting as a lid. However some of these thermal aquifers are overlain by confining layers that keep the hot water under pressure. Reservoir temperatures remain below the boiling point of water at any pressure because the heat source is not strong enough. Surface temperature is not higher than boiling temperature of water at atmospheric pressure. These fields may also occur in areas with normal heat flow. A hot water field is of economic interest if the reservoir is found at a depth of less than 2 km, if the salt content of the water is lower than 60 g/kg, and if the wells have higher flow-rates, 150 t/h or higher.

Wet steam fields contain pressurized water at temperatures exceeding 100°C and small quantities of steam in the shallower, in the lower pressure reservoir sections. The reservoir dominant phase in is liquid, being also the phase that controls the pressure inside the reservoir. Steam is not uniformly present occurring in the form of bubbles into the liquid water, and does not noticeably affect fluid pressure. An impermeable cap-rock lid generally exists to prevent the fluid from escaping to the surface, thus keeping it under pressure. However, at any depth below the water bears its own hydrostatic pressure. When the fluid is brought to the surface and its pressure decreases, a fraction of fluid is flashed into steam, while the larger part remains as boiling water. Once a well penetrates such reservoir, the pressurized water rises into the well due to higher pressure. The surface manifestations of such fields include boiling springs and geysers. The heat source is large and commonly of magmatic origin. The produced water often contains large quantities of chemicals (up to 100 g/kg, in some fields even up to 350 g/kg), mainly chlorides, bicarbonates, sulfates, borates, fluorides, and silica. These chemicals may cause severe scaling problems to pipelines and plants. More than 90% of the hydrothermal reservoirs exploited on an industrial scale are of the wet steam type. Electricity generation is their optimal utilization. One important economic aspect of wet steam fields is the large quantity of water extracted with the steam. Owing to its high chemical content this water has to be disposed of through reinjection wells drilled at the reservoir margins.

Vapor-dominated reservoirs produce dry saturated or superheated steam at pressures higher than atmospheric pressure. They are geologically similar to wet steam fields however the heat transfer from depth is much higher. Usually their permeability is lower than in wet steam fields, and the cap-rock presence is of critical importance. Water and steam co-exist, steam being the dominant phase, regulating the reservoir pressure, which is practically constant throughout the reservoir. These fields are called dry or superheated fields, producing superheated steam, with small quantities of other gases, mainly $CO_2$ and $H_2S$. The production mechanism in these fields is when wells penetrate the reservoir and production begins, depressurized zones form at well-bottom. The pressure drop produces liquid water boiling and vaporization in the surrounding rock mass. A dry area, without liquid water, forms near the well-bottom and steam flows here, expand and cool, but the addition of heat from the very hot surrounding rocks keeps steam temperature above the vaporization level for the pressure existing at that point. In this way, the well produces superheated steam with a degree of superheating which may reach 100°C, for example with wellhead pressures of 5–10 bar and a steam outlet temperature of more than 200°C. Surface geothermal activity associated with vapor-dominated fields, either dry or superheated, is similar to the activity of wet steam fields. About half of the geothermal electric energy generated in the world comes from six vapor-dominated fields. Hundreds of hydrothermal systems that have been investigated, less than 10% are vapor-dominated, 60% are wet steam fields (water-dominated) and 30% produce hot water. In water-dominated geothermal systems the most common type of fluid found, the primary water type is of near-neutral pH, a sodium-chloride brine (1000–10,000 mg/kg of Cl) containing gas, mainly $CO_2$. The proportion of magmatic volatiles ultimately determines the salinity of the reservoir waters.

Once a reservoir is identified and characterized, the power plant and related infrastructure is designed and equipment selected to optimize the resource use. The goals are to construct an energy efficient, low cost, minimal environmental impact plant. Geothermal fluids, hot, sometimes salty, mineral-rich liquid and/or vapor, are the carrier medium that brings geothermal energy through wells to up the surface. This hot water and/or steam is withdrawn from the underground reservoirs, isolated during production, flowing up wells and converted into electricity in a geothermal power plant or direct used for heating, cooling and hot water into buildings and industrial processes. Once used, the water and condensed steam is re-injected into the reservoir to be reheated. It is separated from groundwater by encased pipes, making the facility virtually free of water pollution. While several other geothermal resource types exist, all U.S. geothermal plants are using hydrothermal resources. Utilization of natural steam for electricity generation is not the only possible application. Hot waters, which are present in large parts of the all continents, can also be exploited for the space heating and industrial processes. The distribution of geothermal energy for nonelectric applications is divided as: (a) 40% for bathing and swimming pool heating, (b) 25% for space heating, (c) 12% for the use in GHPs, (d) 9% for greenhouse heating, (e) 5% for industrial applications, and (f) 9% for fish farm pond, agricultural drying, snow melting, air conditioning, or other uses.

Direct-use of geothermal energy is the oldest, the most versatile and the common forms of utilizing geothermal energy. Unlike geothermal power generation, in which heat energy is converted to electricity, direct-use applications use heat directly to accomplish a broad range of purposes. The temperature range of these applications is from about 10°C to about 150°C. Given this temperature range in the shallow subsurface, these application types of geothermal energy have the potential to be installed everywhere that has sufficient fluid available. Approximately $5.4 \times 10^{27}$ J of thermal energy is available on continents, and nearly a quarter is available at depths less than 10 km. In order to be useful, the heat must be well above ambient surface temperatures and easy to be transferred efficiently. Conditions have traditionally been satisfied in places where hot springs emerge at the surface or in locations where high thermal gradients allow shallow drilling to access heated waters. Such sites are relatively restricted in their distribution, concentrated in regions where volcanic activity has occurred recently or where continental rifting has occurred. For these reasons, a small fraction of the large amount of heat contained within the continents can be economically exploited for direct-use applications. The amount of the heat, readily available is not well known because thorough assessment efforts to quantitatively map the distribution of such resources have been limited. As drilling technology improves and fluid circulation to support heat harvesting at depth improves, the amount of the continental thermal resource will significantly expand. As of 2010, approximately 122 TWh/yr of thermal energy was used for direct-use purposes, value derived from the installed capacity of 50,583 MW. The growth in installed capacity of direct-use applications reflects a rapid growth in international development of this type of systems.

Hydrothermal resources of low to moderate temperature (20°C–150°C) are utilized to provide direct heating in residential, commercial, and industrial facilities. These are including space heating, water heating, greenhouse heating, aquaculture, food dehydration, laundries, and textile processes. These are commonly used in Iceland, the United States, Japan, and France. Geothermal energy resources are also used for agricultural production, to warm greenhouses to help in cultivation. For example in Hungary, thermal waters provide 80% of the energy demand of vegetable farmers. There are also geothermal greenhouses in Iceland and in the western United States. The heat from geothermal water is used for other industrial purposes, such as drying fish, fruits, vegetables and timber products, wool washing, cloth drying, in paper, and in milk industries. Geothermal water can be piped under sidewalks and roads to keep them from icing. Thermal waters are also used in gold and silver mining or for refrigeration and ice-making applications. GHPs have the largest energy use and installed capacity worldwide, accounting for 70.95% of the installed capacity and 55.30% of the annual energy use. The installed capacity is 49,898 MWt and the annual energy use is 325,028 TJ/yr, with a capacity factor of 0.21 (in the heating mode). Most of the GHP installations occur in North American, Europe, and China. However, the number of countries with GHP installations increased to 48 in 2015 from about 26 in 2000. The energy use reported for the GHPs is deduced from the installed capacity (if not reported), based on an average coefficient of performance (COP) of 3.5, which allows for one unit of energy input (usually electricity) to 2.5 units of energy output, for a geothermal component of 71% of the rated capacity [i.e., (COP-1)/COP = 0.71]. The cooling load was not considered as geothermal as in this case, heat is discharged into the ground or groundwater. Cooling, however, has a role in the substitution of fossil fuels and in emission reductions.

### 7.2.3 Improving a Geothermal Resource through Human Intervention

The terms hot dry rock (HDR) or hot fractured rock (HFR) refer to a family of technologies that are not yet commercially available. The objective is to establish any missing factors of a commercial geothermal resource (often permeability or water) where a heat source already exists. HDR experiments have been undertaken as research projects in United States,

Europe, and Japan. The problems are daunting and the costs are high. The HDR and HFR concept is to drill a well into the hot rock and then pump water at very high pressure, causing the rock to fracture. The fractures significantly enhance the heat transfer surface and flow path allowing water to be pumped from the surface into the well, circulated through the man-made fractures, recovered in production wells and used. Either the theory is straightforward; the implementation is still far from commercial viable developments. If HDR and HFR technology is developed and implemented in the future, the geothermal reservoir management strategies and the energy conversion technologies are much the same as the ones discussed. Between the naturally occurring resource and the potential man-made resource base of HDR/HFR is the enhanced geothermal system. EGS technologies seek to supplement a naturally occurring geothermal resource primarily by the addition of more liquid, or by stimulation of wells to tie into a larger, naturally occurring fracture network. The goal of EGS is to extend the life or capacity of existing fields, rather than the creation of an entirely new resource. From a thermodynamic point of view, it may seem that the only true renewable geothermal development would be one in which the extraction rate is the same as the natural heat influx rate into the system. However, reservoir simulations and field observations frequently reveal that the natural heat influx rate increases as production occurs, due to pressure changes that allow more hot liquid to flow into the system. Therefore, operating at a "nonrenewable" level increases the ultimate energy extraction from the resource. More importantly, a true "renewable" level of energy extraction would very often be sub-economic, and is therefore of little interest in the development of geothermal resources for society's benefit. Another definition argues that, as long as the power or heat usage from the resource continues at a constant level for hundreds of years, it approximates a true renewable resource. Again, this is an interesting theoretical discussion, but not one that actually is put into practice in the development of most geothermal resources, for both the reasons of economics and an inability to know what this actual level would be. More valuable than the theoretical discussion of whether geothermal energy is renewable, sustainable, both, or neither, are to look at the history of geothermal development. For example geothermal use at Larderello, Italy, is over 200 years old, starting with mineral extraction in the early 1800s and almost 100 years of commercial power generation. Geothermal power facilities only very rarely cease operation.

## 7.2.4 Resource Assessment and Analysis

Heat can be transferred by three processes: *conduction*, *convection*, and *radiation*. Conduction governs the thermal conditions in almost entire Earth solid layers, playing a very important role into the lithosphere. Convection dominates the thermal conditions in the zones where large quantities of fluids (usually molten rocks) exist, governing the heat transport in the fluid outer core and the mantle. On geological time scales, the mantle behaves like a viscous fluid due to the existence of high temperatures. Convection, involving heat transfer by the movement of mass, is a more efficient heat transport into the Earth compared to the conduction. However, into the Earth interior processes, both conductive and convective heat transfer play important roles. Radiation is the least important mode of heat transport in the Earth. However, the Earth internal heat losses, through the continental and oceanic lithosphere take place primarily by conduction, except near the mid-oceanic ridges where convection through hydrothermal circulation is significant. Cooling of magmatic intrusive bodies inside the crust and the upper mantle takes place through both processes. In anomalous geothermal areas, near geysers, hot water springs and fumaroles, convective heat transfer through circulating hot waters into the shallow subsurface levels far exceeds the background heat conduction. Conduction governs the temperature distribution into the continental lithosphere and the effects of sedimentation, burial, uplift and erosion processes on the subsurface temperature distributions. Convection dominates heat transfer into the deep mantle and outer core. Temperature, described as the object property, which determines the sensation of hotness or coldness felt from contact with it. More unambiguously, using the Zeroth Law of Thermodynamics, temperature of a system is defined as the property that determines whether or not that system is in thermal equilibrium with others system with which it is put in thermal contact. When two or more systems are in thermal equilibrium, they are said to have the same temperature. Temperature is commonly measured in the *Celsius* ($^oC$), *Fahrenheit* ($^oF$), and *Kelvin* (*K*) scales. The first two scales are based on the melting point of ice and the boiling point of water. Heat is defined as the energy transfer between two systems at different temperatures. Heat originates from other kinds of energy according to the first law of thermodynamics. It is important to distinguish between temperature and heat. Temperature is a property of matter, while heat is the energy that is flowing due to a temperature difference. Heat storage is the change of enthalpy or heat content of a medium in the heat transfer path. In accord with the first law of thermodynamics, change of enthalpy ($\Delta H$) occurs as a result of a time change in the temperature ($\Delta T$) of the medium. The amount of heat that a body is able to store as a result of change in temperature, called free energy, depends upon its specific heat capacity (*C*).

It is widely accepted that most geothermal fields are localized in areas of young tectonism and volcanism, and along the active plate boundaries. In fact, large volcanic-related hydrothermal systems only occur in areas where magma comes close to the surface, and this occurs at tectonic plate boundaries and over mantle hot spots. Present technology and economic factors restrict extraction of geothermal energy to the upper few kilometers of the Earth's crust. Geothermal wells, to date, are usually drilled to less than 5 km depth. As in the search for any natural resource, a strategy for geothermal energy exploration must be defined and followed. Once a geothermal area is identified, the next step is to use various exploration techniques to locate the most interesting geothermal sites and identify suitable fluid production targets. It is also necessary to estimate temperature, reservoir volume and permeability, and to predict whether wells are producing steam or only hot water. Ideally the chemical composition of the fluid to be produced is also estimated. To obtain this information, there are several exploration techniques available: inventory and survey of surface manifestations, geological, hydrogeological, geochemical, and geophysical surveys and the exploratory wells. To reduce the cost of exploration, a prescribed sequence of steps, altering the order from time to time depending on our prior knowledge of the area in question is recommended. In some cases, high costs lead to the elimination of some steps. For any geothermal field a series of resource assessments are made during the exploration and development processes. As more data become available the resource assessment becomes more certain. It is important that at each stage the uncertainties are fully estimated and the level of detail is appropriate without giving a false impression of precision. The main requirements for a good resource are: higher temperatures, large quantity of stored heat for longer resource life, with a low rate of liquid production per unit of energy. Other requirements may imply: reinjection well sites available at a lower elevation production for disposal under gravity; produced fluids with a near-neutral pH for low corrosion rates, adequate permeability to ensure adequate outputs from individual wells, low tendency for scaling in pipelines and wells, low elevation and easy terrain for access roads, lower risk of volcanic or hydrothermal eruptions and proximity to electrical grid or loads.

During the early exploration stage, prior drilling, resource assessment is largely qualitative. Once a few wells are drilled, it is possible to undertake a more accurate resource assessment. Resource assessment during production is more in the realm of reservoir engineering. Once production data are available, accurate estimates the resource quality, size and if additional resources can be found or the plant to be decommissioned are possible. The production sustainability from geothermal resources is a topic that has received little attention, leaving the question open to conjecture. As geologic phenomena, hydrothermal systems in the continental crust can persist for tens of thousands of years. However systems lifetimes can be foreshortened by artificial surface production during geothermal energy extraction. Geothermal project feasibility studies typically deal only with developing a certain sized power plant to be run for an arbitrary period, usually 30 years. Such limited studies fail to capture a true measure of the useful energy that can be produced from a resource. Geothermal utilization is divided into two categories, i.e., electric energy production and direct uses. Conventional electric power production is limited to fluid temperatures above 150°C, but considerably lower temperatures can be used in binary cycle systems (organic Rankine cycles, with the outlet temperatures of the geothermal fluid that are above 85°C). The ideal temperature of thermal waters for space heating is about 80°C, but larger radiators in the houses or the use of heat pumps or auxiliary boilers means that thermal water with much lower temperatures can be used.

In important part of geothermal thermal resource assessment consist of the field measurements. It is not uncommon to find more than one kind of fluid at a geothermal system. It is useful to find any relationship between the different fluids and to interpret from the chemistry of the geothermal fluids, their chemical compositions at production depth and the physical reservoir environment. Geo-thermometers are subsurface temperature indicators, using temperature dependent geochemical and/or isotopic composition of hot spring waters and other geothermal fluids under certain favorable conditions. The equilibrium between minerals or mineral assemblages and a given water chemistry is temperature dependent, and the relationship between chemical composition and temperature is predictable for certain parameters or ratios of parameters. Therefore, these parameters or ratios of parameters can serve as geo-thermometers. A commonly used geo-thermometer where geothermal waters that are coming from high-temperature environments (>180°C) is the atomic ratio of sodium to potassium (Na/K). The ratio decreases with an increase in temperature. The Na/K geo-thermometer fails at temperatures lower than 120°C and gives high temperatures for solutions with high calcium contents. A drawback of this thermometer is the preferential absorption of potassium by certain hydrothermal clay minerals, such as montmorillonite, present in the top few hundred meters in geothermal areas. All geo-thermometers have certain limitations. Therefore, the geo-thermometers should be applied with caution, utilizing all preliminary information available about the geological and hydrological setting as well as the chemistry of the fluids.

Geophysical surveys are providing the only means of delineating deep subsurface features, other than drilling, and are covering in short time larger areas, at lower cost when compared to drilling. In addition to major structural features, both shallow and deep, are addressing important questions relevant to the geothermal exploration. These include the heat source, reservoir areal extent, zones of fluid up-flows and highly permeable pockets, and the geothermal energy resource potential assessment. Geophysical techniques can provide valuable inputs to understanding the movement of fluids in response to production or re-injection during the exploitation of a geothermal field, helping in proper management for sustainable production. Therefore, geophysical investigations constitute an essential part of any exploration, in conjunction with geological, hydrological, and geochemical surveys. Geophysical anomalies associated with a geothermal prospect are usually caused by contrasts between physical properties of rocks and fluids inside or near the reservoir. The most common physical properties that are targets of geophysical exploration are temperature, resistivity, density, porosity, magnetic susceptibility, and seismic velocity.

The techniques applied to a geothermal prospect are primarily influenced by preexisting information such as the geological and tectonic setting, the hydrothermal system type, and area hydrological characteristics. Some of the geophysical methods, such as thermal, electrical resistivity, gravity, seismic refraction and well logging, successfully used in several geothermal exploration projects are now well established. Other methods using controlled source electromagnetic, magneto-telluric, self-potential, seismic reflection, and radiometric techniques are being developed and tested. The geophysical techniques used are specific to a particular geothermal resource. The technique choice is mainly governed by the temperature and the physic-chemical properties of reservoir rocks and the fluids contained. Therefore, before deciding on a set of geophysical surveys, it is important to have some ideas about the various geophysical targets that a particular hydrothermal system presents. Exploration work in geothermal areas is aimed at delineating the geometry of shallow geothermal reservoirs with up-flow zones causing geothermal manifestations. However, deeper systems exist without any up-flow zones and surface manifestations. The most suitable geophysical methods for exploring shallow reservoirs differ from those used for deep reservoirs. The fluid thermal state in the reservoir adds further complications. The physical properties of porous rocks, such as resistivity and density, are different in rocks filled with steam and those filled with hot water. A critical target of exploration programs is to locate highly permeable zones, eventually controlling the exploitation by locating suitable drill sites. Locating subsurface permeable zones within a reservoir is essentially an indirect procedure. Therefore, in an exploration the utilization of the available geological, hydrological and geochemical information relevant to permeability in a reservoir is needed to decide on techniques used to delineate highly permeable locales within the hydrothermal systems. The exploration methods are classified according to the depth of investigation, and as direct and indirect methods. Direct methods include thermal exploration techniques, aiming at geometrically mapping anomalous thermal zones. Indirect methods include techniques for investigating geological structures and highly permeable zones that control the fluids' accumulation and motion. Geophysical techniques are used for exploration and to map the extent of the geothermal reservoir, mass changes by fluid withdrawal or injection, and changes in saturation and gas content during the production stage. Microgravity, micro-seismic, and resistivity surveys are often used to monitor an exploited field.

## 7.3 Direct Use of Geothermal Energy

Geothermal heat is used directly for a variety of applications such as space heating and cooling, food preparation, hot spring bathing, balneology, agriculture, aquaculture, greenhouses and industrial processes. Uses for heating and bathing are traced back to ancient Roman times. From about $5.4 \times 10^{27}$ J of the Earth continents' available thermal energy about a quarter is estimated to be available at depths shallower than 10 km. In order to be useful directly, the thermal energy temperature must be above ambient surface temperatures and to be easily and efficiently transferred to the designed premises. Such conditions have are satisfied in places where there are warm or hot springs and in locations where high thermal gradients exist. Such sites are quite restricted in their distribution, being concentrated in regions where there is recent volcanic activity or where the rifting of continents occurs. For these reasons, a relatively small fraction of the heat contained within the continents can be economically employed for geothermal direct use applications. The geothermal energy resources that are readily available are not well assessed and their complete distribution map of such resources is not available. The main geothermal energy direct uses are (1) swimming, bathing and balneology; (2) space heating and cooling including district heating; (3) agricultural, aquaculture, and industrial applications; and (4) geothermal (ground-source) heat pumps. The annual growth rate for direct-use is about 8.3%, with the largest annual energy increase in the GHP technology. The overall global geothermal energy direct use is estimate close to 75 GWt, while the contribution of shallow reservoirs for small commercial, industrial or domestic applications is quite difficult to estimate. The capacity factors for such applications are in the range of 15%–75%. Direct use of geothermal energy in U.S. is about 12.5 GWt, China, 32.0 GWt, and UE is about 20 GWt. The oldest direct use of the geothermal energy is for bathing and therapy by using the hot spring or the surfacing heating hot water. Another traditional direct use of the geothermal energy is the space heating. The hot water, typically at 60°C or higher from geothermal reservoirs, is pumped into the building heating system, through the heat exchanger to provide heat to the building. The water is then re-injected into the geothermal reservoir for reheating. For direct use of geothermal energy a well is drilled into the geothermal reservoir, and there are pumps to bring the hot water to the surface (see Figure 7.4).

Direct use of geothermal energy includes the hydrothermal resources of low to moderate temperatures, providing direct heating in residential, commercial and industrial sector, include among others: space, water, greenhouse, and aquaculture heating, food dehydration, laundries and textile processes. These applications are commonly used in many countries. Unlike geothermal power generation, direct-use applications use heat directly to accomplish a broad range of purposes. The temperature range of these applications is from 10°C to about 150°C. Given the ubiquity of this temperature range in the shallow

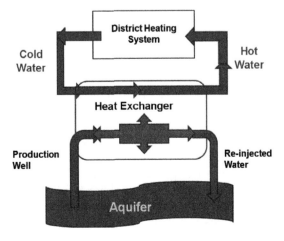

**FIGURE 7.4**
Block diagram of a district heating system.

subsurface, these types of geothermal applications have the potential to be installed almost everywhere. Geothermal resources are also used for agricultural production, to warm greenhouses, to help in cultivation or for industrial purposes, including drying fish, fruits, vegetables and timber products, washing wool, dying cloth, manufacturing paper and in milk industry. Geothermal heated water can be piped under sidewalks and roads to keep them from icing, during cold weather, extracting gold and silver from ore, and even for refrigeration and ice-making. Geothermal, ground-source heat pumps have the largest energy use and installed capacity worldwide, accounting for 70.95% of the installed capacity and 55.30% of the annual energy use. The installed capacity is 50,000 MWt and the annual energy use is 325,028 TJ/yr, with a capacity factor of 0.21 (in the heating mode). The energy use reported for the heat pumps was deduced from the installed capacity, based on an average coefficient of performance (COP) of 3.5, which allows for one unit of energy input (electricity) to 2.5 units of energy output, for a geothermal component of 71% of the rated capacity. The cooling load was not considered as geothermal effect; however, it has a significant role in the use of fossil fuels and pollutant emission reductions.

Thermal equilibrium is achieved when coexisting systems are reaching the same temperature, heat being spontaneously transferred from a hot to a colder body. This fundamental principle upon which all direct use applications rely is also the mechanism of the unwanted heat losses. The ability to manage heat transfer by minimizing unwanted losses and maximizing useful heat is required to set and operate an efficient direct use application. The heat transfer processes are conduction, convection, radiation, and evaporation. Heat transfer by conduction occurs at microscopic level through the exchange of vibrational energy by the atoms and molecules, while at the macroscopic level this process is manifest as changes in temperature when two bodies at different temperatures, are placed in contact with each other at a certain time: $t_1$. If $T_1$ and $T_2$ represent the initial temperatures of the bodies 1 and 2, respectively, and $T_3$ is the equilibrium temperature they eventually achieve at time $t_2$. Note that $T_3$ is not half way between $T_1$ and $T_2$, reflecting the effect of heat capacity of each body material. In this example, the heat capacity of body 1 must be higher than that of body 2. Conductive heat transfer is described by the relationship:

$$\frac{dQ_{cnd}}{dt} = k \cdot A \cdot \frac{dT}{dx} \tag{7.2}$$

where $dQ_{cnd}/dt$ (J/s or W) is the rate at which heat transfer occurs by conduction over the area $A$, $k$ is the thermal conductivity (W/m·°K), and $dT/dx$ is the temperature gradient over the distance $x$ (m). Equation (7.1) is Fourier's law of heat conduction. Equation (7.2) indicates that the heat transfer rate increases by increasing the area over which heat transfer occurs or by decreasing the distance. The thinner the plate, the greater is the temperature gradient across the plate and hence the greater the heat loss rate. Table 7.1 lists the thermal conductivities of some the common materials that may be used in direct use applications. Notice, that there is a difference of more than two orders of magnitude in the rate of heat loss, but that, for many common materials, the difference is more than four orders of magnitude. Although large temperature changes are not associated with most direct-use applications, it must be worth to mention that thermal conductivity is a temperature-dependent material characteristic. Thus, accurate computation of heat transfer rates must take into account the temperature dependence of $k$. It is so important to know the thermal conductivities and spatial geometry of the materials that are used for a direct-use application, or which are encountered when constructing a facility. Incorrect data on these parameters result in seriously under-sizing thermal insulation, inadequately sizing

**TABLE 7.1**

Thermal Conductivities of Selected Materials Used in Direct Use of Geothermal Energy

| Material | Thermal Conductivity (W/m K) |
|---|---|
| Aluminum | 202 |
| Copper | 385 |
| Iron | 73 |
| Carbon steel | 43 |
| Marble | 2.90 |
| Magnesite | 4.15 |
| Quartz | 6.50 |
| Glass | 0.78 |
| Concrete | 1.40 |
| Sandstone | 1.83 |
| Air | 0.0240 |
| Water | 0.5560 |
| Water vapor | 0.0206 |
| Ammonia | 0.0540 |

piping, and underestimating the rate of heat loss to the environment, all of which can seriously compromise the efficient operation of a direct use system.

**Example 7.5:** A spherical tank reservoir of 2 m diameter, having the wall thickness of 10 cm is filled with liquefied natural gas (LNG) at –150 C. The tank is insulated with a 4 cm thickness of insulation ($k$ = 0.015 W/m°C). The ambient air temperature is 20 C. How long takes for the LNG temperature of the LNG to decrease to –120 C. Neglect the thermal resistance of the steel tank and assume only conduction thermal heat transfer to the tank insulation. The density and the specific heat of LNG are 420 kg/m³ and 3.48 kJ/kg°C, respectively.

**Solution:** The LNG reservoir heat loss is:

$$\Delta Q = mC\Delta T = \rho \frac{\pi D^3}{6} C \left( T_{final} - T_{initial} \right) = 420 \times \frac{3.14 \times 2^3}{6} \times 3480 \left( -120 - (-150) \right) = 183.577 \times 10^6 \text{ J}$$

The time is the estimated from Equation (7.2), for a spherical reservoir as:

$$\Delta t = \frac{\Delta Q}{k \cdot A \cdot \dfrac{\Delta T}{\Delta x}} = \frac{183.577 \times 10^6}{0.015 \times 3.14 \times 2^2 \dfrac{20 - (-150)}{0.04}} = 229,270.6 \text{ sec or } 63.686 \text{ days}$$

Heat transfer by convection is a complex process that involves the movement of mass that contains a quantity of heat. Convection represents the heat transfer due to the bulk motion of a fluid. Convective heat transfer also occurs at interfaces between materials, as when air is in contact with a warm pool of water or is forced to flow at high velocity through a heat exchange unit. In such cases, buoyancy effects, the flow characteristics, the boundary layers, the effects of momentum and viscosity, and the surface properties and shape of the geometry of the flow pathway influence the heat transfer. For example, we assume that a cool air at temperature $T_2$ moves over a body of warm water at temperature $T_1$. Viscous and frictional forces act to slow the air movement near the water surface, forming a boundary layer, which is a region where a velocity gradient develops between the interface and the main air mass that has velocity $v$. At the interface the velocity approaches zero. The characteristics of the boundary layer are dependent upon the fluid properties, velocity, temperature and pressure. Heat is transferred by diffusive processes from the water surface to the fluid at the near-zero velocity boundary layer base, causing its temperature to approach that of the water, $T_1$, resulting in a temperature gradient, in addition to the velocity gradient, between the main mass of moving fluid and the water–air interface. This thermal gradient becomes the driving force behind thermal diffusion that contributes to heat transfer through this boundary layer. Advective transport of heated molecules provides an additional mechanism for heat to move through the boundary layer and into the main air mass, resulting in an increase in the temperature of the air. The rate at which convective heat transfer occurs follows Newton's law of cooling, which is expressed here as:

$$\frac{dQ_{cnv}}{dt} = h \cdot A \cdot dT \tag{7.3}$$

where $dQ_{cnv}/dt$ is the rate at which heat transfer occurs by convection, $h$ is the convection heat transfer coefficient (J/s-m$^2$-K), $A$ is the exposed surface area (m$^2$), and $dT$ is the temperature difference between the warm boundary and the overlying cooler air mass, $(T_1–T_2)$. Values for $h$ are strongly dependent on the properties of the materials involved, the pressure and temperature conditions, the flow velocity and whether flow is laminar or turbulent, surface properties of the interface, the geometry of the flow path and the orientation of the surface with respect to the gravitational field, $h$ being variable and specific to a given situation. Determining values for $h$ requires geometry-specific experiments, or access to functional relationships that have been developed in analog cases. For example, convective heat loss from a small pond over which air is flowing at a low velocity can be reasonably accurate represented by:

$$\frac{dQ_{cnv}}{dt} = 9.045 \cdot v \cdot A \cdot dT \tag{7.4}$$

Here $v$ is the velocity of the air and the effective units of the coefficient, 9.045 are kJs/m$^3$·h°C. In the ideal case, radiation heat transfer is represented by considering a so-called ideal black body. A black body radiator emits radiation that is strictly and absolutely dependent only on temperature, so the wavelength of the emitted radiation is strictly inversely proportional to the temperature. At room temperature, for example, an ideal black body would emit primarily infrared radiation while at very high temperatures the radiation would be primarily ultraviolet. However, real materials emit radiation in more complex ways that depend upon both the surface properties of an object and the physical characteristics of the material of which the object is composed. In addition, when considering radiative heat transfer from one object to another, the geometry of the heat source, as seen by the object receiving the radiation, must also need to be taken into account. For most considerations involving radiative heat transfer in direct geothermal energy applications, interfaces are commonly flat plates or enclosed bodies in a fluid, the geometrical factor of minimal importance, and the following relationship is used:

$$\frac{dQ_{rad}}{dt} = \varepsilon \cdot \sigma \cdot A \cdot \left(T_2^4 - T_1^4\right) \tag{7.5}$$

where ε is the emissivity of the radiating body (ε equals 1.0 for a perfect black body), at temperature $T_1$, σ is the Stefan – Boltzmann constant, equals $5.669 \times 10^{-8}$ W/m²K⁴, $T_1$ and $T_2$ are the respective temperatures of the two involved objects, and A is its effective surface area.

Heat transfer through evaporation can be an efficient energy transport mechanism. The factors that influence evaporation rate are temperature and pressure of the vapor, overlying the evaporating fluid, the exposed area, the fluid temperature, the equilibrium vapor pressure and the wind velocity. Each of these properties are relatively simple to formulate individually, however, the evaporation process is affected by factors similar to those that influence convective heat transfer becoming quite complex. A complicating factor is that the boundary layer behavior with respect to the partial pressure varies both vertically away from the interface and along the interface due to turbulent flow mixing, so the ambient vapor pressure is not represented rigorously. Moreover, the evaporation rate is affected by the temperature gradient above the interface, which is affected by the boundary layer properties, which in turn, influence the equilibrium vapor pressure. Since the gradient is the driving force for diffusional processes, the rate at which diffusion transports water vapor from the surface is affected by the local temperature conditions as well as the fluid velocity. These complications have led to an empirical approach for establishing evaporation rates in which various functional forms are fit to data sets that span a specific range of conditions. One such common heat rate relationship due to evaporation used in direct geothermal energy use is:

$$\frac{dQ_{evp}}{dt} = \frac{a \times \left(P_{water} - P_{air}\right)^b \times H_w}{2.778 \times e^{-7}} \tag{7.6}$$

where $P_{air}$ and $P_{water}$ are the water vapor saturation pressures (kPa) at the water and the air temperatures, respectively, $H_w$ is the, $a$ and $b$, two empirical constants, determined by the velocity of the fluid moving over the interface (m/s), that often are calculated as:

$$a = 74.0 + 97.97 \times v + 24.91 \times v^2$$

$$b = 1.22 - 0.19 \times v + 0.038 \times v^2$$

**Example 7.6:** Estimate the convection, conduction, radiation, and evaporation heat transfer rates for a rectangular water pool, constructed of concrete walls 15 cm in the ground, having the dimensions, 15 m by 40 m and 2 m depth, assuming the external air temperature 20°C, the wind speed is 5 m/s and the pool has the temperature of 37°C.

**Solution:** Applying Equations (7.2), (7.4), (7.5), and (7.6) the heat transfer rates due to conduction, convection, radiation, and evaporation are computed as follow. The total area for conduction heat transfer is calculated considering the lateral walls and the bottom:

$$A_1 = 2 \times \left(15 + 40\right) \times 2 + 15 \times 40 = 880 \text{ m}^2$$

$$\frac{dQ_{cnd}}{dt} = 1.4 \frac{W}{m \cdot K} \times 880 \text{ m}^2 \times \left(\frac{17}{0.2 \text{ m}}\right) = 104{,}720 \text{ W(J/s)}$$

Convection and radiation heat losses are taking place only on the pool upper (open) surface.

$$\frac{dQ_{cnv}}{dt} = 9.045 \cdot 5(\text{m/s}) \cdot \left(15 \times 40\right) \text{m}^2 \cdot \left(37 - 20\right)(°C) = 461,295 \text{ J/s (W)}$$

$$\frac{dQ_{rad}}{dt} = 0.99 \times 5.669 \times 10^{-8} \frac{W}{m^2 K^4} \times 600 \text{ m}^2 \left(310^4 \text{ K}^4 - 293^4 \text{ K}^4\right) = 62,807.1 \text{ W(J/s)}$$

And the rate of evaporation heat loss, assuming the partial pressure to 3.7 kPa and 1.23 kPa is then:

$$a = 74.0 + 97.97 \times 5 + 24.91 \times 5^2 = 1186.6$$

$$b = 1.22 - 0.19 \times 5 + 0.038 \times 5^2 = 1.22$$

$$\frac{dQ_{evp}}{dt} = \frac{1186.6 \times \left(3.7 - 1.23\right)^{1.22} \times 1}{2.778 \times e^{-7}} = 14,411,639 \text{ W(J/s)}$$

### 7.3.1 Assessing Feasibility of Direct Use Applications

The amount of heat, $Q_{Load}$, required to operate the function of the designed installation depends upon the specific process and the operation size and the external conditions. Assuming that load, $Q_{Load}$ is constant over time, then the geothermal resource must be of sufficient temperature and flow rate to satisfy:

$$\frac{dQ_{Geotherm}}{dt} > \frac{dQ_{Load}}{dt} + \frac{dQ_{Losses}}{dt} \tag{7.7}$$

From the discussion of heat transfer mechanisms the total heat losses, $Q_{Losses}$ that are needed to be accounted for in any application are the sum of all relevant heat loss mechanisms, conduction ($Q_{cnd}$), convection ($Q_{cnv}$), radiation ($Q_{rdt}$), and evaporation ($Q_{evp}$), and is expressed as:

$$\frac{dQ_{Losses}}{dt} = \frac{dQ_{cnd}}{dt} + \frac{dQ_{cnv}}{dt} + \frac{dQ_{rad}}{dt} + \frac{dQ_{evp}}{dt} \tag{7.8}$$

This equation represents the heat loss that is assumed for the application operating conditions. The heat, $Q_{Load}$, required to perform the function of the designed installation (load) depends upon the specific processes and the operation size. For most applications it is likely that seasonal variability influences the total heat losses through changes in the air temperature and other seasonal weather variables, such wind or radiation. For this reason the design load concept was developed, the most severe condition set a facility is likely to experience, maximizing the heat losses. Hence, when evaluating the feasibility of a potential direct use project it is important to establish whether the resource is sufficient to meet the maximum demanding conditions that are likely to be encountered. In instances, where an abundant resource is available, it may be suitable to size the facility in such a way to meet all probable energy demands. In other instances, possibly for reasons of economics, it may turn out to be sufficient to design the facility such that the geothermal resource is meeting the demand of some maximum percentage of probable events, while the remainder is addressed with other supplemental energy sources. However, direct heating is more efficient than electricity generation and places less demanding temperature requirements on the heat resources. Heat may come from co-generation with a geothermal electrical plant or from smaller wells or heat exchangers buried in shallow ground. As a result, geothermal heating is economical over a much greater geographical range than geothermal electricity. Where natural hot springs are available, the heated water can be piped directly into radiators. If the ground is hot but dry, Earth tubes or downhole heat exchangers can collect the heat. But even in areas where the ground is colder than room temperature, heat can still be extracted with a geothermal heat pump more cost-effectively and cleanly than it can be produced by conventional furnaces. These devices draw on much shallower and colder resources than traditional geothermal techniques, and they frequently combine a variety of other functions, including air conditioning, seasonal energy storage, solar energy collection and electric heating. Geothermal heat pumps can be used for space heating essentially anywhere in the world.

**Example 7.7:** For the pool of Example 7.3 estimate the geothermal inflow rate from a 50°C reservoir to keep the pool temperature at 37°C.

**Solution:** The total heat losses, due to the conduction, convection, radiation, and evaporation are given by Equation (11.8):

$$\frac{dQ_{Losses}}{dt} = 104,720 + 461,295 + 62,807 + 14,411,639 = 1,638,929 \text{ W(J/s)}$$

The geothermal inflow rate is computed as:

$$q_{inflow} = \frac{dQ_{Losses}/dt}{C_P\left(T_{Geothermal} - T_{Water}\right)} = \frac{1,638,929}{4183.3 \times \left(45 - 37\right)} = 489.7 \text{ kg/s}$$

Notice the evaporation heat transfer losses are the most important mechanism for heat losses for the open pools or ponds.

## 7.3.2 District Heating

Approximately $5.4 \times 10^{27}$ J of geothermal energy is available, of which nearly a quarter is available at depths shallower than 10 km. In order to be useful directly, this heat must be significantly above ambient surface temperatures and easy to be transferred efficiently. The basic district heating requirements are a source of warm geothermal fluid, a network of pipe to distribute the fluid, a control system, and a disposal or reinjection system. These conditions have traditionally been satisfied in places where warm or hot springs emerge at the surface, or in locations where high thermal gradients allow shallow drilling to access heated waters. Such sites are relatively restricted in their distribution, being concentrated in regions where volcanic activity has occurred recently or where rifting of continents has happened. For these reasons, a relatively small fraction of the large amount of heat contained within the crust can be economically employed for geothermal direct use applications. However, for example the space heating accounted for about 60,000 TJh/yr of the total 273,372 TJh/yr of energy consumed through direct use applications, which is the third largest user of geothermal fluids for direct use, worldwide. The vast majority of these heating systems involve district systems in which multiple users are linked into a network that distributes heat to users. District heating systems distributes hydrothermal water through piping system to blocks of buildings. Such system design requires matching the distribution network size to the available resource. The resource attributes are the sustainable flow rate (usually this is required to be between 30 and 200 kg/s, depending on the size of the district heating system and the temperature of the resource) and the temperature of the resource. There are three typical components of a district heating system: a production facility, a mechanical system and a disposal system. A production system is the well(s) needed to bring the hydrothermal water/heat energy from the geothermal reservoir. A mechanical system is a system that delivers the hydrothermal water/heat energy to the process. A disposal system is a medium that receives the cooled geothermal fluid. It can be a pond, river or an injection fluid system. The geothermal power ($Q_{Geo}$) that must be provided from a resource:

$$Q_{Geo} > m \times C_P \times \left( T_{Geo} - T_{Rtn} \right) \tag{7.9}$$

Here $m$ is the mass flow rate (kg/s), $C_P$ is the constant pressure heat capacity of the fluid (J/kg-K), $T_{Geo}$ and $T_{Rtn}$ are the temperature (K) of the water from the geothermal source and temperature of the return water after it has been through the network, respectively.

District heating systems distributes hydrothermal water through piping system to blocks of buildings. There are three typical components of a district heating system: a production facility, a mechanical system and a disposal system. A production system is usually a well to bring the hydrothermal water/heat energy from the geothermal reservoir. A mechanical system is a system that delivers the hydrothermal water/heat energy to the process. A disposal system is a medium or area that receives the cooled geothermal fluid, such as a pond, river or an injection well. District heat basic requirements and components are a source of geothermal fluid, a network of pipe to distribute the heated fluid, a control system, and a disposal or reinjection system. System design requires matching the size of the distribution network to the available geothermal resource. The geothermal water from the production well(s) passes through heat exchangers, so its heat is transferred to a cleaner secondary fluid, often fresh water, which is distributed via insulated pipes to the district, where it transfer the energy to the group of houses and/or commercial facilities. Fluid valves control the amount of geothermal water at heat exchangers and balance the heating demands with the supply. Geothermal water does not exit the heat exchangers, dissolved gases or solids are re-injected into reservoir, with almost any environmental impact. The resource characteristics are a sustainable flow rate (usually this is required to be between 30 and 200 kg/s, depending on the size of the district heating system and the temperature of the resource) and the resource temperature of the resource. The geothermal power, $P_G$, provided from a resource is estimated from:

$$P_G = q_m C_P \left( T_{GW} - T_{RW} \right) \tag{7.10}$$

where $q_m$ is the mass flow rate (kg/s), $C_P$ is the constant pressure heat capacity of the fluid (J/kg-K), and $T_{GW}$ and $T_{RW}$ are the temperature (K) of the water from the geothermal source and temperature of the return water after it has been through the network, respectively. The heat demand, $Q_{Load}$, which is imposed on the system, is a complex function of time. During the day the load can vary by up to a factor of 3, and is affected by the seasonal climatic variability. From above equation is clear that the only variable that can be controlled, affecting $P_{GW}$ is the return water temperature, $T_{RW}$ since the other variables are set by the properties of the natural system. Maximizing the temperature drop across the network thus becomes a means to increase the power output of the system. However, how this is addressed depends upon the operating mode that can be employed for the system.

> **Example 7.8:** Assuming that a geothermal well is providing a flow rate of 300 kg/s with the difference of temperature of water between the geothermal source and return of 75°C, what is the geothermal power of this application? If the total losses are 30% pf the geothermal power what is the maximum load power?
>
> **Solution:** The geothermal power of the well is given by the Equation (7.10), as:
>
> $$P_G = 300 \times 4180 \times 80 = 100.32 \text{ MW}$$
>
> The maximum usable power is:
>
> $$Q_{Load(Max)} = (1 - 0.30) \times 100,320,000 = 70.224 \text{ MW}$$

## 7.3.3 Geothermal Heat Pumps

Heat pump technology is considered to be one of the most sophisticated and beneficial engineering accomplishments of the twentieth century. Heat pumps are devices that use a compressor-pump system to extract heat from a low-temperature reservoir and reject it a higher temperatures. They are, in essence Carnot cycle applications, operating at highest efficiency levels, by employing transport heat that already exist, without generation, for both heating and cooling, while using only a small fraction of energy amount that they move. These devices can be used in low-temperature geothermal schemes to maximize heat extraction from fluids, while their specific function in any scheme depends on the used fluid temperature. A geothermal heat pump (GHP) is a heat pump that uses the Earth thermal capacity as an energy source to provide heat to a system or as an energy sink to remove heat from a system, cooling the system. There are three major GHP types: ground-coupled GHPs, ground water GHPs and hybrid GHPs. The ground-coupled GHPs are of two types: vertical closed-loop and horizontal closed-loop, based on the heat pump piping system shape. In the case of a moderate fluid temperature range (50°C– 70°C) the extracted heat depends primarily on the heat exchanger, and how the heat pump is connected to extract additional heat from the geothermal fluid. For low temperature range (less than 50°C, usually 40°C or lower), heat extraction is almost impossible and the heat pump is connected to ensure all heat transfer. Geothermal heat pumps (GHPs) take advantage of the Earth's relatively constant temperature at depths of about a few meters to about 100 m (or 10 ft to about 300 ft). GHPs can be used almost everywhere in the world, as they do not have the requirements of fractured rock and water as are needed for a conventional geothermal reservoir. GHPs circulate water or other liquids through pipes buried in a continuous loop, either horizontally or vertically, under a landscaped area, parking lot or any number of areas around the building, being one of the most efficient heating and cooling systems available. To supply heat, the system pulls heat from the Earth through the loop and distributes it through a conventional duct system. For cooling, the process is reversed;

the system extracts heat from the building and moves it back into the Earth loop. It can also direct the heat to a hot water tank, providing another advantage, free hot water. GHPs reduce electricity use 30%–60% compared with traditional heating and cooling systems, because the electricity which powers them is used only to collect, concentrate and deliver heat, not to produce it.

Direct geothermal systems for heating and cooling typically comprise a primary circuit that exchanges heat with the ground, a heat pump that exchanges and enhances heat transfer between the primary circuit and the secondary circuit, and a secondary circuit that circulates heat within the building. Space heating systems usually require higher temperatures than that of the ground. At first glance, the use of the cooler ground to heat a building may appear a contravention of the second law of thermodynamics (heat flows from hot to cold). Heat pumps overcome this apparent restriction by enhancing the ground-sourced energy with electrical or mechanical work. Refrigerators are examples of common, everyday heat pump. Figure 7.5 gives a schematic of a geothermal heat pump diagram and the basic operating principle. Heat transfer occurs in fluids when they change temperature and/or phase, while the heat transfer associated with phase change is much greater than that corresponding to only temperature change, and heat pumps make use of the properties of refrigerants (which can change phase at suitable operating temperatures and pressures) to achieve efficient heat transfer.

The heat pump principle and operation consist of a cooled, liquid refrigerant that is pumped into the heat exchanger (evaporator), where it absorbs thermal energy from the ambience as a result of the temperature differential, then through the compressor to a condenser where is released the heat, as is shown in Figure 7.6. During this process, the refrigerant then changes state and becomes a gas and then he gaseous refrigerant is recompressed in the compressor, resulting in a temperature increase. A second heat exchanger (condenser) transports this thermal energy into the heating system and the refrigerant reverts to a liquid form, and the refrigerant pressure is reduced again in the expansion valve. Basically, the liquid refrigerant absorbs heat from a heat source and evaporates, the refrigerant, cooler than the heat source, has boiling point below the heat source temperature, the refrigerant gas then passes through compressor, which increases its pressure and temperature and the hot and high pressure refrigerant gas from the compressor is hotter than the heat sink (so that heat flows from the refrigerant to the heat sink). At this higher pressure, the refrigerant gas condenses at a higher temperature than at which it boiled. Thus when the

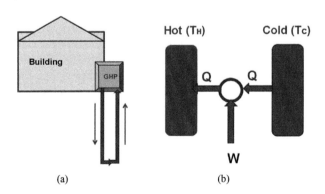

(a)                                                    (b)

**FIGURE 7.5**
(a) Geothermal heat pump and (b) operating principle.

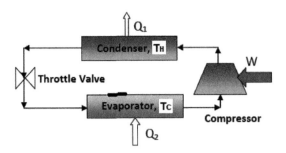

**FIGURE 7.6**
Heat pump schematic and operating diagram.

refrigerant gas reaching the condenser condenses and releases heat. The hot, high-pressure liquid refrigerant then passes through an expansion valve, which returns the pressure and temperature of the liquid to its original conditions prior to the cycle. For GHPs in heating mode, refrigerant evaporation occurs where the heat pump joins the primary circuit and condensation occurs where the heat pump joins the secondary circuit. Refrigerant evaporation cools the circulating fluid in the primary circuit is then re-heated by the ground. This process is reversed in cooling mode as refrigerant condensation heats the circulating fluid in the primary circuit, which is re-cooled by the ground. It is important to note that GHPs require energy input (the compressor and to the pumps that circulate fluid) to move heat around the system. However, this energy input required is small compared to the heat output. A GHP produces about 3.5 kW, even up to 5.5 kW of thermal energy for every 1 kW of electricity used. Heat pump efficiency also increases as the temperature difference between the heat sink and source decreases. GHPs are thus more efficient than air source heat pumps the seasonally averaged ground temperature is closer to the desired ambient building temperature than the air is. In order to transport heat from a heat source to a heat sink, heat pumps need external energy. Theoretically, the total heat delivered by the heat pump is equal to the heat extracted from the heat source, plus the amount of drive energy supplied to it. It is simply a heat engine running in reverse mode, as shown in Figure 7.6. It accepts heat $Q$ from the sink at $T_C$ (at lower temperature), rejecting heat $Q$ into the source which is at a higher temperature $T_H$ in doing so it consumes work $W$, from an external source. From thermodynamic point of view a heat pump thermal cycle is a revere thermal engine cycle, so the input work provides the heat transfer to rise the temperature. The basic relationships governing the heat pumps are the laws of thermodynamics ad are independent of the working fluid, cycle type and the form of heat transfer. The heating efficiency of a heat pump is given by is coefficient of performance ($COP$) defined as the ratio of heat provided per energy input:

$$COP = \frac{\text{Heat Output}}{\text{Input Work}} = \frac{Q_H}{W} < \frac{T_H}{T_H - T_C} \qquad (7.11)$$

However, it is usually to define overall efficiencies for actual (real) heat engines and heat pumps, according to the Carnot cycle, as:

$$W = \eta \cdot Q_H$$

with $\eta < \eta_C$, for heat engines, and $\eta > \eta_C$ for heat pumps, where $\eta_C$ is the Carnot efficiency, operating between the same temperatures. Previous equation is usually rewritten for heat pumps as:

$$W = \frac{Q_H}{COP_{HP}} = \frac{Q_C}{COP_{RF}} \qquad (7.12)$$

Here $COP_{HP}$ is the so-called "coefficient of performance" based on heat output $Q_H$ from a heat pump or vapor recompression system, and $COP_{RF}$ is the coefficient for a refrigeration system based on heat absorbed $Q_C$ from the process, The coefficients of performance are defined for real (actual) systems, with temperature expressed in Kelvin degrees (the absolute temperature) as:

$$COP_{HP} = \frac{\text{Desired Output}}{\text{Required Input}} = \frac{\text{Heating Effect}}{\text{Input Work}} = \frac{Q_H}{W} = \eta_{mech} \frac{T_H}{T_H - T_C} \qquad (7.13a)$$

And

$$COP_{RF} = \frac{\text{Desired Output}}{\text{Required Input}} = \frac{\text{Cooling Effect}}{\text{Input Work}} = \frac{Q_C}{W} = \frac{Q_H - W}{W} = COP_{HP} - 1 \qquad (1.13b)$$

Hence, the heat pump as a low temperature lifting device ($T_C \rightarrow T_H$) gives higher $COP_{HP}$, which is inversely proportional to the temperature stretch and a large amount of upgraded heat per unit power. However, lower values of $T_C$ reduce $COP_{RF}$, so refrigeration systems need more power per unit for upgraded heat as the absolute temperature falls. With the above understanding, one can start analyzing the GHP problems. Theoretically, heat pumping can be achieved by many more thermodynamic cycles and processes. These include Stirling cycle, single-phase cycles (e.g., with air, $CO_2$, or noble gases), solid-vapor sorption systems, hybrid systems (combining the vapor compression and absorption cycle), and electromagnetic and acoustic processes. Some of these systems are close to entering the market or have reached technical maturity, and can become significant in the future. Almost all heat pumps fall on two categories, i.e., either based on a vapor compression, or on an absorption cycle. Heat pumps are used for two purposes, for refrigeration/cooling below-ambient temperature, or as a heat recovery system. However, the equipment used in both cases is very similar.

**Example 7.9:** Compare the heating efficiencies (maximum *COP*) of the same heat pump installed in New Orleans, Louisiana, and in Cleveland, Ohio.

**Solution:** In New Orleans, the climate is milder with higher average temperatures we can assume that $T_H$ (summer) is about 20°C (or 70°F) and that $T_C$ (winter season) is about 5°C (or about 40°F). In Cleveland, assume that $T_H$ is the same, but that $T_C$ (the outside temperature) is much lower, on average is about −10°C (or about 14°F). Since the heat pump is used as a heater, after the conversion to absolute temperatures, the maximum *COP* at each of the two locations is calculated as:

$$\text{Cleveland: } COP_{max} = \frac{T_H}{T_H - T_C} = \frac{293}{293 - 263} = 9.77$$

$$\text{New Orleans: } COP_{max} = \frac{T_H}{T_H - T_C} = \frac{293}{293 - 278} = 19.53$$

The most important benefits of GHPs are that are using about one third or even over half less electricity than the conventional heating and cooling systems, having a good potential for energy savings in any area. The cooling efficiency is defined as the ratio of the heat removed to the input energy, or the energy efficiency ratio (EER). Good GHP units must have a *COP* of 3 or greater and an EER of 13 or greater.

## 7.4 Electricity from Geothermal Energy Sources

The world geothermal electrical capacity installed in the year 2015 was 13.5 GWe, and is expected to reach about 18.0 GWe by 2020. In the industrialized countries, where the installed electrical capacity reaches very high figures, geothermal energy is unlikely, in the next decade, to account for more than one percent of the total. In developing countries, with an as yet limited electrical consumption but good geothermal prospects, electrical energy of geothermal origin could, on the contrary, make quite a significant contribution to the total. The efficiency of the generation of electricity from geothermal steam ranges from 10% to 17%, about three times lower than the efficiency of nuclear or fossil-fuelled plants. Geothermal plants have the lowest efficiency values due to the low temperature of the steam, usually below 250°C. Furthermore, geothermal steam has a chemical composition that is different from pure water vapor, containing non-condensable gases, tending to reduce the overall system efficiency. There are three main technologies for production of electricity from geothermal resources: dry steam-, flash-, and binary power plants. Flash power plants represent 60% of the world's total capacity (with 209 running units) while dry steam plants represent around 26% (with 61 units) and binary plants 10% (with 266 units). The reason for the binary power plants accounting for a small share of the capacity despite the high number of units is the small power per unit production. It is the property and temperature of the geothermal fluid or vapor that determines which techniques that should be used to extract maximum electricity from the resource. Geothermal power plant operation cycles follow laws of thermodynamics, mainly the conservation of energy. The state of the geothermal fluid at different points in the power plant's working cycle and the processes it undergoes is described on a thermodynamic state diagram. The geothermal well is basically a vertical pipe transporting fluid between the reservoir and the surface. The conservation of mass is also important when analyzing these cycles. The most important parameters when considering these working cycles are the fluid properties, such as pressure, temperature, specific volume, enthalpy and entropy. In the operation cycle, the working fluid undergoes phase changes, existing a liquid phase, a gas phase and often a two phase mixture of liquid and gas at some point in the cycle. In a geothermal power plant working cycle, different thermodynamic processes occur in various components of the plant. These processes are assumed to be ideal in most of the geothermal power plant analysis, i.e., neglecting losses. The thermal efficiency of a power plant is derived from the First Law of thermodynamics and is defined as the net power output of a plant divided by the rate of heat supplied to the plant. Thermal efficiency $\eta_{th}$, is a measure of the process quality and is defined as the ratio between the work output and the heat flow into the process:

$$\eta_{th} = \frac{\dot{W}}{\dot{Q}_{IN}} = \frac{\dot{Q}_{IN} - \dot{Q}_{OUT}}{\dot{Q}_{IN}} \tag{7.14}$$

From thermal engineering principles one of the most important concepts governing the operation of a power plant is that the efficiency of the process is determined by the temperature difference between the boiler, $T_H$ and the condenser, $T_C$, expressed in Kelvin degrees. The theoretical efficiency, $\eta_{TCE}$ of the cycle may be calculated by using the following relationship:

$$\eta_{th} = \frac{T_H - T_C}{T_H} \tag{7.15}$$

**Example 7.10:** A geothermal power plant is operating in connection with a reservoir with temperature of 350°C (the highest reservoir temperature limit) and the steam is released into the atmosphere, calculate the plant theoretical efficiency.

**Solution:** Assuming the atmospheric steam release the condenser has an average temperature of 35°C, after converting the temperatures into Kelvins, the plant theoretical efficiency is:

$$\eta_{th} = \frac{(350+273.15)-(20+273.15)}{(350+273.15)} = \frac{330}{623.15} = 0.53 \text{ or } 53\%$$

**Discussion:** This means that, in theory, 53% of the energy contained in the steam would be converted to mechanical energy in the turbine. In a real power plant, due to efficiency losses in equipment and heat transfer processes, only about 1/3 of this theoretical efficiency is achieved, or about 18%, or lower As a result, of the energy contained in the steam, only about 18% would be converted to mechanical energy in the turbine and the remaining 82% is rejected to the atmosphere as waste heat. Beyond the losses in the cycle itself, there is an efficiency associated with the boiler/combustion process and the generator, lowering the actual plant efficiencies. Moreover, in conventional power plants, the temperature at the boiler can be adjusted to specific values that the designer may choose, within the equipment capabilities, allowing the plant to be designed for optimum efficiency. In contrast in geothermal power plants, the resource determines the maximum temperature at which the cycle can operate and thus to a large extent, the cycle efficiency. Geothermal resources, even the highest temperature ones, produce temperatures far less than those at which conventional power plant cycles operate at. Plant efficiency impacts the feasibility of producing geothermal power in two important ways. First, as the efficiency is lower as a result of lower resource temperature for example, the quantity of heat input required to produce a given output increases resulting in higher costs for resource development. At the same time the quantity of heat input required is increasing, the percentage of that input that must be rejected as waste heat is also rising. This increases the cost of the cooling portion of the plant and increases the so-called parasitic load (the energy required to run fans, pumps and other needed operation equipment) as well.

Geothermal power plants use the natural hot water and steam from the Earth to turn turbine generators for producing electricity. Unlike fossil fuel power plants, no fuel is burned in these plants, so there are almost no emissions. Geothermal electricity can be used for the base-load power and for the peak-load power demands, and has become competitive with conventional energy sources in many world regions. Main types of geothermal power plants are briefly discussed here. There are five geothermal power generating systems: *dry steam power plants, single-flash steam power plants, double-flash steam power plants, binary cycle power plants,* and *advanced geothermal energy conversion systems.* Dry steam power plants are the simplest and most economical technology, and therefore are widespread, being suitable for sites where the geothermal steam is not mixed with water. Geothermal wells are drilled down to the aquifer, and the superheated and pressurized steam (180°C–350°C) is converted into electricity by the steam turbine-generator units. In simplest power plants, the low pressure steam output from the turbine is vented to the atmosphere. This improves the turbine efficiency and avoids some of the environmental problems. Italy and United States have the largest dry steam geothermal resources. In single flash steam technology, hydrothermal resource produces heated fluid, in a liquid phase, sprayed into a flash tank, held at much lower pressure than the fluid, causing it to vaporize (flash) rapidly to steam. Then the steam is then passed through a turbine-generator unit producing electricity. To prevent the geothermal fluid flashing inside the well, the well is kept under high pressure. Flash steam plant generators range from 10 to 55 MW, with a standard size of 20 MW in use. Binary cycle power plants are used where the geothermal resource is insufficiently hot to produce steam, or where the resource contains too many chemical impurities to allow flashing. In addition, the fluid remaining in the tank of flash steam plants can be utilized in binary cycle plants (e.g., Kawerau in New Zealand). In the binary cycle process, the geothermal fluid is passed through a heat exchanger, where a secondary fluid (e.g., isobutene or pentane), having a lower boiling point than water is vaporized and expanded through a turbine to generate electricity. Then the working fluid is condensed and recycled for another cycle, and the geothermal fluid is then re-injected into the ground in a closed-cycle system. Binary cycle power plants can achieve higher efficiencies than flash steam power plants and allow the utilization of lower temperature geothermal resources, while the corrosion problems are also avoided. However, the simplest and cheapest of the geothermal cycles used to generate electricity is the direct-intake non-condensing cycle. Such cycles consume about 15–25 kg of steam per kWh generated. Non-condensing systems must be used if the content of non-condensible gases in the steam is very high (50% or higher in weight), being often used in connection to the condensing cycles for gas contents exceeding 15%, because of the high power that is required to extract these gases from the condenser. In power plants where electricity is produced from dry or superheated steam (vapor-dominated reservoirs), steam is fed directly from the wells to the steam turbines. This is a well-developed, commercially available technology, with typical turbine-size units in the 20–120 MWe capacity range. Vapor-dominated systems are less common, steam from such fields has the highest enthalpy, energy content, close to 2800 kJ/kg. Such systems are found in Indonesia, Italy, Japan and the United States. These fields produce about half of the world geothermal electricity. Water-dominated fields are much more common. Flash steam plants are used to produce energy from these fields that are not hot enough to flash a large proportion of the water to steam in surface equipment, either at one or two pressure stages.

If the geothermal well produces hot water instead of steam, electricity can still be generated, provided the water temperature is above 85°C, by means of binary cycle plants. These plants operate with a secondary, low boiling-point working fluid (Freon, isobutane, ammonia, etc.) in a thermodynamic cycle known as the organic Rankine cycle. The working fluid is vaporized by the geothermal heat in the vaporizer, expands as it passes through the organic vapor turbine. The exhaust vapor is subsequently condensed in a water-cooled condenser or air cooler and is recycled to the vaporizers. The efficiency of these power plant cycles is quite low, 5.5% or often less. Typical unit size is 1–3 MWe. However, the binary power plant technology has emerged as the most cost-effective and reliable way to convert large amounts of low temperature geothermal resources into electricity. These large low-temperature reservoirs at accessible depths exist almost anywhere in the world. Notice that the power rating of geothermal turbine-generator units tends to be smaller than in conventional thermal power stations. Most commonly the units are 55, 30, 15,

5 MWe or even smaller. One of the advantages of geothermal power plants is that they can be built economically in much smaller units than conventional power plants. In developing countries with a small electricity market, geothermal power plants with units from up to 30 MWe can be more easily adjusted to the annual increase in electricity demand than a 100 or 200 MWe hydropower plants. The reliability of geothermal power plants is very good, the annual load factor and availability factor are usually about 90%, and geothermal fields are not affected, by annual or monthly weather fluctuations, since the essentially meteoric water has a long residence time in geothermal reservoirs.

The basic types of geothermal power plants are determined primarily by the nature of the geothermal resource at the plant site. The direct steam geothermal plant is applied when the geothermal resource produces steam directly from the well. The steam, after passing through separators to remove small sand and rock particles is fed to the turbine-generator units. These are the earliest power plant types developed in Italy and in the United States. However, such geothermal resources are one of less common of all the geothermal resources, existing only in few places, not being suitable to the low-temperature resources. Flash steam plants are employed in cases where the geothermal resource produces high-temperature hot water or a combination of steam and hot water. The fluid from the well is delivered to a flash tank where a part of the water flashes to steam, which is directed to the turbine. The remaining water is directed usually to the disposal. Depending on the temperature of the resource it is possible to use two stages of flash tanks, in which the water separated at the first stage tank is directed to a second stage flash tank where lower pressure steam is separated. Remaining water from the second stage tank is then directed to disposal. The double flash power plant delivers steam at two different pressures to the turbine. Again, this type of plant cannot be applied to low-temperature resources. The third mentioned type of geothermal power plants are the binary plants. The name derives from the fact that a second fluid in a closed cycle is used to operate the turbine rather than the geothermal steam. Geothermal fluid is passed through a heat exchanger (boiler or vaporizer) or in some power plants, two heat exchangers in series, first a preheater and the second a vaporizer where the geothermal heat is transferred to the working fluid. In the past the working fluids in low temperature binary plants were CFC (Freon type) refrigerants. Current machines use hydrocarbons (isobutane, pentane, etc.) of HFC type refrigerants with the specific fluid chosen to match the geothermal resource temperature. The working fluid vapor is fed to the turbine where its energy content is converted to mechanical energy, and through the generator into electricity. The turbine vapor exits through the condenser are converted back into liquid. In most power plants, the cooling water is circulated between the condenser and a cooling tower to reject the heat into the atmosphere. An alternative is to use so called dry coolers or air cooled condensers which reject heat directly to the air without the cooling water, eliminating any water use by the plant for cooling. Dry cooling is operating at higher temperatures than cooling towers, resulting in lower plant efficiency. Liquid working fluid from the condenser is pumped back to the higher pressure pre-heater/vaporizer to repeat the cycle. The binary cycle is the type of plant which would be used for low temperature geothermal applications. Basically the processes in binary geothermal power plant are very similar to ones in steam turbine based power generation. Currently, off-the-shelf binary equipment is available in modules of 200–1000 kW.

The process of generating electricity from a low temperature geothermal heat source (or in a conventional power plant) involves a thermal process engineers refer to as a Rankine Cycle, which in a conventional power plant include boilers, turbines, generators, condensers, feed water pumps, cooling towers and cooling water pumps. Steam is generated in the boiler by burning a fuel (coal, oil, gas or uranium) is fed to the turbine, converted into mechanical energy, and through the generator into electrical energy. After passing through the turbine the steam is converted back to liquid water in the condenser of the power plant. Through the process of condensation, heat not used by the turbine is released to the cooling water. The cooling water, delivered to the cooling tower where the cycle *waste heat* is often rejected to the atmosphere. However, the modern power plants are using advanced *heat recovering technologies* and *combined heat and power generation* to increase the overall system efficiency. In summary, a power plant is simply a cycle that facilitates the energy conversion from one form to another. In this case the geothermal energy is converted into mechanical energy (turbine) and finally to electricity (generator). We have mentioned previously, that there are geothermal resources that cannot be effectively tapped to produce electricity by using the four basic systems described earlier. Therefore, the advanced geothermal energy conversion systems are used for such geothermal resources, because the systems can be designed to fit the resource conditions. The systems are developed by combining the four basic systems. EGS systems are referring to the creation and the setting of artificial conditions at a location where a reservoir having the potential to produce geothermal energy to become commercially viable. A geothermal system requires heat, permeability, and water, so an EGS corrects any of the reservoir deficiencies. EGS technologies enhance existing rock fracture network, introduce a working fluid, building on a geothermal reservoir that is difficult or impossible to derive energy from using only conventional technologies.

## 7.5 Summary

Geothermal resources can be considered renewable on the time-scales of technological/societal systems and do not require the geological times of fossil fuel reserves such as coal, oil and gas. The recovery of high-enthalpy reservoirs is accomplished at the same site from which the fluid or heat is extracted. Moreover, truly sustainable production can be achieved in doublet and heat pump systems. Of the basic types of geothermal energy, the resource base in United States enhanced geothermal systems is two orders of magnitude higher than in the other types combined, and the same is likely to be true for the world. Commercial geothermal energy exploitation is primarily a heat mining operation rather than tapping an instantly renewable energy source, such as, solar, wind or biomass energy. At the current annual energy consumption rate, geothermal heat mining can theoretically supply the world for several millennia. If a commercial geothermal exploitation project is operated for a typical life of 30 years and then shutdown, the resource would be naturally replenished and available for exploitation again in about a century; with such a scheme a geothermal energy project could be made entirely renewable, and therefore, practically inexhaustible. Between the years 2010 and 2050, geothermal power capacity in the world it is estimated to increase from about 11,000 MW to perhaps as high as 58,000 MW. Rate of growth in power capacity can be higher if adequate commercial incentives are offered by governments and international agencies.

Due to the steady heat flow from the inner parts of the Earth, geothermal resources can be regarded as renewable. A geothermal system can in many cases be recharged as a battery. Utilizing the natural flow from geothermal springs does not affect them. Exploitation through drilled wells and by them application of down-hole pumps always leads to some physical or chemical changes in the reservoir and/or its near vicinity, which could lead to a reduction or depletion of geothermal resources so far as a particular utility is concerned. Geothermal energy has a high availability and capacity factor of about 80%–90%. In comparison with wind, solar and tidal energy, geothermal is clearly an advanced energy source with 61% of the total installed capacity and 86% of the renewable electricity production in recent years. The relatively high share in the electricity production reflects them reliability of geothermal plants. The generation reliability also demonstrates one of the strongest comparative points of geothermal energy. Unlike solar energy, geothermal is available day and night throughout the year and is not dependent on climatic conditions like in the case of wind energy. Geothermal energy is a powerful natural resource with the potential to shape the future of human society. It is relatively clean and environmentally friendly when compared to present energy sources such as fossil fuels. Its possible applications are numerous and its ubiquitous nature means that it is not limited to any physical location for extraction or use, especially given that methods are being developed to access non-ideal systems such as hot dry rock systems.

Geothermal systems can be classified by a variety of criteria, such as the nature of heat transfer (conductive vs. convective), the presence or absence of recent magmatism or volcanic activity, the particular geologic setting (e.g., type of volcanic environment) or tectonic setting (e.g., type of plate boundary or intra-plate geologic hot spot), and, of course, temperature (low-, moderate-, and high-enthalpy systems). Other criteria include fluid chemistry, such as acidic or near pH-neutral systems; vapor- vs. liquid-dominated systems; and how the system is used (power, direct use, or geo-exchange). It is not uncommon for the different classifications to overlap. Ultimately, the classification of any geothermal system is based on study, playing a significant role in how that system is developed. For example, a liquid-dominated system, based on temperature and mass flow rate confirmed by drilling, can be used for flash or binary power generation, combined power and heat, or direct use, whereas, the rare vapor-dominated systems are used for power generation, being the most economic and efficient resource use. Although power generation captures much of the attention of the geothermal energy field, direct-use is much more widely applied as sub-power generating temperatures of fluids are widespread and can be developed with less cost than power plants. Indeed, one of the important attributes of geothermal systems is their wide use range over a cascading range of temperatures. Even where no hot fluids or rocks are present, the Earth acts like a thermal bank, where heat can be stored during the summer and withdrawn during the winter. In regions characterized by hot summers and cold winters, geo-exchange systems can significantly reduce energy consumption using conventional fossil fuel sources.

## Questions and Problems

1. How do direct use of geothermal energy applications work?
2. How much geothermal energy is used internationally?
3. How much geothermal energy is used in the United States? How much is the potential of using geothermal resources in the United States? How does geothermal energy benefit local economies?
4. How does geothermal energy benefit developing countries?
5. What geological areas are most likely to have higher heat flows?
6. On the map of your country or of a U.S. state, select the most suitable site for a geothermal energy development; describe the tectonic setting and likely types of rocks and geologic structures that would be found. Justify your conclusions.

7. What regions in your country are best suited for geothermal power facilities? Why? What regions are the least suited?

8. What is the global average heat flow at the surface of the Earth? What is its range?

9. What controls heat flow?

10. How does a conventional geothermal power plant work?

11. What is a base-load resource?

12. What factors influence the cost of a geothermal power plant?

13. How do geothermal heat pumps work?

14. What are the factors determining the geothermal resource sustainability?

15. What non-conventional technologies are used or will be used in the next decade for geothermal energy generation?

16. List the most promising technologies that will expand geothermal energy uses in the near future.

17. List the most important potential environmental benefits of using geothermal energy.

18. How does geothermal energy benefit the U.S. economy?

19. Identify locations in the United States that are suitable for geothermal energy.

20. Ground effect heat pumps require: (a) large heat exchanger surfaces, (b) large sources of fuel, (c) an earthquake zone, (d) extensive forest fires, or (e) none of the above.

21. The crust of the Earth is: (a) closer to the Earth's surface than the core, (b) farther from the Earth's surface than the core, (c) is between the inner and outer cores, (d) below the mantle, or (e) none of the above.

22. It is generally held that geothermal fluids must exceed 150°C in order to be practical for generating electricity. Allowing a surface temperature of 25°C, how deep must one drill to hit this temperature at a location having a normal gradient (see the chapter for the mean temperature gradient).

23. Discuss three environmental considerations that must be addressed when developing a direct-use application.

24. Describe the considerations that would have to be addressed if a district heating system were to be developed using ground source heat pumps.

25. For a direct-use fish pond application, what are the two most significant sources of heat loss? What methods or strategies could be employed to reduce the heat losses?

26. Assume a geothermal water resource of 50°C was available. Plot the flow rate necessary to prevent the pool temperature from dropping more than 5°C.

27. Discuss the most important environmental considerations that must be addressed when developing a direct-use application.

28. How does a conventional geothermal power plant work?

29. What non-conventional technologies are used for geothermal production?

30. If we are assuming that the U.S. geothermal energy sources generated some 15 billion kWh of electricity. How many conventional 1000-MW power plants are needed to produce this electricity if they operate 9000 hours per year?

31. It is accepted that geothermal fluids must about or exceed 200°C in order to be practical for electricity generation. Assuming a surface temperature of 25°C, how deep must one drill to hit this temperature at a location having a normal temperature gradient?

32. The average geothermal heat flux is about 0.050208 J/m²s in the continental regions. Assuming that three-fourths of this is attributed to the crust, by using the average thermal conductivity for rock, calculate the temperature at 5000 m.

33. If a geothermal well of a good thermal aquifer is producing a flow rate of 250 kg/s of saturated water at 225°C, what is the maximum power that this well may produce, assuming the surface temperature of 25°C and the overall conversion efficiency of 36%?

34. An average house in a certain mid-latitude town needs 2.4 kW in winter months. A geothermal district heating is proposed for this town, consisting of 1500 houses. How much heat power is needed? If the geothermal water is entering in the district heating system at 75°C and is re-injected at 45°C, how much flow rate in kg/s is required for this system?

35. A drilling in a certain location intercepts fluid at 250°C at 1 km depth with a very good flow rate. What types of geothermal applications are suitable for this site? Justify your answer.

36. If a GHP has a COP of 3.50 estimates the electrical energy need for heating (mid-November to mid-April) for a 200m² house in Detroit, Michigan.

37. Repeat the problem above for cooling season (June-August).

38. Estimate the available geothermal power from a well having the following characteristics, 250 m³/h flow rate, the well bottom and out (top) temperatures are 170 C and 120 C, respectively.

## References and Further Readings

1. L. J. P. Muffler and R. Cataldi, Methods for regional assessment, *Geothermics*, Vol. 7, pp. 53–89, 1978.
2. T. J. Ahrens, *Rock Physics and Phase Relations* (ed.), AGU Press, Washington, DC, 1995.
3. E. Barbier, Nature and technology of geothermal energy: A review, *Renewable and Sustainable Energy Reviews*, Vol. 1 (1–2), pp. 1–69, 1997.
4. E. Barbier, Geothermal energy technology and current status: An overview, *Renewable and Sustainable Energy Reviews*, Vol. 6, pp. 3–65, 2002.
5. A. V. da Rosa, *Fundamentals of Renewable Energy Processes*, Academic Press, Cambridge, MA, 2005.
6. J. Twidell, T. Weir, and A. D. Weir, *Renewable Energy Resources*, Taylor & Francis Group, Abingdon, UK, 2005.
7. R. DiPippo, *Geothermal Power Plants: Principles, Applications and Case Studies* (2nd ed.), Elsevier Advanced Technology, Oxford, UK, 2008.
8. J. Tester, E. Drake, M. Driscoll, M. Golay, and W. Peters, *Sustainable Energy: Choosing Among Options*, The MIT Press, Cambridge, MA, 2005.
9. R. Bertani, World geothermal energy production, *GeoHeat Center Bulletin*, Vol. 28, pp. 8–19, 2007.
10. B. K. Hodge, *Alternative Energy Systems and Applications*, John Wiley & Sons, Hoboken, NJ, 2010.

11. E. E. Michaelides, *Alternative Energy Sources*, Springer, Berlin, Germany, 2012.

12. V. Nelson, and K. Starcher, *Introduction to Renewable Energy* (2nd ed.), CRC Press, Boca Raton, FL, 2015.

13. W. E. Glassley, *Geothermal Energy: Renewable Energy and the Environment* (2nd ed.), CRC Press, Boca Raton, FL, 2015.

14. D. R. Boden, *Geologic Fundamentals of Geothermal Energy*, CRC Press, Boca Raton, FL, 2017.

15. S. K. Sanyal, and S. J. Butler, Feasibility of Geothermal power generation from petroleum wells, *Transactions Geothermal Resources Council*, Vol. 33, pp.673–680, 2009.

16. S. K. Sanyal, and S. J. Butler, Geothermal power from wells in non-convective sedimentary formations—An engineering economic analysis, *Transactions Geothermal Resources Council*, Vol. 33, pp. 865–870, 2009.

17. I. W. Johnston, G. A. Narsilio, and S. Colls, Emerging geothermal energy technologies, *KSCE Journal of Civil Engineering,*, Vol. 15, pp. 643–653, 2011.

18. G. Boyle, *Renewable Energy—Power for a Sustainable Future*, Oxford University Press, Oxford, UK, 2012.

19. B. Everett, and G. Boyle, *Energy Systems and Sustainability: Power for a Sustainable Future* (2nd ed.), Oxford University Press, Oxford, UK, 2012.

20. C. Luo, Huang, L. Gong, Y., and W. Ma, Thermodynamic comparison of different types of geothermal power plant systems and case studies in China, *Renewable Energy*, Vol. 48, pp. 155–160, 2012.

21. R. A. Dunlap, *Sustainable Energy*, Cengage Learning, Boston, MA, 2015.

22. V. Nelson and K. Starcher, *Introduction to Renewable Energy (Energy and the Environment)*, CRC Press, Boca Raton, FL, 2015

23. V. Nelson, and K. Starcher, *Introduction to Renewable Energy (Energy and the Environment)*, CRC Press, Boca Raton, FL, 2015.

24. W. E. Glassley, *Geothermal Energy: Renewable Energy and the Environment* (2nd ed.), CRC Press, Boca Raton, FL, 2015.

25. D. R. Boden, *Geologic Fundamentals of Geothermal Energy*, CRC Press, Boca Raton, FL, 2017.

# 8

## Hydropower and Marine Energy

---

## 8.1 Introduction, Water Energy, Basics

Water energy and hydropower, like wind and solar energy, have been used for centuries. Initially bits were used for sailing, or as source of mechanical power for grinding grain, for sawing wood or in primitive textile shops. Like most of the renewable energy sources, all types of water energy are driven ultimately by the solar energy. Unlike most other renewable energy sources, hydropower large or small is a major contributor to world energy supplies. Hydropower currently represents worldwide a significant source of electrical energy and compared to fossil and nuclear fuel, hydro resources are found almost everywhere. Hydropower plants are providing close to 20% of the world's electricity supply. Historically the hydropower technology started with the wooden waterwheels since Romans times. Various waterwheel types had been in use in many parts of Europe for more than 2000 years, mostly for milling grain, but also to provide mechanical power to small primitive industrial workshops. However, by the time of the Industrial Revolution, waterwheel technology had been developed to a fine art, and device efficiencies approaching 70% were being achieved for the many of systems that were in regular use. Improved engineering designs during the nineteenth century, combined with the needs to develop smaller and higher speed devices to generate electricity, led to the development of modern hydropower turbines. The first modern hydropower turbine was designed in France in the 1820s by Benoît Fourneyron who called his invention a hydraulic motor. Toward the end of nineteenth century many mills were replacing their waterwheels with hydroturbines, and engineers and entrepreneurs started looking how hydroturbine could be used for large-scale electricity generation. The golden age of hydropower was the first half of the twentieth century, before oil took over as the dominant force in energy supply. Europe and North America built dams and hydropower stations at a rapid rate, exploiting up to 50% of the technically available hydro energy potential, while the equipment providers sprung up to supply this booming market. Whereas the large hydropower manufacturers have since managed to maintain their business on export markets, in particular to developing countries, the small hydropower industry has been on the decline since the 1960s. A few countries, notably Germany have boosted this sector in the last decades with attractive policies favoring "green" electricity supply, but small hydropower generally cannot compete with existing fossil fuel or nuclear power stations so that, without government incentives to use non-polluting power sources, there has been no firm market for small hydropower in developed countries for several decades.

Hydropower, large and small, remains by far the most important of the "renewables" for electrical power generation worldwide, being available in a broad range of projects, scales, and types. Projects are usually designed to suit particular needs and specific site/location conditions. They are classified by project type, head, purpose, and installed capacity. Sizewise categories are different worldwide due to varying development policies in different countries. The hydropower types are: run-of-river, reservoir-based, and hydropower pumped energy storage. Hydropower impacts, negative to positive, are well known both from environmental and social perspectives. Experience gained during past decades in combination with new sustainability guidelines, innovative planning based on stakeholder consultations, and scientific know-how is promising to achieve a high sustainability performance for future hydropower projects. The world's technically feasible hydro potential is estimated at 14,370 TWh/yr, which equates to 100% of today's global electricity demand. The economically feasible proportion of this is currently considered to be 8080 TWh/yr. The hydropower potential exploited in 2000 was about 2700 TWh/yr, about 20% of the planet's electricity. All other renewables combined provided less than 2% of global electricity consumption. North America and Europe have developed most of their hydropower potential, but huge resources remain in Asia, Africa, and South America. Small hydro (up to 10 MW) currently contributes over 40 GW of world power capacity. The global small hydropower potential is believed to be in excess of 100 GW. In many countries hydropower is one of the largest contributors to electricity generation. It is not uncommon in developing countries that a large dam to be the main generating electricity source. Brazil, Canada, China, Russia, and the United States currently produce more than half of the world's hydropower. Since 2000, world hydropower has grown at an average annual rate of 2%. However, in developed countries it has been declining at an annual rate of 0.3%. In South America, the highest hydropower potential is in Brazil, where it exceeds 2200 TWh/yr. In Africa, Democratic Republic of the Congo has the highest hydro-energy potential, and Norway's potential resources are the highest in Western Europe. Other countries with substantial potential include Canada, Chile, Colombia, Ethiopia, India, Mexico, Paraguay, Tajikistan, and the United States. China's hydro-energy resources are the largest of any country with an estimated theoretical potential of more than 6000 TWh/yr, almost double of the current world hydropower generation, with an economically feasible potential of more than 1750 TWh/yr. According to the International Energy Agency (IEA) for the world hydroelectricity market, the world hydroelectricity generation is projected to increase to 4680 TWh in 2030, at an average annual rate of 1.8%.

The oceans cover a little more than 70% of the Earth surface, being the world's largest solar energy collector and energy storage system. Ocean energy can be divided into six types of different origin and characteristics: ocean wave, tidal range, tidal current, ocean current, ocean thermal energy, and salinity gradient. Currently, all ocean energy technologies except tidal range are at early development stages from conceptual up to demonstration stage. Ocean wave and tidal current energy are the two of most advanced ocean energy technologies, being expected to significantly contribute to the future energy supply. On an average day, the 60 million square kilometers of tropical seas absorb an amount of solar energy equal in heat content to about 250 billion oil barrels. If less than 0.1% of this stored solar energy is converted into electricity, it would supply more than 20 times the today total daily consumed electricity in the United States. IEA has estimated that wave energy alone has the potential to meet 10% of the world's current electricity demand. Calculations of the total wave energy in the world yield to average values between 1–10 TW. However, estimates of the level of energy that descends on the world coasts, and how much of this can be technically and economically harvested, vary greatly in the literature. Some of the wave energy positive features are: relatively high utilization, worldwide resource magnitude, low energy flux variability compared to wind energy, relatively high energy density, free energy compared to fuel-based energy sources, installations that likely to have positive artificial reefing effects and degree to which wave energy would pollute the environment or add to greenhouse gas emissions is insignificant compared to the fossil fuels. OTEC systems convert solar radiation into electric power, by using the ocean natural thermal gradient, the temperature difference between the warm surface water and the cold deep water layers below 600 m, which is about 20°C, so they can generate significant amount of power. The cold seawater used in the OTEC process is also rich in nutrients and it can be used to culture both marine organisms and plant life near the shore or on land. OTEC concept has existed for over a century as fantasized by Jules Verne in 1870 and conceptualized by French physicist, Jacques Arsene d'Arsonval, in 1881. However, an operating OTEC power facility was developed only in the late 1920s.

Harnessing ocean energy is truly a challenge spanning over many areas of physics, marine sciences, and engineering, such as hydrodynamics, mechanics, electromagnetism, electrochemistry, electronics, energy conversion, or marine biology. If all of the above challenges are to be met, then a holistic perspective is critical to the ocean energy system designers. Ocean energy is only used to generate electricity and hence the primary ocean energy consumption is the same as fuel inputs to power generation. However, world ocean energy use decreased at an average annual rate of 1.4% from 2000 to 2010, accounting for only a very tiny proportion of energy consumption. Tidal energy has been utilized on a commercial scale to date only in developed countries and on limited developments. Either tidal energy is predictable, cannot be used to generate electricity at consistent levels constantly. Twice in every 12.42 hours (24 hours in some locations) the tidal current speed and hence the electricity generation capability falls to zero. If tidal energy is required to produce a sustained base-load for a grid, energy storage and/or power back-up are needed. Ocean waves, driven primarily by the wind and ocean currents, a complex interactions between solar heating, winds near the Equator, tides, water density and salinity are ocean mechanical energy forms. However, waves are rarely of consistent length or strength, the energy levels may vary considerably from wave to wave, from day to day, and from season to season, due to the variations in local and global wind conditions. This inherent variability needs to be converted to a smooth electrical output to be a reliable electricity supply source. Moreover, there are sites on the western and southern coastlines where regular storms generate consistent swells with periods of wave energy failure both of low frequency and short duration. Higher level forecasting, grid management, or possibly energy storage systems are needed to smooth out such supply variability. OTEC, on the other hand, is potentially suitable for base load electricity generation, as the ocean temperatures on which it relies show only slight seasonal variations.

### 8.1.1 Hydropower Basics

The main components of a hydroelectric system may be classified into two groups: (1) *the hydraulic system* components that include the turbine, the associated conduits-like penstocks, tunnel, and surge tank and its control system; and (2) *the electric system* components formed by the electrical generator, usually synchronous type, and the control system. Hydraulic power can be captured wherever a flow of water falls from a higher level to a lower level. Hydraulic turbines are defined as prime movers that transform the kinetic energy of the falling water into mechanical energy of rotation and whose primary function is to drive a electric generator. The vertical fall of the water, the *head*, is essential for hydropower generation; fast-flowing water on its own does not contain sufficient energy for useful power production except on a very large scale, such as offshore marine currents. Hydropower is essentially a controlled method of water descent usefully utilized to generate power.

Hydroelectric plants utilize the energy of water falling through a head that may vary from a few meters to up to 1500 or even 2000 m. To manage this wide range of heads, many different kinds of turbines are employed, which differ in their working components.

Two quantities are required for hydropower: a water flow rate $Q$, and a head $H$. However, it is better to have more head than more flow, since this keeps the equipment relatively smaller. The *gross head*, $H_g$, is the maximum available vertical fall in the water, from upstream to downstream level. The actual head seen by a turbine is less than the gross head due to losses incurred when transferring the water into and away from the turbine. This reduced head is known as the *net head*. In a hydropower plant, the water potential energy is first converted to equivalent amount of kinetic energy. Thus, the water height is used to calculate its potential energy; this energy is converted to water speed at the turbine intake and is calculated by balancing the water potential and kinetic energy. Turbines are devices designed to extract energy from a flowing fluid. The geometry of turbines is such that the fluid exerts a torque on the rotor in the direction of its rotation. The generated shaft power is then available to derive generators or other devices. Based on the conservation of energy, and laws of dynamics, the hydropower energy conversion is summarized in Figure 8.1. However, a thorough study of each conversion level is needed in order to explain how hydropower is converted into electricity, to estimate the power output, conversion efficiency, and how it can be used as sustainable energy source. The performance of hydraulic turbines is strongly influenced by the hydraulic system, topography characteristics of water conduit that feeds the turbine, including the water inertia effect, water compressibility, and penstock characteristics. The energy in water can be potential energy from a height difference, used to generate electricity, which is what most people think of in terms of hydroenergy from water stored in dams. However, there is also kinetic energy due to water flow in rivers and ocean currents. The potential energy of a mass of water stored in a reservoir, having a volume $V$, having a head $H$, is expressed as:

$$P.E. = W = \rho g V H \tag{8.1a}$$

**Example 8.1:**  Find the potential energy for 5000 m³ of water at a height of 100 m.

**Solution:** By Equation (8.1a), the potential energy is:

$$P.E. = \rho g V H = 10^3 \cdot 9.806 \cdot 5000 \cdot 50 = 4.903 \times 10^9 \text{ J}$$

The specific energy of a hydropower plant is the quantity of potential and kinetic energy, which one-unit mass of water delivers when passing through the plant from an upper to a lower reservoir. The expression of the specific energy is Nm/kg or J/kg and is designated

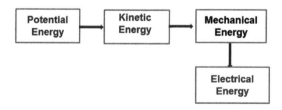

**FIGURE 8.1**
Hydropower energy conversion process.

as [m²/s²]. In a hydropower plant, the difference between the level of the upper reservoir $z_{res}$ and the level of the tail water $z_{tw}$ at the penstock exit to the turbine is defined as the gross head, $H_g = z_{res} - z_{tw}$. This is one of the most hydropower fundamental parameters, the difference between the maximum and minimum water heights, between the upstream side of the turbine and downstream of the turbine at the draft tube outlet. Potential energy of the water is direct proportional to the reservoir head level. A high head level would mean that the potential energy for that system is higher. The corresponding gross specific available energy is then:

$$E_g = gH_g \tag{8.1b}$$

These considerations lead to the definition of the net head across the turbine as described in Equation (8.1b), according to the IEC41 standard, were the draft tube outlet loss is regarded to be a power plant loss and not a turbine loss. Note that the flow rate $Q = A \cdot v$ (m³/s) where $A$ is the cross section of draft tube outlet and $v$ is the flow velocity. To express the energy for the hydropower system we are using the Bernoulli Equation, expressed as:

$$E = gH = \left( gh_0 + \frac{v_0^2}{2} + gz_0 \right) - \left( gh_{tw} + \frac{v_{tw}^2}{2} + gz_{tw} \right) \tag{8.2}$$

Here "0" refers the penstock intake (reservoir level) while "tw" to the turbine input level. When a water discharge (flow rate) $Q$ [m³/s] passes through the hydropower plant, the delivered power is determined as in the following. The effective head is the difference between the energy head at the turbine entrance and the energy head at the exit of the draft tube. When the volume of waters moves from the maximum level at $dV_1$ to the minimum level of $dV_2$ for a height of $H$, the flow rate or discharge is the difference between them, a mechanical work is produced and is given by the equation:

$$W_g = \rho g V H_g$$
$$V = dV_1 - dV_2 \tag{8.3}$$

By using the equation for work, theoretical power output of the hydropower system can be calculated:

$$P_g = \frac{dW_g}{dt} = \rho g Q H_g \tag{8.4}$$

In Equations (8.1) through (8.4), $g$ is acceleration due to gravity (9.806 m/s), $\rho$ is the water density (1000 kg/m³), $H_g$ is the gross head (m), while $Q$ is the volumetric flow rate through the turbine. Power is measured in units of Watts. However, the net head, $H$, defined in Equation (8.2), is used to estimate the theoretical energy of which the majority will be utilized and transformed to mechanical energy by the turbine. The rest of the energy will be converted to losses i.e., mainly increased heat energy in the water and a very small negligible part as a heat flux to its surroundings. The available power for the turbines in a hydropower plant project depends on the available flow $Q$ m³/s and is defined by:

$$P = \rho E Q \tag{8.5}$$

The available energy stored in a reservoir is determined by multiplying the mass of water (i.e., the stored volume of water multiplied by the density) with the net head. The equation for stored energy of a water volume $V$ and a net head $H$, i.e., $E = g \cdot H$, yields:

$$W = \rho \cdot V \cdot E \qquad (8.6)$$

**Example 8.2:** For a reservoir with a volume of $V = 500 \cdot 10^6 \, m^3$ and a head of $H = 400$ m, calculate the available storage of energy which can be utilized by the turbine when ignoring head losses in tunnels and the losses in turbines and generators:

**Solution:** From Equation (8.5) the stored available energy is

$$W = \rho \cdot V \cdot E = 10^3 \times 500 \cdot 10^6 \times 9.806 \times 400 / \left(3600 \cdot 10^3\right) = 10.9 \cdot 10^8 \, kWh = 1.09 \, TWh$$

The actual power output is the theoretical power output time the efficiency, $\eta$, expressed as:

$$P = \eta \rho g Q H \qquad (8.7)$$

**Example 8.3:** Estimate the power output from a reservoir with net head of 80 m and the volume flow rate of 30 m/s and efficiency 90%.

**Solution:** From Equation (8.7), the power output is:

$$P = \eta \rho g Q H = 0.90 \times 9.806 \times 1000 \times 30 \times 80 \approx 21.181 \, MW$$

The hydraulic efficiency of a turbine excludes friction losses on the outside of the runner, leakage loss of water that does not pass through the runner blades and mechanical friction losses. The hydraulic efficiency of a well-designed turbine is 98–99% and can be developed as shown in the following description.

**Example 8.4:** Estimate the flow rate required for an 85% hydropower plant efficiency with a net head of 60 m to satisfy the electrical needs of a load of 15 kW.

**Solution:** From Equation (8.7) the flow rate can be found as:

$$Q = \frac{P}{\eta \rho g H} = \frac{15,000}{0.85 \cdot 1000 \cdot 9.80 \cdot 60} = 0.030 \, m^3/s$$

As the water hits the impulse vanes of hydroturbine, a dynamic force will exist in order for the vanes or buckets to start rotating. The rotation of the vanes converts the potential energy to kinetic energy. The force on the moving vane or bucket by a jet of water is derived as the equation of force, and the torque that is exerted on the vanes by the jet of water, the product of the force, can be computed. Once the value for torque has been obtained, it is now possible to calculate the theoretical power that is exerted by the wanes. Power is simply the product of torque and angular velocity of the runner in rad/s.

$$P = T \cdot \omega \qquad (8.8)$$

If the water weight is low, it will reduce the force on the bucket, torque and power. If the buckets are placed too far apart, water would flow through and only very little energy will be extracted from the system. In order for the runner to perform efficiently, the water should leave the runner in an axial direction. However, this will not be possible as to do so would mean obtaining completely axial flow at all the gate openings. Hence, the absolute velocity of the water is considered as the water exits from the runner to be equal to the water discharged divided by the area of the draft tube. Hydropower turbines are used to extract energy from a fluid and by this decrease the total energy of the fluid. The total energy in the fluid is measured by a total head that composes of various forms of energy as follows:

$$H_{tot} = H_{pr} + H_{st} + H_{vl} + H_{fr} \tag{8.9}$$

Terms in the Equation (8.1) are defined below.

$$H_{pr} = \frac{p_2 - p_1}{g\rho} \quad - \text{ Pressure Head}$$

$$H_{st} = H_{res} - H_{tw} \quad - \text{ Static Head}$$

$$H_{vl} = \frac{v_2^2 - v_1^2}{2g} \quad - \text{ Velocity Head}$$

$$H_{fr} \quad - \text{ Friction Head}$$

The friction head reflects the losses in a system and is commonly expressed in meters, while the static head is basically the earlier defined gross head. A turbine system denotes a system, in which a turbine is used to extract energy from a fluid. The system consists of pipes (or ducts) on the pressure and the discharge side of the turbine as well as eventually valves, reservoirs, or other devices. A turbine decreases the total head in a system. This implies that there is high-energy fluid available at the inlet of a turbine and that the fluid leaves the turbine with reduced energy content. Depending on the application, the primary contribution of the high-energy source might be different:

1. Hydroturbine driven by high-velocity fluid which results from a great difference in elevation: The primary high energy source is static head that is transformed into velocity head by flow acceleration.
2. Hydroturbine that is driven by the flow in a river: The primary high-energy source is velocity head.

Hydropower generation involves the water storage, conversion of the hydraulic (potential) energy of the fluid into mechanical (kinetic) energy, using hydraulic turbines, and conversion of the mechanical energy to electrical energy in electric generators. Hydroelectric units have been installed in capacities ranging from a few kilowatts to nearly 1 GW. Hydroelectric power plants are of three major types. (a) Run-of-the-river, in which small amounts of water storage is used to generate electricity, with a very little control of the flow through the plant. (b) Hydropower plants with a large storage, consisting of an artificial basin (created by a dam on a river course) that allows storing water and thus controlling the flow through the plant on a daily or seasonal basis, having several hydroturbine-generator units. (c) Pumped hydropower storage designed to operate during off-peak hours water is pumped (by means of reversible pump-turbines or dedicated pumps) from a lower reservoir to an upper reservoir, so that the energy is thus stored for later production during peak hours. Net head is lower than gross head due to energy losses in the penstock:

$$H = H_g - Losses = H_g - H_{Losses} \tag{8.10}$$

Penstock efficiency is the ratio of net and gross head, and head losses, $H_{Losses}$, is given by:

$$\eta_p = \frac{H}{H_g} = 1 - \frac{H_{Losses}}{H_g} \tag{8.11}$$

We are defining the efficiency of a pump or turbine to impart energy to or extract energy from water, as the ability of hydraulic structure or element to conduct water with minimum energy losses. The hydraulic efficiency is defined as the ratio of power delivered to the runner, $gH$, to the power supplied at the inlet, $W$, expressed as:

$$\eta_y = \frac{W}{gH} \tag{8.12}$$

The volumetric efficiency is the ratio of the net flow rate (discharge) to the input flow rate, as:

$$\eta_v = \frac{Q_{net}}{Q_{in}} \tag{8.13}$$

The input power in the turbine-generator unit is then given by:

$$P_{in} = \eta_y \eta_v \rho g Q_{net} H \tag{8.14}$$

Overall hydropower turbine efficiency is computed by multiplying the efficiency of turbines, efficiency of generators, and is on average for most of the turbine-generator units in range 60%–70%. Net or output power hydropower turbine output, the difference in gross power and mechanical and auxiliary losses of the turbine-generator unit, and is expressed as:

$$P_{out(el)} = P_{in} - P_{mech} - P_{aux} \tag{8.15}$$

The overall efficiency of the hydropower generation system is

$$\eta_{gen} = \frac{P_{out(el)}}{P_{in}} = 1 - \frac{P_{mech} + P_{aux}}{P_{in}}$$

Net power output of the hydropower generation unit is:

$$P_{out(el)} = \eta_y \eta_v \eta_{gen} \rho g Q H_g = \eta_t \rho g Q H_g \tag{8.16}$$

Here, $\eta_t$ is the total turbine-generator unit efficiency. The overall plant efficiency is the product of the overall generation unit efficiency and the penstock the efficiency, and is expressed as:

$$\eta = \eta_p \eta_t = \frac{P_{out(el)}}{\rho g Q H_g} \tag{8.17}$$

Hydraulic turbines extract mechanical energy from water which has a high head. The power of a rotary hydraulic machines, such as a hydropower, depends upon the working fluid density, $\rho$, speed of rotation, $N$, the characteristic diameter, $D$, the change in the head, $\Delta H$, the flow rate or discharge, $Q$, and acceleration due to the gravity, $g$. The functional relationship of power is expressed as:

$$P = f(\rho, N, D, \Delta H, Q, g) \tag{8.18}$$

There are two main types, reaction and impulse, the difference being in the manner of head conversion. In reaction turbines, the water fills the blade passages and the head change or pressure drop occurs within the impeller. They can be of radial, axial, or mixed flow configurations. Impulse turbines convert first the high head through a nozzle into a high velocity jet that strikes the blades at one position as they pass by. Reaction turbines are smaller because water fills all the blades at one time. For hydraulic impulse turbines, the pressure drop across the rotor is zero, the all pressure drop across the turbine stages occurs in the nozzle row. The Pelton wheel is the classical example of impulse turbines. A high-speed jet of water strikes the Pelton wheel buckets and is deflected, while the water enters and leaves the control volume surrounding the wheel as free jet. Ideally, the fluid enters and leaves the control volume with no radial component of velocity. For impulse turbines, the total fluid head is converted into a large velocity head at the exit of the supply nozzle. The pressure drop across the bucket (blade) and the fluid relative speed change across the bucket are negligible, while the space surrounding the rotor is not completely filled with fluid. The individual jets of fluid striking the buckets are generating the torque, eventually transferred to the generator. For reaction turbines, there is both a pressure drop and a fluid relative speed change across the rotor. Guide vanes act as nozzle to accelerate and turn the flow in the appropriate direction as the fluid enters the rotor, and the pressure drop occurs across the guide vanes and across the rotor, being best suited for higher flow rate and lower head such as are often encountered in hydroelectric stations associated with a dammed river. The working fluid completely fills the passageways through which it flows. The fluid angular momentum, pressure, and velocity decrease as it flows through the turbine rotor, and the turbine rotor extracts energy from the fluid.

In order to estimate the water power potential fluid mechanics principles and laws are applied. For any stretch of a watercourse or river, characterized by a difference in level (head) of $H$ meters, conveying a flow rate or discharge of $Q$ (m³/sec), the theoretical (potential) power, $P_{st}$, can be expressed by:

$$P_{st} = \rho QH = 1000QH\left(\frac{\text{kg} \cdot \text{m}}{\text{s}}\right) = 13.405QH(HP) = 9.786QH\left(\text{kW}\right) \qquad (8.19)$$

If the rate of flow changes along a stretch, the mean value of the discharges or flow rates pertaining to the two terminal sections of the stretch or current is to be substituted in the Equation (8.7). The theoretical power resources of any river or river system are given by the total of the values computed for the all individual stretches or flow currents, and is given by the expression:

$$P_{theor} = 9.786\sum QH\left(\text{kW}\right) \qquad (8.20)$$

Potential water power resources can be characterized by different values according to the discharge taken as basis of computation. The conventional discharges or flow rates, used to characterize hydropower potential of a river are listed here. Minimum potential power, or theoretical capacity of 100%, is the term for the value computed from the minimum flow observed, $P_{teor100}$. Small potential power represents the theoretical capacity of 95% can be derived from the discharge of flow rate of 95% duration as indicated by the average flow duration curve, $P_{teor95}$. Median or average potential power is the theoretical capacity of 50%

can be computed from the discharge or flow rate of 50% duration as represented by the average flow duration curve, $P_{teor50}$. Mean potential power represents the value of theoretical mean capacity can be ascertained by taking into account the average of mean flow. The average of mean flow is understood as the arithmetic mean of annual mean discharges for a period of 10–30 years. The annual mean discharge is the value that equalizes the area of the annual flow duration curve.

An important aspect of the hydropower assessment and analysis consists of good measurements on the hydropower site and surrounding areas, secondary rivers, etc. For power extraction purpose from a stream or river, it is important to measure the water flow rate and head. Head can be measured as vertical distance (feet or meters) or as pressure (e.g., lb/sq ft, N/m²). Regardless of the size of the stream, higher head will produce greater pressure and therefore higher output at the turbine. An altimeter can be useful in estimating head for preliminary site evaluation, but should not be used for the final measurement. Low-cost barometric altimeters can reflect errors of 150 ft (46 m) or more, GPS altimeters are often less accurate. Topographic maps can be used to give an estimate of the vertical drop of a stream. But two methods of head measurement are accurate for design: direct height measurement, and water pressure. The second major step in evaluating a site's hydropower potential is measuring the flow of the stream. Stream levels change through the seasons, so it is important to measure flow at various times of the year. Three popular methods are used for measuring flow in small hydropower applications: container, float and weir. Determining the turbine performances requires measurement of hydraulic power (what is available to the turbine) consisting of flow inlet water level, tail-water lever, turbine-generator output, which is the one sent to the grid. Absolute flow measurement methods include area-velocity (current meters) method, ultrasonic (transit times of ultrasonic pulses) method and dye dilution (change in concentration of an injected tracer) method. Dye dilution method requires long length of conduit (penstock) for mixing, or injection manifold, careful handling and preparation of dye and equipment calibration. Once the procedure is underway, testing is relatively quick, about 15 minutes per operating point, and has no impact to plant operation (no dewatering, etc., to implement). Ultrasonic method is suitable for long and short penstocks, requiring site-specific installation. It is a relatively fast measurement, has impact to plant operation (dewatering, etc., to install), and once installed, future testing is easy to perform. When assessing a micro-hydro site, we are interested in quantifying the available head and the flow rate, since both are necessary in determining power. Of the two attributes, head is usually considered more desirable because it results in smaller diameter pipes and fittings, reducing overall system costs. However, when working with a resource as site specific as hydropower, one must accept whatever is available.

### 8.1.2 Ocean Energy Assessment and Forecasting

One of the most important initial steps toward market deployment of ocean energy systems is the one involving assessment, characterization, and mapping of ocean energy resources. For example, the assessment of wave energy resources includes the identification of areas with high wave energy, the quantification of average energy resources (e.g., total annual wave energy yield), and the description of the resource by using parameters such as significant wave height, wave energy period, and mean wave direction. Precise estimates and description of available wave energy resources at high spatial and temporal resolution are needed for proper planning and the optimization of the design of ocean energy converters. The current state of technology development ultimately determines how much of the resource can be exploited with the main technical parameters to be improved being device efficiency and capacity factor. Reducing uncertainties concerning the available resources also increase the confidence of investors, allowing a better determination of the value of investments and minimizing risks. During the last decades, ocean wave energy resources have been assessed for various regions. The first wave energy resource assessments have been made using buoy data limited to local conditions. The second generation of assessments included buoy data in combination with deep water numerical models which can assess offshore wave resources which helped overcoming the limitations of first generation assessments, the limited time period of the buoy measurements and the uncertainties of extrapolating local data to other locations. Recent tools incorporate radar measurements and allow modelling the wave generation and propagation in coastal regions. Usually, wind and bathymetry data are used as an input for these models. Typical output parameters are: wave height, mean wave period, peak wave period, and mean wave direction. Ocean wave energy resource assessments are performed on global, regional and local levels. Wave forecasting is performed by statistical techniques or physics-based models, such as WAM of the European Centre for Medium-Range Weather Forecasts or WAVEWATCH-III of the National Oceanic and Atmospheric Administration. Statistical approaches include neural networks, regression-based techniques, and genetic algorithms. Statistical models are more accurate for short-time horizons (up to 6 h) while physics-based models perform better for longer time horizons, while a combination of both methods leads to more overall accurate results. Statistical models might be sufficient for to be used for electricity utilities for generating, operating, and trading.

Tidal current energy resources have been assessed, for a number of years, often, through direct measurements, performed on-site. In the last decades, 2-D and 3-D modelling techniques have been employed to assess tidal current energy resources by modelling current velocities. Recent publications assess also the hydrodynamic effects of power extraction and consider for example change to the flow field, change in water surface elevation, or disturbances in tidal dynamics. For example, tidal current energy is calculated as function of sea water density, velocity, velocity availability factor, neap/spring factor and peak spring-tide velocity. However, it is not possible to convert all tidal current energy power due to Betz' limit and mechanical losses in the turbines. These limitations are accounted via the power coefficient. Tidal energy resource assessments have been performed for many regions and coastal areas of the world. Tidal current forecasts are usually readily available. For example, NOAA Current Predictions allows forecasts up to 48-h horizon, one week and annual predictions, which are available online. The German Federal Maritime and Hydrographic Agency provides current predictions, up to 3 days . The model used is a 3-D model that takes into account meteorological forecasts for the North Sea and Baltic Sea provided by the German Weather Service (DWD), tides and external surges entering the North Sea from the Atlantic, as well as river runoff from the major rivers. The forecasts for the sea computed by the operational circulation model cover 48 h. A number of commercial offers are available, mainly aiming at navigation (e.g., MaxSea). They are mainly based on data available from public institutions, such as NOAA or Meteorological Services.

## 8.2 Small and Mini Hydropower

Hydroelectric power comes from water at work, water in motion. It can be seen as a form of solar energy, Sun powering the hydrologic cycle that gives water to the Earth. In this cycle, atmospheric water reaches the Earth's surface as precipitation. Some of this water evaporates, but much of it either percolates into the soil or becomes surface runoff. Water from rain and melting snow eventually reaches ponds, lakes, rivers or oceans where evaporation is constantly occurring. Water vapor passes into the atmosphere by evaporation then circulates, condenses into clouds, and some returns to Earth as precipitation, completing the water cycle. Nature ensures that water is a renewable resource. Hydropower is the most economical way to generate electricity today, no other energy source, renewable or not, is comparable to it. Producing electricity from hydropower is inexpensive because, once a dam or the equivalent has been built and the equipment installed, the energy source, the flowing water, is free. Hydroelectric plants also produce power at a minimal cost due to their sturdy structures and simple equipment. From a thermal perspective, some of the gravitational energy associated with the decrease in height is not converted to hydropower. That energy is converted into heat, which increases the temperature of the water and the surroundings. The maximum temperature rise is approximately one Celsius degree per 400 m of height decrease. Existing hydro plants have been described in three categories: small, micro, and pico. Small hydro represents hydroelectric power on a scale serving a small community or an industrial facility. The definition of a small hydro project varies, with a generating capacity of up to 10 MW generally accepted as the upper limit of what can be termed small hydro. Micro-hydro is a term used for hydroelectric power installations that typically produce up to 100 kW of power. These installations can provide power to a small community or may be connected to electric power networks. Pico-hydro is a term used for hydroelectric power generation of under 5 kW. It is useful in small, remote communities that require only a small amount of electricity. The key advantages of small hydro are:

1. High efficiency (70%–90%), by far the best of all energy technologies;
2. High capacity factor (typically >50%);
3. High level of predictability, varying with annual rainfall patterns;
4. Slow rate of change; the output power varies only gradually from day to day (not from minute to minute);
5. A good correlation with demand, i.e., output is maximum in winter;
6. It is a long-lasting and robust technology; systems can readily be engineered to last for 50 years or more; and
7. It is also environmentally benign.

Small hydro is in most cases "run-of-river"; in other words, any dam or barrage is quite small, usually just a weir, and little or no water is stored. Therefore, run-of-river installations do not have the same kinds of adverse effect on the local environment as large-scale hydropower systems. Small-scale, mini, micro, or pico hydroelectric systems can be constructed in many options, such as on dam-toe, canal drops, and return canals of thermal power stations and also in the flowing small river as well as small revolute which are flowing usually nearby villages. Area required for the construction work is small as canal already exists. It requires very small gestation period and such power stations can be ready for generation within 3 years, in contrast to the large hydro schemes. Small hydropower plants (SHPs) are usually up to capacity of 25 MW whereas mini-hydro plants are above 100 kW but below 1 MW either stand-alone scheme or more often feeding into the grid. Micro-hydroelectric plants (MHPs), ranging from a few hundred Watts for battery charging or food processing application up to 100 kW usually provided power for small community or rural industry in remote area away from the grid. The most important steps in establishing a small-scale hydropower are consisting of:

1. *Site Selection*: MHP are to be situated in hilly areas where there are natural falls, on the canal drops, or at the dam-toe; long range studies are not required for such site selection. Wind energy conversion system should be located preferably in the areas where the wind are strong and persistent, where daily wind flow is variable but monthly and annual average speed should be remarkably constant from year to year.
2. Similarly energy harvesting by other renewable energy systems small hydropower is very much customized by site specific characteristics.
3. *Grid Connection Issues*: In MHP, input power is almost constant. Quantum of power fed to the grid remains constant in a season. Operation of grid connected small or mini hydroelectric station is rather smooth comparing with other renewable energy conversion systems, without facing many problems like production of harmonics, abnormalities in voltage and frequency, etc.
4. *Operation, Maintenance, and Control Issues*: Operation of small, mini, or micro-hydropower systems is smooth, maintenance free, and easy to operate, whereas for wind or wave power, there are problems such as noise pollution, teething troubles, power fluctuations, poor performance due to operation, and maintenance. However, the development of appropriate instrumentation for signal conditioner, computer interfacing mechanism, and software for different aspects of system operation is a major challenge.

A small-scale hydropower facility generates power through the kinetic energy of moving water as it passes through a turbine. Most small-scale hydropower facilities are "run-of-river," meaning that the natural flow of the river is maintained, and that a dammed reservoir is not created in order to generate power. Without a permanent dam to block river flow, nor a large reservoir to flood arable land and disrupt river temperature and composition levels, many of the negative riverine effects of traditional large-scale hydropower are avoided with a small-scale hydropower plant. However, some of the small, mini, or micro-hydroelectric power system may consist of a small reservoir, a section of a stream river, or an irrigation canal, used to provide water to turbine, governor, generator, and power electronic interface and eventually energy storage. The water is passed from reservoir to turbine through penstock. When water strikes at the blades of the turbine, it converts hydraulic energy into mechanical energy. Head is described as the vertical distance, or as a function of the characteristics of the channel or pipe. Most SHP sites are categorized as a low or high head. Low head refers to a change in elevation of less than 10 ft (3 m). A vertical drop of less than 2 ft (0.6 m) will probably make a small-scale hydroelectric system unfeasible. The net head ($H_0$) of an SHP can be created in quite number of ways, being the most known the following two types: building a dam across a stream in order to increase the water level just above the plant, or diverting part of the stream, with a minimum of head loss, to just above the plant. Figure 8.2 shows the main components of a hydropower scheme. The basic hydropower principle is based on the conversion of a large part of the gross head, $H_G$ (m), (i.e., net head $H_0$ (m)) into mechanical and electrical energy:

$$H_0 = H_G - \Delta H_{AB} \tag{8.21}$$

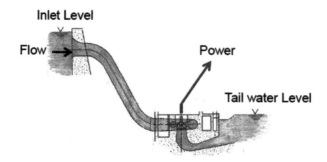

**FIGURE 8.2**
Flow-penstock's configuration.

Here head losses along the total conversion system are expressed by $\Delta H_{AB}$ (m).

**Example 8.5:** A small hydropower facility has a head of 100 m. Determine: (a) the water speed entering the turbine, (b) if the volume flow rate is 25 m³/s determine the penstock diameter, and (c) if the head loss due to the friction is 10% of the static head, determine the actual velocity approaching the turbine and the required penstock diameter.

**Solution:**

a. The water velocity entering the hydropower turbine is given by:

$$v = \sqrt{2gH} = \sqrt{2 \cdot 9.80 \cdot 100} \approx 44.3 \text{ m/s}$$

b. Applying continuity equation for the given flow rate, the penstock diameter is then:

$$D = \sqrt{\frac{4\pi v}{Q}} = \sqrt{\frac{4 \times 3.14 \times 44.3}{25}} \approx 4.72 \text{ m}$$

c. The actual head for 10% friction losses, using Equation (8.10) is:

$$H_{effective} = H - 0.1H = 0.9H = 90 \text{ m}$$

$$v_{eff} = \sqrt{2gH_{effective}} = \sqrt{2 \cdot 9.80 \cdot 90} = 42 \text{ m/s}$$

And the required penstock diameter is

$$D_{req} = \sqrt{\frac{4\pi v_{eff}}{Q}} = \sqrt{\frac{4 \times 3.14 \times 42}{25}} \approx 4.60 \text{ m}$$

The head in Equations (8.8), (8.10) or (8.11) is the gross head, which does not account for the losses due to turbulence and friction in the piping. The effective head, which is the head at the turbine inlet in the form of hydraulic pressure, is the gross head minus the head losses. Head losses are a function of the pipe length, diameter, surface texture, flow-rate and the number and type of fittings between the intake and the turbine. Typically pipe losses are separated into two parts: the losses due to the pipe itself and the losses due to the fittings. For a straight pipe, friction is proportional to the velocity of the water and to the ratio of the pipe's length with respect to its diameter. This relationship is expressed mathematically by the Darcy Equation, given by:

$$H_{Lmajor} = f \frac{L \cdot v^2}{2g \cdot D} \tag{8.22}$$

where $H_{Lmajor}$ represents the head loss due to friction, $L$ is the length of the pipe or penstock, $D$ the pipe or penstock diameter, $v$ the average flow velocity, $g$ acceleration due to gravity, and $f$ the friction factor. The friction factor is specific to the material and construction of the pipe (i.e., whether its inner surface is rough or smooth) and the characteristics of the flow (laminar or turbulent). The most widely used method of obtaining the friction factor is through the use of the Moody diagram. The Moody diagram shows the friction factor plotted against the Reynolds number, with a series of parametric curves related to the relative roughness of the pipe. The Darcy Equation (8.22) can be used to calculate energy loss due to friction in long, straight sections of round pipe. Pipe losses due to turbulence, such as from fittings and valves, are called minor losses, and are calculated using the following relationship:

$$H_{L\,minor} = \frac{k_L \cdot v^2}{2g} \tag{8.23}$$

where $H_{Lminor}$ represents the head loss due to turbulence, and $k_L$ is the loss coefficient. The effective head is therefore the gross head minus the head losses, expressed as:

$$H_{eff} = H_G - H_{Lmajor} - H_{Lminor} \tag{8.24}$$

One should always attempt to minimize the length of the pipe as well as the number of elbows, valves, and other fittings in the flow path, as their combined losses can be significant. Any reduction in the effective head will reduce the power output proportionately. It is not uncommon for a micro-hydro system to have an effective head that is as much as 30% less than the gross head. However, the pipe losses are not the only losses that must be considered. The efficiencies of the turbine, generator and electronics also rob power from the system. Small water turbines rarely achieve efficiencies greater than 80%. Combined with losses in the generator and power electronics, the overall efficiency is likely to be closer to 50%. To account for these equipment losses, the power equation is modified by the total efficiency coefficient, so:

$$P_{Out} = \eta_{tot} \cdot \rho \cdot g \cdot Q \cdot H_{eff} \tag{8.25}$$

This equation provides a reasonable estimate of the power output of a hydroelectric system regardless of its size or construction. The relationship in this form will be used throughout this analysis.

An SHP design should be the result of the work of a multi-disciplinary engineering or multi-specialist team including hydrologic, hydraulic, structures, electric, mechanical, geologic, and environmental experts. They cannot be scaled down from large hydroelectric projects. SHP layout schemes are usually characterized through different intakes and diversion structures depending on the type of the conveyance system (e.g., total pressurized or mixed). The SHPs include some essential components are penstock, power house, tailrace, generating plant, and allied equipment. Figure 8.3 shows small

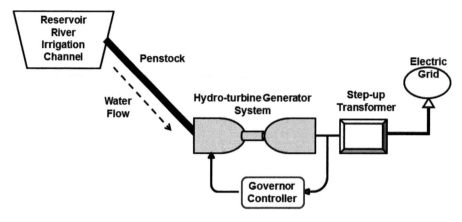

**FIGURE 8.3**
Schematic diagram of a mini or small hydroelectric power plant.

hydro system. The two small-scale hydropower systems that are being discussed in this section are the sites with capacities below 100 kW (referred to as micro-hydropower (MHP) systems) and sites with capacity between 101 kW and 1 MW (referred to as SHP systems). MHP systems, which use cross flow turbines and Pelton wheels, can provide both direct mechanical. Main function of governor is to control generator speed so that its frequency remains constant. Gate position of turbine is controlled through servomotor, which adjusts water flow to produce power according to the load connected. Hydropower turbines convert water pressure into mechanical shaft power, which can be used to drive an electricity generator, or other machinery. The general formula for a hydro system's energy generated over a certain interval of time ($\Delta t$) are given by the following relationship:

$$E_{hydr} = P \cdot \Delta t = \eta \cdot \rho \cdot g \cdot Q \cdot H \cdot \Delta t \qquad (8.26)$$

where $P_{hydr}$ is the mechanical power produced at the turbine shaft (W), h is the hydraulic efficiency of the turbine, $\rho$ is the density of water (kg/m³), $g$ is the acceleration due to gravity (m/s²), $Q$ is the volume flow rate passing through the turbine (m³/s), and $H$ is the effective pressure head of water across the turbine (m). The potential energy in water is converted into mechanical energy in the turbine as a result of the water pressure which applies a force on the face of the runner blades and then decreases as it passes through the reaction turbine. The relation between the electrical, mechanical and the hydraulic powers can be obtained by using the hydraulic turbine, power-train and generator efficiencies, $\eta_h$, $\eta_{mech}$, and $\eta_t$, expressed in the following relationship:

$$P_{electric} = \eta_{mech} P_{mech} = \eta_{mech} \eta_h P_{hydro} = \eta_t P_{hydro} \qquad (8.27)$$

The best turbines can have hydraulic efficiencies in the range 80% to over 90% (higher than most other prime movers), although this will reduce with size. Micro-hydro systems tend to be in the range 60%–80% efficient. Water is taken from the river by diverting it through an intake at a weir. The weir is a man-made barrier across the river which maintains a continuous flow through the intake. Before descending to the turbine, the water passes through a settling tank or forebay in which the water is slowed down sufficiently for suspended particles to settle out. The forebay is usually protected by a rack of metal bars which filters out water-borne debris which might damage the turbine such as stones, timber, or man-made litter. The mini-hydro or micro-power power plants mainly consist of a small reservoir or irrigation canal, governor, turbine, and generator. The water is passed from reservoir to turbine through penstock. When water strikes at the blades of turbine it converts hydraulic energy into mechanical energy. Currently mini hydro schemes employ conventional equipment which have resulted them an un-economical option. In order to make mini-hydro schemes a cost effective technology different new designs have been proposed in almost every component of mini-hydropower plant. The new designs include penstock, hydraulic turbines, generators and governor controller. In a run-of-river or diversion-type small hydropower systems, the power generating capacity of a water mass, flowing through the river at a velocity, $v$, is computed as the kinetic energy over time:

$$P_w = \frac{1}{2}\left(\frac{m}{t}\right)v^2 = \frac{1}{2}\rho Q v^2 \tag{8.28}$$

For the flow rate through an opening with the area, $A$ ($Q = A \cdot v$), the power density (power per unit of area) is expressed as:

$$\frac{P_w}{A} = \frac{1}{2}\rho v^3 \tag{8.29}$$

This equation is similar to the one of wind power density per unit rotor area, but the water density is substantially much higher than the air density. The converted electric power depends on the turbine efficiency, turbine blades power coefficient, and generator efficiency:

$$P_{el} = P_w\left(C_P \eta_{tr} \eta_{gen}\right) \tag{8.30}$$

Here, $C_P$ is the turbine blades power coefficient, $\eta_{tr}$, $\eta_{gen}$, are turbine and generator efficiencies.

**Example 8.6:** A run-of-river system installed on a small river, 25 m wide and 4.5 m deep, with water flowing at 3.20 m/s, diverts 25% of the flow. If overall efficiency is 75%, what is the power output?

**Solution:** The river volumetric flow rate is:

$$Q = A \cdot v = 25 \times 4.5 \times 3.25 = 360 \text{ m}^3/\text{s}$$

The flow diverted to run-of-river system is the:

$$Q_{r-r} = 0.25 \times 360 = 90 \text{ m}^3/\text{s}$$

From Equations (8.9) and (8.10), the power output is:

$$P_{electric} = \eta_t P_{hydro} = 0.85 \cdot 0.5 \cdot 90 \cdot 1000 \cdot (3.2)^2 = 391.68 \text{ kW}$$

## 8.2.1 Small Hydroelectric Power Technology

The role of the hydropower plants is to capture the energy in flowing water and convert it to usable energy. The hydropower potential depends on the availability of suitable water flow and then, where the resource exists, these plants can provide cheap, clean, and reliable electricity. Moreover, small-scale hydropower plants, when designed taking into account of surroundings without interfere significantly with river flows, have minimal negative environmental impacts; also because they don't need a reservoir, being in large part run of the river, respect to the large hydropower systems. A hydroelectric turbine converts the energy from falling water into rotating shaft power. The selection of the best turbine for any particular hydro site depends upon the site characteristics, the dominant ones being the head and flow available. Selection also depends on the desired running speed of the generator or other device loading the turbine. Other considerations, such as whether the turbine will be expected to produce power under reduced flow conditions, also play an important role in the selection. All turbines have a power-speed characteristic and an efficiency-speed characteristic. They will tend to run most efficiently at a particular speed, head, and flow. Hydroelectric power plants are of three major types:

1. *Impoundment*: a large hydroelectric power system that uses a dam to store river water in reservoir. Water stored in the reservoir is then used to generate electricity.

2. *Diversion*: a diversion facility channels a portion of a river through a canal or penstock. This system may not require the use of a dam.

3. *Run-of-river*: the system uses water with in the natural flow range and it requires a little or no impoundment.

Hydroelectric turbines can be crudely classified as high-head, medium-head, or low-head machines. However, this is relative to the size of machine: what is low head for a large turbine can be high head for a small turbine; for example, a Pelton turbine might be used at 50 m head with a 10 kW system but would need a minimum head of 150 m to be considered for a 1 MW system. The main reason that different types of turbine are used at different heads is that electricity generation requires a shaft speed as close as possible to 1500 rpm to minimize the speed change between the turbine and the generator. Turbines used in hydroelectric systems have runners of different shapes and sizes. There are two main categories of hydroelectric turbines in use: **impulse and reaction turbines**. Figure 8.4 is showing the schematic diagrams of the impulse and reaction hydropower turbines. The selection of any type of hydropower turbine for a project is based on the head and the flow or volume of water at the site. However, other deciding factors include how deep the turbine must be set, efficiency and cost.

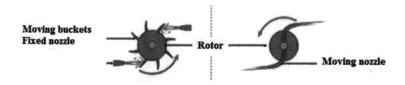

**FIGURE 8.4**
Schematic diagrams and the operation principles of the impulse turbine (left panel) and the reaction turbine (right panel).

The speed of any given type of turbine tends to decline with the square-root of the head, so low-head sites need turbines that are faster under a given operating condition. The reaction turbine rotor is fully immersed in water and is enclosed in a pressure casing. The runner blades are profiled so that pressure differences across them impose lift forces, akin to those on aircraft wings, which cause the runner to rotate. In contrast an impulse turbine runner operates in air, driven by a jet (or jets) of water, and the water remains at atmospheric pressure before and after making contact with the runner blades. Turbines used in hydroelectric power systems have usually runners of different shapes and sizes. The selection of any type of hydropower turbine for a specific application is based on the head and the flow or volume of water at that site. However, other deciding factors include how deep the turbine must be set, efficiency and cost. The choice of turbine for hydro plant depends upon the available water potential and local conditions. Turbine comes in wide variety of designs and sizes. The hydroelectric turbines are also classified according to the available energy at the turbine inlet, flow direction through vanes, head at the turbine inlet and specific turbine speed. According to specific turbine speed, there are three major classes: (1) low-specific speed and high-head turbine (Pelton); (2) medium-specific speed and medium-head turbine (Francis); and (3) high-specific speed and low-head turbine (Kaplan and Propeller). Depending upon specific site conditions, turbines are designed as Dam Base, Canal Fall, Run-of-River, and Hilly Region.

In order to improve the overall system efficiency the pipe length, the number of elbows, valves, and other fittings in the flow path, as their combined losses can be significant need to be minimized. Any reduction in the effective head reduces the power output proportionately. It is not uncommon for a micro-hydro system to have an effective head that is as much as 30% less than the gross head. However, the pipe losses are not the only losses that must be considered. The turbine efficiencies, generator, and electronics must be considered. Small-water turbines rarely achieve efficiencies greater than 80%. Combined with losses in the generator and power electronics, the overall efficiency is usually closer to 50%. To account for these equipment losses, the power equation is modified by the total efficiency coefficient, so:

$$P_{Out} = \eta_{tot} \cdot \rho \cdot g \cdot Q \cdot H_{eff} \tag{8.31}$$

This equation provides a reasonable estimate of the power output of a hydroelectric system regardless of its size or construction. The relationship in this form is used throughout the small hydroelectric system analysis. For a reservoir type small hydroelectric power, the analysis of the potential electricity are taking into account the input power (the potential energy), the losses into the penstock, turbine blades power coefficient, and the turbine and generator losses. Equation (8.21) can be re-written in this case to estimate the output electric power, as:

$$P_{Out} = g Q_{mass} H_{eff} \times \left( C_P \eta_{pen} \eta_{tr} \eta_{gen} \right) \tag{8.32}$$

### 8.2.1.1 Impulse Turbines

Impulse turbine uses the water kinetic energy to drive the runner and discharges to atmospheric pressure, being moved by the water jet at atmospheric pressure before and after making contact with the runner blades. Water that falls into tail after striking the buckets has little energy remaining, so the turbine has light casing serving only for the purpose of protecting the surroundings. Impulse turbines are usually used in systems with high head and low flow. There are three common types of impulse turbines: the Pelton, the Cross-flow, and the Turgo hydropower turbine. Pelton turbine consists of a wheel with a series of split buckets (vanes) set around its rim, and a high velocity water jet is directed tangentially at the wheel, hits each bucket, and is split in half, so that each half is deflected back almost through 180°. Nearly all the water energy goes into propelling the bucket and the deflected water falls into a discharge channel. The jets are through nozzles, each with an axis in the runner plane and a needle (or spear) valve to control the flow. To stop the turbine, in case the turbine approaches the runaway speed due to load rejection, the jet is deflected by a plate, no longer impinging on the buckets, so the valve is closed slowly keeping the over-pressure surge to an acceptable minimum. The water kinetic energy leaving the runner is lost and so the buckets are designed to keep exit velocities to a minimum. This turbine does not require draft tubes since the runners are positioned above the maximum tail water to permit operation at atmospheric pressure. Pelton turbines are usually applied in systems with large water heads. Unlike the Francis turbine, Pelton and cross-flow turbines can operate at high efficiencies even when running below their design flow.

A Turgo turbine is similar to the Pelton but the jet is designed to strike the plane of the runner at an angle (typically 20°) so that the water enters the runner on one side and exits on the other. Therefore the flow rate is not limited by the discharged fluid interfering with the incoming jet (as for Pelton turbines), so Turgo turbines can have smaller diameter runner than a Pelton for an equivalent power. Turgo turbines have also different shape of the buckets, with the water jet entering the runner through one side and exits through the other side. A Turgo turbine has a higher running speed which makes a direct coupling of turbine and generator more likely, thus increasing the overall efficiency and decreasing maintenance costs. Turgo turbines operate effectively in systems with large water heads. Turgo turbine based system is capable of handling varying seasonal flows and can operate efficiently in variety of different heads. These small turbines are ideal for connecting to ranchers existing gravity-fed irrigation systems, streams and creeks, for remote home-sites to reduce electrical consumption through net metering.

The cross-flow turbine has a drum-like rotor with a solid disk at each end and gutter-shaped "slats" joining the two disks. A jet of water enters the top of the rotor through the curved blades, emerging on the far side of the rotor by passing through the blades a second time. The shape of the blades is such that on each passage through the periphery of the rotor the water transfers some of its momentum, before falling away with little residual energy. A cross-flow turbine has a drum-like rotor and uses an elongated, rectangular-section nozzle which is directed against curved vanes on a cylindrically shaped runner. Cross-flow turbines are less efficient than the modern-day turbines, but can accommodate larger water flows and lower heads. A jet of water enters the turbine, thus gets directed through the guide-vanes at a transition piece upstream on the runner which is built from two or more parallel disks connected near their rims by a series of curved blades. The flow is directed to a limited portion of the runner by the guide vane at the entrance to the turbine. The turbine allows water to flow twice through the blades. In the first stage, water flows from the outside of the blades to the inside; in the second stage, the water passes from the inside back out. The flow leaves the first stage attempts to cross the open center of the turbine but as the flow enters the second stage, a compromise direction is achieved which causes significant shock losses.

### 8.2.1.2 Reaction Turbines

This type of hydropower turbine generates electricity from the mutual action of pressure and by moving water. Reaction turbines exploit the oncoming water flow to generate hydrodynamic lift forces to propel the runner blades. They are having a runner that always functions within a completely water-filled casing. The reaction turbine operates when the rotor is fully submerged in water and is enclosed in a pressure casing. The runner blades are profiled so that pressure differences across them impose lift forces, akin to those on aircraft wings, causing the runner to rotate. Reaction turbines are generally appropriate for sites with lower head and higher flow rates compared with the impulse turbines. The runner blades are profiled so that pressure differences across them impose lift forces, akin to those on aircraft wings or wind turbine blades, which cause the runner to rotate. Reaction turbines are generally appropriate for sites with lower head and higher flows compared with the impulse turbines. Typical reaction turbine types are Propeller, Francis, and Kinetic ones. Reaction turbines have a diffuser known as a "draft tube" below the runner through which the water discharges. The draft tube slows the discharged water and reduces the static pressure below the runner and thereby increases the effective head. The two main types of reaction turbines with a few variants are the propeller (with Kaplan variant) and Francis turbines.

Propeller-type turbines are similar in principle to the propeller of a ship, but operating in reversed mode. Various configurations of propeller turbine exist; a key feature is that for good efficiency the water needs to be given some swirl before entering the turbine runner. With good design, the swirl is absorbed by the runner and the water that emerges flows straight into the draft tube with little residual angular momentum. Methods for adding inlet swirl include the use of a set of fixed guide vanes mounted upstream of the runner with water spiraling into the runner through them. Another method is to form a "snail shell" housing for the runner in which the water enters tangentially and is forced to spiral into the runner. When guide vanes are used, these are often adjustable so as to vary the flow admitted to the runner. In some cases, the blades of the runner can also be adjusted, in which case the turbine is called a Kaplan. The mechanics for adjusting turbine blades and guide vanes can be costly and tend to be more affordable for large systems, but can greatly improve efficiency over a wide range of flows.

A propeller hydroelectric turbine generally has an axial flow runner with three to six blades depending on the designed water head. For higher efficiency the water needs to be given some swirl before entering the turbine runner. Propeller turbines are suitable for systems with low water heads. There are several different types of propeller turbines: bulb, Kaplan, Straflo, and tube turbine. The Kaplan turbine has adjustable blade pitch, and it can achieve high efficiency under varying power output conditions. The methods used for adding inlet swirl include fixed guide vanes mounted up stream of the runner and snail-like shell housing for the runner, in which the water enters tangentially and is forced to spiral into the runner. In the case of the Kaplan turbine, the blades of the runner are adjusted. Adjustment of the turbine blades and guide vanes can greatly improve efficiency over a wide range of flows; however, it is costly and so can only be economical in larger systems. The unregulated propeller turbines are commonly used in micro-hydro systems where both the flow and head remain practically constant.

The Francis turbine is essentially a modified form of propeller turbine in which water flows radially inward into the runner and is turned to emerge axially. The runner is most commonly mounted in a spiral casing with internal adjustable guide vanes. Reaction turbines require more sophisticated fabrication than impulse turbines because they involve the use of more intricately profiled blades and profiled casings, making them less attractive for use in micro-hydro in developing countries. Nevertheless, because low head sites are far more common and often more closer to where the power is needed, work is being undertaken to develop propeller machines which are simpler to construct. Francis type is the most common type of hydropower turbine in use. This turbine generally has radial or mixed radial/axial flow runner, which is most commonly mounted in a spiral casing with internal adjustable guide vanes. Water flows radially inward into the runner and emerges axially, causing it to spin. In addition to the runner, the other major components include the wicket gates and the draft tube. The runners with smaller diameter are made of aluminum bronze casting, while the larger runners are fabricated from curved stainless steel plates that are welded to cast steel hub. Francis turbines are applied in hydroelectric systems with medium head size and their efficiency can be above 90% but tend to have higher cost.

### 8.2.1.3 *Pump as Hydroelectric Turbine*

Alternative options to the conventional hydroelectric turbines are the use of pumps, in reverse mode of operation as prime mover. The initial cost of small, mini, or micro-hydropower plants largely depends upon the cost of equipment. One way to reduce this cost of equipment is to use centrifugal pump as a turbine (PAT). PAT can be considered as a cost-effective alternative and viable option for small, mini, micro, or pico hydropower generation especially in rural hilly and complex terrain areas or in agricultural land areas. PAT being mass produced all over the world is a standard product available in variety of sizes of different head and flow rates. These are cheaper and easily available in the market. Their repair and maintenance can be easily carried out by local technicians and in local workshops. The PAT cost is about 50% less than the cost of conventional hydroelectric turbine of similar characteristics. PATs are available in wide variety of power from about 1.7–160 kW range. PAT is basically a pump which can operate in a turbine mode if the direction of flow is reversed, with higher efficiency in turbine mode operation. With suitable conditions, PAT can cover the range of multi-jet turbines, cross-flow turbines, and small Francis hydroturbines. Standard pumps can be easily used for electrical power generation operated in the reverse mode. Axial, radial, and mixed centrifugal pumps operating in reverse mode can operate as hydroelectric turbines. Despite of having many advantages over conventional turbine, PAT has one major drawbacks, prediction of turbine characteristics of the centrifugal pump is very difficult. PAT selection for a specific mini-hydro site is a major problem, while the turbine characteristics of pumps cannot be generalized. However, despite intensive research on adapting methods for determining the optimum behavior of pumps from analytical analysis, simulation, experimental work, computational studies optimum results still have not been found. One of the reason is that manufacturers of pumps do not provide their characteristics curves and it is very essential to know about the characteristics of pumps for successful operation of pumps as a hydroelectric turbine.

### 8.2.2 Generators and Control

The power generation industry almost exclusively uses large synchronous generators, as they have the advantage of a variable reactive power production, i.e., voltage control. There are two most common used types of generators: synchronous and asynchronous electric machines. However, DC (direct current) electric generators are sometimes used, especially for pico-hydroelectric applications. An electric generator is a device that converts mechanical energy to electrical energy. A generator forces electric current to flow through an external circuit. The source of mechanical energy may be a reciprocating or turbine steam engine, water falling through a turbine or waterwheel, an internal combustion engine, a wind turbine, a hand crank, compressed air, or any other source of mechanical energy. Generators provide nearly all of the power for electric power grids. An electric generator uses the rotor shaft speed and torque to convert mechanical energy to electrical energy with the use of electromagnetic fields. The main concern with generators in hydroelectric turbines is that they must produce electricity compatible with that in the electrical grid, if the system is grid connected for the given site. In America, the grid is at 60 Hz, while in most parts of Europe the grid runs at 50 Hz. The electricity must have the same characteristics, be of sufficient quality, and be connected in such a way as to not interrupt the existing current flow. The function of an electrical generator is providing a means for energy conversion between the mechanical torque from the hydro rotor turbine, as the prime mover, and the local load or the electric grid. Different types of generators are being used with hydroelectric system, induction (asynchronous), synchronous, or occasionally DC electric generators.

Synchronous generators are standard in electrical power generation and most commonly used in most power plants. They are used especially in large power plant applications either there is significant research for their applications in lower power range. However, more often in renewable energy induction generators are employed. Asynchronous generators are more commonly known as induction generators. A great deal of research has been carried out, over the last two decades regarding the application of the induction generator as an alternative option for electricity generation over the synchronous generator in small, mini, or micro-hydropower schemes for making these schemes cost effective. The induction generators have lower cost per unit of KWh as compared to synchronous generators. Besides, induction generators are more robust and have easy starts and control mechanisms, self-protection against faults. It has the ability for generating power at changing speeds and can be used in an off-grid or in connected mode with synchronous generator for load sharing. It can also be operated as generator when its stator winding is connected with capacitor and rotor is driven by prime mover. In that case the magnetizing lagging reactive power is provided by the capacitor. This capacitor establishes the air-gap flux. This configuration enables the induction machine to work as a self-excited induction generator. Induction generators excited with capacitor are emerging as a suitable candidate for renewable energy power generation operating in stand-alone mode. Despite all these advantages, induction generators have encountered problem in maintaining the frequency and voltage within its range.

The induction generator is a standard three-phase induction motor, wired to operate as a generator. Capacitors (C) are used for excitation, by connecting unequal excitation capacitance across the windings of the motor, converting a three-phase motor into a single-phase generator. This is cost-effective and has been popular for smaller off-grid systems below 10–15 kW. They have the advantage of being rugged, cheaper than synchronous generators, robust and widely available and can withstand over-speed and overload. For MHP, induction generators, with capacitive VAR controllers, are used for both standalone and grid connected mode. Synchronous generators are also used preferably for standalone mode. Although most induction generators in operation are employed in wind power, such machines have also been used in medium-size hydro and thermal plants. However, most distributed generation systems employ synchronous generators, which can be used in thermal, hydro or wind power plants. In the electromagnetic transient simulations, the synchronous generators were represented by an eight-order model, which was reduced to a sixth-order model in the transient stability simulations. Usually, synchronous generators connected to distribution networks are operated as constant active power sources, so that they do not take part in the system frequency control.

Induction generators need very little auxiliary equipment and can be run in parallel with generator without hunting at any frequency. For IG speed variation of prime mover is less important. A self-protective feature of this is that if there is terminal short circuit, excitation fails and so does the generator output. Its disadvantage is that it draws considerable amount of lagging KVAR from the supply for excitation, the efficiency is comparatively poor and can operate only at leading power factor. An induction generator controller for micro-hydropower systems using the induction generator has also been developed. ELG is not suitable for an induction generator. This is the generator of choice if micro-hydropower system is to supplying to the grid. Induction generators are increasingly used these days because of their relative advantageous features over conventional synchronous generators. These features are brushless and rugged construction, low cost, maintenance and operational simplicity, self-protection against faults, good dynamic response, and capability to generate power at varying speed. The later feature facilitates the induction generator operation in stand-alone/isolated mode to supply far flung and remote areas where extension of grid is not economically viable; in conjunction with the synchronous generator to fulfill the increased local power requirement, and in grid-connected mode to supplement the real power demand of the grid by integrating power from resources located at different sites. The reactive power requirements are the disadvantage of induction generators. This reactive power can be supplied by a variety of methods, from simple capacitors to complex power conversion systems.

Synchronous generators are the most commonly used machine for generation of electrical power for commercial purpose is the synchronous generator or alternator. Typical alternators use a rotating field winding excited with direct current, and a stationary (stator) winding that produces alternating current. Since the rotor field only requires a tiny fraction of the power generated by the machine, the brushes for the field contact can be relatively small. In the case of a brushless exciter, no brushes are used at all and the rotor shaft carries rectifiers to excite the main field winding. Synchronous generators driven at low speeds by prime-movers like water turbines will have salient pole construction having large number of projected poles. Synchronous generators in standalone mode: When synchronous generator is independently supplying load, increasing its excitation will increase the no load voltage. This will raise the whole load characteristics and increase the value of terminal voltage. Permanent magnet synchronous generator is a solution that is appreciated in small wind turbines but it cannot be extended be extended to large-scale power because it involves the use of big and heavy permanent magnets. However, the synchronous generators, used in the large hydropower plants are not well suited for renewable and nonconventional energy generation. New technologies and configurations are under research of in testing for commercial application. One of them is represented by the six-phase synchronous generator, used initially for uninterruptible power supply. In the last decade, the application of six-phase generators for renewable energy started to increase. However, in the past none has adapted six-phase synchronous generator in small, mini, or micro-hydropower plants to show its economic viability. Several studies were conducted to find additional advantages possessed by six-phase synchronous generator over conventional three-phase generator. Usually, the three-phase synchronous generator was formed the six-phase by adapting a phase belt splitting. It should be noted that electrical displacement is of great interest because it allows recombination of three-phase power in step-up transformer bank. Experiments were carried out on constant voltage as well as on constant frequency or speed operation. Six-phase has an advantage that it can supply two three-phase loads separately. Another advantage of using this type of generator is that by employing a transformer having six-phase to three-phase windings, output of two three-phase winding can be supplied to a single three phase loads. The use of transformer has also the advantage that in case of failure in any of three-phase winding set, the system will not lead to shut down but load continued to be supplied through remaining healthy winding thus improving system reliability and robustness, while reducing the cost.

*DC generators* or dynamos are electrical machines that produce direct current with the use of a commutator. Dynamos were the first electrical generators capable of delivering power for industry, and the foundation upon which many other later electric-power conversion devices were based, including the electric motor, the alternating-current alternator, and the rotary converter. Today, the simpler alternator dominates large-scale power generation, for efficiency, reliability and cost reasons. A dynamo has the disadvantages of a mechanical commutator. Also, converting alternating to direct current using power rectification devices (vacuum tube or more recently solid state) is effective and usually economic.

Small-scale hydropower plants due to their complex nature and non-linear behavior of hydraulic turbine encountered very often frequency and voltage variation in the system. Different control techniques are proposed and needed for maintaining voltage and frequency within the prescribed range. Voltage can be controlled by varying the excitation, while frequency can be controlled by making generation equal to load through governor control scheme. In the past, mechanical hydraulic governors were used, today being replaced by electrohydraulic, PI/PID governors, and control schemes. Fundamental mathematical models of all these governors, control schemes and turbines suitable for hydro-electric power plants are intensively discussed in the literature and are beyond the scope of this book. Different governing control techniques for frequency stabilization have been discussed and proposed over the years, covering the classical PID controller to the intelligent control techniques. PID controllers have better response in dealing governor control because it provides three functions to control the system. Control for water flow is also important for a number of applications as in hydropower, tank water level, etc. Turbine gate opening and water flow plays an important role for generation and in maintaining the system frequency. At present, water is controlled by conventional control techniques. However, they can be controlled by using robustly approaches with fuzzy logic control schemes.

## 8.3 Tidal and Wave Energy

### 8.3.1 Tidal Energy

The tides are cyclic or periodic variations in the level of seas and oceans, representing the planetary manifestation of the potential and kinetic energy fluxes present in the Earth-Moon–Sun system. This results in some regions of the world possessing substantially higher local tidal variation than others. Tidal energy is energy generated from tidal movements, containing both potential energy, related to the vertical fluctuations in sea level, and kinetic energy, related to the horizontal motion of the water column. Tidal periodicity varies according to the lunar and solar gravitational effects, respective movements of the Moon and Sun, and other geographical peculiarities. The mean interval between conjunctions of the Sun and Moon has a cycle of 29.53 days, which is known as Synodic month or lunation. There are three different types of tidal phenomena at different Earth locations. Semidiurnal tides with monthly variation has a period of 12 hours 25 minutes, due to the Earth rotation relative to both Sun and Moon, consequently the tidal phenomenon occurs twice every 24 hours 50 minutes 28 seconds, each landmass is exposed to two high tides and two low tides during each period of rotation. Tide amplitude varies according to the lunar month, with higher tidal range at full Moon and new Moon, when Sun and Moon are aligned. Neap tides occur during half-moon as the resultant gravitational pull is at its minimum. However, one of the tides has greater range than the other, having a higher high and a lower low, therefore, a greater tidal flow while water is coming in and going out during the period between high and low level. Furthermore, the tidal output peaks and troughs four times a day as the tide comes in and out twice daily. Diurnal tides with monthly variation are found in China Sea and Tahiti, having the tidal period is of 24 hours 50 minutes 28 seconds. During each Earth rotation, a point of the Earth surface pass through different parts of the equilibrium tide envelope and therefore experience a diurnal variation in tide levels. Mixed tides combine the characteristics of diurnal and semidiurnal tides. Moreover, they can also display monthly and bimonthly variation. They are found in the Mediterranean Sea. Moreover, there are several periodic phenomena that are affecting the tidal behavior.

Tidal currents occur in coastal areas and in places where the seabed forces the water through relatively narrow boundaries. Thus both high tidal ranges and narrow channels are generally required to cause significant tidal stream currents. The range of a spring tide is commonly about twice the neap tide range. The common tidal range is about 50 cm in the open ocean. However, the tidal amplitude can be increased by several local effects such as shelving, funneling, reflection and resonance. The shelving effect consists on increasing the deep water tidal wave height as the wave slows down when entering in shallow water areas. The tidal amplitude can be further increased due to a funneling effect, which occurs when the tidal bulge progresses into a narrowing estuary. Moreover, tidal wave can also be reinforced by reflections of the waves by the coastline. At some sites, the tidal flow can be heightened to more than 10 m by resonance effects, i.e., Bay of Fundy, in Canada, where the greatest tides in the world can be found, and the Severn Estuary in England. The resonance effect it takes place when the tide at the mouth of the estuary can resonate with the natural frequency of tidal propagation up the estuary. Tidal currents can flow in two directions; the current moving in the direction of the coast is known as flood current and the current receding from the coast is known as ebb current. The current speed varies from zero to a maximum. The zero current speed occurs between the ebb and the flood current, slack period; while the maximum speed is reached halfway between the slack periods. All tidal variations rise and fall and flood and ebb current can be utilized to generate electricity. The generation of electrical power from ocean tides is very similar to conventional hydroelectric generation.

The tidal energy resource is vast and sustainable, and has a great potential for electricity generation. There are two different means to harness tidal energy. The first is to exploit the cyclic rise and fall of the sea level using barrages the second is to harness local tidal currents, analogous to wind power also called marine current turbine. An advantage of both tidal range and current energy is that they are highly predictable with daily, biweekly, biannual, and even annual cycles over a longer time span of a number of years. Energy can be furthermore generated both day and night and tidal range is hardly influenced by weather conditions. Tidal stream is slightly more affected by the weather, but the fluctuations in the long run are lower than, for example, wind and solar. Tidal energy is one of the oldest forms of energy used by humans. Indeed, the earliest evidence of the use of the oceans' tides for power conversion dates back to about 900 A.D., but it is likely that there were predecessors lost in the anonymity of prehistory. Much later, American colonists built tidal-powered mills in New England, consisting of a storage pond, filled by the incoming tide trough a sluice and emptied during the outgoing (ebb) tide through a waterwheel. The tides turned waterwheels producing mechanical power to mill grain, the power was available during 2–3 hours, usually twice a day. However, the economically exploitable resource is currently small because of the considerable costs associated with energy extraction and the environmental impacts of some tidal energy technologies, notably barrages and lagoons (tidal pools). There are few estimates of the world tidal energy resource potential. It can be harnessed using two main technologies:

1. Tidal barrages (or lagoons) are based on the rise and fall of the tides, which are generally consisting of a barrage that encloses a large tidal basin. Water enters the basin through sluice gates in the barrage and is released through low-head turbines to generate electricity.

2. Tidal stream generators are based on tidal or marine currents, which are free-standing structures built in channels, straits or on the shelf and are designed to harness the kinetic tide energy. They are essentially turbines that generate electricity from horizontally flowing tidal currents (similar to wind turbines).

## 8.3.2 Tidal Barrage Methods

Currently several places in the world are producing electricity from tides, with the biggest tidal barrage power plant, located in La Rance, France, operating since 1966, generating 240 MW. Other operational barrage sites are in Nova Scotia (20 MW), near Murmansk, Russia (0.4 MW) and the Eastern seaboard of China (3.2 MW). An estuary or bay with a large natural tidal range is artificially enclosed with a barrier. Electrical energy is produced by allowing water to flow from one side of the barrage to the other. To generate electricity, tides go through low-head turbines with a variety of operation modes, as shown in Figure 8.5, similar to conventional hydropower. There are two common configurations, single basin schemes and multiple basin schemes. The simplest ones are the single basin tidal energy systems, requiring a single barrage across an estuary, involving a combination of sluices which when open can allow water to flow relatively freely through the barrage and gated turbines to generate electricity. Ebb generation depends on the height of the tides. At high tide, water is retained behind the barrage by the sluices. At low tides, water flows reverse out through the turbine. Double-basin systems allow for storage (adjusting the power output to demand of consumers), one basin is filled at high tide and the other one is emptied at low tide. The turbines are placed between the basins. The main basin behaves like the ebb generation mode. Two-basin schemes offer advantages over normal schemes in that generation time can be adjusted with high flexibility and it is also possible to generate almost continuously. In normal estuarine situations, however, two-basin schemes are very expensive to construct due to the cost of the extra barrage. There are some favorable geography locations that are well suited to this type of schemes.

 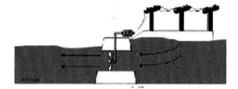

**Flow in (High Tides)**          **Flow out (Low Tides)**

**FIGURE 8.5**
Tidal power system configuration and operation.

Tidal barrages use the potential energy of the tides to generate electricity. Given a basin, the theoretical potential energy can be calculated as:

$$PE = \rho g A \int z dz = 0.5 \rho g A H^2 \tag{8.33}$$

$PE$ is tidal potential energy (J), $g$ the acceleration of gravity (9.806 m/s²), $\rho$, the seawater density (approximately 1022 kg/m³), $A$ the sea area (m²), $z$ the vertical coordinate of the ocean surface (m), and $H$ the tide amplitude (m). For seawater, the product ($g \cdot \rho$) is equal to 10.0156 kNm⁻³. Tidal energy technology is similar to conventional hydropower plants, but in this case, the current flows in both directions, meaning, that tidal barrages are unable to produce electricity at a constant rate, as they have to wait for sufficient hydrostatic head between both sides of the dam. However, tidal electricity production is totally predictable allowing for ease operation. A tidal barrage is generally a dam placed across an estuary with a tidal range of about 5 m or higher. The basic tidal barrage elements are turbines, sluices gates, embankments, caissons, and ship locks. Caissons, very large concrete blocks, house sluices, turbines, and ship lock; any section of the barrage not containing sluices or turbines may be completed using blank caissons. The dam can either be placed at the entrance of channels where ocean water gets inside the land via a bay or between the main land and an island or just in between two islands. Embankments function is to seal the basin, where it is no sealed by caissons. The sluices open when there is an adequate difference in the water elevation on the barrage sides, the hydrostatic head that is created causes the water flow through the turbines, turning an electric generator to produce electricity. When deploying a tidal barrage, technical evaluation of the following aspects will be necessary before the final form of the project is decided: (a) type of structure: single basin or double basin, (b) dam and plant location, (c) operating mode: single or double action with or without pumping, (d) unit power of turbines and generators and (e) total power output. If the basin has a surface area $A$ and the tide a range $R$, the mass of water trapped in basin is then $\rho AR$ for this geometry. At low tide this mass of water will drop by a height $R/2$, so the maximum potential energy available per tide is then, using Equation (8.33) as:

$$PE = 0.5 \rho g A R^2 \tag{8.34}$$

Potential energy averaged over the tidal period $\tau$ yields to the average potential power:

$$\bar{P} = \frac{\rho A R^2 g}{2\tau} \tag{8.35}$$

However, the tide range is not constant over the month, having a sinusoidal variation. If we have a mean range $R_{Mean}$ and $R$ varying between $R_{max}$ and $R_{min}$ then the average power is given by:

$$\bar{P} \simeq \frac{\rho A g}{2\tau} R_{Mean}^2 = \frac{\rho A g}{4\tau} \left( R_{max}^2 + R_{min}^2 \right) \tag{8.36}$$

**Example 8.7:** For a conversion efficiency of 25%, calculate the available tidal energy during one tidal cycle for an estuary with area of 360 km², tide amplitude of 7.5 m, and sea water density of 1022 kg/m³.

**Solution:** Using Equation (8.34) the average tidal energy is:

$$E = 0.5 \rho g A R^2 = 0.5 \cdot 1022 \cdot 9.80 \cdot 360 \cdot 10^6 \left(7.5\right)^2 \simeq 101.408 \times 10^9 \text{ J}$$

Multiplying this result with the efficiency we can get the available electrical energy

$$E_{ele} = 0.25 \cdot 101.408 \times 10^9 = 25.352 \times 10^9 \text{ J}$$

Barrage or dam-type tidal energy systems are the best suited for inlets connecting a lagoon or an reservoir to the open sea. At the end of channel a dam is constructed to control the tidal water flow in either direction. Turbine-generator units are installed in the conduits connecting the dam sides to convert the water kinetic energy into electricity. At high tides the water flows from the open sea to the lagoon or reservoir, generating the electricity, while at low tides the process is reversed and the water flows in opposite direction to the turbine, generating electricity. The amount of the potential energy is determine by the difference in hydraulic heads (Equation 8.23), and is given by the hydrostatic potential energy:

$$PE = mg\Delta H = mg \frac{H_{high} - H_{low}}{2} \tag{8.37}$$

The output electric energy that can be generated by the dam-type tidal energy system is determined by the power coefficient of the turbine blades, $C_P$, the turbine efficiency, $\eta_{tr}$ and the generator and electronics efficiency, $\eta_{ge}$:

$$WE = \eta_{tr}\eta_{ge}C_P \cdot PE \tag{8.38}$$

**Example 8.8:** A dam-type tidal energy is constructed between a semi-circular lagoon and open sea, having a diameter of 1800 m. Compute the output electric energy is the average head difference between high and low tides is 8 m, the power coefficient is 35%, the turbine efficiency is 85%, and the generator-electronics-transmission efficiency is 93%.

**Solution:** The volume of water flowing through the turbine through high tide is:

$$m = 0.5 \cdot \pi \left( \frac{D^2}{4} \right) \cdot \rho_{sea\ water} \cdot \Delta H = 3.31128 \times 10^9 \text{ kg}$$

Using Equations (8.26) and (8.27) the available potential energy and the output electricity are:

$$PE = 3.31128 \times 10^9 \cdot 9.806 \cdot 8 = 2.59733 \times 10^{11} \text{ J}$$

$$WE = 2.59733 \times 10^{11} \cdot 0.35 \cdot 0.85 \cdot 0.93 = 7.187 \times 10^{11} \text{ J}$$

Theoretically tide energy can become one of the major renewable energy sources, because has a long-term predictability, simple to estimate available power and the generation technology is well established and mature electric generation technology. However, among the major limitations of the tide energy systems are: it is an intermittent energy source, depending on the tidal cycle, very high initial system investment, longer system investment return than of the conventional power generation systems, and the dam may have negative effects on sea are ecosystem and navigation.

### 8.3.3 Marine Current Power Systems

Tidal stream generators harness energy from currents generally in the same way as wind turbines. The higher density of water, about 800 times the air density (exactly 832 times higher), means that a single generator can provide significant power at low tidal flow velocities (compared with wind speed). Given that power varies with the density of medium and the cube of velocity, it is simple to see that water speeds of nearly one-tenth of the speed of wind provide the same power for the same size of turbine system. However, this limits the application in practice to places where the tide moves at speeds of at least 2 knots (1 m/s), even close to neap tides (tide range is at a minimum). Ocean current resources are likely limited to the Florida Current, which flows between Florida and the Bahamas. Estimates of the energy present in the Florida Current date back to the mid-1970s, when the use of this resource for electricity generation was first proposed. While these early studies indicate an energy flux potential of as much as 25,000 MW through a single cross-sectional area, the amount of energy that could be extracted is uncertain, primarily because of concerns that reducing the energy in this portion of the Gulf Stream could have negative environmental consequences. Early modeling suggested that an array of turbines totaling 10,000 MW of capacity would not reduce the current's speed by more than what has been observed as its natural variation, and thus might be feasible. Further investigation is required to determine the magnitude of the technically available resource. Since tidal stream generators are an untested technology, no commercial scale production facilities are yet routinely supplying power; no standard technology has yet emerged. A large variety of designs are being experimented with, some very close to large-scale deployment. Several prototypes have shown promise with many companies, but they have not operated commercially for extended periods to establish performances and rates of return on investments. Tidal current turbines extract the kinetic energy from the moving unconstrained tidal streams to generate electricity. Currents have the same periodicity as vertical oscillations, being thus predictable, although they tend to follow an elliptical path. The ideal kinetic energy is calculated as:

$$E = mv = \rho V v$$

where $m$ is the mass of water $m = \rho V$, and $\rho$ is the seawater density (approximately 1022 kg/m³), and $V$ is the water volume. The ideal power for a mass of water passing through the rotor with a cross sectional area, $A$, can be expressed as follows:

$$P_T = 0.5 C_P \rho A v^3 \qquad (8.39)$$

Here, $P_T$ is the power developed by the rotor (W), $A$, the area swept out by the turbine rotor (m²), $v$, the stream velocity (m/s), and $Cp$, and the power coefficient of the turbine, which is the percentage of power that the turbine can extract from the water flowing through it. According to the studies carried out by Betz, the theoretical maximum amount of power that can be extracted from a fluid flow (water or air) is about 59%, which is referred to Betz limit. Tidal current technology extracts the kinetic energy in a similar way to harvest wind energy from air. However, there are several differences in the operating conditions. Operating under similar conditions, water is 832 times denser than air, while the water flow speed is much smaller. Due to the density difference between air and water, the power intensity in water currents is significantly higher than air streams. Consequently, water current turbine can be built considerably smaller than an equivalent powered turbine. In contrast to atmospheric air flows, the availability of tidal currents can be predicted very accurately. Another specific advantage of tidal current devices is the limited environmental impact as their installation requires minimum land use, and fully submerged devices will not affect optically or acoustically their surroundings. Finally, submerged marine current converters are considered to operate in safe environment, disturbances caused by extreme weather conditions are significantly attenuated to depths, where these devices are placed. However, since tidal current turbines operate in water, they experience greater forces and moments than wind turbines. In addition, tidal current turbines must be able to generate during both, ebb and flood currents, and be able to withstand the structural loads when not generating electricity.

**Example 8.9:** Calculate the diameter of tidal current turbine with power coefficient of 0.36, installed in a current of 3.85 m/s that generate 5.4 MW.

**Solution:** From Equation (8.4) the rotor area can be calculated as:

$$A = \frac{P_T}{0.5 C_P \rho \cdot v^3} = \frac{5.4 x 10^6}{0.5 \times 0.35 \times 1022 (3.85)^3} = 529.08 \text{ m}^2$$

The rotor diameter is then:

$$D = \sqrt{\frac{4A}{\pi}} = \sqrt{\frac{4 \times 529.08}{3.14}} = 25.96 \text{ m}$$

In the simplest form, a tidal current turbine is constituted by a number of blades mounted on a hub, a gearbox, and a generator. The hydrodynamic effect of the water, flowing through the blades causes the rotor to rotate, turning the generator to which the rotor is connected through the gearbox. The gearbox is used to transform the rotational speed of the rotor shaft to the desire power input of the generator. The electricity produced is transmitted to land through cable. Turbines are mounted to a support structure required to withstand the harsh environmental conditions. The foundation choice depends mainly on geographical conditions, water depth, seabed conditions, streams, and the type of turbine to be installed.

Moreover, it is an important aspect concerning feasibility and profitability of the devices. There are three main support structures. The first one, gravity structure consists on a big mass of concrete and steel, which is attached to the base of the structure to provide stability. The second option is known as piled structure which is pinned to the seafloor using one or more steel or concrete beams. And finally, the third option is the floating foundation; its structure is usually moored to the seafloor using chains or wire. In this case, the turbine is fixed to a downward pointing vertical beam, which is fixed to a floating structure.

Tidal current turbines are designed to extract the kinetic energy from the currents to generate electricity. Currently, there are mainly two types of turbines: horizontal axis and vertical axis tidal current turbines.

First, horizontal axis tidal current turbines are the ones where the blades rotate about a horizontal axis which is parallel to the direction of the water flow. They are arrayed underwater in rows, similar to some wind farms. The optimum operating point of the turbines is for coastal currents speeds between 4 and 5.5 mph. In those currents, a 15 m diameter tidal turbine can generate as much energy as a 60 m diameter wind turbine. The ideal locations for tidal turbine farms are close to shore in water depths of about 20–30 m. Horizontal turbines have slightly higher efficiency than vertical turbines. However, as they depend on the current direction, a mechanism to make the blades rotate is needed, generally, they are very complex. Following, some real examples of this type of turbines are mentioned. Sea-flow turbines that have about 11 m diameter rotor, with full span pitch control. It is mounted on a steel tubular pile, 2.1 m in diameter, set in a hole that was drilled in the seabed and tall enough to always project above the surface of the sea. They are installed in a mean depth of 25 m, 1.1 km off the nearest landfall in North Devon, UK. Under favorable conditions it has exceeded its 300 KW rated power with a 15 rpm rotor. It is not grid-connected, dumps its power into resistance heaters.

Current turbines (MCT) was developed by Hammerfest Strom. This turbine can be installed on the seabed offshore or near shore depending on the tidal current strength. The blades of 15–16 m are able to rotate on their own axes, allowing the turbine to be optimized to current conditions and also operate in both directions of the tide. A 300 KW system was tested, however a larger design is being developed that provide 750–1000 KW of power. Vertical axis tidal current turbines are cross-flow turbines, with the axis positioned perpendicular to the direction of the water flow. Cross flow turbines allow the use of a vertically oriented rotor which can transmit the torque directly to the water surface without the need of complex transmissions systems or an underwater nacelle. The vertical axis design permits the harnessing of tidal flow from any direction, facilitating the extraction of energy not only in two directions, the incoming and the outgoing tide, but making use of a full tidal ellipse of the flow. Moreover, the blades are easily built and their span can be easily increased. However, this type of turbines experiment a lot of vibrations, as the forces exerted on the bladders are very different, consequently, it is difficult to reach stability. In vertical axis turbines as in horizontal axis ones the rotation speed is very low, around 15 rpm. Other two vertical axis projects are introduced, and they are discussed here. Kobold turbine, at which the main feature is the high starting torque that makes it able to start spontaneously even in loaded conditions. A pilot plant is moored in Messina, Italy, in an average tidal current of about 2 m/s. With a current speed of 1.8 m/s, the system can produce a power of 20 KW. Blue energy project are using four fixed hydrofoils are connected to a rotor that drives an integrated gearbox and an electrical generator assembly. The turbine is mounted in a concrete caisson which anchors the unit to the ocean floor. The generator and the gearbox are placed above water surface and are readily accessible for maintenance and repair. A unit turbine output power is expected to be about 200 KW, for large-scale production, multiple turbines are linked in series.

### 8.3.4 Wave Energy

Today, still wave power technologies are almost absent from the world energy market. This is an area of multi-physics where conventional solutions do not exist, being a defining characteristic of wave power research over the years, and this is also the case today. Wave power development has faced many difficulties and a multitude of solutions. Some of the major challenges are the survivability of parts exposed to the forces of the ocean, investment costs associated with large structures, excessive over-dimensioning needed to handle mechanical overloads, long life mooring difficulties, transmission of energy to shore, and the transformation of wave motion into high-speed, rotating generator motion.

The waves referred here are wind generated ocean or sea waves, and in essence they concentrated solar energy. Due to unevenly solar radiation distribution, the Sun energy is transferred to and concentrated into the wind. When the wind blows over a stretch of water, waves are created, and the wind energy transferred to the waves is concentrated once again. It starts with small pressure differences on the ocean surface, due to turbulence in the wind, that create small irregularities or small waves on the ocean surface. Resonance between the vertical wind pressure and these small waves, together with sheer stress due to higher wind speeds at the crests compared to the troughs, then act upon the waves and make them grow. When they are big enough other processes take over, the friction of the wind on the water and the pressure differences created by the sheltering effect of the lee side of the wave compared to the wind side of the wave causes the waves to continue to grow. Energy is continually transferred from the winds to the waves throughout all of these processes. The size that they ultimately reach depends on three things, the wind speed, the time during which the wind blows, and the distance over which the wind is blowing, *the fetch*.

Ocean waves, created by the wind in the above way, are called wind waves, and as long as they are under the wind generating influence, corresponding physical state of the ocean is called wind-sea. Waves are however, not dependent on winds to exist, after they have been created, can propagate unaided by the wind. Waves that are no longer under the generating influence of the wind are called free or swell waves. A general term for ocean physical state in terms of waves is sea state, representing a shorter time period, and can be thought of as having a typical wave height and period, while the wave climate is the long term description that contains the combined information found in many years of sea states. The wind sea typical appearance is very chaotic, and waves are appearing and disappearing with various heights and wave lengths and in various directions in an indiscernible pattern. Mathematically, a combination of sine waves of various heights, lengths, directions and positions will recreate the chaotic-looking sea state. An interesting property of ocean waves is that the speed, $c$, with which they move is proportional to its wave length as described by:

$$c = \frac{gT}{2\pi} \tag{8.40}$$

where $g$ is the acceleration of gravity and $T$ is the wave period. The consequence is that when the wind has ceased its generating influence on the wind sea, and as the waves continue to propagate across the ocean, then the original chaotic sea state will gradually come apart due to the different periods of the wave components that combine to make up the wind sea. Longer waves travel at the highest speeds while the converse is true for the shorter waves. After a long stretch of ocean, the separation of waves will be almost complete and if at this time the first long waves arrive at a beach somewhere then the beach dwellers will see long, almost perfect waves rolling in and hitting the shore. The linear theory for ocean waves is based on two fundamental equations, the continuity and Navier-Stokes equations. The continuity equation, or the conservation of mass, basically states that the change in density of a fluid within a volume equals the flow of mass into or out of the volume. Navier-Stokes equation is basically Newton second law stating that force equals mass times acceleration, $F = ma$. The left side of the equation represents the acceleration of the fluid, and the right side represents the force per mass where the three components represent forces associated with the pressure gradient, the viscosity, and external forces.

The total ocean or sea wave energy can be calculated, as energy per unit of wave front width by summing up the potential and kinetic energy. The potential energy of a wave of length $L$ is generated by the displacement of the water away from the mean sea level. The kinetic energy of a wave is a result of combined horizontal and vertical water particle motions. The total potential and kinetic energy of an ocean wave is expressed as:

$$E = \rho g A^2 \qquad (8.41)$$

where $g$ is the acceleration of gravity (9.806 m/s²), $\rho$ is the water density (e.g., 1022 kg/m³ for sea water), and $A$ is the wave amplitude (m). Using the speed of wave propagation:

$$c_g = \frac{\lambda}{2T}$$

The average energy flux or power over a wave period, computed by multiplying the total energy $E$ with the speed of wave propagation, expressed as:

$$P_W = \rho g A^2 \frac{\lambda}{2T} = \frac{\rho g^2 T A^2}{4\pi} \qquad (8.42)$$

where $T$ is the wave period (s) and $\lambda$ is the wavelength (m). The dispersion relationship was used to describe the wave period $T$ and the wavelength $\lambda$ as:

$$\lambda = \frac{g T^2}{2\pi} \qquad (8.43)$$

Instead of using the wave amplitude, wave power can also be rewritten as a function of wave height, $H$. Considering that the wave amplitude is half of the wave height then the wave power is expressed as:

$$P_W = \frac{\rho g^2 T H^2}{16\pi} \qquad (8.44)$$

Or using the sea water density of 1023 kg/m³, the power in kilowatts per meter, related to the wave height and period can be expressed as:

$$P_W = 1.96 H^2 T \qquad (8.45)$$

**Example 8.10:** An ocean wave has the period 12 seconds and a height of 3.2 m. What is the available power for one nautical mile of wave front?

**Solution:** From Equation (8.34) the power in kW/m of wave front is:

$$P_W = 1.96H^2T = 1.96 \cdot (3.2)^2 \cdot 12 = 233.5 \text{ kW/m}$$

A nautical mile converted in m is equal to 1852 m, so the available power for a wave front of one nautical mile is:

$$P_{nm} = 1852 \times 233.5 = 432,442 \text{ kW} \approx 432.4 \text{ MW}$$

Wave energy is an intermittent energy source. This means that the available power from the source varies in time in a way that is uncontrollable. It may be possible to predict the power of waves and time of occurrence at a particular location a few days ahead of time, but it is not possible to control the waves themselves. The intermittency of these sources also make them practically incapable of supplying all of the energy to a society because, as long as there is no means of large electric energy storage in existence, the energy consumed in a society has to be produced at the time of use. The share of the total power supply to an electric grid that may be made up of intermittent sources, while remaining stable, depends on the amount of spare capacity that is rapidly available to the grid. Many systems have been proposed that aim to divide the myriad of existing wave energy conversion concepts into a few general categories, see e.g. Here, a very basic division is presented where the converters are arranged into three categories: wave activated bodies, overtopping devices and oscillating water columns, as shown in Figure 8.6.

There are many technologies, designed to convert wave energy into electricity, with the most promising ones summarized below. The general term for technologies that harness the energy of ocean waves is wave energy converter (WEC). The Pelamis is a device that falls into the wave activated bodies category. It is composed of semi-submerged cylinders linked by hinged joints. The waves induce motions in the joints which are resisted by hydraulic rams that pump a high-pressure fluid through hydraulic motors. The hydraulic motors drive electric generators. The Wave dragon is a large, floating, overtopping device that has arms which focus the waves up a ramp and into a water reservoir. The water is then led from the elevated reservoir back to the ocean, driving turbines in the process. Limpet is a traditional OWC located onshore on the island of Islay, off the west coast of Scotland. It was installed in year 2000 and produces power to the electric grid. The advantages of oscillating body converters include their size and versatility since most of them are floating

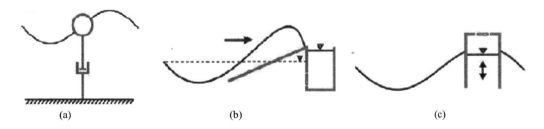

(a)          (b)          (c)

**FIGURE 8.6**
General categories of wave energy converters: (a) wave activated bodies, (b) overtopping devices, and (c) oscillating water columns.

devices. A distinct technology has yet to emerge and more research, to increase the PTO performance and avoid certain issues with the mooring systems, needs to be undertaken.

Overtopping converters (or terminators) consist of a floating or bottom fixed water reservoir structure, and also usually reflecting arms, which ensure that as waves arrive, they spill over the top of a ramp structure and are restrained in the reservoir of the device. The potential energy, due to the height of collected water above the sea surface, is transformed into electricity using conventional low head hydroturbines (similar to those used in mini-hydro plants). The main advantage of this system is the simple concept—it stores water and when there is enough, lets it pass through a turbine. Key downsides include the low head (in the order of 1–2 m) and the vast dimensions of a full scale overtopping device.

Wave activated bodies include all concepts where the motion of the ocean is transferred directly to the motion of the device. A subcategory of wave activated bodies that is worth mentioning is the point absorbers. The main characteristic of point absorbers is that they are small in the horizontal dimensions compared to the length of the waves from which they aim to capture energy. In other words, they take up a relatively small area of the ocean surface. Overtopping devices consist of a ramp or a tapered channel that forces the water of incoming waves to rise up and spill into a pool or reservoir. In this way, since the water surface of the reservoir is elevated relative to the ocean surface, the energy of the waves has been converted to potential energy. In a manner resembling hydropower plants the water is lead back into the ocean through a turbine. Oscillating water columns have an oscillating pillar of water that pumps air through a turbine. This motion of the water pillar is achieved by e.g., taking a pipe and placing it partly submerged in the sea. It consists of a partially submerged structure, forming an air chamber with an underwater opening that allows sea water to flow into the chamber. The waves that roll against the pipe will make the internal water surface oscillate. This oscillation is normally used to pump air which drives a turbine. A useful expansion of these categories is to add device location information. A typical set of definitions is to say that the device may be located onshore, near shore, or offshore. Near shore are relatively shallow areas and offshore sites have depths were the waves are not affected by the bottom. Most energy can be found offshore since the waves have yet to lose energy in friction against the seabed.

## 8.4 Ocean Thermal Energy Conversion

Ocean thermal energy conversion generates electricity indirectly from solar energy by harnessing the temperature difference between the sun-warmed surface of tropical oceans and the colder deep waters. The idea behind OTEC is the use of all natural collectors, the sea, instead of artificial collector. A significant fraction of solar radiation incident on the ocean is retained by seawater in tropical regions, resulting in average year-round surface temperatures of about 23°C. Deep, cold water, meanwhile, forms at higher latitudes and descends to along the seaboard, or toward the equator. The warm surface layer, which extends to depths of about 100 or 200 m, is separated from the deep cold water by a thermocline. The temperature difference, $T$, between the surface and thousand-meter depth ranges from 10°C to 25°C, with larger differences occurring in equatorial and tropical waters, as depicted in Figure 8.7. To establish the limits of the performance of OTEC power cycles, as a rule-of thumb is that a differential of about 20°C is necessary to sustain viable operation of an OTEC facility. OTEC power systems operate as cyclic heat engines. They receive thermal energy through heat transfer from surface sea water warmed by

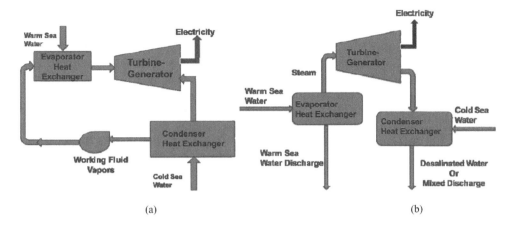

**FIGURE 8.7**
Block diagram of (a) closed-loop OTEC system and (b) open-loop OTEC system.

the Sun, and transform a portion of this energy to electrical power. A portion of the heat extracted from the warm sea water must be rejected to a colder thermal sink. The thermal sink employed by OTEC systems is sea water drawn from the ocean depths by means of a submerged pipeline. A steady-state control volume energy analysis yields the result that net electrical power produced by the engine must equal the difference between the rates of heat transfer from the warm surface water and to the cold deep water. The limiting (i.e., maximum) theoretical Carnot energy conversion efficiency of a cyclic heat engine scales with the difference between the temperatures at which these heat transfers occur. For OTEC, this difference is determined by $T$ and is very small; hence, OTEC efficiency is low. Although viable OTEC systems are characterized by Carnot efficiencies in the range of 6%–8%, state-of-the-art combustion steam power cycles, which tap much higher temperature energy sources, are theoretically capable of converting more than 60% of the extracted thermal energy into electricity. The low energy conversion efficiency of OTEC means that more than 90% of the thermal energy extracted from the ocean's surface is "wasted" and must be rejected to the cold, deep sea water. This necessitates large heat exchangers and seawater Sow rates to produce relatively small amounts of electricity.

OTEC is quite an old concept aiming to tap solar energy stored as sensible heat in the upper mixed layer of tropical oceans. Deep cold seawater originally formed at polar margins provides the low temperature needed for an appropriate working fluid, such as ammonia to complete a thermodynamic cycle, e.g., Rankine cycle; the mechanical work produced is easily convertible to electricity. OTEC is operating by using warm water that is collected on the surface of the tropical ocean and pumped by a warm water pump. Because practical temperature differences are only of the order of 20°C, with much of this resource needed in the process heat exchangers, the cycle thermodynamic efficiency is of the order of 3%. As a result, several cubic meters per second of seawater are necessary to produce just 1 MW of net electricity. Such facts have thus far prevented OTEC and some of its by-products from being economically competitive. The water is pumped through the boiler, where some of the water is used to heat the working fluid, usually propane or some similar material. If it is cooler you can use a material with a lower boiling point like ammonia. The propane vapor expands trough a turbine which is coupled to a generator that generating electric power. Cold water from the bottom is pumped through the condensers, where the vapor returns to the liquid state. The fluid is pumped back into the boiler. Some small fraction of the power from the turbine is used to pump the water through the system and to power other internal operations, but most of it is available as net power.

There are three types of OTEC designs: open cycle, closed cycle, and hybrid cycle. In an open cycle, seawater is the working fluid. Warm seawater is pumped into a flash evaporator where pressure as low as 0.03 bar cause the water to boil at temperatures of 22°C. This steam expands through a low-pressure turbine connected to a generator to create power. The steam then passes through a condenser using cold seawater from the depths of the ocean to condense the steam into desalinized water. In a closed cycle, a low boiling point liquid such as ammonia or another type of refrigerant is used as the working fluid in a Rankine cycle (common steam cycle). The heat from warm seawater flowing through an evaporator vaporizes the working fluid. The vapor expands through a turbine, flows into a condenser where cold seawater condenses it into a liquid. A hybrid cycle is a combination of both closed and open cycles where flash evaporator seawater is used as the closed cycle working fluid. OTEC systems can be sited anywhere across about 60 million square kilometers of tropical oceans—anywhere there is deep cold water lying under warm surface water. This generally means between the Tropic of Cancer and the Tropic of Capricorn. Surface water in these regions, warmed by the Sun, generally stays at 25°C or above. Ocean water more than 1000 m below the surface is generally at about 4°C. It would not be profitable, for example to use an OTEC power plant in the Baltic Sea, because the average temperature is about 8°C–10°C. The temperature difference between the warm surface water and the cold deep water has to be more than 20°C (68°F), to have an efficiently operated OTEC system, capable of producing a significant amount of power. It is clear that to extract higher powers, higher temperature difference is required. OTEC has little impact on the surrounding environment. Another attractive feature of OTEC energy systems is that not only the end products but also several other by-products, such as desalinated water, include energy in the form of electricity. The OTEC system is similar to a heat engine, operating between the hot and cool water ocean levels. The OTEC system uses a fluid that circulates in a cycle, and with the aid of a heat exchanger, the heat of the hot water is absorbed by the working fluid. The fluid gets expanded after taking the heat and drives a turbine, which is mechanically coupled to an electric generator. After rotating the turbine, the working fluid needs to be condensed to continue the cycle. The block diagram of a general closed cycle OTEC system, as in Figure 8.7 is showing the evaporator, condenser, and turbine, used in an OTEC. The drawback of an OTEC system is the low temperature difference between the hot and cold ends of the engine, which results in a very large volume of water requirement. The maximum possible efficiency of such system is given by the Carnot cycle efficiency:

$$\eta = \frac{W}{Q_H} = 1 - \frac{T_C}{T_H} \tag{8.46}$$

**Example 8.11:** Calculate efficiency of a non-ideal OTEC system operating at 21°C difference temperature, if the surface layer has a temperature of 25°C.

**Solution:** From above equations, the theoretical efficiency is:

$$\eta = 1 - \frac{T_C}{T_H} = 1 - \frac{273.15 + 4}{273.15 + 25} = 0.07 \text{ or } 7.0\%$$

However, the efficiency of a non-ideal OTEC system, operating between a hot water layer ($Q_H$) and a cold water layer ($Q_X$) is lower than the ideal Carnot cycle operating between the same temperatures:

$$\eta = 1 - \frac{Q_C}{Q_H} < 1 - \frac{T_C}{T_H} \tag{8.47}$$

**Example 8.12:** An OTEC system is operating in an area where the warm water surface temperature is 25°C and the cold layer is at 4°C. The temperature of the evaporator is 20°C and the one of condenser is 12°C. Calculate the ideal energy gain from 1 m³ of this sea water.

**Solution:** The energy available from 1 m³, operating between 20°C and 12°C is:

$$Q = mC\Delta T = 1 \times 1023 \times 3930 \times (20 - 12) = 32.163 \text{ MJ/m}^3$$

For an OTEC system operating between the surface warm layer and cold layer the ideal efficiency calculated in Example 8.10 is 7%. The net energy extracted is:

$$E_{net} = 0.07 \times 32.163 \times 10^6 = 2.251 \text{ MJ/m}^3$$

Theoretical, it is possible to convert the energy in a 23° temperature difference at an efficiency of 7%–8%. In actual practice, it is possible to do this at slightly more than 3% efficiency, which may mislead. About the energy converted, as is the value of Example 8.10. The efficiency is not influenced by the amount of power available is small, or that power generated for this source need be expensive. This energy is equivalent to the same amount of water passing through a hydroelectric dam with a water height of 56 m, an OTEC plant needs to handle no more water than a hydroelectric plant of the same capacity. This temperature difference is constantly renewed by the action of the Sun and the ocean currents, and is therefore inexhaustible. The amount of water constantly available for this use is enough to provide at least 300 times world's total electricity usage. One notice, the steam locomotives, which were used during the middle of the nineteenth century, had a thermal efficiency of only about 3%. To count this efficiency you can use an equation, which is called the Carnot factor, and can be presented like this:

$$W = \frac{T_S - T_D}{T_S} Q \tag{8.48}$$

where $W$ is the work, $T_S$ is the surface (worm) water temperature (K), $T_C$ is the deep (cold) water temperature (K), and $Q$ is the thermal value. If we are using the values, surface temperature 27°C and deep temperature 4°C, the Equation (8.24) looks like:

$$W = \frac{T_S - T_D}{T_S} Q = \frac{27 - 4}{27 + 273.15} Q = \frac{23}{300.15} Q \approx 0.076Q \tag{8.49}$$

OTEC Advantages and Disadvantages

Advantages

1. OTEC uses clean, renewable, natural resources. Warm surface seawater and cold water from the ocean depths replace fossil fuels to produce electricity.
2. Suitably designed OTEC plants will produce little or no carbon dioxide or other polluting chemicals.
3. OTEC systems can produce fresh water as well as electricity. This is a significant advantage in island areas where fresh water is limited.
4. There is enough solar energy received and stored in the warm tropical ocean surface layer to provide most, if not all, of present human energy needs.
5. The use of OTEC as a source of electricity will help reduce the state's almost complete dependence on imported fossil fuels.

Disadvantages

1. OTEC-produced electricity at present would cost more than electricity generated from fossil fuels at their current costs.
2. OTEC plants must be located where a difference of about 20°C occurs year round. Ocean depths must be available fairly close to shore-based facilities for economic operation. Floating plant ships could provide more flexibility.
3. No energy company will put money in this project because it only had been tested in a very small scale.
4. Construction of OTEC plants and lying of pipes in coastal waters may cause localized damage to reefs and near-shore marine ecosystems.

When two extensive currents of water, one warm and one cold, exist in close proximity to one another, it is possible to operate a power plant utilizing this temperature differential. This energy may be extracted wherever a temperature difference driving force exists. The extraction of such energy becomes more difficult, costly, and less efficient as the temperature difference between the high- and low-temperature layers decreases. The costs are not yet competitive where the technology for this idea has been shown to work. If this energy source were developed, there would also be a number of environmental impacts to be considered. The large-scale mixing of warm and cold water could have significant impacts on the ocean, biota and climate. The large surface areas in the heat exchangers are subjected to the flow of corrosive seawater, and metallic elements will therefore be introduced into the seawater. Loss of working fluid might also be a problem if leaks into the system, or if there are unexpected spills. Other problems include the impacts of techniques used to inhibit bio-fouling and corrosion, the impacts of coastal zone facilities associated with the operations of the offshore plants, and the installation and operation of electrical distribution systems.

## 8.5 Summary

The chapter discusses some of the background topics of the water energy conversion. Hydroelectricity is the term referring to electricity generated by hydropower. The production of electrical power "arises" through the use of the gravitational force of falling or flowing water. It is the most widely used form of renewable energy. Once a hydroelectric complex is constructed, the project produces no direct waste and has a considerably lower output level of the greenhouse gas carbon dioxide ($CO_2$) than any of the fossil fuel powered energy plants. The major advantage of hydroelectricity is the elimination of the cost of fuel. The cost of operating a hydroelectric plant is not affected by increases in the cost of fossil fuels such as oil, natural gas, or coal, and no imports are required. These plants also have long lives, with some plants still in service after 50+ years. There is also tremendous energy in waves. Waves are caused by the wind blowing over the surface of the ocean. In some areas of the world, the wind blows with enough intensity and force to produce large waves. The west coasts of the United States and Europe and the coasts of Japan and New Zealand are excellent candidates for harnessing wave energy. There are no large commercial wave-energy plants, but there are a few small ones. They have been used to power the lights and whistles on buoys. This resource might produce enough energy to power local communities. Wave and tidal energy sources are non-depletable resources, which are increased use of the resources does not affect resource availability. However, estimates of resource availability may change over time as new measurement methods become available. In addition, the quantity of the resource that can be utilized will change over time as new technology developments allow increased exploitation of ocean resources. OTEC system is an energy-generating technology that takes advantage of the temperature difference between the ocean's shallow warm water and cold deeper water. OTEC uses the heat stored in warmer surface water to rotate the steam-driven turbines. Therefore, the energy is produced from the natural thermal gradient in the seawater layers gradient which is a self-replenishable source of energy. The OTEC system is similar to a heat engine, which operates between the hot and cool water levels of the ocean.

## Questions and Problems

1. Calculate the speed of a surface wave on deep water of wavelength of 120 m.
2. What is the approximate world capacity of ocean energy (waves, tides, and ocean thermal energy)?
3. Give classification of hydroelectric power plant.
4. Write advantages, disadvantages and application of hydroelectric power plant.
5. A Pelton turbine works with a head of 100 m and a water flow equal to 10 m³/s. What is the power output if the efficiency of the plant is 67%?

6. A village needs 20 m³/day for water, collected from a waterfall with a head of 18 m. Estimate the hydraulic power.

7. Discuss the factors for site selection for a hydroelectric plant.

8. The Three Gorges Dam that spans the Yangtze River in China is the world's largest power station in terms of installed capacity. (a) In a normal year the river flow rate is equal to 30,000 m³/s. The dam has a head of 80 m. Calculate the power output if the overall efficiency of the power plant is 90%. (b) In a drought year, the river flow rate can decrease to 50% of its normal value, and the head to 75 m. Calculate the dam power output if the hydropower plant efficiency remains the same.

9. List the top ten countries in the world and their installed capacity for large hydropower.

10. A water turbine converts power from a river to the rotation of a shaft which is used to run an electric generator. If the efficiency of the turbine is $\eta_t = 80\%$ and that of the generator is $\eta_g = 90\%$, what is the overall efficiency $\eta_{tg}$ of the turbine-generator combination? If 100 kW of electrical power is produced, what is the rate of energy loss?

11. Water falls from a head 85 m at a flow rate of 30 m³/s, and the overall conversion to electricity is 70%. How many typical single-family homes can this hydropower system provide with electricity?

12. An artificial reservoir is 20 km long, 1.8 km wide and has an average depth of 80 m. 1.5% of the reservoir volume drops 100 m and passes through a hydroelectric conversion system with the overall efficiency of 0.82%. What is the electrical power output?

13. Give the classification of hydraulic turbines.

14. What are the main parts of a Pelton turbine?

15. Define the following terms: (a) Gross head; (b) Net head; (c) Hydraulic efficiency; (d) Volumetric efficiency; (e) Mechanical efficiency; and (f) Overall efficiency.

16. Explain the impulse turbine.

17. Explain the reaction turbine.

18. A reservoir-type small hydropower system has a penstock of 1.8 m diameter, the water speed at the penstock exist is 12 m/s. Assuming the power coefficient of 36%, turbine efficiency 85% and the generator-electronics efficiency 93%, compute the generator output electric power. If the penstock efficiency is 93% what is the water head?

19. Consider a turbine system as the one depicted below. The turbine is connected to an upper reservoir on the pressure side and discharges to a lower reservoir. The following is given, static pressure $p = 1\ 80$ kPa, volume flow rate at position 2 is 10 m³/s, height $H = 400$ m, and pipe diameter at turbine inlet $d_2 = 0.6$ m, pipe diameter $d_3 = d_2$, static pressure $p_3$ is 105 kPa. The gravitational constant is $g$ 9.81 m/s². Working medium is water with a density of 1000 kg/m³. Friction is neglected. Determine the following: flow speed at turbine inlet, total head at turbine inlet, difference in total head over turbine, and maximum possible power produced.

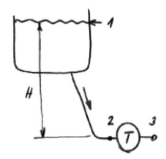

20. A site has a head of 100 m with a variable flow rate, as shown in table below. Calculate the output power if the efficiency is (a) 100% and (b) 50%. Plot the results.

| Month | Jan | Feb | Mar | Apr | May | Jun | Jul | Aug | Sep | Oct | Nov | Dec |
|---|---|---|---|---|---|---|---|---|---|---|---|---|
| Flow Rate (m³/s) | 10 | 12 | 100 | 80 | 35 | 28 | 25 | 17 | 22 | 60 | 45 | 14 |

21. List the major advantages and disadvantages of the onshore and offshore wave energy generation.

22. What is total tidal energy at 85 and efficiency during a falling tide from a basin of 250 km² with a tidal range of 7.0 m?

23. Estimate the power density of a water turbine with 16 m diameter is placed in channel with a tidal current moving with a speed of 3.8 m/s.

24. Assuming a power coefficient of 45%, water speed of 3.5 m/s, the rotor swept area diameter of 0.6 m, calculate the output electrical power and the energy generated in one month (assuming the steady flow at 3.5 m/s), if the turbine and generator efficiencies are 80% and 95%, repectively.

25. A basin of 189 km² with a tidal range of 10.8 m drains trough an opening of 250 m wide. What the average water velocity during the falling tide? What tis the available tidal energy during one tidal period?

26. An ocean wave has a height of 3 m and a period of 11.5 seconds. What tis the available power per 1 km of wave front?

27. In storms the waves may reach a height of 12 m and periods of 15 seconds. What is the available power per 100 m of storm wave front?

28. A wave, traveling at a speed of 8 m/s and has a height of 2.4 m is incident on an energy conversion device that has efficiency of 25%. How large needs to be the device to generate 1500 kW of electricity?

## References and Further Readings

1. F. Bueche, *Introduction to Physics for Scientists and Engineers*, McGraw-Hill, New York, 1975.
2. V. Quaschning, *Understanding Renewable Energy Systems*, Earthscan, London, UK, 2006.
3. R. A. Ristinen and J. J. Kraushaar, *Energy and Environment*, Wiley, Hoboken, NJ, 2006.

4. J. Andrews and N. Jelley, *Energy Science, Principles, Technology and Impacts*, Oxford University Press, Oxford, UK, 2007.
5. E. L. McFarland, J. L. Hunt, and J. L. Campbell, *Energy, Physics and the Environment* (3rd ed.), Cengage Learning, Mason, OH, 2007.
6. F. M. Vanek and L. D. Albright, *Energy Systems Engineering: Evaluation and Implementation*, McGraw-Hill, New York, 2008.
7. J. Twidell, and T. Weir, *Renewable Energy Sources* (2nd ed.), Taylor & Francis Group, London, UK, 2006.
8. B. Sorensen et al., *Renewable Energy Focus Handbook*, Academic Press, London, UK, 2009.
9. M. A. El-Sharkawi, *Electric Energy—An Introduction*, CRC Press, Boca Raton, FL, 2009.
10. A. Vieira da Rosa, *Fundamentals of Renewable Energy Processes*, Academic Press, London, UK, 2009.
11. B. K. Hodge, *Alternative Energy Systems and Applications*, Wiley, Hoboken, NJ, 2010.
12. H. J. Wagner, and M. Jyotirmay, *Introduction to Hydro Energy Systems—Basics, Technology and Operation*, Springer, Berlin, Germany, 2011.
13. G. Boyle, *Renewable Energy—Power for a Sustainable Future*, Oxford University Press, Oxford, UK, 2012.
14. B. Everett, and G. Boyle, *Energy Systems and Sustainability: Power for a Sustainable Future* (2nd ed.), Oxford University Press, Oxford, UK, 2012.
15. R. Belu, Hydroelectric power systems, small, in *Encyclopedia of Energy Engineering & Technology* (Online) (eds, A. Anwar et al.), Taylor & Francis Group, Vol. 2, 2014.
16. R. A. Dunlap, *Sustainable Energy*, Cengage Learning, Stamford, CT, 2015.
17. V. Nelson, and K. Starcher, *Introduction to Renewable Energy* (*Energy and the Environment*), CRC Press, Boca Raton, FL, 2015.
18. E. E. Michaelidis, *Alternative Energy Sources*, Springer, New York, 2012.
19. H. J. Wagner, and M. Jyotirmay, Introduction to hydro energy systems—Basics, technology, *KSCE Journal of Civil Engineering*, Vol. 15(4), pp. 643–653, 2010. doi:10.1007/s12205-011-0005-7.
20. F. R. Spellman, *Environmental Impacts of Renewable Energy*, CRC Press, Boca Raton, FL, 2014.

# Appendix A: Common Parameters, Units and Conversion Factors

**TABLE A.1**

Physical Constants in SI Units

| Quantity | Symbol | Value |
|---|---|---|
| Avogadro constant | N | $6.022169 \cdot 10^{26}$ kmol$^{-1}$ |
| Boltzmann | k | $1.380622 \cdot 10^{-23}$ J/K |
| First radiation constant | $C_1 = 2 \cdot \pi \cdot h \cdot c$ | $3.741844 \cdot 10^{-16}$ Wm$^2$ |
| Gas constant | R | $8.31434 \cdot 10^3$ J/kmol K |
| Planck constant | h | $6.626196 \cdot 10^{-34}$ Js |
| Second radiation constant | $C_2 = hc/k$ | $1.438833 \cdot 10^{-2}$ mK |
| Speed of light in a vacuum | c | $2.997925 \cdot 10^8$ m/s |
| Stefan-Boltzmann constant | $\sigma$ | $5.66961 \cdot 10^{-8}$ W/m$^2$K$^4$ |
| Speed of light | c | 299,792.458 m/s |
| Elementary charge | e | $1.602176 \cdot 10^{-19}$ C |

**TABLE A.2**

Multiplication Factors

| Multiplication Factor | Prefix | Symbol |
|---|---|---|
| $10^{12}$ | Terra | T |
| $10^9$ | Giga | G |
| $10^6$ | Mega | M |
| $10^3$ | Kilo | K |
| $10^2$ | Hecto | H |
| 10 | Deka | da |
| 1 | N/A | — |
| 0.1 | Deci | D |
| 0.01 | Centi | C |
| $10^{-3}$ | Mili | M |
| $10^{-6}$ | Micro | M |
| $10^{-9}$ | Nano | n |
| $10^{-12}$ | Pico | p |
| $10^{-15}$ | Femto | f |
| $10^{-18}$ | Atto | a |

**TABLE A.3**

System of Units and Conversion Factors

| U.S. Unit | Abbreviation | SI Unit | Abbreviation | Conversion Factor |
|---|---|---|---|---|
| Foot | ft | Meter | m | 0.3048 |
| Mile | mi | Kilometer | km | 1.6093 |
| Inch | in | Centimeter | cm | 2.54 |
| Square feet | ft$^2$ | Square meter | m$^2$ | 0.0903 |
| Acre | acre | Hectare | ha | 0.405 |
| Circular mil | cmil | $\mu$m$^2$ | — | 506.7 |
| Cubic feet | ft$^3$ | Cubic meter | m$^3$ | 0.02831 |
| Gallon (U.S.) | gal(U.S.) | Liter | l | 3.785 |
| Gallon (UK) | gal(UK) | Liter | l | 4.445 |
| Cubic feet | ft$^3$ | Liter | l | 28.3 |
| Pound | lb | Kilogram | kg | 0.45359 |
| Ounces | oz | Gram | g | 28.35 |
| US ton | ton (U.S.) | Metric ton | ton (metric) | 0.907 |
| Mile/hour | mi/h | Meter/second | m/s | 0.447 |
| Flow rate | ft$^3$/h | Flow rate | m$^3$/s | 0.02831 |
| Density | lb/ft$^3$ | Density | kg/m$^3$ | 16.020 |
| lb-force | lbf | Force | N | 4.4482 |
| Pressure | lb/in$^2$ | Pressure | kPa | 6.8948 |
| Pressure | bar | Pressure | Pa | 10$^5$ |
| Torque | lb.force ft | Torque | Nm | 1.3558 |
| Power | ft.lb/s | Power | W | 1.3558 |
| Power (horsepower) | HP | Power | W | 745.7 |
| Energy | ft.lb-force | Energy | J | 1.3558 |
| Energy (British thermal unit) | Btu | Energy | kWh | 3412 |

**TABLE A.4**

Common Energy Conversion Factors

| Energy Unit | SI Equivalent |
|---|---|
| 1 electron volt (eV) | $1.6021\ 10^{-19}$ J |
| 1 erg (erg) | $10^{-7}$ J |
| 1 calorie (cal) | 4.184 J |
| 1 British thermal unit (Btu) | 1055.6 J |
| 1 Q (Q) | $10^{18}$ Btu (exact) |
| 1 quad (q) | $10^{15}$ Btu (exact) |
| 1 tons oil equivalent (toe) | $4.19 \cdot 10^{10}$ J |
| 1 barrels oil equivalent (bbl) | $5.74 \cdot 10^{9}$ J |
| 1 tons coal equivalent (tce) | $2.93 \cdot 10^{10}$ J |
| 1 m$^3$ of natural gas | $3.4 \cdot 10^{7}$ J |
| 1 liter of gasoline | $3.2 \cdot 10^{7}$ J |
| 1 kWh | $3.6 \cdot 10^{6}$ J |
| 1 ft$^3$ of natural gas (1000 Btu) | 1055 kJ |
| 1 gal. of gasoline (125,000 Btu) | 131.8875 kJ |

**TABLE A.5**

Specific Work (Mechanical Work per Unit of Mass) in kJ/kg, of
Major Geothermal Energy Resources

| Resource Type | Specific Work (kJ/kg) |
|---|---|
| Geothermal steam (280°C, 15 atm) | 955 |
| Saturated steam (220°C) | 930 |
| Saturated water (liquid at 200°C) | 158 |
| Saturated water (liquid at 160°C) | 97 |
| Geo-pressured liquid water (140°C, 25 atm) | 75 |

**TABLE A.6**

Refractive Index of Some of the Common Materials (at 20°C)

| Material | Index of Refraction | Material | Index of Refraction |
|---|---|---|---|
| Diamond | 2.419 | Benzene | 1.501 |
| Fluorite | 1.434 | Carbon disulfide | 1.628 |
| Fused quartz | 1.458 | Carbon tetrachloride | 1.461 |
| Glass (crown) | 1.520 | Ethyl alcohol | 1.361 |
| Glass (flint) | 1.660 | Glycerin | 1.473 |
| Ice | 1.309 | Oil, turpentine | 1.470 |
| Polystyrene | 1.590 | Paraffin (liquid) | 1.480 |
| Salt ($NaCl_2$) | 1.544 | Water | 1.333 |
| Teflon | 1.380 | Air (0°C, 1 atm) | 1.000293 |
| Zircon | 1.923 | Carbon dioxide (0°C, 1 atm) | 1.00045 |

**TABLE A.7**

Properties of Dry Air

| Temperature (K) | $\rho$ (kg/m³) | $C_P$ (kJ/kg K) | k (W/m K) |
|---|---|---|---|
| 293 | 1.2040 | 1.0056 | 0.02568 |
| 300 | 1.1774 | 1.0057 | 0.02624 |
| 350 | 0.9980 | 1.0090 | 0.03003 |
| 400 | 0.8826 | 1.0140 | 0.03365 |
| 450 | 0.7833 | 1.0207 | 0.03707 |
| 500 | 0.7048 | 1.0295 | 0.04038 |
| 600 | 0.5879 | 1.0551 | 0.04659 |
| 700 | 0.5030 | 1.0752 | 0.05230 |
| 800 | 0.4405 | 1.0978 | 0.05779 |
| 900 | 0.3925 | 1.1212 | 0.06279 |
| 1000 | 0.3525 | 1.1417 | 0.06752 |

*Source:* U.S. National Bureau Standards (U.S.), Tables of Thermodynamic and Transport
Properties, Circular 564, 1955.

*Symbols:* T absolute temperature, degrees Kelvin; $\rho$ density; $C_P$ specific heat capacity;
$\mu$ viscosity; $\nu = \mu/\rho$; and k thermal conductivity. The values of $\rho$, $\mu$, $\nu$, k, and $C_P$
are not strongly pressure-dependent and may be used over a fairly wide range
of pressures.

**TABLE A.8**

Properties of Water

| Temperature (°C) | ρ (kg/m³) | $C_P$ (kJ/kg K) | k (W/m K) |
|---|---|---|---|
| 0.00 | 999.8 | 4.225 | 0.566 |
| 4.44 | 999.8 | 4.208 | 0.575 |
| 10.0 | 999.2 | 4.195 | 0.585 |
| 20.0 | 997.8 | 4.179 | 0.604 |
| 25.0 | 994.7 | 4.174 | 0.625 |
| 100.0 | 954.3 | 4.219 | 0.684 |

**TABLE A.9**

U.S. Standard Units

| Quantity | Type | Symbol | SI Conversion |
|---|---|---|---|
| Foot | Length | ft | 0.305 m |
| Inch | Length | in | 0.0254 m |
| Mile | Length | mi | 1609.34 m |
| Second | Time | s | 1 s |
| Pound | Mass/weight | lb | 0.454 kg |
| Ounce | Mass/weight | oz | 0.02835 kg |
| Gallon | Volume | gal | 3.7854 l |
| Fluid ounce | Volume | fl oz | 29.574 ml |
| Fahrenheit | Temperature | °F | (9/5) °C + 32 |
| Ampere | Current | A | 1 A |
| Volt | Potential/voltage | V | 1 V |

# Appendix B: Design Parameters, Values and Data

### TABLE B.1

Conversion Coefficients for Insulation

| Parameter | SI | American Engineering System |
|---|---|---|
| Solar constant | 1367 W/m² | 430 Btu/ft²/hr. |
| Solar constant | 4870 kJ/m²/hr. | 1.520 HP/sq. yd. |
| Solar constant | — | 1.94 Langley/min. |
| Solar radiation | 1000 W/m² | 317 Btu/ft²/hr. |
| 1 Langley | — | 1 Cal/cm² = 3.69 Btu/ft² |
| | 11.35 kJ/m² | 1 Btu/ft² |

### TABLE B.2

Temperature Conversion Formulas

| | Degree Celsius (°C) | Degree Fahrenheit (°F) | Kelvin Degree (K) |
|---|---|---|---|
| Degree Celsius (°C) | — | $\frac{9}{5}(^\circ C + 32).$ | K − 273.15 |
| Degree Fahrenheit (°F) | $\frac{5}{9}(^\circ F - 32)$ | — | $1.8 \times K - 459.67$ |
| Kelvin degree (K) | °C + 273.15 | $(459.67 + ^\circ F)/1.8$ | — |

### TABLE B.3

Terrain Categories

| Terrain Category | Terrain Description |
|---|---|
| 0 | Smooth, flat terrain, small water body |
| 1 | Open, flat, no trees or obstructions or near/across water or desert, snow cover |
| 2 | Open, flat rural areas with no trees obstructions, few and low obstacles, e.g. farmland, semiarid areas |
| 3 | Open, flat or undulating with very few obstacles (open grass, farmland with few trees, hedgerows, other barriers), tundra, prairie |
| 4 | Build-up areas with some trees and buildings (suburbs, small towns, woodlands, shrubs, broken country with large trees, small fields with hedges) |

**TABLE B.4**

Solar Collector Important Angles

| Azimuth | Tilt | Θ | Orientation |
|---|---|---|---|
| N/A | 0° | 90°–β | Horizontal (Flat) |
| — | 90° | Varies | Vertical wall |
| 0° | 90° | Varies | South facing vertical |
| −90° | 90° | Varies | East facing wall |
| +90° | 90° | Varies | West facing wall |
| A | 90°–β | 0° | Tracking system |

**TABLE B.5**

Standard Test Condition (Solar Energy)

| Parameter | Value |
|---|---|
| Solar radiation | 1000 W/m² |
| Solar spectrum | AM 1.5 |
| Temperature | 25°C |

**TABLE B.6**

Monthly Day Number and Recommended Average Days

| Month | Day of Month (n) | Date | Monthly Average Day (n) | Declination (δ–degrees) |
|---|---|---|---|---|
| January | $I$ | 17 | 17 | −20.9 |
| February | $31 + i$ | 15 | 47 | −13.0 |
| March | $59 + i$ | 16 | 75 | −2.40 |
| April | $90 + i$ | 15 | 105 | 9.40 |
| May | $120 + i$ | 15 | 135 | 18.8 |
| June | $151 + i$ | 10 | 162 | 23.1 |
| July | $181 + i$ | 17 | 198 | 21.2 |
| August | $212 + i$ | 16 | 228 | 13.5 |
| September | $243 + i$ | 15 | 258 | 2.20 |
| October | $273 + i$ | 15 | 288 | −9.60 |
| November | $304 + i$ | 14 | 318 | −18.9 |
| December | $334 + i$ | 10 | 344 | −23.0 |

**TABLE B.7**

Solar Thermal Collectors

| Collector Type | Temperature Range (°C) | Concentration Ratio |
|---|---|---|
| Flat-plate collector | 30–80 | 1 |
| Evacuated-tube collector | 50–200 | 1 |
| Compound parabolic collector | 60–240 | 1–5 |
| Fresnel lens collector | 60–300 | 10–40 |
| Parabolic trough collector | 60–250 | 15–45 |
| Cylindrical trough collector | 60–300 | 10–50 |
| Parabolic dish reflector | 100–500 | 100–1000 |
| Heliostat field collector | 150–2000 | 100–1500 |

**TABLE B.8**

Mid-Season Monthly Average Daily Extraterrestrial Radiation ($MJ/m^2$)

| Latitude | January | April | July | October |
|----------|---------|-------|------|---------|
| 90° N | 0.00 | 19.3 | 41.2 | 0.00 |
| 80° N | 0.00 | 19.2 | 40.5 | 0.60 |
| 70° N | 0.10 | 23.1 | 38.7 | 4.90 |
| 60° N | 3.50 | 27.6 | 38.8 | 10.8 |
| 50° N | 9.10 | 31.5 | 40.0 | 16.7 |
| 40° N | 15.3 | 34.6 | 40.6 | 22.4 |
| 30° N | 21.3 | 36.8 | 40.4 | 27.4 |
| 20° N | 27.0 | 37.9 | 39.3 | 31.6 |
| 10° N | 32.0 | 37.9 | 37.1 | 35.0 |
| 0° | 36.2 | 36.7 | 34.0 | 37.3 |
| 10° S | 39.5 | 34.5 | 29.9 | 38.5 |
| 20° S | 41.8 | 31.3 | 25.2 | 38.6 |
| 30° S | 43.0 | 27.2 | 19.9 | 37.6 |
| 40° S | 43.1 | 23.3 | 14.2 | 35.5 |
| 50° S | 42.3 | 16.8 | 8.40 | 32.4 |
| 60° S | 41.0 | 10.9 | 3.10 | 28.4 |
| 70° S | 40.8 | 5.00 | 0.00 | 24.0 |
| 80° S | 42.7 | 0.60 | 0.00 | 20.6 |
| 90° S | 43.3 | 0.00 | 0.00 | 204 |

**TABLE B.9**

Thermal Properties of Some Common Materials

| Material | Density ($kg/m^3$) | Heat Capacity CP ($kJ/kg \cdot K$) | Temperature Range ($\Delta T$ in °C) |
|----------|--------------------|------------------------------------|--------------------------------------|
| Water | 1000 | 4.190 | 0–100 |
| Ethanol | 780 | 2.460 | −117–79 |
| Glycerin | 1260 | 2.420 | 17–290 |
| Oil | 910 | 1.800 | −10–204 |
| Synthetic oil | 910 | 1.800 | −10–400 |
| Engine oil | 888 | 1.880 | 0–160 |
| Propanol | 800 | 2.500 | 0–97 |
| Brick | 1600 | 0.840 | 20–70 |
| Concrete | 2240 | 1.130 | 20–70 |
| Sandstone | 2200 | 0.712 | 20–70 |
| Granite | 2650 | 0.900 | 20–70 |
| Marble Clay sheet | 2500 | 0.880 | 20–70 |

**TABLE B.10**

Average Clear Day Solar Radiation Calculation Coefficients, Declination and Equation of Time for the 21st Day of Each Month (SI and American Engineering Systems)

| Month (21$^{st}$ Days of the Month) | A (W·m$^{-2}$) | A (Btu/hr·ft$^2$) | B | C | δ (deg.) | Equation of Time (hours) |
|---|---|---|---|---|---|---|
| January | 1230 | 390 | 0.142 | 0.058 | −20.05 | −0.19 |
| February | 1215 | 385 | 0.144 | 0.060 | −10.50 | −0.23 |
| March | 1186 | 376 | 0.156 | 0.071 | −0.050 | −0.13 |
| April | 1134 | 360 | 0.180 | 0.097 | 11.85 | 0.02 |
| May | 1105 | 350 | 0.196 | 0.121 | 20.32 | 0.06 |
| June | 1190 | 345 | 0.205 | 0.134 | 23.45 | −0.02 |
| July | 1185 | 344 | 0.207 | 0.136 | 20.55 | −0.10 |
| August | 1107 | 351 | 0.200 | 0.122 | 12.35 | −0.04 |
| September | 1151 | 365 | 0.178 | 0.092 | 0.000 | 0.13 |
| October | 1192 | 378 | 0.160 | 0.073 | −10.55 | 0.26 |
| November | 1222 | 388 | 0.150 | 0.063 | −19.85 | 0.13 |
| December | 1233 | 391 | 0.142 | 0.057 | −23.45 | 0.03 |

**TABLE B.11**
Electric Power Common Terms

| Term | Definition | Standard Unit | Common Used Unit |
|---|---|---|---|
| Power density | Power per unit volume | W/m$^3$ | kW/m$^3$ |
| Specific power | Power per unit mass | W/kg | kW/kg |
| Electric power | | | |
| Output | Power × time | J | kW$_e$h |
| Rated power | Power output of a power plant at nominal operating conditions | MW | MW |
| Hate rate (HR) | Thermal BTUs required to generate 1 kW$_e$h (3412 Btu = 1 kW$_t$h) | W$_t$h/W$_e$h | kW$_t$h/kW$_e$h |
| Capacity factor (CF) | Average power/rated power (per a specific time period) | — | — |
| Load factor (LF) | Average power/maximum power (per a specific time period) | — | — |
| Availability factor (AF) | Fraction of time that power generation system (unit) is available (usually non-dispatchable units) | — | — |

# Index

Note: Page numbers in italic and bold refer to figures and tables, respectively.

9781032337944